# ARCTIC INSTITUTE OF NORTH AMERICA

## ANTHROPOLOGY OF THE NORTH:

### TRANSLATIONS FROM RUSSIAN SOURCES

*Editor:* HENRY N. MICHAEL

T0256684

ARCTIC INSTITUTE OF NORTH AMERICA

ANTHROPOLOGY OF THE NORTH:

TRANSLATIONS FROM RUSSIAN SOURCES / NO. 5

# The Archaeology and Geomorphology of Northern Asia: Selected Works

*Edited by* HENRY N. MICHAEL

PUBLISHED FOR THE ARCTIC INSTITUTE OF NORTH AMERICA
BY UNIVERSITY OF TORONTO PRESS

ANTHROPOLOGY OF THE NORTH: TRANSLATIONS FROM RUSSIAN SOURCES
is supported by National Science Foundation grant GN-212

# EDITOR'S PREFACE

THE EIGHTEEN ARTICLES appearing in this, the fifth, number of *Anthropology of the North: Translations from Russian Sources*, were mostly published between the years 1957 and 1963. The exceptions are S. I. Rudenko's "The culture of the prehistoric population of Kamchatka," published in 1948, and A. P. Okladnikov's "Paleolithic remains in the Lena river basin," published in 1953. Thirteen of the articles deal with the archaeology and five with the geomorphology of selected areas of northern Asia. Dr. Chester S. Chard of the University of Wisconsin analyzes the contents and meaning of these articles in his Introduction to the book and fruitfully correlates them with other sources which have been made available to the English-reading specialist over the past few years. In the Notes and References attached to each article, editorial reference has sometimes been added about the availability in English translation of a cited article.

As in previous issues of this series, we have followed the transliteration system recommended by the United States Board on Geographic Names, with the exception that the Russian "soft sign" is not rendered as an apostrophe. Names of non-Russian authors who have published in Russian, or whose works have been translated and published in Russian, are transliterated according to the above system. Thus Shrenk rather than Schrenk, Grintsevich rather than Hryncewicz, and others. However, well-established spellings of Russian place-names in English have been preferentially retained. Thus Moscow, not Moskva; Caucasus, not Kavkaz; Tien-shan, not Tyan Shan; and so on.

Throughout these articles, any words or sentences appearing in square brackets are those of the Editor unless specified as those of the Translator. Any words or sentences appearing in parentheses are those of the author of the article.

Many of the articles describe archaeological sites located in the Soviet Far East (often referred to by the authors as just "the Far East"). This area occupies the entire Pacific littoral of the Russian Soviet Federated Socialist Republic and is limited to the west by the crests of the Stanovoy, Dzungzhur, Kolyma, and associated ranges. Some of the general names for the various types of administrative areas have been transliterated rather than translated since no ordered or generally accepted translation of them exists. A *kray* denotes a very large administrative unit. At present there are nine *krays* in the Soviet Union, six of them in the R.S.F.S.R. and three recently established ones in the "pioneer" regions located in the north of Kazakhstan. Each *kray*, except the Maritime (Primorskiy), contains subdivisions called *oblasts*. These may be *autonomous oblasts* containing an ethnic group, as in the case of the six *krays* of the R.S.F.S.R., or they are simply *administrative oblasts*, as in the case of the newly established Kazakhstan *krays*. Additionally, Krasnoyarsk *kray* in western Siberia contains two *national okrugs*, very large areas sparsely inhabited by a small ethnic group with limited autonomy. Several of the administrative *oblasts* of the various Union republics also contain *national okrugs*. (Only ten of the fifteen Union republics have oblasts.) Oblasts are further subdivided into *rayons*. Those republics which do not have oblasts are directly subdivided into *rayons* and other, usually smaller, administrative units. In addition to this fairly complicated administrative division with its specific terminology, some of the older works, particularly ethnographic ones, use administrative terms no longer in existence, such as *guberniya, uprava, uyezd, ulus, volost,* or *sloboda.* Whenever these occur, they are transliterated rather than translated.

A goodly number of the articles—the five on geomorphology and Okladnikov's "The Shilka cave"—were ably translated by Dr. David Kraus. Mrs. Penelope Rainey, whose translations have appeared in past issues of this series, has skillfully done the translations of Okladnikov's "Paleolithic remains in the Lena river basin" and Larichev's "Neolithic remains in the upper Amur basin at Ang-ang-hsi in Tungpei." Miss Helen Frenkley has translated Gerasimov's "The Paleolithic site Malta" and Arutyunov, Levin, and Sergeyev's "Ancient burials on the Chukchi peninsula." Mrs. Lillian Ackerman has translated the three short articles dealing with Eskimo archaeology on the Chukchi peninsula, namely, Levin's "Fieldwork on the Chukchi peninsula in 1957" and "An early Eskimo cemetery at Uelen" and Levin and Sergeyev's "The penetration of iron into the Arctic." She has also translated the lengthier article by Okladnikov, "Ethnic and cultural connections of middle Yenisey tribes during the Neolithic." The translator of Larichev's "Ancient cultures of northern China" was Mrs. Lilian Volins; of Andreyev's "Certain problems relating to the Shell Mound Culture," Miss Helen O'Connell; and of Rudenko's "The culture of the prehistoric population of Kamchatka," Mr. A. Karpovich. Several of the maps accompanying the articles had to be redrawn *de novo* and this was carefully done by Mr. Edward Schumacher of the Smithsonian Institution.

I am very much indebted to several individuals who have helped me with the editing of this lengthy work. Dr. Henry B. Collins of the Smithsonian Institution has critically examined the articles dealing with Eskimo archaeology and has made valuable suggestions concerning terminology. Dr. Schuyler Cammann of the University of Pennsylvania has read Larichev's articles on the archaeology of northeastern China and has adjusted the translation of the numerous place-names and other terms contained in these articles. Mrs. Natalie Frenkley of the *Arctic Bibliography* staff has edited the multilingual bibliographies attached to many of the articles and has also drawn from her rich experience to make valuable suggestions concerning several difficult passages. To all go my profound thanks.

HENRY N. MICHAEL

*November 1964*

# CONTENTS

EDITOR'S PREFACE iii

MAP OF SITES vi

INTRODUCTION. CHESTER S. CHARD ix

THE ARCHAEOLOGY OF NORTHERN ASIA: SELECTED WORKS

List of Abbreviations 2
The Paleolithic Site Malta: Excavations of 1956–1957. M. M. GERASIMOV 3
Paleolithic Remains in the Lena River Basin. A. P. OKLADNIKOV 33
Ethnic and Cultural Connections of Middle Yenisey Tribes during the
    Neolithic: The Origins of the Samoyedic Peoples. A. P. OKLADNIKOV 80
The Shilka Cave: Remains of an Ancient Culture of the Upper Amur River.
    A. P. OKLADNIKOV 112
Neolithic Remains in the Upper Amur Basin at Ang-ang-hsi in Tungpei.
    V. YE. LARICHEV 181
Ancient Cultures of Northern China. V. YE. LARICHEV 232
Certain Problems Relating to the Shell Mound Culture. G. I. ANDREYEV 249
The Culture of the Prehistoric Population of Kamchatka. S. I. RUDENKO 265
Fieldwork on the Chukchi Peninsula in 1957. M. G. LEVIN 296
An Early Eskimo Cemetery at Uelen: A Preliminary Report on the Excavations
    of 1958. M. G. LEVIN 305
The Penetration of Iron into the Arctic: The First Find of an Iron Implement
    in a Site of the Old Bering Sea Culture. M. G. LEVIN and D. A. SERGEYEV 319
New Finds in the Old Bering Sea Cemetery at Uelen. S. A. ARUTYUNOV and
    D. A. SERGEYEV 327
Ancient Burials on the Chukchi Peninsula. S. A. ARUTYUNOV, M. G. LEVIN, and
    D. A. SERGEYEV 333

THE GEOMORPHOLOGY OF NORTHERN ASIA: SELECTED WORKS

List of Symbols 348
The Quaternary Period in Western Siberia. A. I. POPOV 349
The Main Features of the Quaternary History of the Southwestern Baykal
    Region. S. S. VOSKRESENSKIY 372
Results of a Paleobotanic Investigation of the Quaternary Deposits of the
    Angara Region. M. P. GRICHUK 394
Stratigraphy of the Quaternary Deposits of the Middle Lena and the Lower
    Aldan Rivers. N. S. CHEBOTAREVA, N. P. KUPRINA, and I. M. KHOREVA 452
A Brief Outline of the Vegetation, Climate, and Chronology of the Quater-
    nary Period in the Upper Reaches of the Kolyma and Indigirka Rivers
    and on the Northern Coast of the Sea of Okhotsk. A. P. VASKOVSKIY 464
Bibliography 506

# INTRODUCTION

THE CULTURAL HISTORY of Northern Asia holds much of interest for the student of culture process and human adaptation. The variety of often rigorous environments, the presence or absence of geographic barriers affecting movement and diffusion, the impingement of contrasting ethnic and cultural forces from east and west—to name only a few of the factors involved—make this an area of prime concern to anthropology. Developments here, also, may shed important light on historical problems of Eastern Europe, Central Asia, the Far East, and even the New World. In fact, the prehistory of Northern Asia is attracting growing interest among American archaeologists and anthropologists owing to its significance for basic research, including that into the question of the origins of the indigenous populations and cultures of the New World, a major interest of long standing in American anthropology. The seemingly isolated Western Hemisphere provides a laboratory situation for the testing of many basic theoretical assumptions regarding human behavior, "human nature," and cultural evolution. Determination of the nature and extent of Old World influence on the Americas is vital to the solution of such problems. Since a considerable if not preponderant part of this influence must be attributed to Northern Asia, its identification and evaluation depends primarily on the fullest possible knowledge of the prehistory of this part of the world.

The present volume was conceived as a contribution to provide information on the archaeology and also on the Pleistocene geomorphology of Northern Asia, since the latter is indispensable to an understanding of the earlier phases of human history here. The past thirty years have seen a commendable amount of data collection and research in both fields. Much still remains to be done, but this part of the world is no longer the proverbial "blank spot" of popular imagination. The dearth of information in the Western World is primarily a product of the language barrier, and is thus more apparent than real. From the growing Russian literature, the Advisory Committee of this series has selected for publication here eighteen papers of permanent value or major interest not available elsewhere, which appeared in Soviet sources during the years 1948–63.

As it stands, the volume is primarily directed at the specialist with some knowledge of the area. For those persons approaching the subject for the first time, or with only a limited acquaintance, the following works will provide general orientation and aid in placing the papers in proper perspective.

The outstanding work in English is Henry N. Michael's "The Neolithic Age in eastern Siberia" (*Transactions of the American Philosophical Society*, vol. 48, part 2, 1958). A briefer but important treatment of western Siberia is Demitri B. Shimkin's "Western Siberian archaeology, an interpretative summary" (*Selected Papers of the Fifth International Congress of Anthropological and Ethnological Sciences*, Philadelphia, 1960, pp. 648–661). For northeastern China, the most comprehensive treatment will be found in Kwang-chih Chang's *The Archaeology of Anicent China* (Yale University Press, 1963). Marija Gimbutas's "The prehistory of Eastern Europe, Part I" (*Bulletin of the American School of Prehistoric*

*Research,* no. 20, 1956) and "Middle Ural sites and the chronology of northern Eurasia" (*Proceedings of the Prehistoric Society,* vol. 24, 1958, pp.120–157) contain much that is relevant to Siberian problems. Other general treatments include Chester S. Chard's "An outline of the prehistory of Siberia. Part I. The pre-metal periods" (*Southwestern Journal of Anthropology,* vol. 14, no. 1, 1958, pp. 1–33), "The Neolithic in Northern Asia: A culture area approach" (*Anthropologica,* n.s., vol. 2, no. 2, 1960, pp. 240–248), and "Soviet scholarship on the prehistory of Asiatic Russia" (*Slavic Review,* vol. 22, no. 3, 1963, pp. 538–546).

The following more-specialized discussions focus on particular regions or topics:

CHARD, CHESTER S. Chronology and culture succession in the Northern Kuriles. *American Antiquity,* vol. 21, no. 3, 1956, pp. 287–292.

——— Observations on the Lena Neolithic. *Asian Perspectives,* vol. 1, 1957, pp. 183–198.

——— Mesolithic sites in Siberia. *Asian Perspectives,* vol. 2, no. 1, 1958, pp. 118–127.

——— Organic tempering in Northeast Asia and Alaska. *American Antiquity,* vol. 24, no. 2, 1958, pp. 193–194.

——— Routes to Bering Strait. *American Antiquity,* vol. 26, no. 2, 1960, pp. 283–285.

——— The Old World roots: Review and speculations. *Anthropological Papers of the University of Alaska,* vol. 10, no. 2, 1963, pp. 115–121.

——— Check-stamped pottery in prehistoric Eurasia. In *A Pedro Bosch-Gimpera en el septuagesimo aniversario de su nacimiento,* Mexico, 1963, pp. 95–101.

GAUL, JAMES. Observations on the Bronze Age in the Yenisei Valley, Siberia. *Papers of the Peabody Museum,* vol. 20, Cambridge, 1943, pp. 149–186.

JETTMAR, KARL. The Karasuk culture and its southeastern affinities. *Bulletin, Museum of Far Eastern Antiquities,* no. 22, Stockholm, 1950, pp. 83–126.

——— The Altai before the Turks. *Bulletin, Museum of Far Eastern Antiquities,* no. 23, Stockholm, 1951, pp. 135–223.

MOVIUS, HALLAM L., JR. Late Pleistocene (4th glacial) conditions and Palaeolithic settlement in Soviet Central Asia and Western Siberia. *Actes du IV Congrès International du Quaternaire,* 1955.

QUIMBY, GEORGE I. The prehistory of Kamchatka. *American Antiquity,* vol. 12, no. 3, 1947, pp. 173–179.

TOLSTOY, PAUL. Some Amerasian pottery traits in North Asian prehistory. *American Antiquity,* vol. 19, no. 1, 1953, pp. 25–39.

——— The archaeology of the Lena basin and its New World relationships. *American Antiquity,* vol. 23, no. 4, 1958, pp. 397–418; vol. 24, no. 1, 1958, pp. 63–81.

YOSHIZAKI, MASAKAZU. Prehistoric culture in southern Sakhalin. *Arctic Anthropology,* vol. 1, no. 2, 1963, pp. 131–158.

Malta, the subject of the first archaeological report in the present volume, is the type site for the early stage of the Siberian Palaeolithic; there are no indubitable sites of comparable age with the exception of the Ust-Kanskaya Cave in the Altay, which is thought to represent a still earlier time. Current estimates place the age of Malta at 15,000–20,000 years, which seems not unreasonable in the light of the radiocarbon date from the later site of Afontova Gora II. The initial work at Malta was carried out intermittently between 1928 and 1937; the reports which appeared are not readily accessible for the most part. The present paper deals with the new excavations carried out in 1956–57 by one of the original investigators, M. M. Gerasimov, a famous figure in Soviet science well known for his reconstructions of the features of early human skulls. It is important because the author seems to have no plans for publishing a more definite report on this key site, and because his views and interpretations differ in important respects from his earlier ones. It has been the general view that the early stage of the Siberian Palaeolithic displayed definite western elements, which disappear in the subsequent "classic" Siberian Palaeolithic, the roots of which are now traced to Mongolia (cf. "Northeast Asia," *Asian Perspectives*, vol. 8, no. 1, 1964). Gerasimov here plays down the western aspect, denying, for example, the presence of true burins and real prismatic cores. The paper is of particular interest for its description of Palaeolithic dwellings with clear division into men's and women's sides, and differing structures for summer and winter use. The discovery of actual hafted tools is also noteworthy. Those who, like the present writer, have been privileged to examine the materials recovered can only regret that a fuller report will not be forthcoming. The only other sources on the 1956–57 excavations are to be found in *Kratkiye soobshcheniya Instituta arkheologii*, no. 82, 1961, pp. 128–134, and *Asian Perspectives*, vol. 5, no. 1, 1962, pp. 118–119.

The author of the next three papers, Academician A. P. Okladnikov, is a towering figure in Siberian archaeology who has devoted a lifetime to pioneering investigations in the eastern half of the area and is responsible for the bulk of our information. His report on the Lena valley is still the only work on the Palaeolithic of this important region, describing in detail all possible occurrences, based on his own observations. It is important to bear in mind that the later stages of the Siberian Palaeolithic are clearly postglacial, and it is here that the Lena sites would seem to belong, with the possible exception of Chastinskaya. All are merely temporary campsites with scanty remains. The basis for assigning them to the Palaeolithic is mainly typological, plus in some cases a geological-stratigraphic situation comparable to that of the late Palaeolithic finds on the Angara and Yenisey rivers. Judging by the evidence presented, man was a late-comer in the Lena valley and did not penetrate north of Yakutsk until fairly recent times.

Okladnikov's summary and interpretation of the available data on the Yenisey Neolithic is also the only source on an important region. The Yenisey served both as a major frontier and also as a channel of communication through the heartland of Siberia. Information on the Neolithic stage here is drawn mainly from scattered finds, typologically dated; these are described and discussed, and their significance examined. A few sites are referred to but never described. Cultural parallels and possible imports from the Baykal region, most apparent in the earlier stages, provide a basis for cross-dating. Some links are also seen with the Afanasyevo culture of the steppe and with western Siberia and even the Urals. Okladnikov believes there is evidence indicating the arrival of Uralian tribes from the west towards the close of the Neolithic—most probably the ancestral Samoyedic stock. In general, he sees the area east of the Yenisey as the domain of the Mongoloid Altaic peoples in prehistoric times, and the area to the west as the territory of the Caucasoid Uralians.

The region of the Amur headwaters was a crossroads of cultural influences moving from the Amur basin to the Baykal area and from Mongolia to Yakutia. It is also the possible homeland of the Altaic peoples, where the historic territories of Tungus, Turk, and Mongol meet. Despite its potential importance, the region is almost unknown archaeologically. Hence the significance of Shilka Cave, a prehistoric habitation site excavated in 1952 and 1954, and assigned on typological grounds to the Eneolithic or initial Bronze Age, and possibly overlying a somewhat earlier burial. The thin cultural deposit, probably the result of repeated occupations, is treated as a unit. (The burial context had been disturbed previously.) The bulk of the report consists of a detailed description of all the artifacts with comparative remarks. Cultural parallels are seen primarily with

Trans-Baykal, less with the Lena and Angara valleys, and, to some extent with Manchuria—but very few with the Pacific coast or the Amur valley. The report concludes with an extended discussion of prehistoric cultures in the surrounding regions.

V. Ye. Larichev, a former student of Okladnikov and now his closest collaborator, is one of the very few Soviet archaeologists who reads Chinese and controls the Chinese sources. He has interested himself particularly in the prehistory of Manchuria, or "Northeastern China" (Tungpei) as it is called nowadays. The two papers in the present volume reflect his research on this area. The first is an analysis of the Neolithic and Bronze Age of Manchuria based on the published sources in all languages. He sees three culture areas represented during the Neolithic: the loess area (Yangshao culture), the eastern Chinese plain (Lungshan culture), and the Manchurian Neolithic proper. It should be pointed out that the data for resolving the then-obscure relationship between the Yangshao and Lungshan were not available at the time the paper was written. It has now been demonstrated that Yangshao and Lungshan represent the two major temporal stages of the Chinese Neolithic, embracing all of northern China (see Kwang-chih Chang, *The Archaeology of Ancient China*, referred to above). The Manchurian Neolithic itself Larichev subdivides into four regional complexes: (1) Jehol-Liaoning, the truly indigenous Manchurian culture, agricultural, revealing close ties with northern Korea and the southern Maritime Territory of Siberia; despite Yangshao parallels, this is a distinctive culture. (2) The completely different Lin-hsia region, which is culturally simply the eastern extension of Mongolia. (3) The Ang-ang-hsi–Nonni–Sungari region, culturally a southern extension of Trans-Baykal. (4) The lower Sungari, which belongs culturally with the lower Amur and the interior portion of the Maritime Territory. According to Larichev the essential feature of the Manchurian Neolithic is its mosaic nature —the result of its intermediate position relative to a number of quite distinctive adjacent culture areas. The Bronze Age, on the contrary, is characterized by cultural uniformity over the entire area with the exception of the eastern mountain tribes, and by the northward penetration of Manchurian-Chinese culture for the first time towards the Baykal region. Larichev believes that this "slab grave" culture, as it is commonly labelled, grew out of the indigenous Jehol-Liaoning Neolithic, borrowing bronze metallurgy from the Shang civilization. The sites indicate sedentary life and a mixed farming-stock-breeding economy. These "barbarian" tribes of the steppe had far-flung trade connections to the west and links with the contemporary dolmen culture of Korea and the Shell Mound culture of the Maritime Territory. Larichev hypothesizes that this ethnic continuum represents the ancestors of the Tungus-Manchu.

His second paper describes an important group of Neolithic sites near Tsitsihar, based partly on obscure Manchurian reports and Chinese sources, but primarily on unpublished recent Soviet collections from two of the previously known sites. The Tsitsihar sites are important not only for the abundance of finds, but also because these were recovered *in situ*, whereas most other Manchurian and Mongolian Neolithic sites occur as surface finds devoid of context. For the sake of completeness, a detailed description is given of the total assemblage of artifacts from all sources. Larichev concludes that from the standpoint of the stone and bone industry, the area represents the easternmost outpost of Siberian-Mongolian culture, with which it shares a similar environment. Yet in ceramics, in particular, many southeastern analogies can be discerned. The result is a unique regional culture with a hunting-gathering base, forming the connecting link between China and Siberia.

G. I. Andreyev specializes in the prehistory of the Maritime Territory of Siberia and has been responsible for a considerable part of the recent fieldwork in that area as field director of a section of the Far Eastern Expedition headed by A. P. Okladnikov. The purpose of his paper is to propose a chronological division of the Shell Mound culture of the Vladivostok area, to trace its historical development, and to clarify its economic basis. The significance of the paper lies in the fact that Andreyev presents a counterview in several respects to the picture set forth by Okladnikov in his authoritative *Distant Past of the Maritime Territory,* an English translation of which will soon appear in the present series. Andreyev sees two closely related temporal stages in the Shell Mound culture, and points out the inshore, shallow-water nature of the economy (which Okladnikov had considered a truly maritime one), supplemented by the breeding of pigs, and perhaps of dogs also, for food. Andreyev sees no clear evidence of cultivation either. (The most recent estimates by Okladnikov place the Shell Mound culture in the 11th–9th centuries B.C., based on cross-dating and extrapolation from the few radiocarbon dates for this part of Siberia.)

S. I. Rudenko is one of the most distinguished senior archaeologists in the Soviet Union, best known for his work in the Altay region (Pazyryk frozen tombs) and as author of the first volume in the present translation series ("The Ancient Culture of the Bering Sea and the Eskimo Problem"). His survey of Kamchatka represents a painstaking synthesis of all available information as of 1947—unstudied collections in local museums as well as all published and unpublished reports. Rudenko himself did no fieldwork in the area. Although the author readily admits that his data are inadequate to establish a relative chronology of the known sites, the paper has remained the definitive work on Kamchatka prehistory down to the summer of 1964, when the results of N. N. Dikov's important new excavations were announced, and is still useful for its detailed discussion of artifact types and house remains. Early ethnographic sources are drawn on to amplify the archaeological picture. It should be noted that Rudenko repeats Jochelson's identification of 11th century Japanese coins in the Kurile Lake site, which has since been shown to be erroneous (see *American Antiquity*, vol. 21, no. 3, 1956, pp. 287–292). Kamchatka prehistory has taken on a new dimension with Dikov's report (*Arctic Anthropology*, vol. 3, no. 1, 1965) of his excavations at Ushki on the Kamchatka river, one of the most important sites yet discovered in northeastern Asia, where four successive occupations are represented, separated by layers of volcanic ash. The lowest level, classified as "Mesolithic," displays parallels with the famous preceramic Shirataki site in Hokkaido and has a radiocarbon date of 10,675 years B.P. This is overlain by a core-and-blade complex with burins considered to be typologically "preceramic Neolithic," and ascribed on the basis of palynological data to the postglacial climatic optimum. The succeeding full Neolithic is especially noteworthy for the first indubitable labrets to be found *in situ* in northeastern Asia, and the very large communal pit dwellings. The topmost level at Ushki is a fishing settlement that could be ascribed to the ancestral Kamchadal.

The archaeological section of the present volume concludes with a collection of brief preliminary announcements on the 1956–1960 excavations in the Uelen cemetery on the Siberian shore of Bering Strait. These more ephemeral contributions are included owing to the importance and interest of the finds and the fact that the appearance of a definitive report may be delayed by the unfortunate death of the senior investigator, M. G. Levin. Levin, who was Deputy Director of the Institute of Ethnography, Academy of Sciences of the U.S.S.R., was the leading Soviet authority on the Eskimo problem, and Uelen represents his last major fieldwork. (For an obituary of this distinguished scholar by Henry N. Michael, see *American Antiquity*, vol. 29, no. 4, 1964, pp. 480–483; one of Levin's major works, "Ethnic Origins of the Peoples of Northeastern Asia," was published as no. 3 of the present series.) The Uelen cemetery, discovered in 1955, is certainly one of the major sites of prehistoric Eskimo culture, and in the rich and spectacular nature of the finds it is unsurpassed. The majority of the 76 burials opened are assigned to the Okvik–Old Bering Sea phase; the remaining few to Birnirk or Punuk. Perhaps the most important aspect of the finds is the first Eskimo cranial series from this oldest phase (70 individuals from Uelen and 16 additional from the Ekven cemetery), which proved to resemble the modern eastern Eskimo, thus demonstrating the antiquity of this "classic" type and also providing further substantiation of W. S. Laughlin's hypothesis that the western Eskimo, owing to their large population, have undergone progressive brachycephalization over the course of time, while the original Eskimo type has survived only in the small, isolated communities of the eastern Arctic. The Uelen cemetery also yielded the earliest known iron artifact (a graver) from the Bering Straits area, demonstrating trade contacts with areas where iron was in use—most likely either the lower Lena or the western coast of the Sea of Okhotsk. The spectacular nature of the art finds will be apparent from a glance at the illustrations. It should be noted that the blood group data which Levin mentions having collected among the native population—the first from this part of Asia—have been published in English in *Arctic Anthropology*, vol. 1, no. 1, 1962, pp. 87–92 (the M and MN columns of Table 4 have been transposed). It should also be noted that an English translation of Arutyunov, Levin, and Sergeyev's "Ancient burials of the Chukchi peninsula" has previously appeared in *Arctic Anthropology*, vol. 2, no. 1, 1964. Another translation is included in the present volume for the sake of completeness in the coverage of Uelen. This paper also gives a preliminary account of a new burial site, the Ekven cemetery, of similar age. Further work at Ekven is anticipated, with just as interesting results.

The five papers on geomorphology are from a 1956 symposium on "The Ice Age in the European Section of the U.S.S.R. and in Siberia," and were selected as being of special interest for the early human history of Northern Asia. Popov presents some new material and a generalized Pleistocene history of the lower Ob and the northern part of the West Siberian Plain. He advances as a major thesis that a large area of this northern part was inundated by the sea, the transgression following the maximum glaciation (see also the paper by Movius cited above). Voskresenskiy describes the land forms, Pleistocene deposits, and environmental history of the upper Angara river basin—an area of special interest for the earliest human occupation of Siberia—relating these to archaeological data. Grichuk presents paleobotanical data for the Angara basin in Pleistocene times based on recent pollen studies, and discusses the vegetation history of the region. Chebotareva, Kuprina, and Khoreva characterize the Pleistocene deposits of the Yakutsk-Vilyuy depression; and Vaskovskiy outlines the Pleistocene history of a large portion of northeastern Siberia, hitherto little known, based on his own recent fieldwork. Three glacial and two interglacial stages are identified and described.

The reader who wishes to supplement the coverage provided in this volume is directed to the following published English translations of Russian papers dealing with aspects of the archaeology of Northern Asia:

ABRAMOVA, Z. A. Krasnyy Yar—A New Palaeolithic Site on the Angara. *Arctic Anthropology*, vol. 3, no. 1, 1965.

ANDREYEV, G. I. and ANDREYEVA, ZH. V. 1959 Field work of the Coastal Section of the Far Eastern Expedition in the Maritime Territory. *Arctic Anthropology*, vol. 3, no. 1, 1965.

DIKOV, N. N. Archaeological materials from the Chukchi peninsula. *American Antiquity*, vol. 28, no. 4, 1963, pp. 529–536.

——— The Stone Age of Kamchatka and Chukotka in the light of the latest archaeological data. *Arctic Anthropology*, vol. 3, no. 1, 1965.

DIKOVA, T. M. New data on the characteristics of the Kanchalan site. *Arctic Anthropology*, vol. 3, no. 1, 1965.

FORMOZOV, A. A. Microlithic sites in the Asiatic USSR. *American Antiquity*, vol. 27, no. 1, 1961, pp. 82–92.

LARICHEV, V. E. Neolithic settlements in Cis-Baykal (1957–59 excavations). *Arctic Anthropolgy*, vol. 1, no. 1, 1962, pp. 93–95.

——— On the Microlithic character of Neolithic cultures in Central Asia, Trans-Baykal and Manchuria. *American Antiquity*, vol. 27, no. 3, 1962, pp. 315–322.

——— Neolithic settlements on the lower reaches of the Ussuri river. *Arctic Anthropology*, vol. 3, no. 1, 1965.

LEVIN, M. G. Ethnic origins of the peoples of Northeastern Asia. *Anthropology of the North: Translations from Russian Sources*, no. 3, Toronto, 1963.

MEDVEDEV, G. I. The place of the culture of Verkholenskaya Gora in the archaeological sequence of the Baykal region. *American Antiquity*, vol. 29, no. 4, 1964, pp. 461–466.

OKLADNIKOV, A. P. Ancient cultures and cultural and ethnic relations on the Pacific Coast of North Asia. *Proceedings of the 32nd International Congress of Americanists*, Copenhagen, 1958, pp. 545–556.

——— Ancient cultures in the continental part of North-East Asia. *Actas del XXXIII Congreso Internacional de Americanistas*, Tomo II, San Jose, 1959, pp. 72–80.

——— Archaeology of the Soviet Arctic. *Acta Arctica*, Fasc. XII, Copenhagen, 1960, pp. 35–45.

——— A note on the Lake Elgytkhyn finds. *American Antiquity*, vol. 26, no. 1, 1960, pp. 97–98.

——— The Palaeolithic of Trans-Baykal. *American Antiquity*, vol. 26, no. 4, 1961, pp. 486–497.

——— Palaeolithic sites in Trans-Baykal. *Asian Perspectives*, vol. 4, 1961, pp. 156–182.

——— The temperate zone of Continental Asia. In *Courses Toward Urban Life* (Viking Fund Publication No. 32), 1962, pp. 267–287.

——— The introduction of iron in the Soviet Arctic and Far East. *Folk*, vol. 5, Copenhagen, 1963, pp. 249–255.

——— The history of fishery in North Asia. *Folk*, vol. 5, Copenhagen, 1963, pp. 256–258.

——— Ancient population of Siberia and its culture. In M. G. Levin and L. P. Potapov (editors), *The Peoples of Siberia*, Chicago, 1964, pp. 13–98.

——— An ancient settlement on Pkhusun Bay. *Arctic Anthropology*, vol. 3, no. 1, 1965.

OKLADNIKOV, A. P. and NEKRASOV, I. A. New traces of an inland Neolithic culture in the Chukotsk (Chukchi) peninsula. *American Antiquity*, vol. 25, no. 2, 1959, pp. 247–256.

——— Ancient settlements in the valley of the Main river, Chukchi peninsula. *American Antiquity*, vol. 27, no. 4, 1962, pp. 546–556.

RUDENKO, S. I. The ancient culture of the Bering Sea and the Eskimo problem. *Anthropology of the North: Translations from Russian Sources*, Toronto, 1961.

———— The Ust-Kanskaya Paleolithic cave site, Siberia. *American Antiquity*, vol. 27, no. 2, 1961, pp. 203–215.

SEMENOV, A. V. The ancient culture of the Koryak district. *Arctic Anthropology*, vol. 3, no. 1, 1965.

VASILEVSKII, R. S. Ancient Koryak culture. *American Antiquity*, vol. 30, no. 1, 1964, pp. 19–24.

Other Russian reports are summarized in the following:

*American Antiquity*, vol. 17, no. 3, 1952, pp. 261–262; vol. 20, no. 3, 1955, pp. 283–284; vol. 21, no. 4, 1956, pp. 405–409; vol. 22, no. 3, 1957, pp. 304–305.
*Anthropological Papers of the University of Alaska*, vol. 8, no. 2, 1960, pp. 119–130; vol. 9, no. 1, 1960, pp. 1–10; vol. 10, no. 1, 1961, pp. 73–76.
*Arctic Anthropology*, vol. 1, no. 1, 1962, pp. 84–86.
*Asian Perspectives*, vol. 5, no. 1, 1962, pp. 118–126.
*Southwestern Journal of Anthropology*, vol. 11, no. 2, 1955, pp. 150–177.

Additional English-language sources on the archaeology of Northern Asia include abstracts of Russian reports in the "Arctic" section of the annual *Abstracts of New World Archaeology* (published by the Society for American Archaeology, 1530 P Street, N.W., Washington 5, D.C.); brief abstracts of Russian publications on both archaeology and geomorphology in the various volumes of the *Arctic Bibliography* (published by the U.S. Government Printing Office); complete annotated bibliographies on the archaeology of eastern Siberia and Mongolia, with notices of recent fieldwork, appearing annually in the "Northeast Asia" section of *Asian Perspectives* (Bulletin of the Far Eastern Prehistory Association, c/o Department of Anthropology, University of Hawaii, Honolulu, Hawaii); and the COWA Surveys and Bibliographies for Northern Asia (published by the Council for Old World Archaeology, 11 Divinity Avenue, Cambridge 38, Mass.) which provide selective coverage at intervals of a few years. *Asian Perspectives*, in particular, regularly notes all new translations in English or other Western languages.

CHESTER S. CHARD

*University of Wisconsin, Madison*
*October 1964*

# THE ARCHAEOLOGY OF NORTHERN ASIA:
## SELECTED WORKS

# LIST OF ABBREVIATIONS

| | |
|---|---|
| ESA | Eurasia Septentrionalis Antiqua |
| GAIMK | Gosudarstvennaya akademiya istorii materialnoy kultury |
| GIM | Gosudarstvennyy istoricheskiy muzey |
| IAK | Izvestiya Archeologicheskoy komissii |
| IAN SSSR | Izvestiya Akademii nauk SSSR |
| IGAIMK | Izvestiya Gosudarstvennoy akademii istorii materialnoy kultury |
| IIMK | Institut istorii materialnoy kultury |
| IOAIEKU | Izvestiya obshchestva arkheologii, istorii i etnografii pri Kazanskom universitete |
| Izv. RAI | Izvestiya Russkogo arkheologicheskogo instituta |
| KSIA | Kratkiye soobshcheniya Instituta arkheologii |
| KSIIMK | Kratkiye soobshcheniya o dokladakh i polevikh issledovaniyakh Instituta istorii materialnoy kultury |
| LGU | Leningradskiy gosudarstvennyy universitet |
| LOIIMK | Leningradskoye otdeleniye Instituta istorii materialnoy kultury |
| MAE | Muzey arkheologii i etnografii AN SSSR |
| MAR | Materialy po arkheologii Rossii |
| MGU | Moskovskiy gosudarstvenniy universitet |
| MIA | Materialy i issledovaniya po arkheologii SSSR |
| OAK | Otchet Arkheologicheskoy komissii |
| PIMK | Problemy istorii materialnoy kultury |
| RANION | Rossiyskaya assotsiatsiya nauchno-issledovatelskikh institutov obshchestvennykh nauk |
| RGO | Russkoye geograficheskoye obshchestvo |
| SA | Sovetskaya arkheologiya |
| SMYA | Suomen Muinasmiusto yhdistyksen Aikakausskirya |
| Tr. AS | Trudy Arkheologicheskogo sezda |
| Tr. GIM | Trudy Gosudarstvennogo istoricheskogo muzeya |
| VAN | Vestnik Akademii nauk SSSR |
| VDI | Vestnik drevney istorii |
| ZOOID | Zapiski Odesskogo obshchestva istorii i drevnostey |
| ZORSA | Zapiski otdeleniya russko-slavyanskoy arkheologii Russkogo arkheologicheskogo obshchestva |
| ZRAO | Zapiski Russkogo arkheologicheskogo obshchestva |

M. M. GERASIMOV

# THE PALEOLITHIC SITE MALTA

## EXCAVATIONS OF 1956–1957*

MALTA is rightly considered one of the most interesting Paleolithic sites even outside the confines of eastern Siberia. A great deal has been written about it. Published data concerning the site (its stone- and bone-working techniques, ornamentation, sculpture in the round, design patterns, and so on) are widely used in scientific, popular, and instructional literature.

Investigations of this site were started by the author in 1928.[1] S. N. Zamyatnin participated in the excavations of 1932, and G. P. Sosnovskiy[2] took part in those of 1934. A geological study of the site was made by V. I. Gromov, who visited Malta twice, in 1928 and 1930.[3] Preliminary studies of the fauna and a determination of the species were conducted by the author.

The mammalian fauna from the excavations of 1932–1937 was identified and published by V. I. Gromov in his monograph.[4] Birds and fishes were identified by A. Ya. Tugarinov and V. V. Dorogostayskiy. One of the well-preserved reindeer skulls was studied and described by K. K. Flerov.[5]

The basic part of the archaeological material from the 1928–1931 excavations is stored at the State Hermitage [in Leningrad], and the material of the 1932–1937 excavations in the Museum of Anthropology and Ethnography of the Academy of Sciences of the U.S.S.R. [also in Leningrad]. A small collection of stone tools and casts of ornaments and statuettes is kept in the Irkutsk Museum. A large collection of the fauna deposited with this museum has not yet been studied thoroughly; a significant number of the finds have been lost although they were recorded at the time of discovery. N. I. Sokolov sorted the remainder of the faunal collection and, in oral communication, confirmed [the existence of] a great number of reindeer remains, that is, that more reindeer were actually killed than was indicated in reports dealing with the identification of fauna during preliminary fieldwork.

Malta is regarded by some scientists as a multilayered site. This [view] is a result of insufficient acquaintance with the factual data. Collapsed roofs of dwellings were mistaken for traces of a cultural layer positioned higher than the basic cultural layer of the site. In 1932, a small fire pit was actually found above the base layer of this site. However, it was completely isolated and constituted the remains of a temporary encampment of a very late Paleolithic culture, not connected with the base layer. The notion that the entire area, or a great part of it occupied by the base layer of Malta, is overlain by remnants of an upper horizon is not true to fact.

The basic cultural layer of Malta is represented by a single horizon over 40 m long and approximately 20 m wide, i.e., by an area of 8000 m². The small area marking an upper horizon was found at the very edge of the southeastern periphery of the site; it covers only 30 m², but since it was found to be in an undis-

*Translated from *Sovetskaya etnografiya*, no. 3, 1958, pp. 28–52.

turbed state, its stratigraphic importance is significant. The few implements found there are so characteristic that their archaeological dating presents no problems. The entire complex may be dated to the Baday culture.* The cultural remains lay under a  stratum of humic sandy loam 45–50 cm thick in the upper part of a lower-lying loess-like sandy loam 20–25 cm thick and dark brown in color.

Lower still is a stratum of light-yellow sandy loam up to 50 cm thick, which proved to be completely sterile. Its structural characteristics—unusual porosity and a columnar structure—make the slightest disturbance of the original structure noticeable. Below this sterile stratum is the layer containing the finds of the basic horizon of Malta. In contrast to the upper [strata], this layer is characterized by a completely different structure. It consists of a sandy loam with horizontal stratification and veins of a darker reddish-straw-colored plastic clay, frequently containing barely smoothed small particles of limestone and gravel. The time of formation of the sterile layer of the loess-like sandy loam is archaeologically datable: its upper limits by the Baday culture, its lower limits by the Malta culture (Fig. 1).

Chronologically, the upper horizon Baday culture seems to precede the basic archaeological complex of Verkholenskaya Gora [Mountain] on the Angara.† Sites on the Yenisey river, corresponding chronologically to the Baday culture, are those of the Afontova Gora III type which, in turn, are synchronous with the Azilian-Tardenoisean era of southern Europe. This dating of the Baday culture must still be defined more accurately, but as a preliminary approximation it does not evoke any strong objections from either archaeologists or geologists.

The basic horizon of Malta as yet does not have a precise dating. Yefimenko dates it to the late Solutrean,[6] Sosnovskiy to the Aurignacian-Solutrean.[7] In a number of my reports, I suggested the dating of Early Aurignacian of the Siberian type.[8] Gromov, not disturbed by the time lapse between the Early Aurignacian and late Solutrean, states that all three datings fall approximately within the same period. He is inclined to regard the Malta site as somewhat older than Afontova Gora I and II. He is obviously correct in considering the Paleolithic sites on the Yenisey known at present as later than Malta. Although the faunal complex seems to be confirmatory, it is difficult to show, when the terraces of the Belaya and Yenisey river basins are compared, that the terraces of Afontova Gora and Malta are contemporaneous formations.‡ To deduce an archaeological evolvement of late Paleolithic cultures of the Yenisey or Angara type from a culture of the Malta type is hardly possible; in any case, there is no direct continuity between them.

The Malta site has been visited not only by Gromov, but also by V. N. Maslov, N. I. Sokolov, N. A.  Florensov, N. A. Logachev, and many other geologists. However, there is still no complete geological description of the site. In his extensive monograph describing the Malta site, Gromov includes a geomorphological map of the surrounding area, a cross-section of the left bank of the Belaya where the site is located, a listing of the fauna, drawings of the finds, and so on. He notes the level of the terrace, but hesitates to identify [its height] in relation to the Belaya river basin.[9]

It seems to me that the geological dating of the Malta site on the Belaya river

---

*[Named after an epipaleolithic site located on the left bank of the Belaya, a tributary of the Angara. It was excavated by Gerasimov and designated by him as a type site.—Editor.]

†[An Upper Paleolithic site, 3 km distant from the city of Irkutsk, first excavated by Ovchinnikov in 1897.—Editor.]

‡[The site of Malta is located on the Belaya river.—Editor.]

and the Buret one on the Angara will become significantly more accurate when the correct relationship of the terraces of the two rivers is ascertained. In turn, the determination of synchronous development between the Angara and Yenisey terraces will facilitate the establishment of the time-relationship of the Malta culture and the series of evidently later Yenisey sites of the Afontova Gora types I, II, III, and IV.

During the past few years, I have had occasion to visit numerous Paleolithic sites in Siberia and the European part of the U.S.S.R., including the Crimea. In observing the conditions of the deposition of cultural layers, the floor levels of dwellings, I determined by constant comparison that in the distribution of finds the Malta cultural layer differs sharply from the other sites. It is characteristic of all European and most Siberian sites that the finds never lie in one horizon, that is, the cultural layer is actually of considerable depth. In addition, the finds rarely are positioned horizontally but they often form an accumulation of thickly massed flints and animal bones; moreover, many of the finds lie in a vertical, obviously not the original [i.e., disturbed], position. The exposed floor, that is, the platform beneath the cultural layer, to a greater or less degree is always pitted by depressions, small holes, and large hollows. Nothing of this kind was discovered in Malta during all the years of excavation. The pits or holes where skeletons of polar foxes or fragments of other animal skeletons were found had a completely different aspect. The hiding places also have a different shape and were more often dug in the walls of dwellings rather than in the floors. The platform surface, when cleared of cultural debris, is always more or less horizontal and follows the contour of the microrelief of the ancient terrace surface. As a rule, only the floor of a dwelling was slightly depressed, tamped down; nevertheless, it was always level. There are no pits in the dwellings except for the specially dug fire pits. As a matter of fact, the structural peculiarities of the layer containing the finds are such that any disturbance, even if it happened in antiquity, is immediately noticeable. It should be noted that, unlike most European sites, the Malta site is completely undisturbed by marmot burrows. Only three burrows have been uncovered in the sizable excavated area of over 1000 m² and these did not disturb the cultural layer.

The cultural layer of Malta does not usually have any specific coloring; only in spots do traces of charcoal occur, or small spots of lavender or purple, or more rarely of red and green colored soils. Only in those places that were used for habitation complexes or campfire sites did the layer acquire a more or less distinct coloring explainable either by large accumulations of ashes, charcoal, dyes, or fragmented bones (mostly mammoth tusks), and other organic remains. In the latter case, the layer of sandy loam is darker and more plastic to the touch. Chips of bone and flint are rarely found outside of the living complex; they always lie in a natural, horizontal position. Any deviation [from this position] indicates either that the layer was disturbed or that the artifacts were manipulated by man in antiquity. By close scrutiny of the vertically placed limestone slabs and bones, one can always determine the purpose for which this was done.

As in past years [of excavation], it was noted that all finds were specific to a definite locale of the site—the fire pits or dwellings. Not once was there found an accumulation of bones or other objects of interest outside a dwelling complex; even flint and bone chips were always localized in definite groups depending on the conditions of their production. In a number of places within the site, most often on the edge of the terrace, near the dwellings, specific kinds of debris were found. These were accumulations of kitchen refuse: crushed and charred bones, cemented together by the action of [hot] ashes. Fragments of flint, quartz, pieces of broken and charred stones, and occasional pieces of damaged bone and stone artifacts were found mixed with the bones. The extent of these accumulations was determined by the natural relief. Crowded conditions in the dwellings, the necessity to maintain a fire in the hearth constantly, necessitated special care of the living space. Food and industrial refuse—bones, stones, coal, and ashes— were systematically gathered and dumped in specific places. Large bones, mammoth tusks and rhinoceros horns, and reindeer antlers were almost always used for the construction of walls. It has been noted frequently by visitors to the Malta site that the small number of fragments of reindeer long bones did not correspond to the number of antlers and skull fragments found. This can only be explained by [the fact] that bone chips were the principal fuel. The prevailing belief that only spongy bones burn is incorrect. [However,] the use of bones for fuel should not be taken as evidence that there were no forests in the immediate vicinity of the site. The analysis of charcoal found at the site, carried out by A. F. Gammerman, indicates that birch and, apparently, spruce grew near the site. But stocking wood for fuel presented certain difficulties. Bone, on the other hand, although it is hard to light, burns for a long time and bone coal and ash preserve warmth longer. Furthermore, it was not necessary to stock this fuel, as it was already at the location. There are almost no so-called temporary campfires at the site. As a rule, the fire was kindled in specially constructed hearths, sometimes of rather complex construction, with special windbreaks and ashpits.

In the process of cleaning and sorting the remains of dwellings, it was possible in a number of instances to note details of construction which separated the dwellings into several types. Apparently the most widespread type was semisubterranean with a firmly packed floor at a depth of 50–70 cm; the earthen walls were strengthened by a framework of stone slabs and large bones, often placed vertically. The roof of such a house was apparently covered with hides placed over a framework of wooden poles resting on the walls. The hides were weighted with reindeer antlers and a thin layer of earth. In a number of cases these houses had walls on three sides only with the open side facing the river; the roof on the open side was held up by vertical, thin poles anchored at their base by bones. Additionally, traces of light, tent-like structures, apparently for summer use, were discovered. The basis for the projection that these were specifically summer dwellings is the fact that bones of waterfowl and fish were found in them and in no other dwellings.

In connection with the construction of the Bratsk hydroelectric station the necessity arose to make a geological and paleontological survey of the Quaternary deposits in the Angara river valley and the estuary parts of the Angara's tributaries between Irkutsk and Bratsk. The Geological Institute of the East Siberian branch of the Academy of Sciences of the U.S.S.R. organized a special Angara Geological and Paleontological Quaternary Expedition. Territorially, the Malta site is included in the area of operations of this expedition. A cross-disciplinary investigation of this interesting site could yield a large collection of archaeological remains, and also a sizable number of faunal remains. Besides, [a determination of] the stratigraphy of the site could reveal interesting data about the time and rate of formation of the Belaya river terraces. In connection with [these investigations], the directors of the Institute of Ethnography of the Academy of Sciences of the U.S.S.R. sent me to work under the auspices of the Geological Institute and to renew the archaeological investigations of Malta and of a number of other sites on the lower Belaya river. In the course of two field seasons, 1956 and 1957, new data were gathered concerning this interesting site of the Upper Paleolithic.

Numerous sites of the Upper Paleolithic along the Angara, Belaya, and Yenisey rivers and a number of sites in the Kostenko-Borshchevsk region cling to the banks of ancient gullies and ravines; frequently, they are situated far from a river and, generally, they have a very extensive area of occupancy. In contrast to these, the Malta site occupies a narrow strip, no wider than 20 m, along the very edge of the terrace. Formerly, it was thought that part of the site had been destroyed by erosion of the bank of the terrace. However, as has now been determined conclusively, most of the Malta dwellings were located along the very edge of the present-day escarpment of the terrace. This cusp of the third-flood terrace is supported, along almost the entire length of the site, by a thin [remanent] strip of the second terrace, now largely eroded and disturbed; segments of it in the given area partially rest against a slope towards the steep shelf of a Cambrian stratum. Half a kilometer away from the site, the third terrace turns sharply, receding from the present-day riverbed. At the same time, it gradually becomes lower and merges with the second terrace which, in this area, is wide and slopes towards [the third terrace].

The excavations of 1956 and 1957 were started at the very edge of the terrace. They were a direct continuation of the 1937 excavations. We were interested in uncovering a previously established archaeological complex, and in confirming previous findings in regard to the extent of the cultural layer and the degree of preservation of the cusp of this ancient terrace.

In the process of investigation of the cusp of the third-floor terrace in the area of the site, it was established that the cusp, delimited by the uppermost extent of level "D," was not in the form of a steep escarpment. The plane of inclination of the cusp did not exceed 35°. The surface and cusp of the ancient terrace are covered, as mentioned earlier, by a deposit of loess-like sandy loam, deepened in color by humus in its upper part. The edge of the present-day cusp of the third-flood terrace (and it seems to have been preserved intact in a number of places) forms a steeper escarpment—of inclination up to 45°. The cultural layer, which lies parallel to the cusp of the ancient terrace, thins out [to a wedge] as it comes closer to the edge. Not one fragment of stone or bone was found at the break of the terrace or farther down the slope [of its escarpment].

During excavations at Malta, three heaps of kitchen refuse were found on the terrace slope, in natural depressions; this finding confirms our proposition that the ancient cusp of the third terrace was not destroyed by subsequent erosion. This can also serve as a partial basis for the proposition that the cusp of the terrace was thickly covered with shrubbery during the existence of the site and possibly even later. Until recently, a barrier of literally impenetrable bird-cherry thickets separated the first-flood terrace from the bottomland. Possibly similar conditions prevailed when Paleolithic man settled on the third-flood terrace. This terrace, which has the present height of 18 to 20 m, was not more than 6 or 7 m above water level during ancient times.

At the time of the excavations, the thick humus layer and a large part of the underlying straw-colored loess-like sandy loam had been cut away by road builders, which made it considerably easier to uncover the cultural horizon. In those places where all the layers remained undisturbed, it could be observed that the stratigraphic cross-section of this particular area of the site was altogether analogous to the one which we had repeatedly determined and published. The diagram (Fig. 1) indicates clearly the character and thickness of the layers:

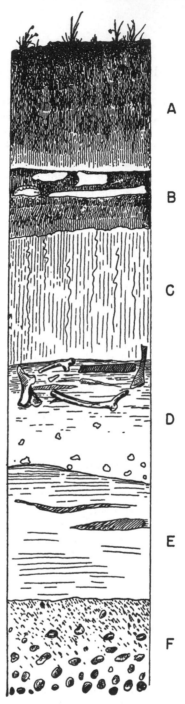

FIGURE 1. A stratigraphic cross-section of the Malta site. A, soil layer, loess-like sandy loam, darkened by humus. B, dark-brown loess-like sandy loam; remains of the Baday culture were found in its upper part. C, light-straw-colored columnar loess-like sandy loam; a sterile layer. D, dark-straw-colored layered loess-like sandy loam; the horizon of the Malta culture lies in its upper part. E, dark-straw-colored clayey loess-like sandy loam containing lenses of rough clayey sand and lumpy raspberry-colored clay. F, brown-colored clayey sand containing pebbles.

Layer A: The soil layer, a sandy loam deep black in color, 40–50 cm thick.

Layer B: A dark-brown to straw-colored sandy loam, 20 cm thick.

Layer C: A light-straw-colored sandy loam, porous, seemingly structureless; when dry it has a tendency to crack vertically, columnarly; rarely, small, often crushed shells of ground mollusks Succinaea (*Pupilla* sp.) are found; there are no traces of animal bones or human cultural remains; [the layer is] 40–50 cm thick.

Layer D: A much darker, straw-colored, layered sandy loam, apparently heavier [more compact], with clearly visible lenses of clayey soil of a brownish-raspberry color. The upper part of this loam, of a darker color, gives the impression of a horizon of buried soil. When freshly cleaned off, this horizon is very clearly observable and appears as a washed-out, thin, dark-straw-colored strip; it should rather be considered as part of the lower than of the upper layer because, when dry, it loses to a great extent its intensity of color and becomes horizontally layered. When the separation of a large block of the top layer of sandy loam is attempted, it always breaks from the underlying loam along the top of the buried soil horizon. This occurrence made it considerably easier to uncover the cultural layer, as it coincides with the horizon of the buried soil.[10] The general thickness of the layer is 30–45 cm. The buried soil is 5–10 cm thick. Poorly smoothed pebbles are occasionally found in the lower part of this layer.

Layer E: A coarse, dark-straw-colored loam, layered with lenses ranging from rough-grained clayey sand to lumpy, plastic, raspberry-hued clay. Its thickness does not exceed 50 cm. Farther down, this layer becomes coarser, brownish-colored clayey sand with pebbles imbedded in it. As the depth increases, so does the number of pebbles, and at a depth of approximately 1 m we find a bed of gravel over 2 m thick, resting on a deep foundation of Cambrian limestone.

The place and size of the excavation started in 1956 were conditioned by the necessity of uncovering, at one and the same time, the entire habitation complex which had been discovered [in part] in 1937. The length of the excavation was 20 m, the width 7 m. For easy localization of the finds, the cleared area was subdivided into 1-m squares. The rows were labeled with letters and the files with figures. As usual, the O line [from which depth was calculated] passed through the upper edge of the cultural stratum.

In this area of 140 m² the complex of cultural remains marked for excavation was stripped. The uncovered platform of the original surface of the terrace was mostly horizontal and only the very edge, forming a narrow strip of 1.5 m to 2 m, sloped noticeably—about 10°–12°—towards the river.

All the large and small animal bones, antlers, tusks, thin and massive limestone slabs, and flint chips that were of interest to us were amassed in a compact area which stood out from its surroundings because of a somewhat darker coloring. This elongated area, shaped like an irregular trapezoid, was centered, it so happened, in the excavated part. The wide side of this area (the base of the trapezoid) faces the river. The peripheral part of the excavation surrounding the area in question was the regular, overall, cultural layer which unites the entire extent of the site. This overall cultural stratum of Malta is poor in finds and has no particular coloring. It follows the contour of the upper part of layer D (the horizon of the buried soil) and contains [occasional] cultural remains throughout the extent of the site. It extends even beyond the limits of the site, but there it contains neither bones nor flint.

The uncovered complex was made up of an accumulation of reindeer tines, large and small bones of various animals, and numerous limestone slabs. In the

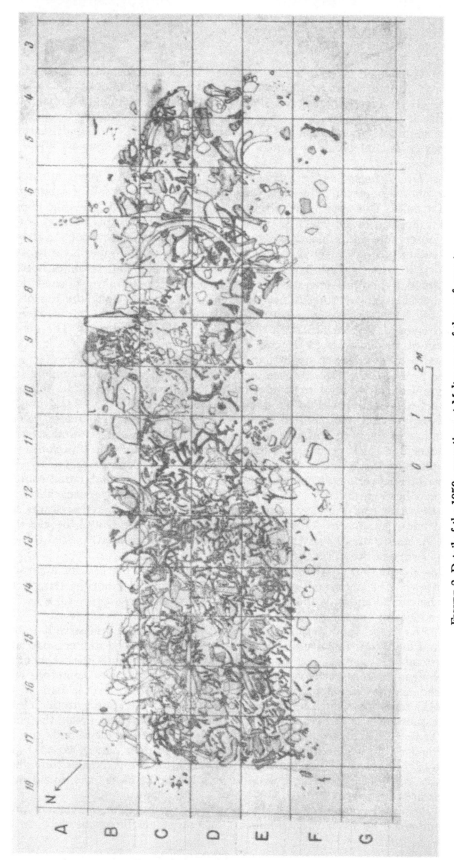

FIGURE 2. Detail of the 1956 excavations at Malta: area of the roof cave-in.

process of clearing this accumulation, it became obvious that its boundaries were delineated by the edges of a depression made in ancient times (Fig. 2). The total length of the heap of debris was 14 m, its width about 6 m; that is, it occupied an area of 84 m². 

The bones and stones were unevenly distributed over the area of the accumulation. The more massive, large slabs lay at the three sides [of the trapezoid], forming a wide arc. Some of them are over 1 m long, 50 cm wide, and 25 cm thick. Particularly massive slabs lay along the boundary of rows B and C. Slabs found within [inside] the arc, in the central [meter-]squares of the accumulation, are quite different. They are thin (up to 2.5 cm), and their dimensions often reach 70 × 50 cm. As one gets closer to the edge of the accumulation nearer the river, the slabs become scarcer and smaller. The last row of squares contained no slabs.

Numerous fragments of reindeer antlers were found, mostly above the slabs, and these were more numerous in the northwestern portion of the heap; in the southeastern part, mammoth shoulder blades and other large bones were more numerous (Fig. 3a). Most of the larger mammoth and rhinoceros bones—pelvis, shoulder blade, jaw, thigh, and bones of the lower leg—served, with the slabs, as a framework for the walls.

On the very top of the debris, flint chips were strewn about irregularly, in heaps, lying over the bones and slabs, in spaces between them, but always within the accumulations of bones in the heap. This was probably the refuse from the stone-tool manufacturing. Not a single complete implement was found among them.

Numerous fragments of reindeer antlers lay mostly in the central part of the accumulation. Their position over the slabs suggests that they could not have been a part of the wall or roof.

A number of complete antlers found along the edges of the depression near the large slabs apparently formed part of the collapsed above-ground section of the dwelling. In removing the antlers and slabs from the surface of the accumulation, we uncovered a number of thin slabs of Cambrian limestone. Their large size and regular distribution on essentially one level provides evidence that they may have been used to weight down the hides which served as the roofing and also themselves formed part of it.

During the excavation of the heap, which apparently was the collapsed roof of a dwelling, the outline of a dwelling which had been dug in the terrace was unearthed. The hollow in the slope of the ancient cusp of the terrace uncovered in this manner was in the form of a wide, level space surrounded on three sides by a low wall. The wide side facing the river was open, i.e., there is not a trace of a wall. The side opposite it was the long side of the dwelling, its back wall. During the time of occupancy the walls of the dwelling were made of stones and bones. Their remnants have collapsed into the hollow. Most of the bones and slabs lay in such a position that reconstructing their original position posed no difficulties. It is interesting that when the slabs and bones were raised, they easily fitted into their original places; also, when placed on edge, the slabs were quite stable (Fig. 3b). After all the slabs and bones of the collapsed roof were removed, what at first appeared to be the bottom of the hollow came into view. However, after careful scrutiny, it became obvious that it was a thin, sterile layer between the roof debris and the floor of the dwelling. This layer was uneven, its greatest thickness of 12–15 cm being near the back wall. As the sterile layer reached out to the edge of the dwelling, it became thinner and tapered off so

FIGURE 3. Excavation of a dwelling, 1956. *a*, partial view of the roof debris; stone slabs and reindeer antlers which settled on the floor of the house pit are clearly visible. *b*, a corner of the dwelling with vertically placed limestone slabs and mammoth bones.

that along the line of row F it was barely 2.5–1.5 cm thick. This sterile layer stood out because of its color, but more so because of its structure. It was of gray color, which differed from the reddish floor of the dwelling as well as from the straw-colored debris of the roof. In addition, it was layered, plastic, and contained quite a number of rough pieces of limestone and other hard particles. Of special interest was the fact that this layer could be easily detached from the floor surface of the dwelling. During the process of cleaning the floor it was noted that this sedimentary layer filled all the spaces between objects lying on the floor. During the final clearing of the floor and walls of the dwelling, several

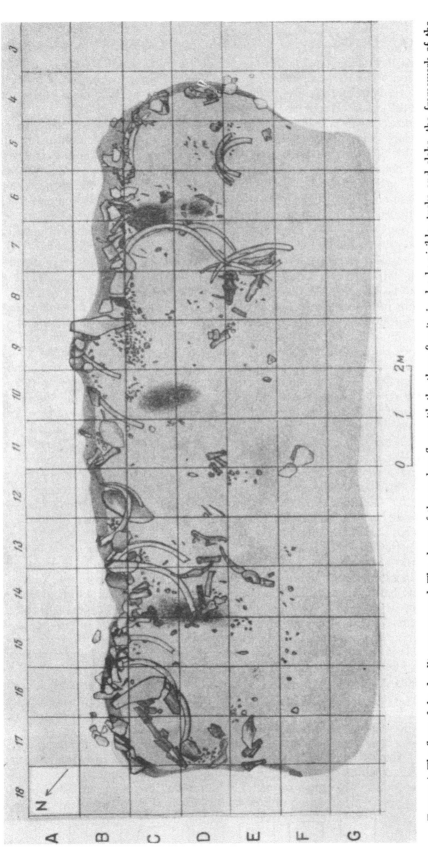

FIGURE 4. The floor of the dwelling uncovered. The shape of the sunken floor with the three firepits is clearly visible; tusks and slabs, the framework of the walls, lie near the walls.

narrow rivulet microchannels were uncovered in which small rivulets had brought deposits onto the terrace from its [higher] bank. The silty mud was brought in by these rivulets very slowly, gradually, layer by layer, and filled in the spaces between the collapsed roof and the floor of the dwelling.

The floor is smooth, firmly packed, and in a number of places retains its original coloring (where it had been powdered with a thin layer of red ochre). It is noteworthy that the collapsed roof did not fall apart into numerous [pieces of] slab and bone—it seems to have lowered itself as a unit, apparently because it was held together by one base. Taking into consideration the size of the dwelling, such a base could have been only a mammoth hide. During the collapse, the free edge of the roof fell to rest immediately on top of the floor of the dwelling, while in those places where the roof rested on the crumbled remnants of the walls, an empty cavity remained which was later filled in by the silty, sterile layer.

When cleared of debris, the floor of the dwelling proved to be more or less horizontal with a barely noticeable incline towards the front edge, i.e., the side facing the river. This is explained by the fact that the floor of the dwelling was dug in the form of a wide hollow at the very edge of the terrace. The natural sides of the hollow served as walls. The height of the back wall was 53–57 cm and was almost even along its entire length. Both side walls, following the configuration of the ancient terrace, gradually decreased from 53–57 cm to 10 cm in height as they came closer to the front edge. The level bottom of the hollow formed the floor of the dwelling. Over the natural earthen walls were constructed walls of vertically placed limestone slabs and bones of large animals, which were part of the framework for the earthen wall superstructure. The entire construction was strengthened by a row of mammoth tusk pieces 1.5–2 m long. Ten such pieces were set into the bottom of the pit, parallel to the back wall. At the time the dwelling was excavated, the tusks lay on the floor, although their position convincingly enough points to their original use (Fig. 4). Three hearths were found in the floor of the dwelling, two of them of rather intricate construction (Fig. 5a, b). On the median line of the dwelling, along row D, bone sockets were found—bases for the poles which had served as roof supports. This dwelling seems to have been an imitation of a rock shelter.

Most of the bone and stone objects found in the dwelling lay in the immediate vicinity of the hearths and walls. A number of objects were located in special hiding places dug into the walls near the hearths. In one such hiding place were found two half-finished bracelets made of mammoth ivory; in another, a likeness of a bird; in a third, a fragment of a female statuette; in a fourth, three stone implements. On the floor there were few, but typical, tools. Most of these were found along the walls. It is especially interesting to note the discovery of two stone tools with bone handles. One of these is a thin lamella with a transverse blade inserted into a handle made of an antler tine of a young deer (Fig. 6a); the other is a massive wedge-shaped blade with a thick antler handle. Both the blade and the handle are polished from lengthy use (Fig. 6b). A similar implement was found at Malta already in 1929, but no such tools have been uncovered in other Siberian or western European Paleolithic sites of this period.

In contrast to earlier artifacts found at Malta, the stone inventory [of this dwelling] shows more archaic features. Disk-shaped cores predominate, degenerate forms of choppers occur, and [there are] many arched scrapers. There are knife-like tools with crude retouching, numerous lamellae, usually shortened, wide, and apparently knapped off disk-shaped cores; there are many perforators, no burins at all, or any really regular prismatic cores; there are comparatively

FIGURE 5. Hearths in the dwelling excavated in 1956.
*a*, the left hearth, the woman's half; the sunken pit lined with slabs is clearly visible: 1, a small hiding place in which a bracelet and an unfinished bracelet were found; 2, a tubular mammoth bone which served as a base for the pole supporting the roof.
*b*, the right hearth, the man's half.

many coarse, scraper-like tools made from large pebbles. The retouch on all the tools is steep, with almost no secondary touch-up or smoothing [creeping] retouch (Figs. 7 and 8).

The assortment of bone implements is typical to the Malta culture and consists of already known types. There are many needles of an early type, without an eye, many awls, massive [projectile] points, daggers, knives, dart heads, and so on (Figs. 9 and 10).

There are few ornaments, but they are very unusual. In addition to beads and pendants, brooches, ornamented plates, stone ornaments, and other artifacts were found (Fig. 11).

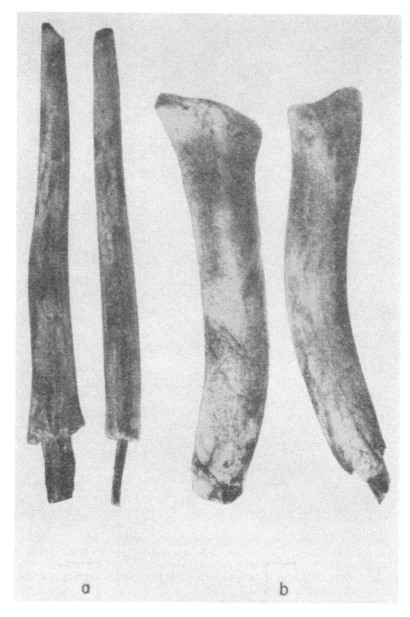

FIGURE 6. Stone implements with antler handles: *a*, retouched flint lamella; *b*, adze-like flint blade.

Of considerable interest are the various statuettes of bone. The representations of waterfowl are magnificent in their simplicity (Fig. 12*a*). The primitive artist achieved near perfection in his depiction of a partridge (Fig. 12*b*).

During the excavations of 1928 to 1937, twenty female statuettes of bone were found. In 1956, four were found. One of these was a fragment. A second (one of the largest) has a number of details not found on statuettes discovered previously: it has a complicated hairdress with a braid thrown over the shoulder onto the chest, and bracelets on the arms (Fig. 13*a*). The third statuette is magnificent

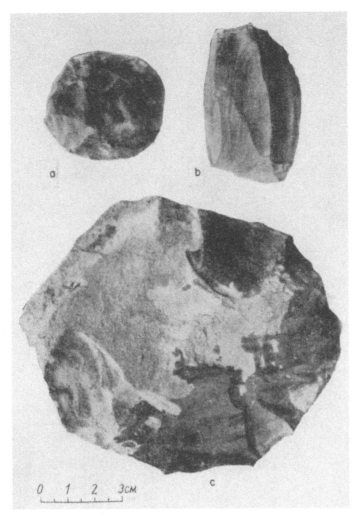

FIGURE 7. Stone inventory: *a*, chopper; *b*, prismatic core; *c*, large discoid core.

in the execution of the face and the hairdress (Fig. 13*b*); the fourth is of a traditional form, but very small (Fig. 13*c*).

It should be noted that in the distribution of the inventory in the dwelling, a certain regularity can be noted. The male inventory lies near the hearth on the right, that is, the hunting implements and representations of birds. Near the left one is the female inventory—knives, scrapers, needles, awls, ornaments, and female statuettes.

In the summer of 1957, the excavations at Malta were continued [and] 348 m² were excavated. This excavation adjoined the 1956 one at its southeastern part. This area, located behind the dwelling uncovered in 1956, turned out to be almost sterile. A few flint fragments, some bones, and small piles of kitchen refuse were found only in the [meter] squares near the dwelling; in a few places limestone slabs were found. Five dwellings of various size and shape were uncovered in the northwestern part of the excavation, at the edge of the terrace. Two

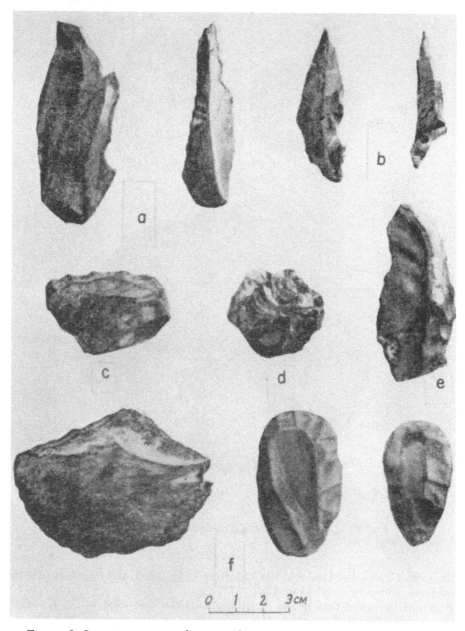

FIGURE 8. Stone inventory: *a*, knives with steep retouch of the Aurignacian type; *b*, perforators [punches]; *c*, scraper with steep retouch; *d*, a small discoid chopper; *e*, a typical lamella; *f*, scrapers.

of these were tent-like summer dwellings of which nothing remained except slightly depressed, saucer-like, packed down, areas with traces of small hearths in the center. Even in ancient times, the foundations of these dwellings were used by man as dumps for kitchen refuse. At the time of excavation they were

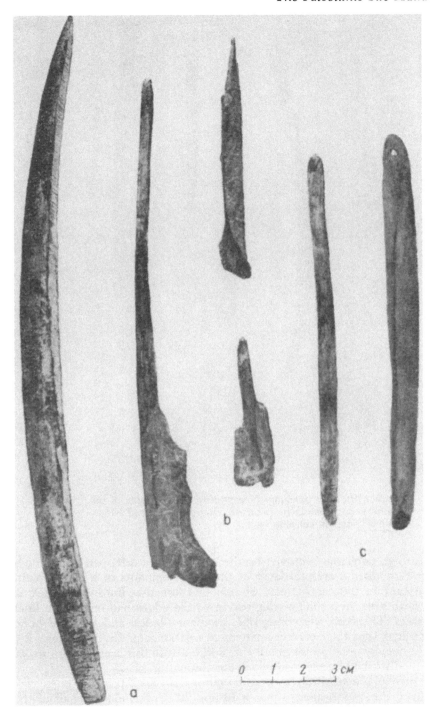

FIGURE 9. Bone inventory (male, from the right half of the dwelling): *a,* dart point with notches on the inside edge; *b,* awls made from the tubular bone of the reindeer; *c,* pressure flakers, perforated for suspension.

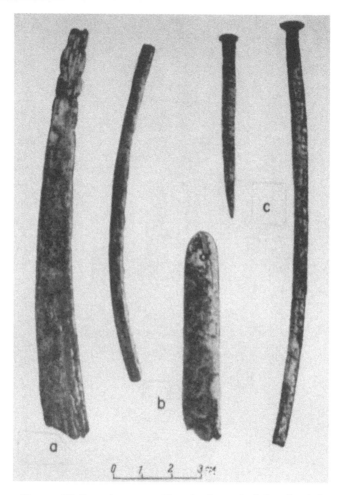

FIGURE 10. Bone inventory (female, from the left side of the dwelling): *a*, a spatula for paints; *b*, fragments of head bands, "diadems"; *c*, pins, hairpins.

severely disturbed, partially destroyed by the roadbed of a dirt road whose deep ruts had cut through the cultural layer of the site. Remnants of a young mammoth were found in the central part of one of these dwellings (Complex 2) under the hearth area, in a shallow depression whose edges had once been lined with small slabs. The skull, vertebrae, ribs, shoulder blades, and shoulder bone, by and large, had kept their correct anatomical relationship. Unfortunately, it is difficult to derive any conclusions about this find because this area of the excavation was badly disturbed by the road which cut through it.

The third dwelling, a semisubterranean dwelling, had been partially uncovered during the 1956 excavation along squares 19 and 20. Unfortunately, half of this dwelling was destroyed by [the construction of] a sod house of Russian settlers in the 18th century. The remaining portion of the dwelling is a rectangular building, 3 × 4 m. Its total length was probably about 9 m (Fig. 14). The roofing of this semisubterranean house was similar in its construction to that of the dwelling uncovered in 1956, but the wooden framework rested on the walls on both sides.

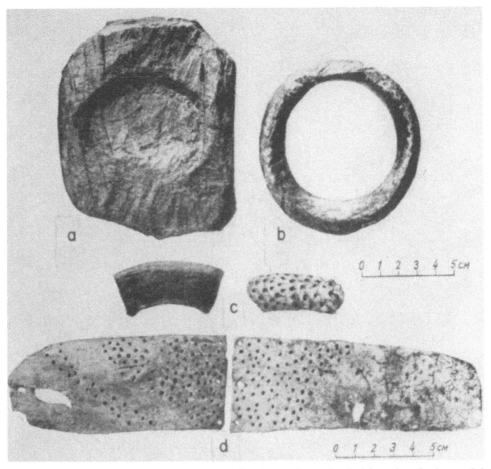

FIGURE 11. Ornaments: *a*, unfinished bracelet; *b*, an almost finished bracelet (both *a* and *b* were found in a small hiding place near the hearth in the woman's half of the dwelling); *c*, bracelet fragments found in the woman's half of the dwelling; *d*, male breast ornament found near the hearth in the man's half of the dwelling

FIGURE 12. Bird carvings found in the right half of the dwelling: *a*, a waterbird in flight (a duck or a loon); *b*, a partridge.

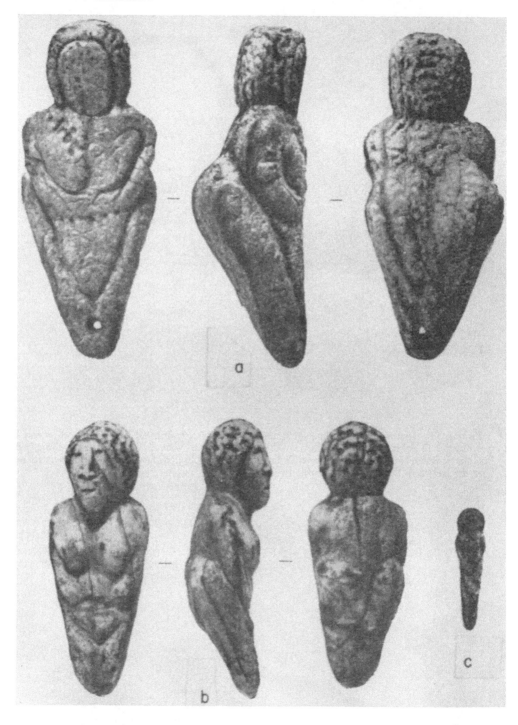

FIGURE 13. Female statuettes found in the left, woman's half of the dwelling: *a*, carving of a stout woman with braided hair and bracelets on her arms (three views); *b*, a carving of a woman with a face (three views); *c*, a miniature female figurine.

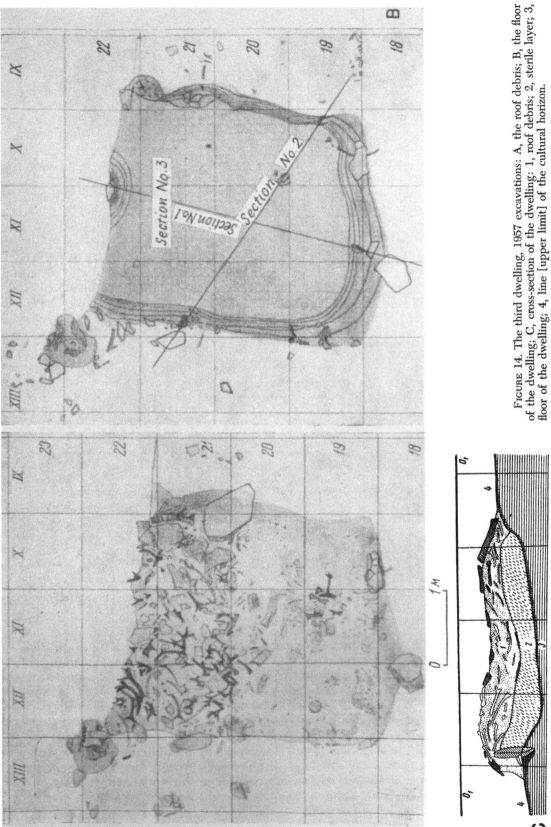

FIGURE 14. The third dwelling, 1957 excavations: A, the roof debris; B, the floor of the dwelling; C, cross-section of the dwelling: 1, roof debris; 2, sterile layer; 3, floor of the dwelling; 4, line [upper limit] of the cultural horizon.

It seems that the roof was also made of mammoth hides over which were placed reindeer antlers and thin limestone slabs. Under the [collapsed] roof, a sterile layer was uncovered with a thickness that reached 20–25 cm near the walls and 8–10 cm in the center. Structurally, this sterile layer was sandy and thin-layered; it contained tiny pieces of bone charcoal and by its bluish-gray color could be definitely distinguished from the straw-colored debris of the roof and the dark, hematite-colored floor. This sterile layer was separated easily from the walls and floor of the dwelling because it was firmly packed. Of the hearth, only a small part of the ashpit remained. It is of interest that the hearth had been raised 20–25 cm above the floor (Fig. 15). During the clearing of the roof, flint fragments, several damaged stone tools, and pieces of two stone dart points were found on its [the roof's] surface. On the floor, not far from the hearth, two pear-shaped beads were found, [also] buttons made from a calcite crystal and a kind of an unusual artifact made from mammoth ivory, which, at first glance, looks like a beetle (Fig. 16*b*). Across from the collapsed entryway, beyond the hearth, in the back wall of the dwelling, a hiding place was uncovered in which a statuette of a flying bird was found. Judging by the long neck, the small head, and the heavy body, it appears to be a representation of a swan (Fig. 16*a*).

FIGURE 15. General view of the exposed floor of the third dwelling; the raised hearths and walls are clearly visible.

The disturbed part of the sod house adjoined the fourth dwelling of which [only] an arced wall, constructed of large [stone] slabs, rhinoceros and mammoth skulls, and their long bones, remained (Fig. 17*a*, *b*). Inside this wall, remnants of the collapsed roof were preserved. The wide "mouth" of the arc faced the river. Two long bones, which served as bases for the poles which held up the roof, were uncovered in the central part of this wide entrance. The roof was probably also made of mammoth hides. Reindeer antlers and thin slabs found on the inner side of the wall were joined structurally to the wall and formed a part of the roof. The floor of the dwelling was not especially sunk [by the removal of the soil], but was firmly packed and charred. There was no hearth. Judging by the charring of the floor, the fire was most often lit near the front edge of the dwelling. After the inhabitants put out the fire, they must have swept the coal

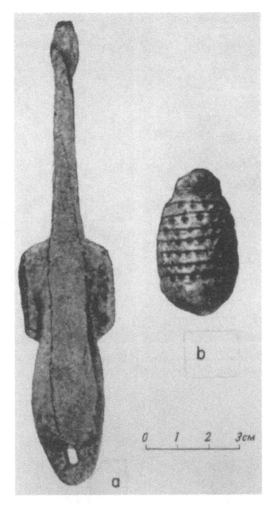

FIGURE 16. Statuettes made of mammoth ivory: *a*, representation of a swan in flight; *b*, a stylized carving resembling a beetle (?).

and hot ash over the entire floor surface, as a result of which most of the burnt bone charcoal and ash collected along the walls inside the dwelling. A sandy, layered sterile deposit was found between the collapsed roof and the floor; it contained a large quantity of small, barely smoothed pieces of limestone and bone charcoal. The deposit was 15 cm thick at the wall, and thinned out at the entrance. It was clearly traceable because of its dark-gray color, and was separated easily from the floor. The shoulder blade of a large bull was excavated at the left [side of the] wall under the roof and on [top of] the sterile layer. At its base, it had a special perforation apparently to secure it to the wall in a horizontal position. On this unusual shelf a reindeer dorsal vertebra was found, the vertebral canal serving as a case for a small female statuette decorated with horizontal lines (Fig. 18*a*). Near the center of the entrance of the dwelling was found an unfinished female statuette (Fig. 18*b*). In the back of the dwelling, near the right [side of the] wall, lay a rhinoceros skull. It is quite obvious that it

FIGURE 17. Walls of the fourth dwelling, 1957 excavations. The walls built of bones of large animals and limestone slabs are clearly discernible: *a*, the left wall; *b*, the right wall.

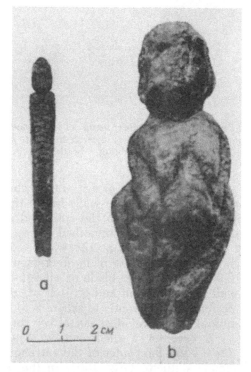

FIGURE 18. Female statuettes: *a*, stylized representation of a clothed woman; *b*, an unfinished representation of a stout woman.

was used as a seat and simultaneously as a hiding place for storing especially precious objects. In this hiding place over 350 small pieces of [mammoth] ivory were found. Out of these it was more or less possible to collect and glue together a representation of some sort of an animal (a hair seal or a bear?).

At the very back of the dwelling, near the wall, the skull and wing bones of some large bird were found. However, these were not the remains of a bird whose flesh had been eaten. Apparently, ancient man saved its skull and wings for some purpose unknown to us.

Thin, double-pointed dart heads were uncovered in two places at the back wall on the outside of the dwelling.

The stone inventory found in this dwelling contained an assortment of implements typical of Malta: large hammering and scraping tools made of pebbles, core-like tools, discoid choppers, arched scrapers, numerous blades with and without retouch, and perforators [punches].

The fourth dwelling was located in an area where the terrace sloped upward rather sharply, and as a result it stood higher than those uncovered earlier. Stratigraphically, however, it was located over the same horizon of buried soil as were the other Malta dwellings. This horizon is the basic layer of the Malta culture.

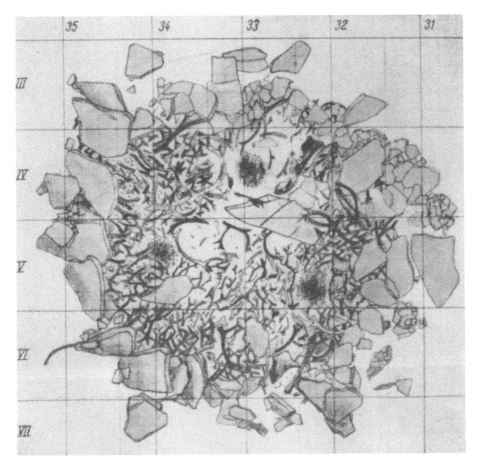

FIGURE 19. The fifth dwelling (excavations of 1957), in the process of being cleared; the walls and roof debris are clearly visible.

In the northern part of the excavation, on the same level as the fourth dwelling, a fifth was uncovered. This small but monumental structure adjoined the right wall of the fourth dwelling. Structurally, it differs from all [dwellings] uncovered earlier; it consists of a ring of massive limestone slabs [and is] about 4.5 m in diameter. Many of the slabs were standing on edge, having retained their original position (Fig. 19). Inside the ring, a large accumulation of reindeer antler tines was found. A certain order was noticed in the distribution of the antlers: they formed a distinct ring directly adjoining the ring of slabs and were connected to it structurally. These were the walls of the dwelling, which had collapsed into the interior. There was no trace of an entrance. Inside the ring, over the slabs and antlers, patches of ash and burned bones—thrown there when the hearth was cleaned—were uncovered in three places.

Under the collapsed roof, as in the other dwellings, a sterile, gray, layered stratum was found, which was separated easily from the floor.

A hearth, the bottom of which was lined with thin limestone slabs, had been dug in the floor, almost in the center (Figs. 20 and 21). Almost all the important artifacts of bone and stone were found close to the walls of the dwelling. The dwelling's male and female sections are clearly distinguishable. At the right side of the hearth—the man's side—pieces of split nephrite, flint lamellar blades, small

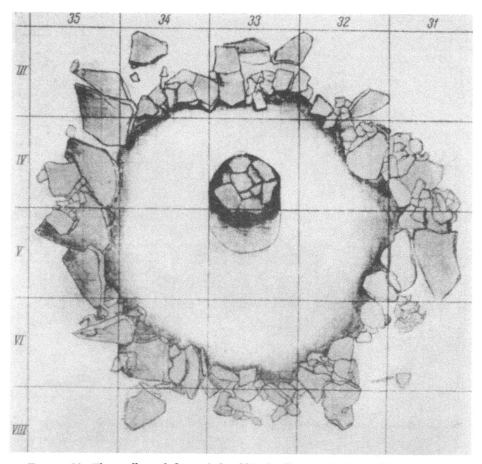

FIGURE 20. The walls and floor of the fifth dwelling with the roof debris removed.

Figure 21. The fifth dwelling after clearing.

choppers, a bone dagger, an unfinished point, and thin chips of mammoth ivory were uncovered; representations of birds (a loon and a swan) were found in the sterile layer above this area (Fig. 22*a*, *b*). At the left side of the hearth—the woman's side—were found beads of mammoth ivory, pendants of calcite crystals, ivory buttons decorated with parallel zigzagging lines, needles, awls, scrapers, and knives. Above these, in the sterile layer, there was a female statuette (Fig. 22*c*).

Thus, as a result of the excavations at Malta during the past two years, data have been assembled which permit a consideration not of separate complexes, but of the entire settlement. However, in this article no conclusions are made about the site [as an entity] since the [analytical] study of the material has only begun. Besides, the excavation of the site is not yet finished, there being an unexcavated area of 250 m². Nevertheless, the significance of this site for the study of the ancient history of Siberia is unquestionably very great.

Indirectly, this material is of interest also to geologists. The condition of the cultural layer in the dwellings, the location of the roofs, the state of preservation of the hearths—all this clearly confirms our earlier observations and proposition that the site has not been disturbed and that the ancient cusp of the terrace has not undergone appreciable changes. It has been confirmed that the cultural layer is connected with the horizon of buried soil. Man settled on this terrace at the time of the completion of its alluvial formation. It is quite obvious that the site was occupied at the same time; no instance of repairs or superimpositions of dwellings was noted. Man lived here for a relatively short time.

A study of the fauna, which is so richly represented at Malta, may answer a number of problematic questions. Conclusions made on the basis of preliminary field evaluations obviously cannot be considered final. Nonetheless, they are of definite interest. In particular, an account of the animal remains will help resolve questions of the seasonal economy of the people at this place and of the length

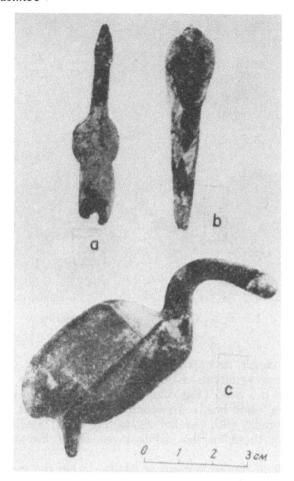

FIGURE 22. Statuettes: *a*, stylized representation of
a duck in flight; *b*, representation of a woman (head-
less); *c*, representation of a standing swan.

of existence of the site. In the analysis of the animals, the first thing to be eva-
luated is the significance of these animals as objects of the hunt and the degree
of their utilization by man.

Most of the bones found were those of the reindeer (*Rangifer tarandus*).
According to preliminary count, 150 pairs of antlers were found. Finds of reindeer bones preserved in [correct] anatomical relationship should also be noted.
This supports [the theory] that the flesh was removed while still raw and then
cured or smoked. The reindeer was fully utilized—its meat for food, sinews for
thread, skins for clothing, antlers for implement handles, and, finally, bones for
fuel.

The mammoth (*Elephas primigenius*) is represented by eight specimens,
apparently young ones. Mammoth meat was eaten, the large bones and tusks
were used as the framework for the dwellings, and most of the bone tools and
ornaments were made from mammoth ivory. The hides served as roofing for
the dwellings.

The woolly rhinoceros (*Rhinoceros antiquitatis*) [is represented by] fourteen specimens. Three rhinoceros skulls possess a peculiar morphological structure which distinguishes them from the typical woolly rhinoceros: despite the fact that these were adult specimens (the teeth were worn, the fifth molar had been cut), they had no nasal septum. Further study will solve the question of the species to which they belong.

There are five specimens of [wild] cattle (*Bos* sp.). One of them is unquestionably a bison (*Bison priscus*). The rest, represented by skull fragments, have a much flattened forehead and long, thin, steeply implanted horns. They have not yet been classified as to species.

The very small number of bones of predatory animals (eighteen specimens of the polar fox, four of the wolverine, one each of the lion and wolf) attest to the fact that hunting for furs had not yet arisen. Large predatory animals were killed by man in self-defense.

It should be noted that the Malta hunter had not yet acquired a constant companion, the dog. The lack of dog bones on the site, and also the lack of gnawing marks on bones of other animals, testify to this. Besides, as has already been noted, animal bones were frequently found in their natural anatomical relationship; this would have been impossible in the presence of dogs.

Now, a few words about the dating of the site. Stratigraphical data and the faunal complex indicate that the Malta site is the most ancient in the Angara basin. That this site belongs to an early phase of the Upper Paleolithic is confirmed first of all by the archaic technique of flint flaking, the absence of pressure-flaking, and the survival of early forms of tools combined with the absence of true burins, stone dart points, and so on. All this indicates the earliest phase of the Upper Paleolithic. There is a point of view which maintains that the ornaments and sculptures of Malta show aspects of a late stage of art development, characteristic of the terminal Paleolithic (Magdalenian). The great variety of forms, the realism of the sculptures and graphic designs, the absence of the commonplace [repetitive], of prescribed standards in the execution of sculptural detail, and also of geometric patterns of ornamentation are undoubtedly signs of a developed, though early, art. The combination of an archaic inventory and developed art is characteristic of the early stage of the Upper Paleolithic. Elements of undoubtedly local Siberian forms have been noted at the Malta site. However, their details are completely different from those of sites belonging to the terminal Upper Paleolithic of Siberia. The sources of the Malta culture are different. All this greatly complicates the dating of the Malta site. Nevertheless, when geological, paleontological, and archaeological data are considered, it must be dated to an early stage of the Upper Paleolithic—the era of the Malta culture, which represents a Siberian variant of the Aurignacian.

# Notes and References

1. M. M. Gerasimov, *Malta—paleoliticheskaya stoyanka* (Malta, a Paleolithic site), Irkutsk, 1931; *idem*, Raskopki paleoliticheskoy stoyanki v sele Malta (Excavations of a Paleolithic site in the village of Malta), *Izvestiya Gosudarstvennoy akademii istorii materialnoy kultury*, no. 118, Leningrad, 1935; *idem*, Obrabotka kosti na paleoliticheskoy stoyanke Malta (The working of bone at the Paleolithic site Malta), *Materialy i issledovaniya po arkheologii SSSR*, no. 2, 1941.

2. G. P. Sosnovskiy, Paleoliticheskiye stoyanki Severnoy Azii (Paleolithic sites of northern Asia), *Trudy II mezhdunarodnoy konferentsii Assotsiatsii po izucheniyu chetvertichnogo perioda Yevropy*, vol. 5, no. 74, 1934.

3. V. I. Gromov, Paleontologicheskoye i arkheologicheskoye obosnovaniye stratigrafii kontinentalnykh otlozheniy chetvertichnogo perioda na territorii SSSR (Paleontological and archaeological bases for the stratigraphy of continental Quaternary deposits within the territory of the U.S.S.R.), *Trudy Instituta geologicheskikh nauk Akademii nauk SSSR*, no. 64, 1948.

4. *Ibid.*

5. K. K. Flerov, Servernyy olen iz paleolita Sibiri (The reindeer of the Siberian Paleolithic), *Izvestiya Gosudarstvennoy akademii istorii materialnoy kultury*, no. 118, Leningrad, 1935.

6. P. P. Yefimenko, Dorodovoye obshchestvo (Pre-clan society), *Izvestiya Gosudarstvennoy akademii istorii materialnoy kultury*, no. 74, Leningrad, 1934.

7. Sosnovskiy, Paleoliticheskiye stoyanki Severnoy Azii.

8. Gerasimov, *op. citati* [Ref. 1].

9. Gromov, Paleontologicheskoye i arkheologicheskoye obosnovaniye. . . .

10. This is a conditional name given to this horizon because all the overlying layers of loess-like sandy loam show obvious traces of diluvial origin, while the underlying stratum of a more plastic loess-like sandy loam is layered and is of alluvial origin.

A. P. OKLADNIKOV

# PALEOLITHIC REMAINS IN THE
# LENA RIVER BASIN*

A. S. UVAROV was the first to suggest the possibility of the existence of Paleolithic sites in northern Siberia. "The only reason that scientists are unaware of the existence of Paleolithic man in these areas," he wrote, "is that the local population pays no attention to chipped tools, mistaking them for ordinary stones and pebbles, whereas the more perfected, polished tools, in particular the very spectacular nephrite axes with their unusually beautiful coloring and often mirrorlike polish, inevitably catch the eye."[1]

I. D. Cherskiy whole-heartedly agreed with Uvarov. According to him, Paleolithic man in Siberia must have lived under considerably better climatic conditions than those existing today, and

found more propitious conditions for living than in the polished Stone Age, when on the Nizhnaya Tunguska [Lower Tunguska] river, for example (at 60° N. lat.), the people who used the nephrite axes which I had found in the area must have withstood the frightfully cold temperatures of this region, which reach −54.8° centigrade. Nevertheless, how far post-Pliocene man spread to the north in Siberia and whether he formed part of the far-northern fauna I have described is not yet known.[2]

It is, therefore, understandable that investigators of the Paleolithic in northern Asia are keenly interested in data occasionally reported in the literature which have a bearing on the distribution of Paleolithic people and sites in Yakutia, including the basin of the Lena and the marine islands far beyond the Arctic Circle.

The first report of the kind is found in the works of Cherskiy himself, who not only was keenly interested in the possibility of discovering new northern Paleolithic sites, but also took an active part in investigating the first Paleolithic site to be found in Siberia—at the military hospital in Irkutsk.[3]

In 1872, Cherskiy, while describing the head of a Siberian rhinoceros brought to Irkutsk most likely from Yakutia, noted on it traces of cuts, some more recent than others, inflicted by some cutting tool near the base of the horns. According to the scientist, his observations in part justified dating the marks of blows on the rhinoceros skull "even to prehistoric times, with the suspicion that the body which was found had been preserved in the ground with the horns already struck off."[4]

Cherskiy, while describing bones of animals excavated in the Tikhono-Zadonsk gold fields in the Olekma district in 1875, again noted the remains of ancient fauna, this time a red deer, with traces of cuts purposefully made by man. This was the left antler of a *Cervus elaphus*, on which, on a "fairly decayed surface," there were evident "clear, deep traces of two blows inflicted by a four-faceted implement, doubtless an arrow."[5]

*Translated from *Materialy i issledovaniya po arkheologii SSSR*, no. 39, 1953, pp. 227–265.

Probable traces of the activity of Quaternary man in eastern Siberia were next reported in 1898 from the Lena valley when N. M. Kozmin published a brief report about traces of the Stone Age in the valley of the Malyy [Little] Patom, a right tributary of the Lena, which joins it near the village of Mach. In his article Kozmin states outright that the traces of man found and studied by him in the valley of the Malyy Patom, contemporary with the mammoth and rhinoceros, were Paleolithic.[6]

Kozmin gives in his report a short account of the remains of a wooden construction, a flooring or *gat*, found by him while working gold-bearing sands, and of his finding nearby, on a high bank terrace, of accumulations of fossil bones of Quaternary animals. He called a *gat* the layer of fossil wood lying at an angle of 20–25° to the plane and consisting of "pieces of timber of different lengths and thicknesses, varying from 1 to 50 cm in diameter." He wrote further: "Considering that the layer of fossil wood was slanted in relation to the plane, while the layers of sand, gravel, and wood above it were horizontal, it is safe to say that it was constructed artificially. The fact that on the southern side it abuts on a projection of the cliff may explain the use of this construction as a ramp from the cliff to the water."[7]

On examining the "pieces of wood from the *gat*, which were beautifully preserved because of the permafrost," the explorer also came to the conclusion that they showed clearly preserved traces of chopping with stone axes "kept intact by the frozen condition of the stratum and turf surrounding layer G, a condition which developed in areas drained of water after the riverbed was deepened through erosion."[8]

Kozmin considers the presence of such sticks hewn with a stone ax as conclusive proof that the *gat* was constructed artificially by Stone Age people. Based on these observations he sets forth detailed speculations about the purpose of the structure and the life of the ancient people who had built it, completing the general picture by a sketch of the landscape of the period. He wrote:

The Stone Age *gat*, as revealed in the course of excavations, abutted on the projection of the cliff, and slanted down to the edge of the terrace, giving easy access to the low riverbank; this was apparently necessary for some purpose, possibly fishing. One may wonder whether the presence of heaps of sticks cut to the same length (about ¼ *arshin**), sometimes lying crosswise, is accidental, or was intended for some construction to help trap fish, or something of the sort.[9]

Near the *gat*, "about 7 *sazhens* [12 m] from it, in the direction of the river," Kozmin reports that he found bones of mammoth and other animals "piled seemingly in heaps, which might only be explained by the existence of Stone Age man's houses, perhaps built over the water like the Swiss pile-dwellings." The "log-path," Kozmin thought, might have connected these dwellings with the bank and at the same time have served the fishing pursuits of the inhabitants. The people might also possibly have lived in caves in the cliffs, as entirely favorable environmental conditions existed there at the time.

The next publication reporting finds which may be dated to the Paleolithic period is that of the geologist M. M. Yermolayev, who was making a special study of the geology of the New Siberian Islands. In 1927–1930 Yermolayev discovered, and we quote, "in strata of the second interglacial period" on Bolshoy Lyakhovskiy, one of the New Siberian Islands, "a piece of mammoth ivory with

*[Ca. 1½ ft.]

traces of work done with stone tools, in soil which was greatly eroded, where it was found together with odd fragments of a mammoth."[10]

In another work, Yermolayev points out that the tusk with traces of working "was apparently lying in its original position" and that the cut on it was made with a stone knife.[11]

The most specific of these reports is that of Kozmin, who not only categorically states that he found indisputable traces of the activity of man living at the same time as the mammoth, but attempts to reconstruct, on the basis of geological and paleontological data, the natural environment and landscape contemporaneous with the ancient inhabitants of the Malyy Patom valley, as well as their settlements and way of life.

We should first of all take into account here, however, the fact that Kozmin himself did not find any stone tools, although he intentionally looked for them and with this idea "examined very attentively all the material obtained." All the more important seems his description of sticks with marks made by stone tools: "The larger and thicker sticks found in stratum G were characteristically shaped like a spindle, which could not be avoided because of the dullness of the ax and the fact that it took two, three, or more strikes of the ax to shape a stick 2–3 cm in diameter."

Yet in this case, too, one must consider that exactly the same type of marks are left on a tree by beavers [gnawing it down for] constructing a dam. The "presence of heaps of sticks cut off to one length," mentioned by Kozmin, is characteristic of the constructions of these animals.[12] The fact that beavers lived in the Lena basin, along its tributaries and neighboring rivers, even in fairly recent times is attested by the studies of A. D. Chekanovskiy and R. K. Maak made during the second half of the 19th century. Chekanovskiy and Maak observed on the Nizhnaya [Lower] Tunguska well-preserved remains of birch *gats* [log-paths] and flooring constructed by beavers. "The founders of both [Upper and Lower] Korelina villages were attracted thither by the abundance of beaver in the area. . . . Today only the 'leavings of the beavers' remain, the birch logs cut by them," he wrote in 1876.[13]

Maak wrote later on the same subject, asserting that the founders of pioneer Russian settlements on the Lena, the villages of Podvolochnoye and Verkhne [Upper] and Nizhne [Lower] Korelina, were attracted there by the abundance of beaver.[14]

The presence of beavers in eastern Siberia not only in most recent times but also in the remote era of the mammoth and rhinoceros was demonstrated by Cherskiy, who observed that a fragment of a beaver's pelvic bone together with remains of a rhinoceros were both stained the same dark-brown.

As for the layer of fossil wood discovered on the Malyy Patom, if this really represents a beaver dam, and not a chance accumulation of pieces of wood deposited by the water, for instance at floodtime, it must be recognized, judging by the stratigraphic conditions, as a very ancient find, invaluable from the paleontological viewpoint, since the remains of the dam on the Malyy Patom are buried in deposits of the old 12–20-m terrace, dated by the remains of Quaternary fauna, including mammoth bones.

It does not at all follow, however, that there were dwellings of Paleolithic people here. The deposits containing the log-path and the fossil animal bones found nearby are undoubtedly of alluvial origin. As Kozmin asserted, they therefore were formed at the bottom of the river. These particular accumulations, or "stores" of bones as Kozmin calls them, in all probability occurred not as a result of the activity of man, but rather of the action of the water which carried them into depressions in the bottom of the riverbed. It is also important that Kozmin apparently found here none of the large accumulations of bones in one spot so typical of the Paleolithic settlements. Usually the bones here did not lie in heaps, but were found by the workmen from time to time in different places. Such isolated finds of faunal remains in alluvial deposits of the Quaternary period are not unusual.[15] Accumulations of mammoth and rhinoceros bones, and the isolated bones from ancient alluvial deposits, have nothing in common with the heaps of bones of Quaternary animals in the real Paleolithic settlements of the open type, that is with kitchen refuse in the strictest sense, particularly if combined with building materials of Stone Age man. Consequently, Kozmin's conception of a log-path or flooring constructed by aboriginal people, with a pile-settlement over the Malyy Patom river, is without foundation.[16]

What is more, if we are to agree with the opinion of Yermolayev that the mammoth tusk from Lyakhovskiy Island had been cut with a "stone knife" in the interglacial period preceding the last (Würm) glaciation, then we must allow that all of Yakutia, up to what today are the islands of the Arctic Ocean, was settled by Paleolithic man. As for the first find described by Cherskiy, if we are to admit that the cuts on the rhinoceros skull were made before it was buried, then this would suggest an even earlier penetration of man into Yakutia, since modern geologists and paleontologists such as V. I. Gromov and I. T. Savenkov consider that the rhinoceros became extinct in Siberia before the mammoth.

Unfortunately, the observations of Cherskiy cannot be verified by new investigations. Actually what must be admitted about his observations is that Cherskiy limited himself to very carefully generalized statements, in which he was undoubtedly right, since not only the conditions under which the skull was found were unknown, but even the place. And, secondly, it is extremely difficult to judge the age of the cuts on the skull; it would have been quite sufficient for the skull to be exposed to air for a short time to have comparatively fresh cuts take on the look of old, or even prehistoric ones. One should take into account also that Cherskiy himself, in his late works on the faunal remains from the New Siberian Islands, when writing about this same rhinoceros skull, and of traces of the work of primitive man in the North, does not again refer to these mysterious cuts. As for the find of Yermolayev in the New Siberian Islands, the tusk in question from Bolshoy Lyakhovskiy was turned over by him to V. I. Gromov for study. The latter's conclusions about the tusk and the supposed traces on it of work by man remained, however, unpublished. Yermolayev himself does not enter fully enough into the reasons for his dating the tusk to the interglacial period. In particular, it is impossible to judge from his communications whether in this case the action of a very important physical factor under arctic conditions can be excluded—so-called thermal erosion, as a result of which bones buried in the frozen layers in the Arctic rise to the surface, then often again sink underground. It is quite possible that the tusk found by Yermolayev went through this process more than once. In this case the cuts found on it might have been made by traders, if not in the 19th, then in the 17th or 18th centuries, or by the local indigenous population.[17]

Thus, the facts recorded in the above-mentioned works of Cherskiy and Yermolayev, as well as the communications of Kozmin, cannot be considered as evidence of the spread of Paleolithic man in the Lena river basin, and the adjacent territory of Yakutia. But from this it certainly does not necessarily follow that there were not or could not have been Paleolithic men, including contemporaries of the mammoth and rhinoceros, in Yakutia and in general in northern Siberia.

Cherskiy, who followed Uvarov in pointing out that all stone tools formerly found in the northern regions of eastern and western Siberia belonged only to the Neolithic period, correctly surmised that "this did not in any way decide the question of Paleolithic man in the negative." "The fact is," he said, "that the crude Paleolithic tools, which do not attract the attention of the local population, can only in the rarest circumstances come into the possession of specialists, who visit these inhospitable places with a haste brought about by the aims and conditions of the expedition on which they are engaged."[18]

Cherskiy's speculations about the possibility of discovering a Paleolithic culture in northern and eastern Siberia have now been upheld as far as the Lena area is concerned. The general nature of the Lena basin, particularly in the upper part between Kachuga and Zhigalovo, is quite favorable for finding Paleolithic remains. As at Krasnoyarsk on the Yenisey or at Irkutsk on the Angara, there are broad river valleys here with their characteristic series of old terraces. It is true that the clayey soil of the upper Lena sometimes differs from that of the Yenisey or Angara in its greater compactness and in its brown or even rosy-red color, which is due to the base rock—red sandstone.[19] Nevertheless, even in the layers of reddish clay along the Lena bones of those animals which coexisted with Paleolithic man on the Yenisey and Angara are found constantly, as has been pointed out by Kropotkin and others after him—the mammoth, rhinoceros, prehistoric ox-bison, and reindeer.

P. A. Kropotkin was the first to discover in the clayey soil of the Lena the ground mollusks *Pupa* and *Succinea* commonly found in the loess deposits. After studying the Quaternary cross-sections of the Lena area, he came to the conclusion, on the basis of these finds of ancient fauna and the general character of the deposits, that the loess-like clayey soils and clays along the Upper Lena "very closely resemble the Rhine and Himalayan loess."[20]

Cherskiy supported the observations of Kropotkin and Chekanovskiy. According to Cherskiy, the loess, with its admixture of loess-like porous clayey soils containing the terrestrial mollusk fauna, is found on the Angara at Irkutsk and on the Lena in the Verkholensk district, where it does not extend too far to the north beyond Verkholensk.[21]

Chekanovskiy also found in the upper layer of red clay at Biryulka village the shells of *Succinea* and *Helix*. He found the same fossils in the red deposits three versts downriver from the village of N.[izhnaya?] Kosogoliskaya on the Lena.[22]

Taking all this into consideration, it is difficult to imagine that in areas so favorable from a geomorphological standpoint and so near to the Cis-Angara region with its numerous Paleolithic sites, there should not have been settlements of the first hunters of the mammoth, rhinoceros, wild horse, bison, and reindeer. The cultural remains of these initial inhabitants should be preserved here also and in similar places as on the Angara and Yenisey, that is, together with the remains of Quaternary fauna in the clayey soil and sandy loam covering the old river terraces. And actually at present there are reasons which allow us to establish indisputably the presence of Paleolithic remains not only on the upper Lena but also on the middle Lena, i.e., in the Yakut A.S.S.R. (Fig. 1).[23] Let us have a look at these sites.[24]

Paleolithic finds on the Lena river were first reported in 1927 above Kachuga, near the village of Biryulka on the right bank of the Biryulka river, 3–4 km downriver from Zalog village, in the Ponomarevo area, on an old terrace not less than 50 m in height.

The eroded terrace revealed horizontal layers of the red Lena sandstone under a thick deposit of reddish clayey soil. In the upper layer of this "century old" clay soil, famous for its fertility, on the surface of old ploughland, bone fragments of large Quaternary animals have occasionally been found in depressions and water

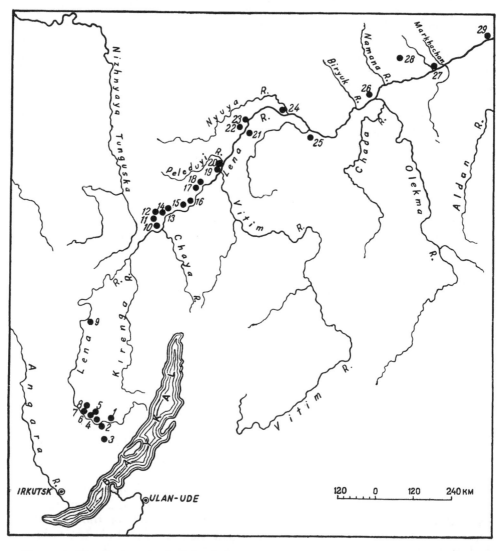

FIGURE 1. Distribution of Paleolithic finds in the Lena river valley. Numbers 1–29 correspond to the numbering of the descriptions of the finds in the text.

holes, and also Paleolithic stone implements, mostly small core tools of the prismatic type, and one lateral flake from a large core

Among the finds there are also two large tools picked up at the site. One of these is a large scraper-like implement with a bow-shaped, convex oval blade and considerably thickened dorsum carefully trimmed from both sides by long and fairly broad parallel flaking. The blade of the instrument is irregular and much blunted, apparently from blows struck vertically on some comparatively hard material. Judging by these traces of use, the implement, in spite of its scraper-like appearance, was used not as a scraper in the strict sense of the word but as a chopping tool.

The second implement also belongs to the category of chopping tools. It is a large, flat core, covered on one side with long, broad, grooves [flutings] left when lamellar blades were struck from it, and with a carefully prepared striking platform. The opposite edge and especially the lower end of the tool were trimmed from both sides with coarse retouching, which in places has almost a Solutrean aspect, with the long and strictly parallel facets characteristic of Solutrean tools forming a slightly wavy blade, massive when seen from above. With such a tool one could easily chop and break either wood or large bones.

Like the cores, these scraper-like chopping tools were prepared from large nodules of white limestone flint of a fairly high quality. Only in one instance was a tool made of dark-gray flint. All the tools are completely covered with a layer of heavy patina, and in addition have a thick limestone crust on one side.

Also worth noting are a cleaver made from an oval pebble, and a large oval scraper made of a flake from a boulder of dark-green jasper. On the lower part of the scraper the smooth surface of the boulder is perfectly preserved, as is often the case with Late Paleolithic scrapers* from Siberia.[25]

The great age of the finds at Ponomarevo is indicated by their shape as well as by the material from which they are made. White flint was very rarely used on the Lena in Neolithic times, while the greenstone jasper-like rock was one of the basic materials used in Paleolithic Siberia for the manufacture of large implements.

2

There are traces of a second Paleolithic settlement at another point, not far from Biryulka village near the mouth of the Manzurka river, on the right, raised bank near Bayraka village in the Khabsagay [Khabtsagay?] region, where in the same year of 1927 an Eneolithic cemetery and Iron Age burials were found. Here the old terrace, about 20 m in height, is easily distinguishable, with its almost perpendicular eroded limestone facing the river. In the yellowish clayey soil covering the limestone are occasional stone chips and flakes, which could belong to the Paleolithic period judging by their stratigraphic positon in the clay layer. In addition to this, a burin of black flinty slate was found in one of the Eneolithic burials. In form and technique of preparation this burin was close to the Paleolithic and quite different from the usual Eneolithic tools, which on the upper

---

*[For a discussion of Late Paleolithic hand-axes and scrapers see: H. N. Michael, The Neolithic Age in Eastern Siberia, *Trans. Am. Phil. Soc.*, n.s., vol. 48, pt. 2 (April 1958), pp. 35 and 36.]

Lena do not include burins at all. It is not impossible that it might have fallen there by chance when the grave was being dug.

*3*

A third Paleolithic find was recorded in 1928 in the Manzurka river valley, on the second floodplain bench at a height of 8–12 m, near the Stepno-Baltay *ulus*. Here separate stone tools of a Late Paleolithic aspect were found, among them a typical oval scraper.

*4*

The fourth Paleolithic site was found in 1941 on the right bank of the Lena near Makarovo village, 15 km below Kachuga, in the area of the unusual burials of the Glazkovo stage and a Neolithic site.

All sites are associated with a well-defined floodplain terrace, 6–8 m high, near to, and downriver from the western limits of Makarovo village. River sand is found at the base of the terrace. Above this is a layer of yellowish sandy clay 30–45 cm deep, and above this is a humus horizon about 20–30 cm thick.

In the humus horizon everywhere along the edge of the terrace are found fragments of clay pots typically Neolithic in ornamentation and general appearance, chips, arrow points, and scrapers. In 1929, a hearth was found here such as was customary in Neolithic settlements, put together from river pebbles and slabs of red sandstone, and around it were found potsherds with impressions of coarse woven cloth or netting characteristic of the Serovo stage of the Baykal Neolithic.

The Paleolithic remains as revealed by the excavation of an ancient hearth found in 1941 lay deeper, in the clayey soil stratum. The hearth was discovered during an examination of the eroded edge of the terrace. The erosion had partially exposed the lowest stones of this construction, which protruded slightly when the turf and clay covering them had been carried away. The uphill part of the hearth and the strata covering it were less disturbed by erosion, and this permitted a fairly accurate correlation of the Paleolithic remains with the deposits of the terrace in question (Figs. 2, 3, and 4).

Clearing of the hearth revealed that it lay beneath the humus horizon, at a depth of 15–20 cm in the clayey soil. The base of the hearth rested on sand, which under the hearth itself was of a reddish shade from the prolonged action of fire. In places the sand was burned to a depth of 10–15 cm (Figs. 5 and 6).

The hearth was very carefully constructed from blocks of red sandstone stood on edge or horizontally positioned, comparatively small in size (20 × 30 cm, 20 × 15 cm, etc.). The general shape of the hearth was that of a bowl. Small stone blocks formed a ring or almost perfect circle around the edge. These were placed vertically, with the lower ends slightly slanted towards the center of the hearth, and were all carefully fitted with their edges touching. Some small slabs of sandstone had been positioned horizontally in the very center of the hearth, in the area delimited by the blocks standing on edge. Some of these slabs were level with the top edges of the blocks framing the hearth, others were lying lower, close to the bottom of the hearth. The bottom was completely paved with horizontally placed slabs of stone. On the very bottom was preserved a little ash and charcoal. The rest of the coals and ash which formerly filled the hearth had seeped out from it through the cracks between the upright blocks, and lay adjacent to the lower part of the hearth blocks further downhill, in the form of a solid black patch, clearly distinguishable against the background of the clay.

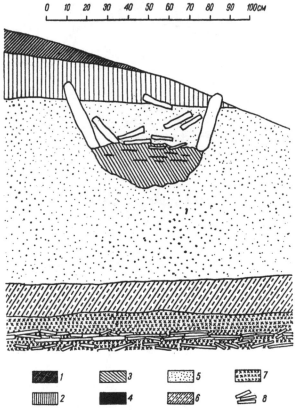

FIGURE 2. Cross-section of the Makarovo site: 1, chernozem; 2, pale yellow loam; 3, burned sand and ash; 4, coal; 5, sand; 6, green-gray coarse sand; 7, red loam; 8, flatstone.

In addition to the charcoal and ash, the hearth contained bits of animal bones, among them a fragment of a metapodial bone and of a rib from a fairly large animal (deer?). Stone tools are represented by one specimen only—a beautifully worked and typical scraper. Its shape is quite common to the Late Paleolithic complexes of Siberia, a massive and broad blade with a convex crescent-like working edge. It is made of a large flake from a water-worn river pebble, of a jasper-like stone, bluish-green in color. Both flat sides, the upper and lower, have been carefully trimmed with broad flaking. On the dorsum, that is the edge of the tool opposite the blade, are remains of the [original] surface of the pebble. The blade is thick and steep. It was carefully retouched twice: the basic facets of the flaking are broad, while below these are small facets of secondary retouching. The curve of the blade is not quite regular. At one end the curve is sharper, and the blade at this end is considerably narrower than at the other. The length of the scraper is 9 cm and the maximum width 5 cm.

Inside the hearth, besides the scraper, there were two fragments of a flat, well-worn river pebble with traces of flaking.

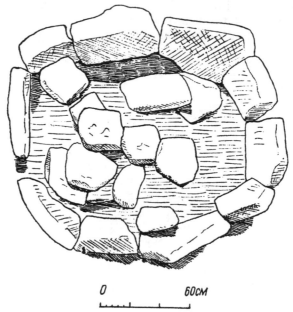

0          60CM

FIGURE 3. Paleolithic hearth of the Makarovo site, view from above.

0   10   20   30   40   50   60   70   80   90   100CM

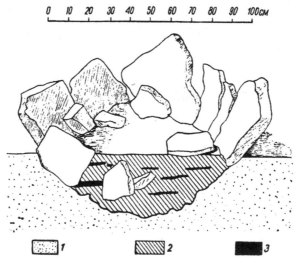

1          2          3

FIGURE 4. Paleolithic hearth of the Makarovo site, cross-section: 1, sand; 2, burned sand and ash; 3, coal.

Near the hearth, on the eroded part of the terrace, there were fairly numerous chips and two tools which should be contemporaneous with the objects found in the hearth.

The first artifact is a small, typical core scraper, 4 cm long and 2.5 cm wide. It was made of a lamellar flake knapped off a pebble of green, jasper-like rock. Like any other articles in this distinctive group of tools peculiar to the Siberian Stone Age, a group composed of a number of special individual types, the core-scraper in question was probably never used as a core, in spite of the well-prepared

FIGURE 5. Hearth of the Makarovo site.

striking platform, but was intended for other purposes. The narrow facets on one side apparently were made only to give a proper shape to the surface of the tool, in place of retouching. The bottom side of the tool preserves the smooth surface of the pebble. The outer side opposite the platform of this tool has a broad, specially prepared concavity, carefully retouched along the edge with fine, secondary flaking so that it forms an arc-shaped, slightly bevelled cutting edge similar to the blades of the grooved Neolithic gouges.

The second tool is a scraper made from a thick longitudinal flake struck from a fairly large, prismatic core of greenish-gray jasper-like stone. Its length is 5 cm, its width 3 cm. The lower part is formed by the surface of the flake, with its pronounced bulb of percussion and rippling. On the upper side there are traces of trimming. There are broad facets along one edge of the scraper, where lamellar blades had been detached from it. The other edge is formed by typical Paleolithic retouching—steep, with broad, deep facets—and forms the convex, bow-shaped cutting edge typical of this type of implement. The retouching also extends to the upper end of the tool, where it is executed with the greatest care and is modified by fine secondary retouching. This tool, like the core-scraper described above, is Paleolithic in character, in both form and retouching, and both implements can therefore be associated with the same period as the finds in the hearth.

Hearths of the Makarovo type have been studied on the Yenisey, in sites at Kokorevo village which have produced valuable material for the clarification of the geologic age and stratigraphy of Late Paleolithic remains.[26] These also looked like ring-shaped constructions made of vertically and horizontally positioned slabs, around which were grouped a few meager cultural remains left at these temporary camps of hunters of reindeer, wild horses, and wild oxen.

Geologically speaking the site at Makarovo village can also be compared in

FIGURE 6. Cross-section of terrace above the floodplain, with Paleolithic layer and hearth. High, ancient terrace in the background.

some ways with other Late Paleolithic sites in Siberia. It came into existence at a time when the present floodplain did not yet exist, and the first alluvial terrace [now], up to 4 m in height, very likely was the floodplain. At that time it was already topped by the [originally] sandy ledge of what is the 6–8-m terrace today, on which had been deposited clayey soil, [during an era] apparently contemporary with the dry, so-called xerothermic period which geologists believe began at the end of the last glaciation. The fact that the settlement of Makarovo by Paleolithic people coincided with the time of the deposition of clayey soil on the 6–8-m terrace is clearly indicated by the position of the finds right in the clay soil. It is further apparent from the contact of the lower stones of the hearth with the alluvial sand which lies below the clayey soil.

Later on as the climate became moister and milder, a humus layer was deposited on top of the clayey soil with its Paleolithic remains, and this layer corresponded to the new forested, taiga period of the new Neolithic culture.

Paleolithic settlements are found on the Yenisey and the Angara in geological conditions identical with those just described. Here they succeed in time the settlements of the Afontova Gora type II (the lower horizon) and Afontova III, when not only the rhinoceros but also the mammoth which survived it had definitely died out.

*5*

In 1941 and 1947 I found three ancient large-scale drawings on the cliffs near Shishkino village; two represented wild horses and one a wild ox. All the drawings may be dated to the Paleolithic period, perhaps in its closing phase.

Information about these drawings and the theories concerning their age have already been reported more than once, so there would be no purpose in repeating them here.

It is the more interesting that in the closest proximity to them, by the same Shishkino cliffs, a Paleolithic settlement was found which yielded substantial material of sufficient quantity and character to date the site. The settlement is associated with specific stratigraphic conditions which support the dating of the Shishkino finds as Paleolithic.

*6*

The Paleolithic site at Shishkino is situated beyond the last houses of the village, upstream on the right bank. This is the place where the bare cliffs called *Shamanki* (these were revered by shamans) begin. Here at the "First Rock" of the Shishkino cliff there is a dry stream-bed and a cape-like prominence of

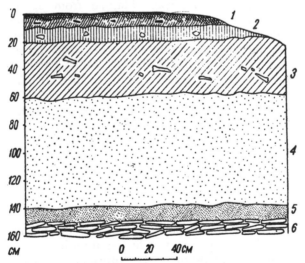

FIGURE 7. Shishkino site, cross-section: 1, turf (Iron Age); 2, dark loam (Neolithic); 3, straw-colored loam (Paleolithic); 4, sand; 5, red sand; 6, water-worn small sandstone slabs.

the 6–8-m alluvial terrace, along the edge of which were found burials of the Glazkovo and Serovo periods.* Test digging at the First Rock promontory showed that in addition to the burials there were here the remains of Iron Age (Kurykan culture) and Neolithic sites, and finally Paleolithic cultural remains associated with the same 6–8-m alluvial terrace (Fig. 7).

*[I.e., Late Neolithic burials.—Editor.]

The very few Iron Age cultural remains lay at the surface, in the turf. These were bones of the domesticated horse and also thick-walled pottery with appliqué bands. Neolithic tools occurred in the humus layer, directly under the turf; these included the characteristic pottery, small and large stone tools, and bone implements, among them a fragment of a harpoon. Well-preserved hearths were also found, made of small slabs of red sandstone and river-worn pebbles. The Paleolithic remains were found still deeper in the upper part of the layer of light yellow clay soil, which in turn rested on fragmented material of slightly worn slabs and boulders of red sandstone. These tools lay at a depth of 35–40 cm from the surface, on the border between the humus and clay layers, in that part of the clay soil which still kept its yellow color but was slightly darker than the lower stratum where limestone was plentiful. It was impossible to separate a well-defined culture-bearing level in the usual meaning of the word. The finds were spread over a considerable area, a few objects grouped together here and there, spottily.

Here first of all were a few very small fragments of animal bones differing from those above by their better state of preservation, since they had suffered less from erosion and plant action, and also were often covered by a protective layer of limestone.

Among the bones were also found chips, flakes, blanks, and stone tools (Figs. 8–11).

The stone tools found in the Paleolithic layer and Shishkino village, in contrast to the Neolithic, were all made from the same material—greenstone rock, a jasper-like flinty argillite. This material was obtained from deposits of well-worn pebbles, as shown by the remains of the pebble surface usually found on the implements. A good many such pebbles were found in this layer, showing flaking scars. In some cases these were undoubtedly crude cores. Thus, for example, two small artifacts of this type covered with flaking scars on both sides are similar to disk-shaped cores. A large disk-shaped core, 9 cm in diameter, is also an object of this type; on its upper side it is covered with broad facets of triangular outline, converging towards the center. There are traces of flaking in one place on the opposite side of the disk. Similar cores, archaic in type, resembling in their shape Mousterian disk-shaped cores, are encountered in other Paleolithic sites, as is well known, for example at Afontova Gora, and always together with prismatic lamellar blades and bone artifacts of Upper Paleolithic shapes.

Among the cores may also be included many-faceted nuclei of irregular form. chipped from all sides (Fig. 8, f).

Among the chipped pebbles and their fragments there are also core blanks of the prismatic type. Such for example is a blank in the form of a large, trapezoidal piece of a pebble. This pebble was first split with a strong transverse blow, which formed the striking platform of the future core. Then the platform was supplementarily shaped by secondary flaking along the edges. After this the core was struck longitudinally, producing two flat and wide scars on the broad sides and two narrow lateral grooves (flutes), corresponding to the direction of the future fractures, which would split off lamellar blades. This core must have had the shape of a flat cone (Fig. 11, f).

Also of interest is a small, thick blank in the shape of an isosceles triangle. On both broad sides the remains of the pebble surface are still preserved, showing that in working the stone the craftsman was particularly careful not to waste any material. The flat bottom surface of the tool was carefully smoothed with broad, shell-shaped flaking. The surface of the upper part was retouched, but only along the edges, with the same type of flaking to form broad, steep cutting edges. This might perfectly well have been a blank for a conical core, or, judging by its small size, rather a blank for a core scraper. The length of the tool is 5 cm, the width at the base 4.4 cm, the thickness 1.6 cm (Fig. 8, *b*).

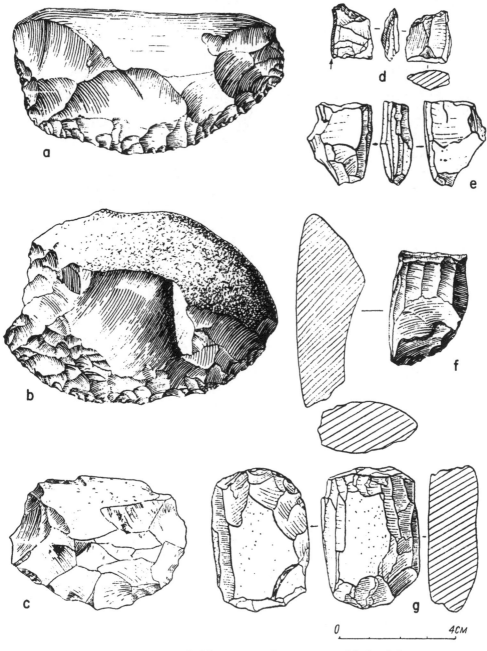

FIGURE 8. Stone tools from the Shishkino site: *a*, *b*, scrapers; *c*, blank; *d*, burin; *e–g*, cores.

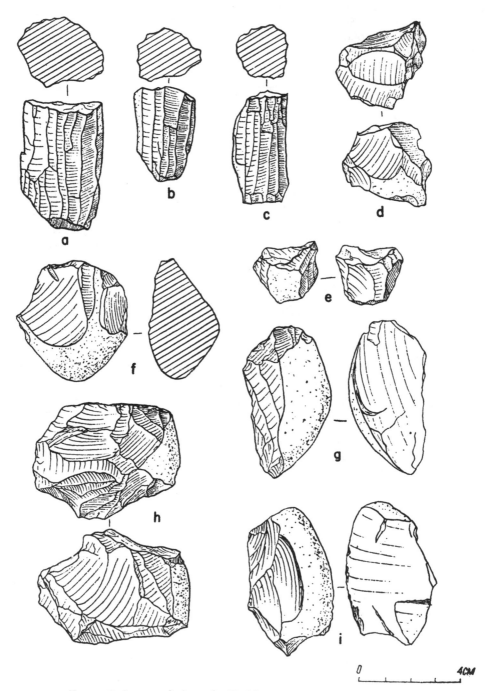

FIGURE 9. Stone tools from the Shishkino site: *a–c*, cores; *d–i*, blanks.

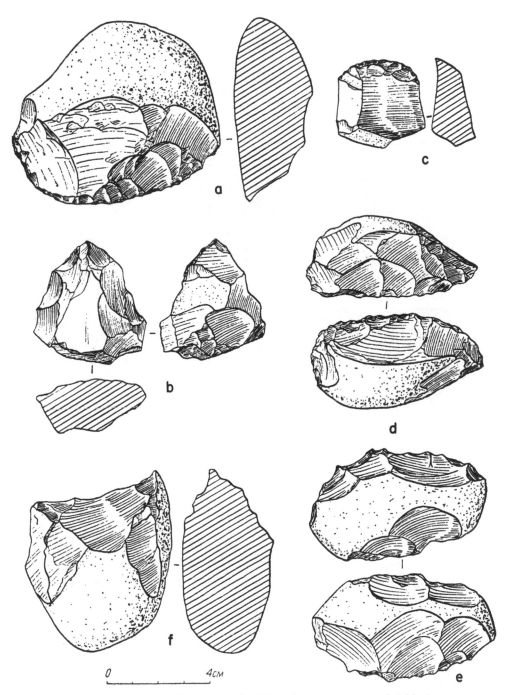

FIGURE 10. Stone tools from the Shishkino site: *a–c*, scrapers; *d–f*, blanks.

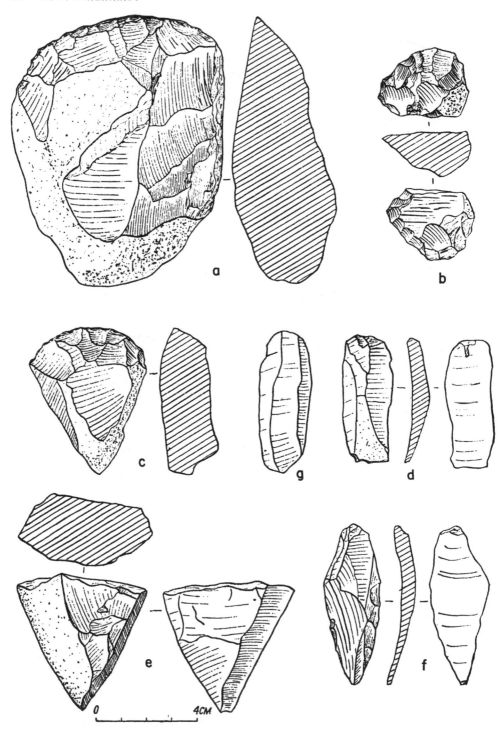

FIGURE 11. Stone tools from the Shishkino site: *a–c*, scrapers; *d*, flake; *e*, core blank; *f*, *g*, flakes.

There is another similar implement, but larger in size.

A massive, oval blank, 7.6 cm in length and 4.8 cm in width, was made differently. Its bottom side is flat; the upper, convex. The lower side is almost completely smoothed with secondary flaking which has left broad, flat facets. The opposite, convex surface of the pebble was also retouched, but in such a way that steep, high, roughly rippled cutting edges were formed along both long ends. The blade on one side is steeper than on the other, and closely resembles the blades usually found on scrapers from Paleolithic sites, not only by its curved, convex shape, but also by the steepness of the retouching (Fig. 8, c). Another blank 6.7 cm long and 3.2 cm wide has a thick, highly arched back showing the original pebble surface. This is another common feature of Late Paleolithic Siberian scrapers. Both sides have been chipped. One side is [essentially] flat, having been formed by one blow, which resulted in a broad horizontal facet, supplemented by others which produced less extensive facets. The other side is convex and also chipped, showing on one end long, narrow facets resembling those used for the shaping of the working edge of [wood] chipping tools or of Aurignacian keeled scrapers. This is especially noteworthy since on the opposite side of the tool the end has a no less characteristic scar forming the fluted, concave blade so usual for chipping tools. It is consequently very likely that this is a blank for a [wood] shaving tool rather than a scraper, which explains its unusual massiveness and the arched back, with the pebble surface (Fig. 10, d).

Other chipped pebbles, both whole and fragmentary, may be classified as blanks for scrapers. This allows us to determine the methods and sequential stages in the manufacture of scrapers from pebbles customary to the inhabitants of the site. In a number of cases it is apparent that such pebble blanks were broken in half, and then flakes were detached in such a way that on one side of the rounded part of the pebble there remained the smooth natural surface, while on the other side a wedge-like, rough, bilaterally chipped blade was formed (Fig. 10, f).

Flakes knapped from cores or chipped from pebbles in general are comparatively few in number. Usually they are large and irregular in outline. Lamellar blades are also scarce, and are usually broad, trapezoidal, with irregular ridges on the dorsum. There are also some more regularly shaped narrow blades struck off typical prismatic cores (Fig. 11, d, f, g).

Completely finished tools are represented by prismatic cores, and large and small scrapers. Of the finished and actually used cores of the prismatic type, two are regularly cylindrical in shape with carefully formed striking platforms at both ends. One of these measures 4.2 cm in length and 2 cm in diameter, the other 4.5 cm in length and 3.5 cm in diameter. The facet scars on them are narrow and strictly parallel. There is also a conical type of core 3.6 cm long, with a striking platform 2.5 cm in diameter.

One prismatic nucleus is distinguished by unusual details. It looks like a core scraper with a broad, curved, convex cutting edge along one long side, extending around the end of the tool opposite the striking platform. The opposite long edge is formed as usual by long and narrow parallel facets. On both broad sides of the nucleus the smooth pebble surface is preserved, showing that almost an entire smooth, flat, water-worn pebble was used in making the tool. At the same time it is quite obvious that the prismatic flakes taken from the edge of this core were not the end-product for which it was worked. When he struck off the thin, narrow flakes from the edge of the core, the craftsman had no thought of using these flakes; it was just his way of giving the pebble the desired shape of a core scraper. It is specially interesting that the convexly rounded, bilaterally retouched blade at the end of the tool opposite the striking platform is thoroughly blunted, apparently from blows against some hard substance. The length of the core is 5 cm, the diameters of the platform 2.6 cm and 1.7 cm.

Scrapers are represented by five large, well-trimmed specimens, of a shape classical for the Siberian Paleolithic.

The first scraper is made from a fragment of a large, flat pebble. It has a thick, tapering, triangular grip with remains of the natural smooth pebble surface. The working edge is steep and massive, convexly curved, prepared with the steep retouching typical of the Late Paleolithic Siberian flaking technique. The length is 5.8 cm, the width 4.6 cm (Fig. 11, c).

The second scraper is 9.2 cm long and 6.1 cm wide and is made of a broad, flat flake from a large pebble. One longitudinal edge retains the smooth pebble surface. The opposite working edge is sharp, massive, and convexly curved. It is trimmed carefully and expertly, in a characteristic and peculiar manner, with long and narrow facets subsequently retouched with fine secondary flaking. On the bottom and sides, the tool is covered with a thick limestone crust (Fig. 8, a).

The third tool of this kind shows an even more characteristic technique of manufacture. It is made from a smooth, flat pebble, completely retaining one rounded, massive side, which forms a well-fitting grip for the hand. The opposite side of the tool is retouched on one side to form a heavy, convex, crescent-like edge. The technique of preparing this tool may have been as follows. First a flat, well-worn pebble of oval shape was chosen and split in two. Then, one edge of the pebble was chipped and smoothed with broad fractures and then trimmed with steep, narrow-faceted retouching; finally it was modified additionally with fine, secondary flaking along the very edge. The other side was left untouched. The length of the tool is 8.4 cm, the width 7 cm (Fig. 8, b).

A fourth scraper was similarly manufactured; it has the same arched, convex blade; its grip is formed by the unworked, rounded end of a flat pebble. The measurements of this tool are, however, appreciably larger than those of the two preceding specimens: its length is 10 cm and its width 9 cm. It is also distinguished by the fact that the working edge is not the long lateral side, but on the narrower transverse end. In this respect the tool in question resembles an ax rather than a scraper. The cutting edge is also appreciably more massive and steeper than in the two tools described previously. It is, however, also convex and crescent-shaped. A trait characteristic of the finishing of this edge is a shallow step formed by the steep secondary retouching (Fig. 11, a).

The last scraper is of great interest; while exactly like those already described in shape and technique of preparation, with the same convex, crescent-shaped edge and the same grip formed of the untouched part of the smooth pebble, it differs from the others by its unusually small size. The length of this miniature side-scraper is only 3.5 cm, the width 3 cm. A distinguishing feature is the bilateral retouching of the edge, which consequently has a serrated, wavy appearance (Fig. 11, b).

There are four [small] scrapers, two of which are end-scrapers. One is made of a comparatively regular, long and broad flake with two curving ridges on its dorsum. The retouching on the narrow, slightly beveled edge is fine enough to serrate it [minutely], and brings to mind the retouching usual for the lamellar end-scrapers of Malta and Buret. The length of the tool is 5.2 cm, the width 1.8 cm.

The second scraper was made of a large flake from a pebble, the original surface of which is still preserved on one side of it. The working edge is slightly slanted and convexly arched, finished with fairly large secondary flaking. The length of the tool is 9 cm, the width 3.5 cm.

The third scraper is one of those typical of Late Paleolithic sites in Siberia. It is made from a short, heavy pebble flake, triangular in cross-section. The working edge is regularly arched in outline like the segment of a circle. The retouching on it is steep, with narrow facets; the characteristic step overhanging from the top results from secondary corrective retouching. The scraper edge is slightly curved on the underside, as though it is fluted. The length of the tool is 3.4 cm, the width 3.6 cm (Fig. 10, c).

The fourth scraper is shaped almost like the one just described, but it is made from a thinner and flatter flake. The blade is also wider.

The last group to be mentioned in describing the inventory from the Paleolithic level of the Shishkino site is that of the gravers [burins].

There is only one authentic burin of the usual type. It is small, thick, of the median type, with a broad cutting edge and three ridges on the dorsum, like some burins from Afontova Gora and Verkholenskaya Gora [Mountain]. It is made of high-quality dark-green jasper.

Two miniature tools may be regarded as burins, though they are unusual in shape and technique of preparation. The first is shaped almost like a core-scraper, since it has lateral ridges of longitudinal flaking. There is, however, no customary retouched edge on the end opposite the striking platform. Yet, on the platform itself there is a broad, burin-type scar. The longitudinal channels on the sides of the tool are also burin-like in character. The length of the implement is 3 cm, the width 2.1 cm. The second tool is even smaller than the first. Its length and width do not exceed 1.8 cm. It accurately reproduces a core-scraper in shape and workmanship. On one side is a broad, arched edge formed by bilateral chipping. The opposite side of the tool is shaped by three narrow microflutings (Fig. 8, d).

Thus, judging from the stratigraphic position and the general geologic data (a 6–8-m terrace, the cultural level in the upper part of the clayey soil), the finds at Shishkino village together with those from Makarovo may be dated to the Paleolithic period, and, more specifically, to one and the same phase of the Late Paleolithic of eastern Siberia.

On the right bank of the Lena half a kilometer downriver from Shishkino village, in a yellow layer of clayey soil on the edge of a 6-m terrace, the remains of stone hearths of the same type and structure as those at Makarovo, with dark, compact interlayers of ash, were found in places. Lack of time prevented me from investigating these in any detail, but it seems more than likely that further search will reveal here also the remains of a Paleolithic habitation.

I also noted earlier a well-defined cultural level at a depth of 2.5–3 m with abundant, finely crumbled charcoal and ash discovered near the village of Kurtukhay [Kartukhay]. This site lies on the right bank of the Lena, just at the mouth and on the right side of the now-dry stream-bed of the Nikolskiy channel, in an artificial cut made for a roadbed and an embankment for a bridge. Here animal bones were found, among them the ribs and long bones of deer, as well as an epiphyseal bone of a fairly large animal, probably an ox. There were two stone tools among the coals and bones; one flake of black flinty slate, and a pebble fragment of dark-green jasper-like rock. All these remains lay embedded in a layer of reddish clayey soil which had apparently slid down the slope and covered them.

The Paleolithic age of these finds, however, was not confirmed by later studies, and the considerable thickness of the underlying layer may be explained by an influx of clay from the steep slope of the original high bank.

## 7

In examining collections of the Irkutsk Museum I discovered among the materials collected by P. P. Khoroshikh in the environs of Verkholensk a series of characteristic flakes, prismatic cores, and a small crescent-shaped, convex scraper, found by him in 1924 and 1925 in blowouts on the 6–8-m terrace, 1.5–2 km above Verkholensk, on the right bank of the Lena in the area of the Sorokinskaya chapel.

These objects give us the right to suppose that here also was a Late Paleolithic settlement, demolished by the wind, with an inventory typical of Siberia.

## 8

In Verkholensk itself in 1951 during the excavation of a Neolithic burial ground, particularly in the northwest part of the excavation, at a depth of 0.5–0.9 m, in a layer of yellowish sandy loam, split pebbles and flakes, as well as crude prismatic cores of black slate analogous to those found at Shishkino, were found separately. It is especially interesting that still deeper, about 1–1.2 m below the surface, there is a well-defined horizon of buried fossil soil. In it are occasionally found isolated pieces of charcoal and fragments of animal bones.

Thus, altogether, in the area between Biryulka village and Verkholensk there are eight ancient sites which may be dated as Paleolithic.

## 9

The next Paleolithic location to be recorded is already at a considerable distance from the Kachuga group described above. This site was found 5 km below Potapovo village, above Ust-Kut, between Omoloy and Boyarsk. Remains of the site were found in 1941 near a promontory on the left side of the now dried-up stream of Vodyanishnyy. Owing to the action of the water at floodtime every year, there is a long stretch here of the eroded vertical wall of a 6–8-m terrace, made up basically of sandy loam and clayey soil. In the place where Paleolithic remains were found the following sequence of strata was recorded:

I. A turf layer with quantities of interwoven and half-rotted tree roots and moss comprising the peat-like substance typical of the thick, damp deciduous taiga prevalent here. Its thickness is 10–15 cm.

II. Light bown, fairly compact and heavy sandy loam, very probably of alluvial origin. Thickness, 25 cm.

III. Brown sandy loam, with small and well-worn river pebbles, lying horizontally the whole length of the natural cut, on one and the same level. Thickness, 10–12 cm.

IV. Pale yellow clayey soil so compact that it is hard to cut with a knife. Here are found shells of land mollusks. Thickness, 25–30 cm.

V. Further down there is a fairly thick layer of pebbles and sand.

VI. Below the pebbles is found layered, sandy soil.

The lower part of the cut is hidden by hillside waste.

The Paleolithic culture remains are contained in the fourth layer from the top, in the clayey soil with land mollusk shells. The remains were associated with a thin hearth layer in which ashes and fine pieces of charcoal were clearly visible. The hearth layer is easily distinguished from the clay soil not only by its darker color, but also by its friability. It is traceable for 1.5 m, and is of varying thickness, in some places thinner, in others, thicker.

Quite small fragments of animal bones were found in the hearth layer, among them the tubular bone of a small animal without epiphyses, and a bone plate, apparently a fragment of the scapula of an animal, with a fine incised line visible on the smooth surface. A small piece of antler was also found, very likely of a reindeer. All bone fragments are dark brown in color as the result of a thin ferruginous incrustation. One or occasionally both sides of the bone fragments have been much corroded by roots, which in modern times reach very rarely to such a depth.

No stone tools were found. Yet, in the hearth layer, together with the bones, there were found rejects: flakes with clearly preserved traces of human workmanship. All flakes are lamellar, sharp-edged, with protruding bulbs of percussion, with percussion scoring, rippling, and even radial splintering on the bottom side, and with ridges from the previously detached flakes on the upper side. Some of the flakes also retain remains of the pebble surface, showing that the material used was principally water-worn river pebbles of jasper-like stone, greenish in color. Two large chips of homogeneous greenstone, with tiny refracting crystals, were also found. This rock was very likely volcanic in origin.

The Paleolithic age of the Vodyanishnyy finds may be considered as indisputable. It is obviously attested to first of all by the stratigraphic position of the cultural remains in the layer of yellow, clayey soil. It is interesting to note in this connection that a seriation of strata such as that at Vodyanishnyy stream is to be seen at a variety of other fairly distant places on the 6–8-m terraces. It corresponds perhaps to the partial changes of a climate that was becoming progressively damper.

It may be inferred that the deposition of yellow clayey soil containing land mollusk shells on top of the sandy loam and shingle of the lower horizons may be connected with conditions arising with a drier climate at the end of the Ice Age. Later, after Paleolithic man had already left this place, the climate apparently became considerably moister, bringing about flooding of the river and the consequent deposition of the sandy soil. The last act in the history of the terrace is connected with the creation of the modern taiga cover and with the thin layer of turf which could not have developed fully in this locality under the conditions of dampness and [therefore] scraggly vegetation.

There is another possible explanation. The geologist N. A. Grave suggests that layers I, II, and III indicated above represent the alluvium, affected in the upper part by the process of soil formation. He thinks that in stratum II there are traces of podzolization, since it is lighter in color than stratum III. The alternation of layers of alluvium and of continental [zonal?] clayey soil, he considers, is connected with a horizontal migration of the river channel, with flooding when the waters were high. At the time of its formation, the clayey soil lay not far from the riverbed. At floodtime and when the course of the river was nearest to the clay soil, the surface of the terrace on which the soil was being formed was inundated and the flow of water deposited gravel and sand. When the water fell and the current slowed down, only sand, i.e., sandy soil, was deposited. The depth of alluvium is not sufficient to suppose that it was caused by climatic change: 25–30 cm could have been deposited during a single spring flooding, such as is currently observed on the Lena. The only evidence in favor of the climatic change theory, according to Grave, is the conformity which has already been noted between this situation and the seriation of alluvial deposits in other areas of erosion. The question can obviously be resolved only by further investigations at the site.

The Vodyanishnyy stream site was apparently a temporary hunting camp; it forms a link between the southern, upper Lena center of Paleolithic culture, and a new center of diffusion of Paleolithic hunters discovered far to the north of Verkholensk.

*10*

The first point north of Kirensk where we can suspect that there are traces of the activity of Paleolithic man lies 1 km above Mironovo village (Fig. 12). After erosion of a 20-m terrace above the floodplain on the left bank of the Lena were found, at a depth of 5 m in stratified sandy loam underlying yellowish, loess-like clayey soil, parts of the skeleton of a grown mammoth, fragmented but concen-

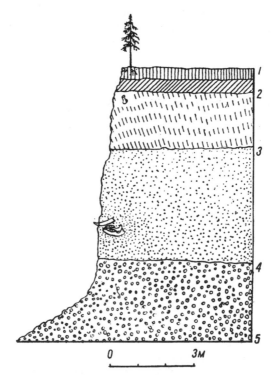

FIGURE 12. Cross-section of the terrace at Mironovo village: 1, humus layer; 2, clayey diluvial deposit; 3, loess-like loam; 4, stratified sandy loam; 5, shingle.

trated in one place and at one level—the lower jaw, pelvis, a tusk about 2.6 m long, scapulae and long bones, these last broken. Together with the mammoth bones were those of a wild horse (or *dzhigetay* [*Equus hemionus*]) and of a small animal, possibly a polar fox, as well as innumerable shells of land mollusks. The accumulations of bones extended along the exposed face in separated groups for about 12 m. Immediately below the layer of finely stratified clayey soil with the mammoth bones, there is the gravel which comprises the base of the terrace.

The general conditions of this find give grounds for supposing that we have here traces of the activity of Paleolithic man. This may represent the remains of a temporary Paleolithic campsite similar to that at Tomsk, or merely the periphery of an ordinary settlement.

## *11*

On the left bank of the Lena, 2.5–3 km above Korshunovo and about 8–10 km below Mutino village, on an eroded 8–10-m terrace, flakes of a Paleolithic aspect and a large scraper of characteristic oval outline, made with the "counter retouching" technique specific for the Late Paleolithic in Siberia, were found. There are also traces of ancient hearths here, built of river pebbles.

## *12*

Traces of a Paleolithic settlement were found about 1–1.5 km above Chastinskaya village, on the precipitous left bank of the Lena which has been considerably undercut by the river (Fig. 13).

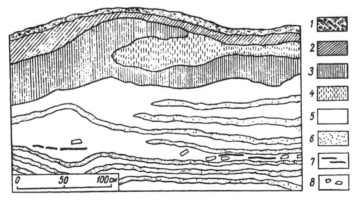

FIGURE 13. Cross-section of part of the excavation at the Chastinskaya site: 1, turf; 2, sandy loam; 3, red loam; 4, friable loam; 5, loam; 6, sand; 7, hearth interlayering; 8, hearthstones.

The place where traces of Paleolithic man were found is geomorphologically characteristic and favorable for the search for Paleolithic remains. Immediately beyond the scenically eroded cliffs winds the narrow and impetuous current of a channel of the Lena, while above it the high bank rises smoothly in tier-like terrace steps to merge with the original bank.

The lowest terrace forms the modern pebbly floodplain with fairly large, little-worn boulders scattered over the surface. The maximal height of this plain is 1.5 m above the [mean] water level. Visible above this is a well-defined terrace about 4 m high. It is composed of silty, compact, clayey soil, dark in color, and is most sharply pronounced at the lower end of the eroded cliff on the left bank. Above it, at the place where the Paleolithic remains have been found, rises still another equally well-defined terrace-like shelf; the height of this is about 10–12 m above the level of the river.

Downstream in the direction of Chastinskaya village the terrace rises slightly, reaching the height of about 15 m above the Lena. The edge of the terrace along its entire length has suffered severely from erosion by the channel current. It is steep and precipitous, covered with fresh slides. The stream of the Krokhalev, which is now dry, and in ancient times also probably phenomena connected with frost action, that is, fractures, seem to have had a most destructive influence. This is the very terrace with which the Paleolithic settlement is associated. The site is situated in an area where the terrace is cut by the small gulley of the Krokhalev streamlet. The stream bed is framed on both sides by the platforms of the 4-m terrace. Above it on the left, jutting out in a sharp triangular wedge, rises a cape-like mound, corresponding to the 10–12-m [high] terrace described above. This mound was severely eroded, the upper part of the deposit destroyed and washed away, so that the height of the exposed part at the place of excavation was only 9.30 m.

The deposit containing the remains reflects the great age and complex past of this terrace. The clearing of the terrace revealed the following:

I. Modern taiga turf fairly friable in structure, brown in color. It is easily lifted as a single layer from the underlying horizon.

II. Light-gray sandy loam, not layered, but extremely fine-grained, almost dust-like. Bits of charcoal are frequent. In one place there was a pile of charcoal, and here the loam was deeply burned. This was apparently what was left of a hearth.

This stratum was clearly defined only in the better-preserved part of the terrace, where the stream depression deepened and was suddenly interrupted by a thickening of a sort of sinter. The sandy soil is probably of aeolian origin. It should be mentioned in passing that in strata of similar light-gray sandy soil near Davydovo village and in other places along the Lena, traces of Neolithic cultures were found.

III. Compact, heavy clayey soil, red, almost cherry-colored. Clinging and sticky when wet, it has a slabby, clotted structure, and can be shoveled only with difficulty. Rarely there are single bits of charcoal, chips, and lamellar blades of non-transparent white flint covered with a deep patina, probably also Neolithic, but older than the remains mentioned above. The thickness is 30–40 cm.

IV. Friable sandy loam of a reddish color gradually merging into yellowish sandy loam, shot through with thin (3–5 cm) layers of clayey soil, greasy, sticky and plastic when wet. Maximum thickness 30 cm, average 20–25 cm.

Such interlayering of clayey soil and sandy loam continues down to a depth of more than 2 m from the surface. The clayey strata do not lie strictly horizontally, but are curved, in places undulating. In their finely clotted structure they follow the structure of the layers of clayey soil lying above them (stratum III). No extraneous intrusions were noted in the numerous clayey soil layerings except for the white, friable, lime inclusions which are fairly abundant in places. A characteristic peculiarity was noted in some of the seams, where under the clayey soil lay a thin, sandy layer which seemed to have been deposited by a perfectly clear [?] stream of river water. The sandy soil is basically light yellow in color, occasionally tinged with gray. Layers of the gray and yellowish river sand can be distinguished, but these do not extend for any distance and soon lens out. The sand is fine-grained, with quantities of mica particles. As in other places, gravels form the base of the terrace.

The culture remains lay at a depth of 1.6 m below the surface, in stratified sandy loam with clayey soil, beneath the fourth and consequently above the fifth thin clayey layer, counting from the top. Under the fifth clay-soil layer were found, in excavating, thirteen more such layers, which were similar but rather thicker towards the bottom. They were lying in sandy soil in just the same way.

The culture-yielding layer in cross-section appeared as a thin seam, dark-gray or in places black in color, depending on the amount of charcoal particles present. Isolated small boulders could be distinguished in this layer, splintered animal bones, and stones worked by the hand of man—flakes, chips, and finished tools.

Owing to the shortage of time the excavation could be conducted only within a narrow strip 9 m long and 2 m wide, along the edge of the bluff; however, the material obtained was quite significant.

When the friable layer covering the culture stratum had been removed, it became apparent that the animal bones, stone and pebble artifacts were distributed in two groups, not isolated, however, but tangent. One group at the western end of the excavation consisted of small river stones, probably from a hearth. Near the stones lay especially heavy accumulations of bits of charcoal, bone fragments from large, less frequently from small animals, and flake chips. Here were also found particles of blood-red mineral coloring matter, ochre, usual for many Paleolithic sites. These did not occur in the second group, but chips and flakes were considerably more numerous, and also many small pieces of animal bones as well as the bones of some small animals, possibly arctic fox.

I quote the relevant passages from my notebook for 1941 to give a more detailed picture of the position of the cultural remains.

During the inspection and clearing of the exposed slide, two stones were found, one round, 20 cm long, the second flat, with a naturally smoothed surface on one side. Further clearing revealed that there were two additional stones lying nearby, in the undisturbed culture layer. One, measuring 21 × 20 cm, was flat; the second, round, of close-grained, jasper-like greenstone, had been split in two with one blow, and had the transverse edge chipped to form a steep, straight cutting edge. A third stone of the same greenstone lay 10 cm away. Near the stones were found large fragments of the splintered bones of a large animal (rhinoceros?), the jawbone of a reindeer, and fragments of antlers.[27] There were also small and thin chips of white quartzite scattered about, as well as chips of green jasper-like rock, and one large, crude flake struck from a greenstone pebble. One chip was conspicuously large. At a distance of 35 cm from it was yet another, smaller, unworked pebble. Two additional ones were found in the same accumulation. Thus, there were seven of these stones in all, counting those that protruded beyond the layer. Not far off a large piece of quartzite pebble was found which had been split with a hard, intentional blow.

The bones and stones were spread on top of a thin interlayer, which was not continuous but intermittent. Here and there small bits of coals were preserved, more often [just] "smears." There were also here a few pieces of ochre. The area containing the cultural remains measured 1 × 1.5 m.

The second accumulation lay 3 m to the east of the first. The cultural remains were similarly associated with a thin seam containing charcoal, lying directly on top of the fifth clay-soil layer, counting from the surface. No large pebbles, but fragments of stones well-worn by the river water were found. Chips were considerably more numerous here, and in addition there were many small pieces of bone and one rhinoceros tooth. Among the bones of small animals were those of arctic fox. There were two large quartzite scrapers among the worked stones and one tool of greenstone, of the Malta half-disk type. A large number of bone fragments are charred.

Consequently, we have before us a picture, typical of Paleolithic settlements in Siberia, of a deposit of varied cultural remains in a well-defined culture horizon.

Some fairly large river pebbles remained unbroken here, probably on the site of a hearth. It is not impossible that some of the stones might have formed part of an enclosure for a hut-like type of light dwelling, covered with skins. At any rate such stones are used even today by northern tribes to weight the edges of the sewed-skin coverings of their dwellings.

Finely splintered animal bones and chips, grouped together at little distance, may also have remained on the site of a second hut, or may simply be a heap of rejects from the manufacture of stone tools and food refuse.

The presence of bits of red mineral coloring and charred animal bones is very characteristic. All this is typical of Late Paleolithic campsites in Siberia.

One detail of the stratigraphy of the terrace deposits is of real significance for an understanding of the conditions in which the inhabitants of the Paleolithic settlement lived. The best-preserved parts of the culture-yielding level were those situated at the extreme ends of the living area, particularly in the eastern part. But towards the western end the whole cultural level deposit, with the seams of clayey soil enclosing it, rises fairly sharply upward and at the same time thins out as if compressed by the layers below and on top of it. At this particular spot the series of clayey interlayers lying above the culture-yielding level disappears. The light-gray sandy loam of the upper horizon (stratum II) also vanishes, and just as unexpectedly the layer of red clayey soil covering it in this place spreads out as though it had swelled and had been thrust sideways. After this the three layers of clayey soil lying above the culture-yielding level break off. In cross-section there they appear as small, curving interrupted bands. The stratigraphy of these layers continues farther on in a normal, undisturbed way.

As for the culture-yielding level, it is broken off vertically into pieces, and the hearthstones are also displaced. One part of the level stayed in its former position, but the other sank 25–30 cm. This brings to mind a phenomenon not unusual in the Far North, where even today such cracks occur, with displacements of sections of the frozen ground. The frost fissure which shattered the continuity of the cultural level at the Chastinskaya site undoubtedly occurred after the inhabitants had left it, which means that the climate prevailing at that time was that which is peculiar to those regions which are in a state of "eternal glaciation," to use the words of the early 19th century scientist Figurin.

The stone artifacts of the Chastinskaya site are not numerous, but typical (Figs. 14, 15). Conspicuous among them are scraper-like tools made from fairly thin lamellar pieces of quartzite pebbles. These pieces still retain a part of the original pebble surface, but their edges are retouched to form durable, heavy cutting blades. The most specific trait of their technical preparation is the retouching, that is, the trimming of the cutting edge by means of detaching small chips with

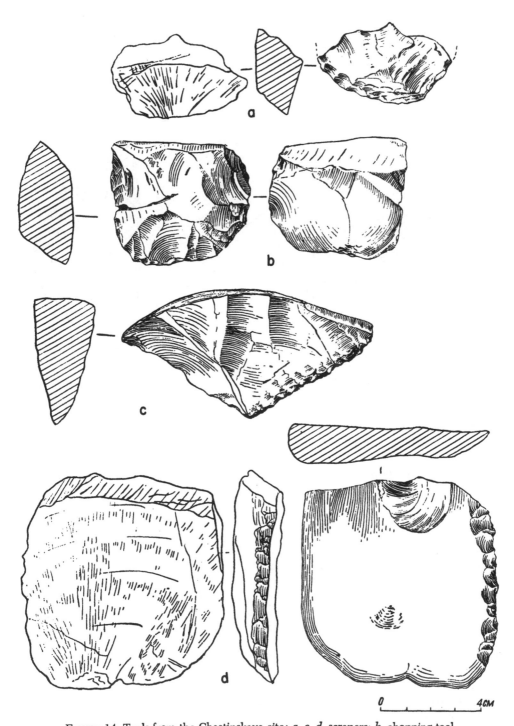

FIGURE 14. Tools from the Chastinskaya site: *a, c, d*, scrapers; *b*, chopping tool.

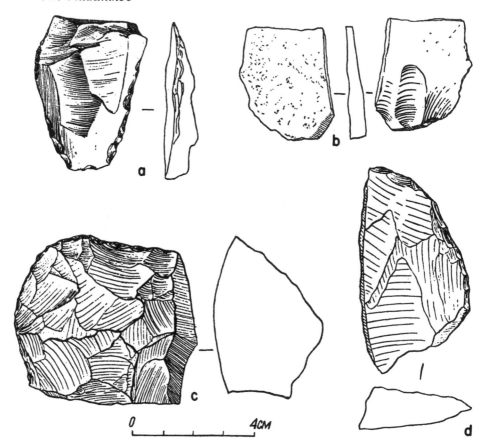

FIGURE 15. Tools from the Chastinskaya site (a, b,) and from Olekminsk (At-Daban) (c, d,): a, point; b, flake; c, scraper; d, retouched flake.

pressure-flaking. This retouching only corrects the edge of the tool from the convex side, but does not extend over the whole surface; neither does it reach the opposite, flat side. Equally crude and simple are the outlines of these tools; they have no regular, geometrically precise, consistent shape.

A total of three such scrapers were found. The first, the largest, is made of a massive, flat flake from a quartzite pebble. The upper side of this scraper is convex and retains the well-worn, smooth pebble surface. There is a cavity on one side made by two flake scars, of which the narrower and deeper one partially covers the other. The lower side, almost flat, retains at one end the characteristic but faintly expressed protuberance of the bulb of percussion. The other end of the flake was broken across with one strong blow.

The working edge of the scraper is on the side. It is almost straight, just slightly convex. The retouching is characteristic, with clear, well-defined facets in spite of the mediocre quality of the material. The blade is fairly steep, but not thick. The side opposite the cutting edge is massive; it is made by the natural curve of the pebble edge, and is well adapted to the hand holding it. The length of the scraper is 8.8 cm, its width 8.2, and its maximum thickness 1.6 cm (Fig. 14, d).

The second scraper is also of quartzite. It was made from a crude, heavy flake or more likely a pebble fragment, and is trapezoidal in shape and triangular in cross-section. One longitudinal edge of the scraper, thick and heavy, still has the pebble surface and might successfully be used as a grip. The opposite edge, thin and straight, is retouched along its entire length. The secondary flaking is fine, sharpening the extreme edge of the tool with shallow, but fairly precise facets. The shape of the tool resembles a crude, unilaterally sharpened end-blade. This impression is strengthened by the presence of a narrow, longitudinal facet at the top of the tool, which is probably fortuitous and altogether dependent on the original shape of the fragment. The length of the scraper is 10 cm, the width 4.9 cm, the maximum thickness 2.8 cm (Fig. 14, *c*).

The third tool, also of quartzite, is a fragment of a fairly large scraper broken by a strong transverse blow. The working edge of the scraper is convex, steep, crudely retouched, with fractures and splintering (Fig. 14, *a*).

The next tool differs from the foregoing scrapers both in outline and in the technique of manufacture and retouching. It is heavy, semidiscoid in shape, of close-grained, dark-green, almost black rock, apparently flinty slate. On the lower side is a large, broad, but slanted bulb of percussion with waves radiating sharply out from it, showing that the blow struck was of great force. The arched, convex edge on the opposite side of the scraper is worked in broad and rather slanting facets, directed from the edge towards the center of the implement. It is typical that the back of the tool, that is the straight, thick edge opposite the cutting edge, was shaped by a long, transverse, flake scar, with the flake being struck from the side. The quartzite scrapers described above are of the same type except that their backs are not formed by flaking [directed] from the side but by one strong and sharp transverse blow (Fig. 14, *b*). The length of the tool is 5 cm, the width 6 cm. In shape and technique of preparation this tool is somewhat similar to the small, disk-like tools characteristic of the [Paleolithic] sites of Malta and Buret.

Just as close to the scrapers from the Chastinskaya site in their type of retouching and crudeness of outline (as though not yet standardized into a common mould) are the large scrapers from Malta and Buret which differ sharply from the standardized "Mousterian-looking" scrapers of the Late Paleolithic Siberian sites.

A fairly large, elongated pebble of greenstone, split by a single, sharp blow, deserves special attention. The even and smooth, almost perpendicular surface of the break was carelessly retouched with minute flaking, and forms a strong, straight cutting edge. The opposite end of the pebble remained unworked, and fits easily into the hand. The implement is very close to the same type of straight-edged chopping tool made from pebbles split by a single blow which are known from sites such as that at Afontova Gora on the Yenisey, in the Upper Paleolithic sites on the Selenga river, and on the Angara, including the materials from Malta and Buret.

The last important object in the Chastinskaya inventory is a broad, flat flake, trapezoidal in form, detached from a greenstone boulder. On its dorsum are the remains of long scars from previous flaking; the opposite side is covered with percussion waves (the bulb of percussion itself has been cut off by a transverse blow). The edge of the flake has been casually retouched with the "counter retouching" technique characteristic of the Siberian Paleolithic, stepped and notched. The length of the flake is 5 cm, the width of the base 4 cm. This tool may be compared to the end points of "Mousterian shape" well known in the Siberian Paleolithic, except for its workmanship, which is cruder and retains the same haphazard kind of shape which is typical of the scrapers from the Chastinskaya site on the one hand, and those from Malta and Buret on the other (Fig. 15, *a*).

Thus, the characteristic traits of the stone inventory at Chastinskaya may justifiably be considered as close in many essential respects to the earliest Siberian Paleolithic sites not known to us, those at Malta and Buret. That the Paleolithic settlement found near Chastinskaya village is particularly close to the Malta and Buret sites on the Angara is also evidenced by the faunal remains. Buret and Malta are distinguished from all the other, later Paleolithic settlements of Siberia by the presence not only of mammoth bones, but also of the bones of the rhinoceros, which died out earlier than the mammoth and in all events did not survive the last (Würm) glaciation. In the Chastinskaya culture-yielding level the remains of rhinoceros were found along with the bones of another representative of glacial fauna, the arctic fox, typical of the Arctic, as well as shells of the cold-climate mollusks usual in the deposits of ancient Lena terraces.

Malta and Buret are remarkable in that they are the oldest among all the Paleolithic settlements discovered in the huge area of Siberia and Mongolia lying east of the Urals and west of the Hwang-Ho [river]. To them must now be added, though with reservations, the site of Chastinskaya on the Lena, which brings the number of known sites of the earliest stage of the Siberian Paleolithic up to three. This is also supported by the faunal remains found there. At the same time, however, there are traits in the stone inventory at Chastinskaya which to a certain extent link it to a later group of Paleolithic settlements in Siberia, that is to the majority of Siberian Paleolithic sites. These traits are indicated first of all by the material from which the tools are made, the technique of retouching, and the shape of the artifacts, which though very archaic are still undoubtedly close to those of the Late Paleolithic tools of Siberia—the large scrapers with convex arched working edges, the end points. All this allows us to consider the Chastinskaya site as an intermediate link between the two basic groups of sites in the Siberian Old Stone Age.

## 13

About 1–1.5 km below Chastinskaya village, beyond the guard post and directly beyond the "Third Rock" on the Lena bluff along the left bank of the Lena river, there extends a well-defined outcropping of the [8–10-m] Lena terrace, described above. In the exposed part of the terrace the following stratification can be observed:

I. Light-gray sandy loam, covered with a thin layer of turf, about 15–20 cm thick.

II. Red, compact clayey soil, 30 cm thick.

III. Yellowish-gray (straw-colored) clayey loam, fairly friable, with many calcareous inclusions, 40 cm thick.

A stratum of sand, or rather sandy loam, follows and below that again clayey loam.

Many hearthstones were found in 1941 in the light-gray sandy loam, some of them cracked by the action of fire; Bronze Age pottery and stone chips were also found in the same place. Numerous white flint chips and one core-like burin were found in the layer of red clayey loam.

In the talus a large, oval, greenstone scraper was picked up, similar in technique of workmanship to one found at Tochilnaya village. This scraper leads one to suppose that a culture layer of the Late Paleolithic period may be found here.

## 14

Paleolithic-type artifacts were found in the outcropping of the 8–10-m terrace 2.5 km below Chastinskaya village, about 2.5–3 km along the left bank of the Lena; these included white quartz chips and a large chopping tool made from an oval quartzite pebble split in two—analogous to that found at the Chastinskaya site, and also to tools from Malta and Buret.

It is interesting that here also there are two well-defined strata. Light-gray sandy loam lies on top of a red clayey-soil layer.

Fairly numerous Neolithic finds coincide with the red clayey soil, among them white, completely patinated chips and flakes of flint of a good quality. Objects of a Paleolithic type were found in the talus.

## 15

Two terraces can be traced near Pyanobykovskaya village on the left bank of the Lena. The first is about 4 m high. The second terrace, composed of friable, sandy loam, is higher. There are two culture-yielding layers on the 4 m terrace, early Iron Age and Neolithic. The early Iron Age stratum is associated with the base of the turf layer, the Neolithic with the sandy loam below the turf. In the talus debris on the high terrace a worked pebble of Paleolithic aspect was found in 1941.

## 16

Eight to nine kilometers below Pyanobykovskaya an 8–10-m terrace extends along the left bank of the Lena, composed (from the top down) of light-gray sandy loam, red clayey soil, and yellowish sandy loam. A layer of large boulders appears at the base of the terrace.

In 1941 the bones of a large animal (wild ox or rhinoceros) and fragments of other bones were found 1 m below the surface in the level of yellowish sandy loam, below the light-gray sandy loam and the red clayey soil. About 1.5 m away from these bones a lamellar blade of black flint was picked up in the talus, and chips of greenstone rock, among them one large piece, a flake from a boulder, crudely chipped, of Paleolithic aspect.

## 17

The next Paleolithic settlement, discovered in 1941 on the Lena, was a site at Dubrovino village, between Pyanobykovskaya and Kureyskaya, that is near the border of the Lenin *rayon* of the Yakut A.S.S.R. Therefore the site should be considered in direct connection with the other Lena sites situated to the north of Kirensk, within the borders of Yakutia.

Near the upper end of Dubrovino village on the left bank of the Lena river there is the outcropping of an old terrace about 25–30 m above the level of the river; it is composed of basic bench gravel covered with sandy loam. The terrace stratification was as follows:

I. Turf, not more than 10 cm thick.

II. Light-yellow sandy loam with reddish inclusions at the bottom, 80 cm thick.

III. Fine, well-worn shingle, lying in red clayey soil, 4.0 m thick.

IV. Coarser shingle, of crystalline rock, about 5.0 m thick.

V. Shingle with large, little-worn boulders (up to 0.5 m in diameter). In places in the upper part of the layer, a seam of conglomerate intrudes, bound together with calcareous cement, 8–9 m thick.

In the severely eroded edge of this terrace, at the very base of the yellow sandy loam and lying almost directly on the layer of shingle, were found a few crude chips of quartzite, quartz, and greenstone. The general aspect of these chips and the material from which they are made contrast sharply with those usual for the Neolithic, where neither greenstone, quartz, nor quartzite were used, but flint, which is plentiful in this area. These chips were spread over an area of not more than 3 m².

At the top of a gulley of a stream which is now dry, 25–30 m from the accumulation of chips, the teeth of a horse were found at the same level, and under identical conditions—on top of the shingle [pebbles], in the base of the sandy loam, at a depth of 1–1.5 m from the surface.

18

In 1941, 6 km above Solyanka village and 1 km above Salzavod, in the outcropping of a 6–8-m terrace and at a depth of 1.3–1.5 m from the surface, a well-defined, dark layer of ancient charcoals and ash was found, sharply contrasting with the light background of the yellowish sandy loam in which it lay. This dark layer is most apparent in one section of this natural cut where a sort of sinkhole is filled with it. The thickness of this layer in the sinkhole is as much as 5 cm, while in other places it is thinner; it can be traced here, intermittently disappearing and reappearing again, over a distance of 6–8 m. Only mollusk shells were found here besides the coals and ash. Still deeper is a distinct layer of river pebbles which extends the whole length of the outcropping in a well-marked, clearly defined horizon.

19

Chipped pebbles, Paleolithic in aspect, were found in 1941, 4–5 km below Solyanka village, in a 5-m terrace on the left bank of the Lena across from an island. At this point the terrace deposit has the following stratification:

I. Light-gray sandy loam, 25 cm thick.

II. Red clayey soil, 30 cm thick.

III. Yellow clayey soil.

A hearth layer [of charcoal and ashes] can be traced for a distance of about 1.5 m in stratum III, at a depth of 1 m from the surface.

*20*

In the talus of the 6-m terrace, 10 km from Kureyskaya and 5–6 km below Solyanka, quartzite and greenstone chips of a Paleolithic character were collected in 1941.

The terrace is composed of the following layers:

I. Light-gray sandy loam, 20 cm thick.

II. Pisolitic red clayey soil with a thin, carbonaceous seam. The thickness of the layer is 35 cm.

III. Straw-colored clayey soil with calcareous inclusions; the thickness of this layer is 1 m.

IV. Sand with a thin carbonaceous seam, 15 cm thick.

V. Non-stratified, yellowish sandy loam, 25 cm thick.

VI. Pebbly sand, 10 cm thick.

VII. Bedded sandy loam, 50 cm thick.

VIII. Sand, 3–5 cm thick.

IX. Bedded sandy loam, 1 m thick.

Iron Age pottery appeared in stratum I. In stratum II were found Neolithic cultural remains including the usual hearthstones. Elk bones were found in this layer, one bone needle, chips, lamellar blades, net-stamped pottery of the Serovo type, and potsherds with wavy-textile ornamentation; also a pestle-hammerstone. These remains lay at a depth of 20–40 cm from the surface. At a depth of 1.75 m, in a sandy seam (stratum IV) there was a layer of hearth refuse, with which chips of the [above-mentioned] Paleolithic type may be associated.

*21*

In 1941, on the right bank of the Lena, some 3–5 km above Khamra village, a large stone tool of Paleolithic type was picked up on the slope of a 12–15-m terrace (Fig. 16).

The terrace is composed of the following deposits:

I. Light-gray, friable sandy loam (almost pure sand). Thickness, 65–80 cm.

II. Reddish-yellow sandy loam, more than 1 m thick.

III. Shingle, with frequent polyhedrons and trihedrons of waltherite, implying the existence here at one time of sandy desert conditions.

At the base of the terrace there is a mass of limestone blocks (1.5–2 m in diameter) brought in by the ice.

In the first level are cultural remains of the Iron Age, including arrow points. Neolithic remains lie in the upper part of the second stratum. A scraper of Paleolithic type was probably upheaved from the bottom of the second level. It is made from a compact pebble of heavy volcanic rock, greenish-black in color, and is oval in outline. The length is 9.3 cm, the width 5 cm. The ventral side is

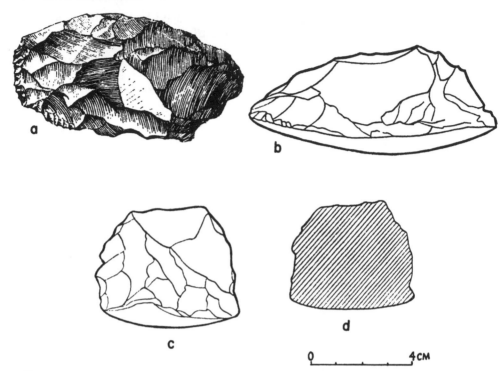

FIGURE 16. Scraper from Khamra village: *a*, top view; *b*, side view; *c*, front view; *d*, cross-section.

smooth, in places scarred with transverse striations, slightly at an angle to the long axis of the implement. The entire dorsum of the tool is crudely chipped. The working edges are convex and very steep, formed by almost vertical flaking, with only occasional corrective flaking along the very edge.

*22*

In Vitim village where the administrative boundary passes between the Yakut A.S.S.R. and Irkutsk *kray* of the R.S.F.S.R., in an eroded part of the 25-m terrace, at the depth of 1.5 m, an old hearth layer in sandy loam was found in 1941, and with it was associated a bone fragment from a large animal, probably a rhinoceros or mammoth.

The structure of the terrace is as follows:

I. Red clayey soil, compact and heavy, characterized by columnar jointing. Thickness of layer, 1.5 m.

II. Light-red friable sandy loam with a carbonaceous seam, 2–3 cm thick. Total thickness of layer, 0.5–0.3 m.

III. A layer of conglomerate; below this, boulders.

At the upper end of the village a mammoth rib was found lying horizontally at a depth of 1 m in the red clayey soil. This indicates that the streak of hearth refuse lying in the sandy loam below the red clayey soil is of Quaternary age. An animal bone was found in the sandy loam.

*23*

Along the left bank of the Lena 1–15 km below Khamra village, there is a 10–12-m terrace, made up of the following layers, starting from the top: (1) light-gray sandy loam; (2) reddish-yellow sandy loam; (3) stratified gray sand with intercalations of brown clayey soil. At the level where the first and second strata meet, Neolithic hearths were found in 1941, and white flint chips.

Adjoining this terrace is a second one, 6–8 m high, which differs substantially from the first in the character of its stratification. A cross-section reveals the following:

I. Humus layer, 50 cm thick.

II. Silty loam of a dark-gray color, with admixture of pebbles and, more rarely, boulders.

III. Light yellow loam.

At the base of stratum II, at a depth of 1.5 m there is a well-defined cultural level, apparently of Neolithic age, with charred and cracked hearthstones and ash.

In stratum III, isolated coals appeared and a well-defined carbonaceous layer, which may be Paleolithic.

*24*

Important among the Paleolithic sites of this area is the site at Nyuya village, discovered in 1941. The location of the site in itself is of interest. At a distance of 1.5–2 km from the village, the valley of the Lena veers sharply from west to east; river terraces descend to the stream at this broad bend. The first of these, the floodplain bench, is low, and stretches up along the bed of a stream which is dry in the summertime. The second terrace attains a height of 6–8 m and has the building of the forestry station on it. The third terrace is the highest, 25–30 m. It is made up of friable deposits. At the base lies greenish-gray limestone marl.

The slopes of the actual banks are slanting, smoothly outlined hills, cut by deep but also slanting valleys. This is the characteristic relief in those regions of the Cis-Baykal which abound in traces of Paleolithic settlements: on the Angara at Irkutsk and near the mouth of the Belaya, on the Lena near Kachuga and also beyond the Baykal, on the Selenga near Kyakhta and Ulan-Bator.

The high terrace of the Nyuya with which the Paleolithic remains are associated is completely covered with thick pine woods. Along the edge of it is a high, sandy embankment of aeolian origin, slanting on the side towards the river and broken off abruptly where it faces the interior of the terrace. The embankment is grown over with trees and is undoubtedly of ancient origin. Where the sand came from is quite obvious—from the slope of the wind-blown terrace facing the river and constituting an unbroken line of blowouts of varying depth.

In the natural erosion-cut of the terrace the strata come in the following sequence:

I. Wind-blown dune sand, 0.5–1.5 m thick.

II. Buried sandy soil, fairly friable, dark brown in color with many fine coals in the sand mass, 15–20 cm thick.

III. Fine, layered yellow-gray aeolian sand containing separate thin seams of pure yellow sand traceable through it. Thickness, 45 cm.

IV. Sand with seams of clayey (two of them are 2–3 cm thick), with which Neolithic cultural remains are associated. Thickness, 20 cm.

V. Pure, yellowish sand, 20–25 cm thick.

VI. Culture layer of the Paleolithic period. This differs from the layers above it in its greater density and darker color. This is sandy loam, rather than sand. From the top down it becomes noticeably yellower, and below it lies sand of a deep yellow-ochre color in which darker and lighter streaks can be seen.

According to all indications level VI is the ancient buried soil, covered by a thickness of stratified aeolian sand. The cultural remains in it lie in one horizontal plane, in the bottom [stratum] of the sandy loam, in contact with the yellow-ochre sand lying below it. The denser structure of this horizon, which in contrast to the friable sand above it abounded in clay particles, is responsible for the presence of a clearly defined escarpment in the profile of the terrace slope. The thickness of the cultural layer is 20–25 cm. Below it again lies friable gray sand, and beneath this a stony foundation bed of greenish-blue limestone.

The cultural remains were associated with a clearly defined small hearth, oval in outline. The upper part of the hearth was dark brown, the lower, black. In and right around the hearth were scattered fragments of split pebbles and chips. The length of the hearth was 1.25 m, the breadth about 40 cm, and the depth 1–15 cm.

Nearby was a second hearth area, 30 cm in diameter and 10 cm thick, and near it also there were some isolated chips. The majority of chips was made not from flint, as has been noted in the Neolithic sites on the same terrace, but from pebbles of greenstone, some of them jasper-like.

There was also a large scraper of decidedly Paleolithic aspect, both in shape and in retouching (Fig. 17, a). This was made from a large flake from a jasper-like river pebble greenish-gray in color, with whitish streaks (Fig. 17, b). The smooth and shining surface of the pebble was preserved on the ventral side of the scraper, with longitudinal striations varying in depth. The dorsal side of the tool was flaked. The working edge is convexly arched, massive, broad and high, and steeply beveled towards the back. The retouch was done with force with notched facets and with subsequent correction of the edge by fine, secondary retouching.

From the description of this tool it is apparent that it is on a par with the most typical tools from the Paleolithic sites on the upper Lena, Angara, Yenisey, and other places in Siberia: this is evidenced by the material itself (a greenstone pebble), its comparatively large dimensions, its oval-convex and steep working edge, and the Mousterian character of the retouch, with typical notching and secondary correction of the massive edge—all traits foreign to the end-scrapers and [hand] scrapers of the Neolithic period.

The same can be said also about another large tool found together with the scraper-like tool just described. This is the largest artifact in the entire collection, and at the same time the most crudely finished, and is made of crystalline greenstone, probably diorite (Fig. 17, b). In outline it is almost an elongated trapezoid, and in general appearance it is similar to a crude, unilateral ax. Its length is 11 cm, the width 5.5 cm.

One side of the tool, which juts out in a steep hump, is chipped all over except for a small section where a piece of the smooth pebble surface remains. The other side is smooth, only occasionally crossed with striations or scratches like

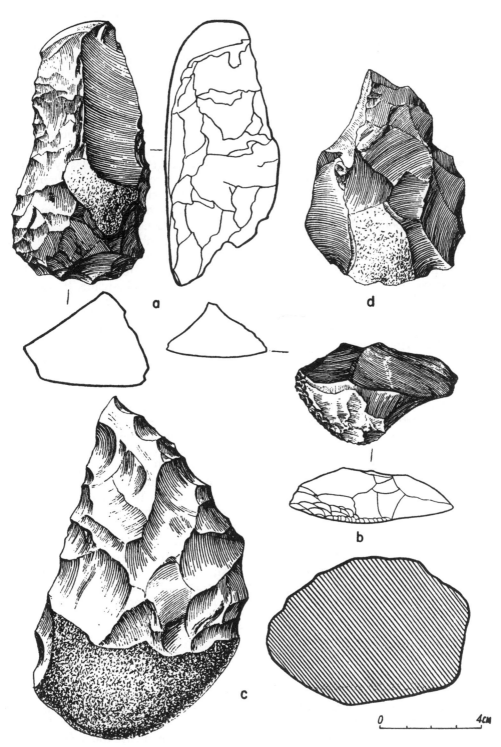

FIGURE 17. Tools from the sites at Nyuya (*a*, *b*), Tochilnaya (*c*), Markhachan (*d*): *a*, ax-like tool; *b*, scraper; *c*, cleaver-like tool; *d*, core-like tool.

those on the scraper described above. The pebble surface here has been left intact and even continues out onto the narrow upper end of the tool, so that it could also be used without a handle, the butt being grasped directly in the hand. The flaking of the tool is crude, as a whole, with splitting and dents. On the blade, however, it is more carefully done, and on a slant. The blade is slightly convex, unilateral, with only a partial corrective flaking done with two or three strokes on the reverse side. This was a genuine, ax-like tool, though still very crude in form and technique, simply flaked and not polished.

If we recall that analogous adze-like and ax-like tools exist in sites of the Afontova Gora type, the presence of this primitive ax, or rather this prototype of the later Neolithic axes at the Nyuya site, appearing together with clearly defined Paleolithic tools, should not surprise us. Many such tools were also found on Verkholenskaya Gora. This tool then fits into the general typology of finds described above, which all have the same Paleolithic age and belong to one and the same long period of the most ancient history of the Lena region.

Together with the scraper and ax-like tool just mentioned there were also flakes and other tools made already of flint rather than greenstone, but nevertheless sharply contrasting in material with what is customary for the Neolithic. For the most part in this area during the Neolithic period a milky-white, translucent or opaque flint was used, and also flint in different shades of rose and gray. In this case, however, the flint was quite unique, probably not of local origin. It was a semi-transparent flint, black or smoky with brown shading, similar to the chalky flint of the Russian plain.

Tools made from flint of similar quality and appearance are found along the lower reaches of the Angara, for example on Kamennyy Island and in other places. It is also characteristic that the flakes and tools of this flint were covered with a typical bluish patina, which is absent on the typically Neolithic flint tools collected nearby; this may serve as a supplementary indirect clue to the great age of tools patinated in this manner. Of the tools made of this sort of flint, one should mention a strongly worked core shaped like a broad cone with flat base, casually evened off with broad, flat, flaking from all sides, equally directed towards the center. Cores of this type are not usually found in the local Neolithic settlements, where only elongated prismatic cores occur. But such cores, shaped like broad cones, are customary in the Paleolithic settlements of Siberia, beginning with the oldest of them, at Malta and Buret.

In addition we should mention a miniature core with this type of side [edge], with a grooved blade at the upper end, executed with one stroke, and with fine corrective retouching. There is a small flake which is also interesting, lightly retouched with fine chipping along one edge opposite the striking platform. This retouching turns the flake into a tool similar to an end-scraper.

25

Below Gatamayskaya village, 3–4 km beyond the island, on the left bank of the Lena, chips struck from greenstone and quartzite boulders were found in 1941. These lay in the talus of the remnant of an old terrace, which now has the appearance in cross-section of a steep-edged embankment. At the same place a core scraper of Late Paleolithic type was found.

The Nyuya site is not the only one in this area. As is usual, the Paleolithic settlements occur in groups, and there is a small group of them here. Particularly interesting finds were those made at Tochilnaya village.

A terrace about 14 m high can be traced on the right bank of the Lena above Tochilnaya. A cross-section of the deposit has the following pattern:

I. Light-gray sandy loam.

II. Light-red sandy loam, probably analogous to the red clayey soil of the higher-lying sections in the Lena valley.

III. Straw-colored clayey soil, compact, finely laminated in structure, lying on top of sand, with a thin carbonaceous seam in it at a depth of 1 m from the surface.

An unusual, crudely chipped tool of Paleolithic aspect was picked up at the foot of the outcropping of this terrace in 1941. It was made from a fairly large pebble of close-grained greenstone, with the pebble surface preserved on the underside, furrowed with scratches and light striations which run transversely and at a slight slant across the implement. In its outline the tool suggests a hand cleaver, as it has a broad heel covered with the pebble surface, and a point. Nevertheless, typologically this is not a cleaver, but a genuine, heavy scraper of Paleolithic type, since it has only one long, very steep broad and convex working edge, shaped by wide and steep flaking. This tool has much in common with the large scraper found 3–5 km above Khamra village and with the large, ax-like tool from the Nyuya river, in respect to both the material from which it is made and also the technique of preparation, even to such small but typical details as the presence on the ventral side of the original pebble surface. As for the scratches and striations, it should be noted that P. A. Kropotkin found boulders covered with glacial scoring on the rivers Nygri and Khomolkho in the neighboring Olekma system. Therefore the striations on Paleolithic tools might be explained by saying that the material used in their manufacture was taken from the glacial boulder and pebble deposits of the Olekma-Vitim upland (Fig. 17, *c*).[28] When they are carefully examined, however, these striations probably have another explanation. According to Kropotkin's observations they correspond most closely to scoring produced by the action of river ice. As in the specimens described by him, they do not have the regularity of direction characteristic of glacial striations —all these scratches are short, very broad, and irregular; they break off, and always bend and twist about.[29] Kropotkin, of course, is describing large boulders, while we are only dealing with fragments of comparatively small stones, but this should not affect the character of the striations, which should be identical in both cases. Other investigators have also observed such a polishing of boulders by river ice, and the scratches and striations produced on their smooth surfaces by [the ice of] Siberian rivers, e.g., I. A. Lopatin, P. Shmidt, A. D. Chekanovskiy,[30] and more recently S. L. Kushev and L. G. Kamanin.[31] The last notes that at the time of the breakup, the river ice carries on "literally before your eyes its task of pseudo-glacial erosion of banks and the rubbly or rolled material found on them."[32]

The tools just described must be dated to the Paleolithic period, and considered as local variations of that tool widely distributed in Siberia and typical for the Paleolithic of northern Asia—the large and technologically crude implement of the scraper, mallet or ax type, principally made from river pebbles and, like the Lena tools, often retaining on the ventral side the original boulder surface.

## 26

Six kilometers below Olekminsk along the left bank of the Lena, and 2 km above the settlement of At-Daban, which has been made famous by the short-story of V. G. Korolenko, there stretches the steep, unbroken ledge of a 15–20-m terrace above the floodplain. The basis of the terrace is a thick layer of gray limestone, strongly eroded by river action. This is shown by the ridged ledges of rock along the banks of the Lena, the remains of cliffs that had been washed away, and the large pieces of water-worn and unworn limestone, strewn about in disarray at the foot of the rocky ledge. Above the limestone foundation of the terrace is a layer of friable deposits of sand and sandy loam. At the bottom is laminated sand, light gray in color, with many ferruginous yellow-ochre bands. Above the sand is a yellowish sandy loam about 1–1.5 m thick, with no sign of lamination.

From the bottom up, the yellow color of the sandy loam becomes progressively richer and stronger, and the consistency more compact, approaching that of clayey soil. In this layer at a depth of 60–80 cm from the surface worked stones were found in 1941. Here there are the usual limestone nodules and wood charcoal in small fragments. The coals lie in groups, in the same horizon, corresponding to an authentic cultural layer.

Above the cultural level lies a more friable, light-yellow sandy loam, and still higher, a dusty, fine-grained sand, probably of aeolian origin. Here again are found bits of coal and distinct hearth lenses, apparently from the Neolithic period, since in other parts at this level there are Neolithic flakes and thin, lamellar blades of white flint characteristic for the Munku site and in general for the middle Lena Neolithic.

The cultural level which lies at a depth of 60 cm from the surface is distinguished by an admixture of bits of coal which give it a darker coloration, and also by the presence of charred and split pieces of river pebbles, which undoubtedly represent the remains of a hearth. There are also greenstone pebbles here which have been intentionally split by man, as well as chips of the same greenstone and of flint. Among the tools encountered are one crude conical flint core, one flint lamellar blade of fairly regular shape and with two flaking ridges on the back, and a crude scraper of typical Paleolithic aspect, made of greenstone. The shape of this tool is rounded, with the step-like retouching typical of the Late Paleolithic of Siberia (Fig. 15, *c*).

All the greenstone flakes have been severely damaged on the upper side by the action of humic acids, while the bottom side is covered by a thick limestone crust. A similar crust covers the bottom sides of the flint tools.

Faunal remains at the site are represented by one small, poorly preserved tubular bone of some large animal.

## 27

A very interesting, altogether original find was recorded in 1940 near Markhachan village in Olekma *rayon* on the heights of Bor-Khaya (or, more precisely, Buor-Khaya).

At this point the original bank comes close to the Lena; breaking [its continuity], the rocky 60-m-high cliff of Buor-Khaya rises as a cape-like promontory of this original bank, bounded on one side by the Lena, and on the other by a small stream which has cut itself a canyon-like path, or gorge, in the limestone of the original bank. The level surface of Buor-Khaya has been cultivated along the edge of the terrace, and the ploughing uncovered a fairly thick layer of pale yellow, sandy loam and loess-like clayey soil, overlying the limestone, in which, in 1940, flakes, several lamellar blades, a core, bones of animals including those of some carnivore (probably a wolf or dog), and distinctive stone tools were found in the deepest furrows.

On reinspection of the ploughland in 1941 similar objects were found, lying in separate, small groups. These artifacts differ sharply from the Neolithic objects known in the central Lena area. The first tool is a large core shaped crudely into a double *rabo* (scraping tool): at each end there is an arched convex, adze-like blade, slightly grooved on the underside. One edge is placed at right angles to the other. Thus one could work with both ends of this very original, double tool. Both blades are wavy, ruffled, jagged along the edge. The material is whitish-gray calcareous flint of fairly high quality. No tools of this sort are known from the Siberian Neolithic, but they do occur in the Paleolithic (Fig. 17, *d*).

The most common occurrence of such scraping tools was in the Upper Paleolithic settlements of the Trans-Baykal, throughout the entire length of the Selenga valley.

The second tool is most suggestive of an oval nucleus of the discoid type. Its broad sides are formed by large, flat flaking facets. At one end the tool has a flattened heel, and at the other a crude cutting edge. Three large artifacts resemble more than anything else the archaic scrapers from the Chastinskaya site. They are made from broad, flat flakes of calcareous flint nodules. The facets of the retouching are fine, just lightly notching the edge of the tool. The shape is irregular, rounded.

At the same place a large prismatic-type core was found.

The local calcareous flint served as material for all these implements. This flint is of a fairly poor quality, but a good deal easier to work than quartzite or the coarse greenstone which the Late Paleolithic inhabitants of Siberia, including those of the Lena valley, preferred to all others for tool material. Nevertheless, we have no reason to doubt the great age of the finds described from the Markhachan Buor-Khaya. Such an age is evidenced by, first of all, the conditions under which the objects were found: the high, ancient terrace of the Lena and the ancient sandy loam deposits, the heavy patina covering these tools, and finally and most important of all, their typological aspect, linking them with the earliest rather than with the late Paleolithic artifacts of Siberia.

In this connection we should emphasize the fact that it is precisely for the early [Upper] Paleolithic sites of the Cis-Baykal and not for the late that the wide use of flint (in this case the local calcareous flint) is characteristic, and not that of coarse [grained] greenstone. This was the case in Malta, and has also been established for Buret. It is therefore very likely that the Buor-Khaya finds at Markhachan may be dated to the very early stage, according to local conditions, of man's settlement of the Lena region. It may be that they are even older than the Chastinskaya site, or contemporaneous with it.

## 28

In 1941 while traveling in the valley of the river Markha (a left tributary of the Lena below Olekminsk), where, 120 km from the mouth of the Markha, there is the remarkable painted cliff of Suruktakh-Khaya, I picked up a crudely chipped scraper-like tool made of a whole, flat pebble. This was at a settlement of sedentary Evenks, bearing the name *Saylyk* ("estival, of the summer"). This tool has a

Paleolithic aspect and is no doubt associated with the outcropping of a high, old terrace nearby, which is composed of sand and sandy loam.

*29*

To complete the survey of the available data, one additional find should be mentioned because it attracts attention by its archaic appearance.

Near the mouth of the stream Balaganaakh Yuryuyete, near the former Petrovskaya village, almost at the very foot of the cliff with the paintings, a crude stone tool of an archaic, almost Chellean chopper was picked up on the surface in 1945, to the right of the mouth of a [tributary], some 150 m from it, on a low ledge of the bank. The tool was made from a large and well-worn river pebble of quartzitic sandstone and had been worked by a continuous series of blows aimed from the edges, which detached broad flakes, leaving the characteristic shell-shaped scars on both broad sides.

The outline of the artifact is triangular. It has a broad heel retaining the original pebble surface, and two blades, heavily waved and serrated, meeting at the end opposite the heel to form a massive point. The side of the artifact which faced downwards is completely covered by a calcareous incrustation, which may be considered an indication of its comparatively great age. It cannot be said with certainty where the artifact came from. One can only suppose that it lay originally on the high, ancient terrace above the rocky ledge with the paintings, and that it was later washed down by water or thrown down by man. Under these circumstances it is impossible to date the find in question with any accuracy.

Incidentally it should be stated that chopper-like tools almost identical in type, proportions, and shape, and made from large river pebbles are occasionally found also in the upper part of the Lena valley on the second terrace above the floodplain. Such, for example, are the crude artifacts found on the right bank of the Lena 5–6 km above Sukhanovo village. This kind of tool is also found on the Angara. A. Linkov found one of them 1 km from Shchukino. He writes: "This was an elongated, egg-shaped ax of quartzite, exceptionally characteristic of the Paleolithic. Fashioned by flaking, it was of elongated shape, with one broad and one pointed end, but the end chipped off, the flaking done on one side, the other side being convex and water-worn. Its length was 6 *vershoks*,* the width 3⅛ *vershoks*."[33]

G. P. Sosnovskiy thought it possible to date the find of Linkov to the early, pre-ceramic Neolithic, and compared it with the early Neolithic artifacts of the Campignian type.[34]

It is interesting that the crude chopper-like artifacts described from the middle and upper Lena and the Angara find of Linkov recall in every detail certain finds on the high shore levels in the north of Scandinavia and the Kola peninsula which correspond in time to the late glacial transgressions of the Arctic Ocean, and date from the so-called Arctic Paleolithic.

Thus a series of finds, Paleolithic in character and in time, have now been established for the Lena region.

The southernmost of these Paleolithic sites were found along the upper reaches of the Lena at Kachuga and Verkholensk, within the Cis-Baykal [region], while the northernmost were at Markhachan village below Olekminsk. They vary in

*[1 *vershok* = 4.45 cm or 1.75 in.]

time within the limits of the Paleolithic period, and are arranged geographically in three groups: the first, near Verkholensk; the second, around Kirensk; and the third, along the Nyuya and the Olekma, entirely within the Yakut A.S.S.R. These three groups as well as the other groups of Paleolithic sites in eastern Siberia (on the Yenisey and the Angara) lie outside the limits of maximum glaciation (Fig. 18).

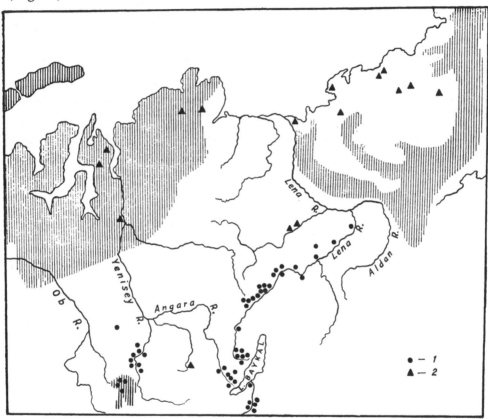

Figure 18. Distribution of Paleolithic remains in eastern Siberia: 1, Paleolithic finds; 2, Quaternary animal remains. The hatching shows the area of maximal glaciation.

One may presume that further work will result in a significant increase in the number of Paleolithic finds on the Lena. We can also hope that finds will be made further down along the Lena below the mouth of the Markhachan, which is the location of the northernmost of the sites so far recorded. In particular, it would be very interesting to look for traces of the Paleolithic along the upper and middle course of the Vilyuy river, where the ancient dunes, or *tukulany* [pl.], are comparable with those along the Nyuya.

## Notes and References

1. A. S. Uvarov, *Arkheologiya Rossii: Kamennyy period* (Archaeology of Russia: The Stone Age), Moscow, 1880.

2. I. D. Cherskiy, Opisaniye kollektsii posletretichnykh mlekopitayushchikh zhivot-nykh, sobrannykh Novo-Sibirskoy ekspeditsiyey 1885–1886 gg. (Description of the

post-Tertiary mammals collected by the Expedition to the New Siberian Islands in 1885–1886), *Zapiski Imp. Akademii nauk*, vol. 65, suppl. 1, 1891, pp. 704, 705.

3. I. D. Cherskiy, Neskolko slov o vyrytykh v Irkutske izdeliyakh kamennogo veka (A few remarks about the Stone Age artifacts excavated at Irkutsk), *Izvestiya VSORGO*, vol. 3, no. 3, 1872, pp. 167–172; *idem*, Zametka ob iskopayemykh ostatkakh severnogo olenya, vyrytykh v okrestnostyakh g. Irkutska, i o sovremennoy im faune (Note on the fossil remains of reindeer uncovered in the vicinity of Irkutsk and on the fauna contemporaneous with them), *ibid.*, vol. 5, no. 2, 1974, pp. 69–78; *idem*, Pribavleniye k iskopayemoy faune okrestnostey g. Irkutska (Additional remarks on the fossil fauna from the environs of Irkutsk), *ibid.*, vol. 5, nos. 3–4, 1874, pp. 108–117.

4. I. D. Cherskiy, Opisaniye golovy sibirskogo nosoroga (Description of the head of a Siberian rhinoceros), *Izvestiya VSORGO*, vol. 10, nos. 1–2, 1879, p. 59.

5. I. D. Cherskiy, Iskopayemyye ostatki s Tikhono-Zadonskogo priiska Olekminskoy zolotonosnoy sistemy (Fossil remains from the Tikhono-Zadonskiy mine of the Olekma gold-bearing gravels), *Izvestiya VSORGO*, vol. 6, nos. 1–2, 1875, pp. 87–88.

6. N. M. Kozmin, Sledy kamennogo veka v doline r. Malyy Patom (Stone Age remains in the Malyy Patom river valley), *Izvestiya VSORGO*, vol. 29, no. 1, 1898, pp. 70–80.

7. *Ibid.*

8. *Ibid.*

9. *Ibid.*

10. M. M. Yermolayev, Geologicheskiy i geomorfologicheskiy ocherk ostrova Bolshogo Lyakhovskogo (Geological and geomorphological sketch of the Bolshoy Lyakhovskiy Island), *Akademiya nauk SSSR, Trudy po izucheniyu proizvoditelnykh sil, ser. yakutskaya*, no. 7, 1932, pp. 147–223.

11. M. M. Yermolayev, Geologiya i poleznyye iskopayemyye Novosibirskogo arkhipelaga (Geology and mineral resources of the New Siberian Archipelago), Yakutskaya ASSR, no. 2 [?], *Geologiya i poleznye iskopayemye* (Geology and mineral resources), Collection of papers edited by V. A. Obruchev, Leningrad, 1933, p. 166.

12. I. B. Arembovskiy, Bobry v Vostochnoy Sibiri (Beavers in eastern Siberia), *Zapiski Irkutskogo nauchnogo muzeya*, no. 1, 1936.

13. A. D. Chekanovskiy, Dopolnitelnyye svedeniya k karte reki Nizhney Tunguski (Supplementary information for the map of the Nizhnaya Tunguska river), *Izvestiya RGO*, vol. 12, no. 5, 1876, p. 411.

14. R. K. Maak, *Vilyuyskiy okrug Yakutskoy oblasti* (The Vilyuy *okrug* of the Yakutsk *oblast*), pt. 1, St. Petersburg, 1883.

15. Cherskiy, Opisaniye kollektsii posletretichnykh . . . , p. 20. I. B. Arembovskiy, Ob ostatkakh deyatelnosti bobra iz postpliotsenovykh otlozheniy r. Bolshoy Patom (Remains of beaver constructions from the post-Pleistocene deposits of the Bolshoy Patom river), *Sbornik trudov Irkutskgo gorno-metallurgicheskogo instituta*, no. 2, 1940; V. N. Skalow, Rechnyye bobry Severnoy Azii (The riverine beavers of northern Asia), *Materialy k pozhaniyu fauny i flory SSSR*, otdel geologicheskiy, n. s., no. 25 (40), 1951.

16. V. A. Obruchev, *Istoriya geologicheskogo issledovaniya Sibiri; period chetvertyy, 1889–1917* (History of the geologic investigation of Siberia; fourth stage, 1889–1917), Moscow–Leningrad, 1937, pp. 278, 296.

17. M. M. Yermolayev, Istoriya otkrytiya Novosibirskogo arkhipelaga, yego issledovaniya i razvitiya ostrovnykh promyslov (History of the discovery of the New Siberian Archipelago, its exploration and the development of insular industries), *Akademiya nauk SSSR, Trudy po izucheniyu proizvoditilnykh sil, ser. yakutskaya*, no. 7, 1932, p. 17; N. V. Pinegin, Ekspeditsii Akademii nauk SSSR na ostrov Bolshoy Lyakhovskiy v 1927–1930 gg. (Expeditions of the Academy of Sciences of the U.S.S.R. to Bolshoy Lyakhovskiy Island in 1927–1930), *ibid.*, p. 93; *idem*, Materialy dlya ekonomicheskogo obsledovaniya Novosibirskikh ostrovov (Contributions to the economic survey of the New Siberian Islands), *ibid.*, p. 249.

18. Cherskiy, Opisaniye kollektsii posletretichnykh . . . , p. 705.

19. For the reddish coloring in the Lena soil, cf. I. V. Nikolayev, *Pochvy vostochnosibirskogo kraya* (Soils of the east Siberian *kray*), Irkutsk, 1934, pp. 54–55.

20. P. A. Kropotkin, *Otchet ob Olekminsko-Vitimskoy ekspeditsii* (Report on the Olekma-Vitim Expedition), St. Petersburg, 1873, pp. 185ff.

21. Cherskiy, Opisaniye kollektsii posletretichnykh . . . , pp. 43, 46; Kropotkin, *Otchet ob Olekminsko-Vitimskoy ekspeditsii*, pp. 186, 188, 189, 272. For the latest data see N. V. Dumitrashko, Geomorfologicheskiy ocherk basseynov verkhnego techeniya rr. Leny i Kirengi (Geomorphological sketch of the basins of the upper Lena and Kirenga rivers), *Trudy Instituta fizicheskoy geografii Akademii nauk SSSR*, no. 23, 1936, pp. 5–61, and in the supplement to this paper, *ibid.*, no. 24, 1937, pp. 105–107; also the article by M. P. Barkhatova, Geomorfologicheskiy ocherk Leno-Baykalskogo vodorazdela (Geomorphological sketch of the Lena-Baykal water divide), *ibid.*, no. 23, 1936, pp. 63–96.

22. A. L. Chekanovskiy, Kratkiy otchet o rezultatakh issledovaniy v leto 1871 goda (Short report on the results of investigations in the summer of 1871), *Izvestiya VSORGO*, vol. 2, no. 5, 1872, p. 38.

23. See the first short reports about these finds in my communications: *Istoricheskiy put narodov Yakutii* (The history of the Yakut peoples), Yakutsk, 1943; *Sakha sirin bylyrgyta* (The distant past of Yakutia), Yakutsk, 1945; Pervaya paleoliticheskaya nakhodka v basseyne Leny: Ponomarevskaya stoyanka (First Paleolithic find in the Lena basin: The Ponomarevo site), *KSIIMK*, no. 12, 1946, pp. 7–9; O pervonachalnom zaselenii chelovekom doliny reki Leny: Obshchiy obzor paleoliticheskikh pamyatnikov (The initial peopling of the Lena valley: General survey of Paleolithic sites), *ibid.*, no. 23, 1948, pp. 3–12; Paleoliticheskiye nakhodki na r. Lene (Paleolithic finds on the Lena river), *Byulleten Komissii po chetvertichnogo perioda, Akademii nauk SSSR*, no. 12, 1948, pp. 97–100; *Istoriya Yakutii, tom. 1: Proshloye Yakutti do prisoyedineniya k Russkomu gosudarstvu* (History of Yakutia, vol. 1: The historical past of Yakutia prior to its incorporation into the Russian state), Yakutsk, 1949; Drevneyshiye naskalnyye izobrazheniya Severnoy Azii (The oldest rock paintings of North Asia), *Sovetskaya arkheologiya*, vol. 11, 1949, pp. 155–170; Osvoyeniye paleoliticheskim chelovekom Sibiri (The occupation of Siberia by Paleolithic man), *Materialy po chetvertichnomu periodu SSSR*, no. 2, 1950, pp. 150–158.

24. The descriptions of sites and localities are numbered to correspond with the numeration in Fig. 1.

25. The artifacts described here are kept in the Irkutsk Museum.

26. G. P. Sosnovskiy, Pozdnepaleoliticheskiye stoyanki Yeniseyskoy doliny (Late Paleolithic sites in the Yenisey valley), *Izvestiya Gos. akademii istorii materialnoy kultury*, no. 118, 1935, pp. 159–164.

27. Identified by V. I. Gromov.

28. Kropotkin, *Ochet ob Olekminsko-Vitimskoy ekspeditsii*, pp. 241–245.

29. *Ibid.*, p. 244.

30. Cherskiy, Opisaniye kollektsii posletretichnykh . . . , p. 650, footnote 1.

31. N. V. Dumitrashko and L. G. Kamanin, Paleogeografiya Sredney Sibiri i Pribaykalya (Paleogeography of Central Siberia and the Cis-Baykal region), *Trudy Instituta geografii Akademii nauk SSSR*, no. 37, 1946, p. 147.

32. L. G. Kamanin, Geomorfologicheskiy ocherk Sredne-Sibirskoy ploskoy ravniny (Geomorphological outline of the Central Siberian Plateau), *Trudy Instituta geografii Akademii nauk SSSR*, no. 29, 1938, p. 104.

33. A. Linkov, Arkheologicheskiye ekskursii. I. Selo Ust-Baley i r. Kitoy (Archaeological excursions. I. The village of Ust-Baley and the Kitoy river), *Sibirskiy arkhiv, Zhurnal arkheologii, istorii i etnografii Sibiri*, Irkutsk, 1911, p. 17.

34. G. P. Sosnovskiy, Paleoliticheskiye stoyanki severnoy Azii (Paleolithic sites in north Asia), *Trudy II Mezhdunarodnoy, konferentsii Assotsiatsii po izucheniyu chetvertichnogo perioda Yevropy*, no. 5, 1934.

A. P. OKLADNIKOV

# ETHNIC AND CULTURAL CONNECTIONS OF MIDDLE YENISEY TRIBES DURING THE NEOLITHIC

## THE ORIGINS OF THE SAMOYEDIC PEOPLES*

IN THE ANCIENT HISTORY of northern Asia, one of the most interesting and important regions of Siberia appears to be the valley of the Yenisey river near Krasnoyarsk. The attention of the very first explorers of Siberia in the 18th century, Gmelin and Messerschmidt, was drawn to this region. They noted that to the east of the Yenisey there begins a new world, in many respects: the natural surroundings change, different people live there, different customs govern. Not less important is the circumstance that along the Yenisey river, which is one of the largest water arteries of Asia, the most convenient route leading from the deep taiga regions of Siberia to its steppe regions in the south, to central Asia, exists. Along this path, cultural influences could penetrate from south to north and vice versa; along this major line of communications, cultural relations could be established more easily. Here could reside, from ancient times, tribes different in culture and origin.

The central position of the region along the Yenisey in Siberia indicates the great importance of its archeological sites for the solution of a series of important historical questions concerning the ancient settlement of all the regions adjacent to it, especially the forested regions.

At the present moment, the Neolithic remains of the countryside west of the Yenisey interest us. Although these remains have been well known for a long time, the general picture of the Neolithic west of the Yenisey remains even now rather unclear. First of all, the viewpoints of investigators on the dating of these Neolithic sites and their relation to the Bronze Age of both the middle Yenisey and its neighboring regions are unclear and contradictory. To this day, for example, the established opinion about the backwardness of the Yenisey tribes, an opinion most clearly formulated by Merhart, has not been examined critically. Merhart, in a number of his works, maintained that the forest tribes of this region were still in the midst of a Stone Age culture during the first half of the 1st millennium B.C., while in the neighboring steppes the imposing burial mounds of the Late Bronze Age were already established. And further to the east, in the Cis-Baykal region, the Neolithic culture continued to exist contemporaneously with a flourishing Iron Age culture in the neighboring regions.[1]

This position was reflected in the viewpoints of local investigators. The investigator of the Cis-Baykal who most consistently voiced it was B. E. Petri.[2] Writing

*Translated from *Sovetskaya arkheologiya*, no. 1, 1957, pp. 26–55.

about the Yenisey region in 1929, V. G. Kartsov, who published the only extensive summary of the archaeological materials of Krasnoyarsk *kray*, wrote that as far as the pace of cultural development and intensity of cultural relations are concerned, the Krasnoyarsk culture area was much retarded in comparison with the neighboring Minusinsk basin: "Being further north, remaining isolated from the arteries of world culture, being hidden among the mountains and forests, it was as if cut off, and because of this, it must have been backward, and grasped the development of world culture, if it may be so expressed, 'second-hand.' "[3] Therefore, judging by the cross-hatching on the inner surface of pottery, the arrangement of the belts of caterpillar ornamentation, and the sharply pointed bottoms of the vessels found at the Yermolayevsk fortified site (which he thought the oldest Neolithic one), he was inclined to date it to the Afanasevo period. The second earliest, in his opinion, was the lower layer of the Ust-Sobakino habitation site on the Yenisey as Krasnoyarsk. On the basis of a copper ring and a vessel similar to those of Karasuk in the Minusinsk region, he dated it to the Karasuk period.[4]

On the whole this was a big step forward compared with the views of Merhart, since the results of S. A. Teploukhov, who had classified the early stages of the Bronze Age in the Minusinsk basin, were brought into play. But the consensus prevailed that the Neolithic of Krasnoyarsk *kray* was a survival culture and coincided in time with the Bronze Age of the Minusinsk region, beginning with the Afanasevo period and ending with the Karasuk, if not later. This conclusion was voiced in full accord with the assumption of isolation and, consequently, also of considerable cultural backwardness of the population of the middle Yenisey in antiquity. The find of a copper celt (later shown by M. P. Gryaznov to be of Karasuk type) near one of the skeletons uncovered by I. T. Savenkov in the valley of the Bazaikha together with other skeletons (about which there were objects of Neolithic aspect) seemed to support such a conclusion.

But with this conclusion there remained unresolved problems. For instance, between the terminal Paleolithic (the same as in the Minusinsk region) and the Karasuk period in the valley of the middle Yenisey—in an archaeological sense, well-studied periods—there still remained an unfilled cultural and chronological gap.

It is especially pertinent to note that at the time, almost thirty years ago, Kartsov correctly recognized that in order to understand the Krasnoyarsk sites of Neolithic aspect fully the relations of the people who had left them with those not only of the steppes but also of the neighboring forest regions of eastern Siberia, and perhaps of western Siberia, were of great significance. In this connection he wrote, in his carefully assembled summary of archaeological materials, that the Krasnoyarsk Neolithic culture in its basic aspects is linked to "the Baykal-Angara cultural area." He thought that the "specific Angara types of artifacts" on the middle Yenisey "possibly could be used to date the (Yenisey—Author) culture." However, he immediately expressed regret that the Angara culture itself remained so far "undated."[5]

At the present time, the Angara, or more correctly the Baykal, Neolithic culture has been studied much more fully and divided into a number of chronological sequences. Consequently, we can again return to this old question, as to whether a comparison of Neolithic finds of the several Angara and middle Yenisey stages will establish definite connections between the populations of these two regions. On this basis it will be possible also to examine more thoroughly the question of dating the Neolithic of the middle Yenisey. And this, in turn, may resolve the question of relations of the forest-zone Neolithic of Siberia and the Bronze Age of the neighboring steppes as well as relations with other Neolithic cultures.

First, we are faced with the undeniable fact that among Neolithic finds of local character there were found objects unquestionably of Angara, or more precisely of Cis-Baykal, aspect and origin, and moreover, they are rather early. This was noted by both Kartsov and Kiselev. The latter in his survey monograph *Drevnyaya istoriya yuzhnoy Sibiri* (The early history of southern Siberia), basing himself on the presence of stone representations of fish in the Yenisey valley (including Bazaikha), supposed that the representations, together with other objects of Neolithic form, indicate for the Yenisey area a Neolithic stage similar to the Serovo on the Angara. On the basis of the distribution of the stone representations of fish he projects the finding of Neolithic Serovo-like sites still further to the west, in the Barabinsk steppe.[6]

The stone sculptured figurines of fish (six have been uncovered along the Yenisey) are so similar to those of the Angara that one might consider them copies of the Angara pattern, or even imported from the Angara region.[7] Such are the representations of fish from the Karaulnaya river (seen by Messerschmidt), from Bazaikha, from the Sisima river, and the villages of Lepeshkino, Korekovo, and Bira.

There are also other artifacts of Neolithic type found on the Yenisey near Krasnoyarsk and upriver, which in both form and material from which they are made at once reveal their Cis-Baykal origin. Such are the artifacts of green nephrite or serpentine, the nearest sources of which are in the Sayan mountains, not far from Irkutsk in the headwaters of the Kitoy and Belaya rivers. Savenkov was the first to note these implements.[8] They have an archaic appearance and are similar to Serovo artifacts of the Cis-Baykal. For instance, the flat adze from the collections of the Minusinsk Museum, found at the village of Karaulniy Ostrog and illustrated by Savenkov, strongly reminds one of many Serovo adzes from Angara graves.[9] The adze made of green serpentine found at Beyskiy village is also distinctive in this respect.[10] Similar finds were encountered also by other investigators. For instance, a nephrite adze found at Bazaikha is also similar in form to the Serovo adzes.[11]

Besides artifacts of nephrite there are also ones of siliceous slate of dark-gray color with a characteristic deep-bluish patina. This material was characteristic for the early stages of the Cis-Baykal Neolithic. Deposits of such slate are found in the Angara valley zone of Angara traprocks. Before the wide distribution of nephrite (i.e., in the Serovo and, partly, the Kitoy periods) this raw material reached the upper Angara and even the Lena. Artifacts made of this slate significantly express the connections of the Yenisey and Cis-Baykal Neolithic periods. Such are the artifacts found in remarkable quantities in the sites near the village of Ladeyki. One of these is a small adze, which is unilaterally convex with a steep back and a thick butt in longitudinal section, and triangular like the adzes of Isakovo* type in cross-section. The retouching, which covers the edge of the implement in its upper part entirely, also correlates it with the Isakovo type, only the cutting edge and adjacent lower part of the adze being polished (Fig. 1, item 1).[12] Another artifact typical of the Serovo in material, shape, and technique of

---

*[A period of the Cis-Baykal Neolithic.—Editor.]

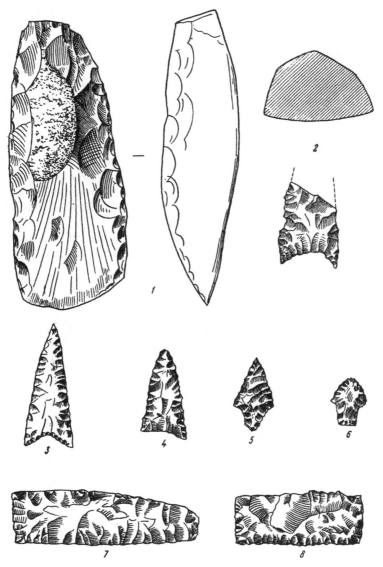

FIGURE 1. Item 1, adze from Ladeyki; 2–6, flint arrowheads from the environs of Krasnoyarsk; 7, 8, side-blades from the environs of Krasnoyarsk (⅔ natural size).

manufacture is a rectangular knife of deep-gray Angara slate.[13] An adze of the Serovo type, from the vicinity of the village of Bateni, was illustrated by Savenkov.[14]

All these artifacts were made on the Angara from local raw materials by craftsmen of the Serovo or earlier Isakovo period, and then found their way to the middle Yenisey.

In the Yenisey region there are still other artifacts which approach in shape or, more exactly, duplicate artifacts that were found in Cis-Baykal habitation sites or graves, although in this case they were made in the Yenisey and not the Angara or Lena valleys. To this complex belong arrowheads with an asymmetrically concave, angular base, [sometimes] like the tail of a swallow (Fig. 1, items 2, 3, 4),[15] tanged arrowheads analogous to those widely used during the Angara Serovo period,[16] bilaterally retouched flakes used as rectangular side-blades in daggers and spear heads, and elongated, triangular end-blades (Fig. 1, items 7, 8).[17] There are also the characteristic, asymmetrically triangular "knives" with a beak-like point which were used as side blades at the point [of a piercing or cutting artifact]. Such artifacts, analogous to those of the Cis-Baykal, were found in the Ust-Sobakino site and in the blowouts at Bazaikha.[19]

Still more important in this respect are the ceramics. The stone artifacts such as adzes, chisels, axes, and knives made of hard-to-get, rare kinds of stone, and in part arrowheads, could spread from one area to another through barter and other forms of contact between neighboring tribes. It is more difficult to propose this for such fragile objects of household use as the thin-walled and often bulky clay vessels.

Vessels with Serovo-type ornamentation were common at Krasnoyarsk in the Yenisey valley during the Neolithic. These vessels were girdled with horizontal

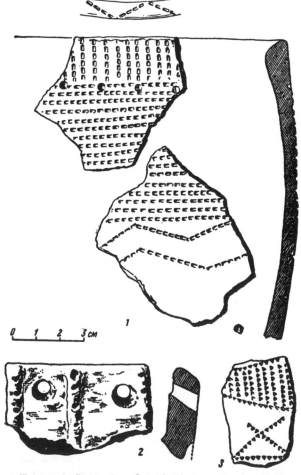

FIGURE 2. Items 1 and 3, fragments of pottery from the Krasnoyarsk region; 2, appliqué lug of a vessel (from Nyasha).

lines, sometimes combined with short vertical or oblique lines. As in the Cis-Baykal, such designs were made by the "teeth of a comb." As a rule such ornamentation was confined to the upper part of the vessel, setting off its rim, and just as in the Cis-Baykal it was combined with a line of depressions [pits], often drilled through, near the very edge of the rim (Fig. 2, item 1).[20]

Rather often complications of the simple, archaic pattern of the Serovo are observed, indicating, if we judge by the chronological scale of the Cis-Baykal, the very end of the Serovo period. In a number of vessels, in many cases, we see an [additional] zigzagging line under the usual wide band of horizontal and vertical lines which delimits the pattern from below.[21] Sometimes the pattern becomes complicated. Thus, in one of the reconstructed vessels from Ladeyki published by Merhart, there are short vertical lines next to the rim, made by impressions of a thin multidentate stamp. The band they form is supplemented with round, perforated holes. Below this is a second band consisting of four horizontal lines made with the same technique. Under it is a zigzagging line from the [lower] angles of which there drop straight, short [dentate] lines (Fig. 3).[22] In other cases there are "dangling" triangles under the horizontal line (Fig. 2, item 3). On the whole, in all these cases the ornamental base is undoubtedly of the Serovo prototype. The vessels are also similar in shape to Serovo vessels. They are either "mitre-like"[23] or semi-ovular.[24]

With a high degree of probability, vessels with appliqué lugs for suspension belong to the Serovo period; these are rather commonly found in the Neolithic sites of Krasnoyarsk. Some of these lugs are rectangular in shape and have vertical perforations (usually two). Others are small round knobs with obliquely drilled [almost] horizontal holes (Fig. 2, item 2).[25]

Most similar to the Cis-Baykal Neolithic are materials from graves at Bazaikha in the Krasnoyarsk region. The finest objects from the graves are the sculptured figurines of elks (Fig. 4). They conform fully in style and spirit to the cliff drawings in the Angara and Lena valleys and sculptured representations from the island of Zhiloy on the lower Angara, and also to the Kitoy finds in the graves of the Tsiklodrom cemetery [near Irkutsk], which were uncovered in 1950, but have not yet been published. Besides the elks, there are other objects, found by Savenkov in the dunes of Bazaikha.[26] With the sixth skeleton there had been

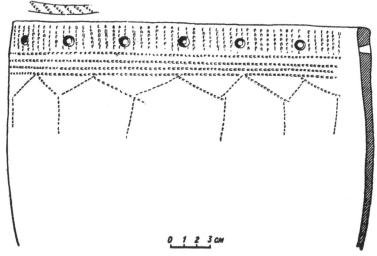

0   1   2   3 CM

FIGURE 3. Vessel from Ladeyki (reconstruction).

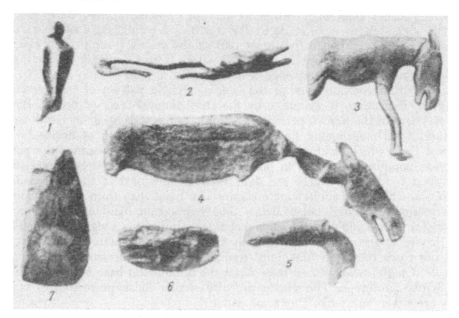

FIGURE 4. Objects from a grave in Bazaikha.

deposited a polished adze (no. 1259–9) made of a raw material—light-gray sili-ceous slate—which undoubtedly originated in the Angara valley (Fig. 5, item 2). The shape of the adze is reminiscent of several sizes encountered on the Angara in the hardware complexes of the transitional period between the Serovo and Kitoy. It also is small, one side slightly convex (in longitudinal section); it is not rectangular in cross-section as are the classical Serovo artifacts of this kind, but almost oval, i.e., it has rounded, narrow edges.

Of the same material and even closer in shape to the late Serovo or early Kitoy period is a second adze-like implement (no. 1269–27) from Bazaikha (Fig. 5, item 7).[27]

An implement of gray Angara slate (now in the Museum of Physical Anthro-pology and Ethnography, cat. no. 1259–25) found in the same grave represents a variant of the usual Serovo knife. Its spine is blunted and it is not polished; the blade is convex, not concave (Fig. 5, item 6).[28]

The flat bone dagger (cat. no. 1259–16) with slots for stone side-blades found in Bazaikha with the fourth skeleton could fit into any Serovo or Kitoy burial on the Angara (Fig. 5, item 1).

The curved bone points (cat. nos. 1259–20, 22, 24) found with the sixth skele-ton resemble in shape musk deer incisors; they duplicate the curved hooks of composite fishhooks well known from the graves of Neolithic age in the Cis-Baykal. Similar artifacts are encountered on the Lena and Selenga in graves of the Serovo period (Fig. 5, items 4, 5). It is interesting that on the Selenga (Fofanovo) and the Lena (Verkholensk cemetery) shanks with broadened heads which look like spikes or tacks with oval perforations were found with these hooks. These were straight shanks of fishhooks to which the short hooks were fastened. Such composite hooks actually joined were found by me in the Verkho-lensk cemetery. The same kind of shank was uncovered by Savenkov at Bazaikha together with the hooks noted above.

The find of a characteristic "Serovo wedge" on the middle Yenisey belonging to

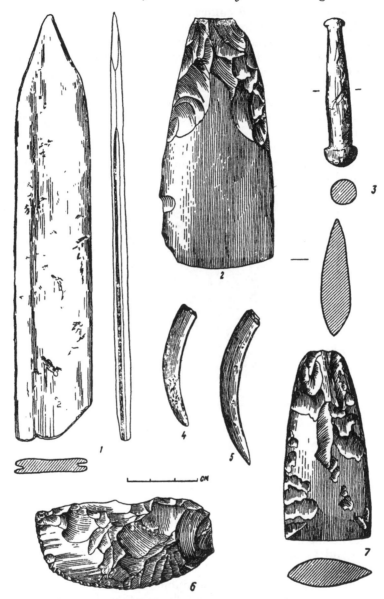

FIGURE 5. Items 1, 2, 4–7, objects from a grave in Bazaikha; 3, Kitoy-type stone shank [weight] from Bazaikha.

a pre-Afanasevo perod helps to elucidate not only the exchange but also the profound connections between the Yenisey and Cis-Baykal peoples.

It cannot be ruled out that this situation represents a direct spread to the west and south (along the upper Yenisey) of tribes whose land of origin was farther to the east, in the lower reaches of the Angara, from present-day Bratsk above the Angara rapids to the village of Strelka, that is, at the confluence of the Angara and Yenisey.

In any case, such penetration to the Yenisey by Angara tribes and their ties with the peoples of the middle Yenisey region could have been facilitated by a common economic structure and cultural level on both the Angara and Yenisey during the Neolithic. This is borne out by materials found in the graves in the environs of Krasnoyarsk, which clearly reflect the presence of a hunting and fishing economy and a way of life which we know existed in the Neolithic settlements of the Cis-Baykal during the Serovo period. Especially interesting in this connection is the coincidence of the inventories of the Yenisey graves not only with those of the Serovo Angara graves but also with the Lena and Selenga remains marked by the common incidence of the particular type of composite fishhook mentioned earlier.

Researches in physical anthropology pertaining to this period of the Yenisey Neolithic do not contradict the archaeological findings. The study of skulls from graves excavated by Savenkov at Bazaikha and V. C. Peredolskiy in the environs

FIGURE 6. Celt from a grave in Bazaikha.

of Krasnoyarsk showed that "with respect to secondary racial types, the Krasnoyarsk graves result in the same picture as do the graves on the upper reaches of the Angara and Lena" although it may be proposed that on the Yenisey "the Mongoloid component was more brachycephalic than in the Cis-Baykal."[29] This is not an essential difference and is of transitional nature.

In connection with this we should briefly mention the attempts to date the graves in the dunes at Bazaikha to the Bronze Age, specifically (as noted above) to the Karasuk period, on the basis of a celt (Fig. 6) uncovered with one of the skeletons. First, on the dunes at Bazaikha, as well as in other places, settlements and graves of peoples that date from the beginning of the Neolithic right up to our times have been uncovered.[30] There is nothing strange, then, in the fact that in the dunes at Bazaikha a grave was found with a Karasuk celt in it, if, of course, the latter did not fall there by chance. Second, and more important, there is now known to us near Krasnoyarsk another grave undoubtedly of the Karasuk period. It has an abundant and typical inventory including two large celts, excellent curved knives, an awl with the "handle" carved to represent the stylized head of a deer (elk?), a tubular needlecase, a needle, a ring, and other artifacts which have nothing in common with the materials taken from Neolithic graves at Bazaikha. Apparently, in the region of Krasnoyarsk there lived during Karasuk times tribes which used excellent metal artifacts of the same sort as had their steppe contemporaries in the Minusinsk basin. This is borne out by the rather numerous finds of Karasuk objects on the middle Yenisey at Krasnoyarsk.[31] Surely the level of culture represented here had nothing to do with that of their remote Neolithic predecessors. Thus, the old hypothesis that the Karasuk age is represented by the basic group of burials at Bazaikha becomes untenable.

Now we can also come to accept the historical reality that the ancient ties of the Cis-Baykal and middle Yenisey areas did not stop after the end of the Serovo period. They continued into the eras that followed, which may be synchronized with the Kitoy and Glazkovo periods of the Cis-Baykal. The characteristic stone shank made of white-veined brown jasper with hemispherical knobs at both ends, for example, testifies to the Kitoy connections. This shank was found at Bazaikha at the natural landmark *Bor* [Fir] and was purchased in 1871 by Korzon.

Also at Bazaikha, a typical Kitoy palette or small mortar for the grinding of red ochre, in the shape of a dish with a hemispherical depression in the center, was found.[32] In other productive sites near Krasnoyarsk, such as Nyasha and Perevoznaya village, there were uncovered typical Kitoy (in terms of the Cis-Baykal) abraders for arrow shafts, erstwhile mistakenly termed "casting molds of the Catacomb type."[33] At Bazaikha a so-called ax (more likely a hoe or an adze) with lobes, also typical of the Cis-Baykal Neolithic, was uncovered.[34] Such implements have been uncovered in the past few years in excavations of Neolithic settlements on the islands and along the shores of the Angara above Irkutsk, where they are dated to the Serovo and Kitoy periods. A small polished knife made of a semitransparent platelet of green nephrite with a convex, unilaterally sharpened blade and of almond shape found at Bazaikha may be dated to the Kitoy period. Fragments of pottery in shape, thickness, color, and ornamentation identical with Cis-Baykal ceramics of the Kitoy or early Glazkovo periods (as known from finds in corresponding layers of Ulan-Khada) may also be so dated. In particular, a large vessel of which almost all of the upper half has been preserved is of this provenience. It was found on the Sobakino river by N. P. Konovalov in 1912 at a depth of 1–1½ m, i.e., in the lower (according to Kartsov) culture-bearing layer of this site. The surface of this vessel is decorated with the usual Baykal stamped ornamentation, forming wide horizontal bands made up of narrower ones, [the wider ones being] joined by slanting double lines (Fig. 7).[35]

The numerous ceramics found on the Yenisey at Krasnoyarsk should be dated to the terminal Kitoy and the Glazkovo periods (on the Cis-Baykal time-scale). These ceramics are richly decorated with stamped ornamentation, forming narrow grooves deeply pressed into the clay, grooves which have stepped "bottoms" because of the repeated pressing of the stamp-paddle. Such fragments belong to vessels of ancient form with almost straight walls and rounded bottoms. They were found, in particular, at the Yermolayevsk site (fortified town site) excavated by Kartsov and also in a number of other places.[36] Among them are rim fragments with a characteristic thickening which forms a sharp rib or cornice and the equally characteristic downward taper of the wide edge of the rim from the external side. The very edge of the rim is shaped like a sharp rib, ornamented with narrow perforations.[37] Such rims are typical of the habitation sites of the Kitoy period in the Cis-Baykal area (Sosnovyy and Lesny islands, upriver from Irkutsk, the finds at Bratsk and near the village of Patrony near Irkutsk).

Fragments of thin vessels are commonly found in the sites at Krasnoyarsk as well as on the Angara, Lena, and along the shores of Lake Baykal. They are stamp ornamented, with the paddle-like stamp leaving a shallow, flat groove with a characteristic, ribbed [stepped] bottom. Such are, for example, the fragments found in Ladeyki.[38] This fine, delicate design is duplicated in tens and hundreds of specimens found in the Cis-Baykal.

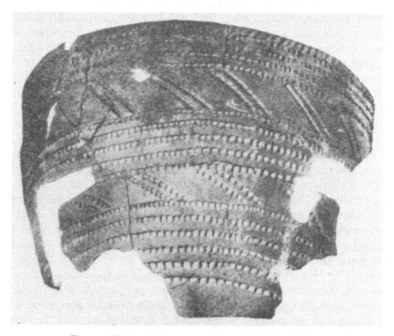

FIGURE 7. Vessel from a site on the Sobakina river.

The finds in the graves on Afontova Gora excavated by A. F. Katkov and by the author point to the existence of definite connections between the inhabitants of the Yenisey valley and of the Cis-Baykal in post-Serovo times. In a number of cases these finds are similar to those from Glazkovo sites of the Cis-Baykal. Such, for instance, is a bone awl with a sculptured human head, flat disk-like mother-of-pearl beads, and a small clay vessel with overall paddle-stamped ornamentation found by me in 1937 (Fig. 8, items a, b).[39] The "small bone rings" and bone blades found by V. C. Peredolsky in graves near the village of Perevoz on the Yenisey also may be cited as being similar to those of the Glazkovo period in the Cis-Baykal.

Thus, in the valley of the Yenisey at Krasnoyarsk at a relatively early time (3rd–2nd millennium B.C.), we see evidence definitely connecting the Neolithic of this and the Baykal regions.

Regarding the ties of the middle Yenisey Neolithic and the steppe cultures, it appears probable that all the Late Neolithic remains along the Yenisey river conforming to those of the Late Kitoy and Glazkovo periods of the Cis-Baykal (and preceding a later period, already attributable to the Bronze Age) belong to the same chronological span as the Afanasevo of the Minusinsk basin.[40] In support of this is the observation of V. G. Kartsov on the similarity of the [middle] Yenisey pottery and that from habitation sites and graves of the Afanasevo era on the [upper] Yenisey. Such similarities are: (1) the pointed bottoms of vessels,

(2) the zonal placement of ornamentation, (3) the presence of caterpillar impressions made with a stamp, (4) the straw-rubbed inner surfaces of the vessels, (5) the absence in this pottery of Andronovo and later modes of ornamentation.

From all this follows a general conclusion that the material culture of the middle Yeniseyan tribes of the 3rd to 2nd millennium B.C. and that of the steppe-dwelling, cattle-breeding Afanasevo tribes to the south were not especially strongly differentiated. The middle Yeniseyans had a few common cultural traits with the steppe people, despite the differences in their modes of life, their economies, and apparently also their origins.

FIGURE 8. Objects from a grave on Afontova Gora: (*a*) representation of a man; (*b*) vessel.

In the materials from Neolithic sites of the Krasnoyarsk region there are indications of connections to the west which, although they are not numerous, are sufficiently distinct and very important for the ethnic and cultural history of the middle Yenisey. These connections were, in the first place, with western Siberia* and with the Urals. In the case of the Urals both the eastern and the western (i.e., the Kama river region) slopes are involved, and possibly even the more remote regions of the European part of the R.S.F.S.R.†

Let us first consider the ceramics. Two vessels found near Krasnoyarsk are of great interest.[41] The first vessel was found in a ditch being dug for a water pipe on Uzenkaya street, "at a depth of 3½ *arshins* [about 8 feet]" according to the entry in the registration book of the Krasnoyarsk Museum. Together with the vessels were found: (1) a broken antler point, (2) a fragment of a mother-of-pearl bead, (3) small pieces of calcined bone. All this was found near the pot or under it: "the vessel stood in the earth with bottom up." Evidently, graves existed here, possibly the remains of a cremation. The presence of mother-of-

*[I.e., Siberia west of the Yenisey.—Editor.]
†[Russian Soviet Federated Socialist Republic.]

pearl beads and the general appearance of the vessel of this curious find are similar to those from the graves on Afontova Gora uncovered by Katkov and myself. This vessel, however, differs decidedly from the Neolithic pottery of the Cis-Baykal by its large dimensions, the height of the vessel being approximately 45 cm and its diameter 40 cm. In shape it is, generally speaking, similar to Cis-Baykal vessels since it has a pointed bottom, but it differs from them by its general proportions and the details of its bottom. At a time when the Cis-Baykal pots (with the exception of the Isakovo ones) characteristically had a more or less rounded bottom, here it is abruptly narrowed and pointed in the lower third, i.e., it is conical. Such a shape of base is peculiar both to the Afanasevo (of Minusinsk) and to several vessels from the Cis-Ural* and the European parts of the R.S.F.S.R. The ornamentation is also not of the Cis-Baykal. The rim is decorated along its edge with an oval, caterpillar-like stamp placed sideways with numerous indentations forming a likeness of a thick cord. Below the rim and down to the very point of the bottom of the vessel are placed stripes of short and narrow oval impressions. Within this ornamentation, round pits are placed at intervals of from 2.5 cm to 3 cm in the third belt of ornamentations from the top.

Still more characteristic is the second vessel, found in Bazaikha and deposited in the Museum of Anthropology and Ethnography of the Academy of Sciences of the U.S.S.R. as no. 2385–1. Its height is 28 cm and the [inside and outside] diameters of the neck are 18 cm and 18.5 cm respectively. In shape it is similar to an oblong half-egg which is somewhat wider in the middle part than the upper. The thickness of the wall is 0.4–0.5 cm. The color of the vessel is brownish-yellow, with slight reddish shades. The outer surface is entirely covered with ornamentation, which is shallow, impressed with a two-toothed (with squarish teeth) stamp, forming continuous horizontal stripes around the vessel. The upper tooth was impressed deeper and more clearly; the lower tooth impression is weaker. The first impression of the ornamentation made it seem that it was made not by separate stamping impressions—two teeth at a time—but rather with a serrated wheel [roulette]. This stamped design covers the pot almost completely, but in five places it is interrupted by horizontal zigzag lines incised with a sharpened stick. Besides the horizontal zigzagging lines, evenly spaced vertical zigzagging lines reaching from the rim one-third of the distance towards the bottom are incised (Fig. 9).

What we have before us is a specific, well-defined, zonal pattern of ornamentation, based on the repetition of horizontal ornamented bands formed by comb-stamping and alternating with zigzagging lines.

Overall ornamentation of pots is not characteristic for the early stages of the Cis-Baykal Neolithic. The ornamentation is concentrated on the upper part of the vessels, and the ornamental pattern typically is not zoned, but of various combinations of horizontal and vertical lines with a characteristic rhythmic alternation of the elements of the design. Both the ornamental elements noted above and, in particular, their pattern of distribution—horizontal, with alternating bands —again relate this vessel to the ceramics of the western Urals, of the forest zone of the European part of the Soviet Union and of Finland.

The vessel from Bazaikha also bore a close resemblance to the vessels from habitation sites on the upper reaches of the Lyapin river, a tributary of the northern Sosva, which were illustrated by V. N. Chernetsov. They are represented

*[I.e., eastern slopes of the Urals (as viewed from the Yenisey).—Editor.]

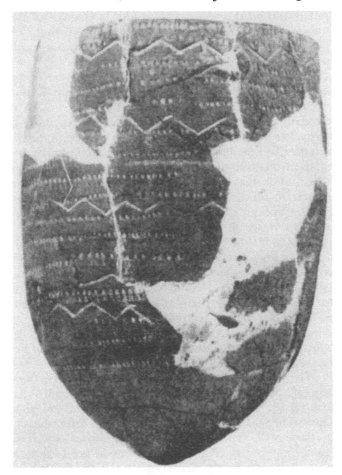

FIGURE 9. Vessel from Bazaikha.

in Chernetsov's collections by pots with egg-shaped bottoms, hand-modeled "and from the rim to the bottom entirely covered with comb ornamentation."[42] The ornamentation consists of a board of horizontal zigzagging at the rim, under which there are round pits filling the space between the additional bands below; the spaces between the bands are filled with three rows of round pits formed by a three-pointed comb stamp.[43]

Pottery similar in ornamentation is close to the Late Neolithic pottery of Karelia and the neighboring regions of Finland (the second group of habitation sites at Lake Onega according to the classification of N. N. Gurina). The Karelian pottery is characterized by the same pointed base and ornamentation of two elements: alternating small pits and short, oblique comb impressions, forming bands that girdle the pot.[44] The central [European] Russian sites of Lyalovo and Balakhna (site no. 5) are characterized by analogous pottery.[45]

Sherds of this type (with zoned, pitted, and comb-impressed ornamentation) at one time were the basis for the assertion of J. Ailio and after him of G. Merhart that these ceramics are encountered far to the east of their region of origin in Finland, Karelia, and the forest zone of [European] Russia. On the whole this view is not correct. No finds supporting a diffusion of the Volga-Oka and Karelian pitted and comb-impressed ceramics to the east of the Urals exist.[46] Generally, pottery similar in ornamentation and shape to that of the west does not reach farther than the Yenisey, and, of course, it does not reach the shores of Lake Baykal in any such form, as Merhart once thought.

Besides, the Krasnoyarsk vessels, even though they have some general resemblance in ornamentation to those of the Volga-Oka and of Karelia, differ from them in one essential feature. The walls of these vessels, as in most Neolithic vessels of eastern Siberia, are not thick, but thin. The pits forming [part] of the ornamentation are therefore not deep but shallow, and relatively small in diameter. This feature, the technique of applying the ornamentation, and the pattern of the ornamentation relate the vessel from Bazaikha and Late Neolithic pottery on the middle Yenisey to that of the Urals, both its eastern and western slopes— in case of the latter to the Kama region. For example, the usual ornamentation of the archaic ceramics from Uralian peat bogs of lightly impressed comb stamps is often duplicated on vessels of the Krasnoyarsk region.[47]

The same may be said about the ceramics of the Kama basin of the Urals, where we also find analogous combinations of this pattern with zigzagging.[48] On the vessel from the Annino Island site illustrated by D. N. Eding, the combination of zigzagging horizontal and comb-impressed bands corresponds, in principle, with what we see on the pot from Bazaikha, although the shape of the pot and the ornamentation are different in details.[49]

The general impression of variance with Cis-Baykal ornamentation and agreement with that of the west, the Ural area, which is given by the two vessels from the Krasnoyarsk region is reinforced by a more detailed study of Neolithic patterns of middle Yenisey ceramics. As an example of western connections, there is Kartsov's "maggot" design already referred to. This element of design, common to the Ust-Sobakino site and elsewhere, is a fairly wide and clearly impressed oval-shaped depression with cross-partitions inside. J. Ailio classifies it with the type of design done with hard-twisted cord, *Wickelschnurornament* (Fig. 10, items 1, 4).[50] Such "caterpillar" stamp impressions in the form of wide ovals are not present on the pottery from the Serovo graves of the Cis-Baykal, where there are only narrow or crescent-shaped comb impressions. In the Krasnoyarsk region they appear repeatedly and in characteristic combination, usually zoned, the impressions forming bands. In other cases vertical zigzagging is placed on the vessel from top to bottom, at the same time forming horizontal bands by the correspondence of the bends (of the zigzags in the same direction). Thus, the surface of the vessel is completely covered.[51] To the west, these designs are known as far as Karelia and Finland.

Characteristic of many specimens of pottery from Neolithic sites on the Yenisey is a design with short, narrow, almost rectangular, slanting impressions, placed close together. This design is common in the Urals.[52] Also close to early Uralian pottery are patterns in the form of triangles, slanting crosses, arc-shaped dotted lines, and others, made with "comb" stamps. Sometimes these join in fanciful complicated combinations (Fig. 11, items 5, 6).[53] The mode of arrangement of such ornamentation is very similar to that seen very early in Uralian pottery, represented by the elongated, egg-shaped vessels of Neolithic aspect.[54]

There is a significant coincidence of other Yeniseyan types of ornamentation with those of the Urals or western Siberia. Thus, for instance, a fragment of a vessel from the Ust-Sobakina site (cat. no. 122–19 in the collections of the Krasnoyarsk Museum) is covered by a design consisting of narrow, rippled grooved bands formed with the stab-and-drag method. The stab intervals are

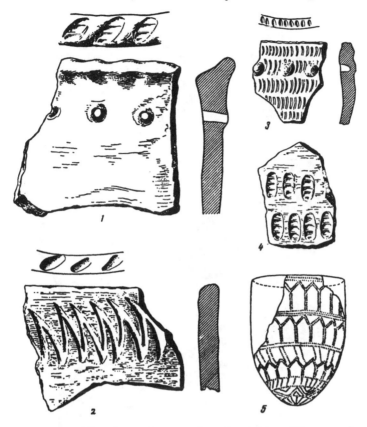

FIGURE 10. Fragment of pottery from the vicinity of Krasnoyarsk and vessel from the Urals. Items 1, 2, Ust-Sobakino site; 3, 4, Ladeyki; 5, Nyasha. (Items 1–4 are ½ natural size.)

short. The design is on the whole simple, but characteristic. It consists of a combination of long and wavy and of short, slanting, also wavy lines. Designs of this pattern were found on other fragments of pottery from Ust-Sobakina sites, at Bazaikha and other places (Fig. 11, items 1–4, 7).[55]

Among many thousands of specimens of Neolithic and later ceramics from the Cis-Baykal, I have not encountered a single fragment with this type of ornamentation. In technique of execution, the Krasnoyarsk vessels, particularly in their wavy, specific ornamentation, reveal a similarity to Neolithic ornamentation found on sherds from western Siberia and the Urals and also from the Kelteminar habitation sites around the Aral Sea.

In describing this ornamentation in his monograph on the Shigir culture, P. A. Dmitriyev divides it into two groups. The "undulating design" known "almost at all sites" belongs to the first group. The second is the "ripple design" (A. E. Teploukhov called it scaly, squamous), which consists of many rows of wavy lines parallel to each other. "Beyond the eastern [slopes of the] Urals," he wrote, "we do not know of such ornamentation and thus the rippling type can be considered characteristic of the Shigir culture."[56] This ornamentation was done, according to Dmitriyev, with an instrument similar to a comb. However, in some cases it was made with the point of a stick.

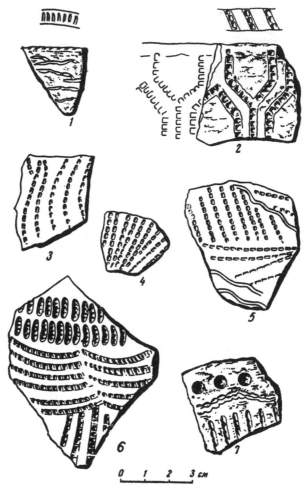

FIGURE 11. Types of pottery from the vicinity of
Krasnoyarsk. Items 1, 2, Ust-Sobakino site; 3–7, Bazaikha.

After an analysis of similar pottery from Uralian Neolithic sites, V. M. Raushenbakh came to the conclusion that it was present even at such early sites as Strelka,
which according to her dates to the end of the Atlantic period, i.e., between the
middle and the end of the 3rd millennium B.C.[57] Similar in age and pottery to the
Strelka site is a Neolithic habitation site studied by O. N. Bader in 1944–1946 on
the Poludenka river near Nizhniy Tagil. The finds on the Poludenka are dated to
the end of the 3rd millennium B.C. At this time there existed, on the Angara and
Lena, the Kitoy stage of the local Neolithic. Ties with the east are indicated by
the find of a lobed axe, which was noted by Bader,[58] as well as by the presence of
green nephrite (which is clearly not of Uralian but of Sayan origin).

Such "axes with ears," whose principal region of diffusion is eastern Siberia
(the Angara, the Lena (including Yakutia), and also the Baykal region, but not
farther to the east), were also found in other places in the Urals as surface finds.[59]
This confirms the significance of the Poludenka finds, which testify to connections
between the peoples of the eastern Urals and those of eastern Siberia even at this
relatively remote time.

To the west, in the Urals and even beyond, another kind of ornamentation is known, which is represented at the Ust-Sobakino site by arched zigzags, placed on one pot in such a way that alternate arches of the zigzag are parallel (Fig. 10, item 2). This ornamentation, formed by "rocking" a curved stamp from side to side while advancing over the surface of the pot, is also completely alien to the Cis-Baykal.[60] Yet, it is usual for the Late Neolithic sites of the Kama region of the western slopes of the Urals, for example, the Levshino site widely known in the literature through the work of A. V. Schmidt and N. A. Prokoshev.[61] It is also found on the Ob. For example, V. N. Chernetsov uncovered it at the Yekaterinovka site in the Omsk region. Exactly the same ornamentation was noted by Slovtsov in the 1880's in the Tyumen region. In both cases, as on the Yenisey, the design was combined with rippling.[62]

The last and very interesting example of ornamentation from the Ust-Sobakino site is the design on a rim fragment of one rather thick vessel.[63] It resembles the net-like design characteristic of the Ob region. The mesh of the net is not square or rhomboid, but in the form of oblong hexahedrons (Fig. 11, item 2).[64] It is remarkable that this design is encountered in the Cis-Ural region in petroglyphs and was noted among the Uralian tribes of the 19th and early 20th centuries as [an element of] tattooing with special significance.[65]

The extensive ties of the Late Neolithic of Krasnoyarsk *kray* are also supported in the realm of art and [religious] beliefs. In the Krasnoyarsk Museum there is a miniature statuette (cat. no. 519) of steatite (agalmatolite), found at Bazaikha; it represents a sitting naked human figure. In profile it depicts the characteristic, smooth curve of the convex back; the buttocks are depicted distinctly enough, but are disproportionately small. There is only one leg, the foot being indicated by a notch (Fig. 12, item 4). The head is comparatively large, the occiput wide and round, the face flat, and the chin sharp. It has no arms.

A glance at this statuette recalls the well-known antler figurine found in Pernov.[66] Both of these have the same characteristic curve of the back which suggests a man in a sitting position, the absence of arms, and the identical, conventional treatment of the face.

The statuette from Bazaikha is therefore, in all respects, an exact, although greatly reduced, copy of the Pernov statuette. The significance of these resemblances is strengthened by the fact that they are expressed in even more clear-cut form in a number of Neolithic sculptures from the European North both in the Soviet Union and in Finland. Such are, for instance, the clay anthropomorphic statuettes from Khonkaniemi in Finland and from Kubenino in the Kargopol *rayon* of Archangelsk *oblast* (Fig. 12, items 6, 7).[67] One of the remarkable wooden anthropomorphic figurines from the Gorbunovo peat bog, made with great care and well preserved, also shows this curved dorsum. This idol was discovered in the second cultural layer in section 6 of the peat bog. A. Ya. Bryusov is of the opinion that the cultural layer in which this figure was found can "with a good deal of certainty be dated to the second quarter of the second millennium B.C." (Fig. 12, item 5).[68]

The range of similarity of the agalmatolite statuette from Bazaikha is not limited to the sculptured anthropomorphic figurines of the Neolithic and Early Bronze ages which were found west of the Yenisey. Besides sculpture in the round there are also petroglyphs.

All the statuettes enumerated above (from Bazaikha, Pernov, Khonkaniemi, Kubenino, and the Gorbunovo peat bog) when viewed in profile bring to mind certain petroglyphs in Karelia. In these, human figures often are shown with the

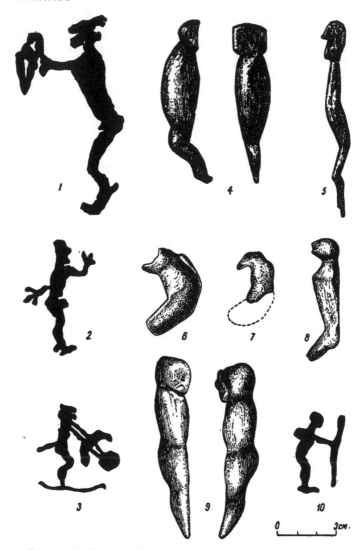

FIGURE 12. Items 1–3, 10, representations of people from the petroglyphs of Karelia; 4, statuette from Bazaikha; 5, idol from the Gorbunovo peat bog; 6, 7, clay figures from Ileksa and Kubenino; 8, bone figure from a grave in Bazaikha; 9, statuette from Pernov.

same odd, invariable curve of the back and legs. They are also depicted as if half-sitting, and are made this way with a definite purpose[69] (Fig. 12, items 1, 2, 3, 10). It is possible that the statuette from Bazaikha either copies in its peculiarities the aspects of the petroglyphs, or that the latter are representations of sculptured statuettes. The probability of such suppositions is supported by the conclusions of S. N. Zamyatnin, who pointed to the general agreement of sculptured (chipped) flint representations and petroglyphs in our North.[70]

In this connection there is yet one more remarkable anthropomorphic statuette, carved from elk antler, uncovered by I. T. Savenkov in the grave which contained the already mentioned elk figurines. This figure has a strange, bird-like head and the same dorsal curvature, which makes it look as if it were half-sitting (Fig. 12, item 8), like the agalmatolite statuettes or the one from Pernov. A carefully made flint representation of a bird (retouched with pressure flaking) also shows this curvature. It was found by Savenkov in the Angara lowland neighboring on the Yenisey region at the well-known habitation site near the Chadobets river. The bird has a long, sharp beak, reminiscent of the beak of the thick-billed nutcracker or the carrion crow, and a long tail. The bird is shown as if in flight, but has very short wings.[71]

Again, we cannot remain silent in view of the similarity between the primitive art of the Yenisey sites and those of northern Europe. M. E. Foss has published an illustration of a clay statuette from the Kinema site. This statuette has the same curved dorsum and bird's beak that we have noted for the Bazaikha statuette. In making a qualitative comparison of the Kubenino statuette Foss introduced the one from Khonkaniemi.[72] She writes that "in spite of the variation in their shapes due to the characteristics of the different materials, we observe an almost complete identity in these sculptures. Both are made without hands and feet, have the characteristic slight bend of the torso and a somewhat elongated crown of the head." The resemblance between the statuettes is, as a matter of fact, so striking that "it seems they came from the hands of one master."[73] This conclusion could be applied in the same measure to the sculptured figurines from the middle Yenisey (Bazaikha). The fact that the Yenisey figurines are not of clay, but of stone and bone, does not lessen the significance of this similarity but, indeed, strengthens it by confirming the presence of a specific traditional style and possibly similar meaning of such figurines. For, to carve such figurines from antler or steatite was more difficult than to fashion them out of soft clay.

Not less interesting is another object (cat. no. 625) also found at Bazaikha; it represents a bear (Fig. 13, item 1). This object has been known for a long time. N. M. Yadrintsev noted that at Bazaikha I. E. Savenkov found "amulets of agalmatolite representing various animals, and a pendant in the form of a stone ring."[74] From this source apparently originates also the bear figurine described above. N. K. Auerbakh noted this find of Savenkov (in the catalogue of his collections) among the artifacts of agalmatolite, which included a head of a wild boar.[75] The bear figurine in question is also a miniature. It is made of reddish-brown soft slate. The head is executed with special care and realistic vitality; it is typical of a bear, with a short, blunt snout and well-expressed ears. The snout of the beast is slightly raised and [gives the impression of] stretching forward with the ears jutting backwards. Paunchy and round in section, the body of the beast has a short stylized "leg," similar to the one-legged human figurine. The leg ends with the ankle. In general form and style, the miniature figure of the bear from Bazaikha startlingly resembles the well-known hammers with animal heads found in the northwest of the Soviet Union and Finland. Their resemblance is so great that the general features of the figure with its broad middle and animal head at one end must point to conscious imitation of the figured hammers (Fig. 13, item 4).

Judging by the photographs in the archives of the Institute for the History of Material Culture,* such bear figurines had been found by Savenkov at Bazaikha.

The second figure is almost analogous to the first (Fig. 13, item 2). I have not been able to locate it in the collections of the Krasnoyarsk Museum.

In considering the figurines from Bazaikha we must mention the find of a sculptured bone figurine of an aquatic bird, apparently a duck. It was uncovered

---

*[Institut istorii materialnoy kultury Akademii nauk S.S.S.R.]

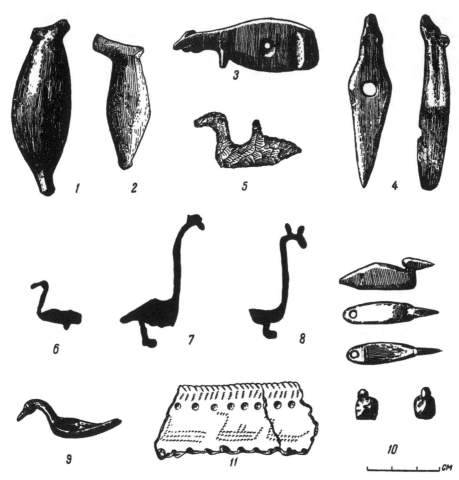

FIGURE 13. Items 1, 2, agalmatolite figures of a bear from Bazaikha; 3, agalmatolite figure from Bazaikha; 4, figured hammer from Karelia; 5, flint figure of a bird from Modlona; 6–8, representations of a bird from petroglyphs in Karelia; 9, 10, figurines of birds from a grave on Afontova Gora; 11, fragment of a vessel from Karelia with representations of birds.

by A. F. Katkov among the contents of the Neolithic burial on Afontova Gora.[76] The duck is depicted simply and schematically, but all the same quite expressively. It is represented as if swimming in water with a raised massive head and long, exaggerated beak. In the tail of the figurine there is a hole for suspension, indicating that it was used as a pendant or as part of a necklace of bone beads and animal teeth which were found next to it in the same grave (Fig. 13, item 10).

Stylized figurines of water birds were widely diffused and were an important element of design in Neolithic and Early Bronze Age ceramics in the northern European part of the Soviet Union and on the eastern slopes of the Urals (Fig. 13, item 11).[77]

Water birds, identical in style and details, also occur among the petroglyphs of Karelia and the flint miniatures (Fig. 13, items 6, 7, 8).[78] A remarkable example of similarity in depicting birds with petroglyphs and in flint is a unique figurine, found by A. Ya. Bryusov during excavation of a habitation site on the

Modlona river. The birds are portrayed with a narrow projection on their backs, which is suggestive of [whipping] wings[79] (Fig. 13, item 5).

Investigators propose that the frequent portrayal of ducks on Neolithic vessels in conjunction with horizontal zigzags signifies that they are shown "together with their element, water, the waves of which are rendered with zigzagging lines."[80] P. P. Yefimenko wrote that the very form of the vessel can be associated with the egg, the symbol of the vital beginning of nature. To him, the horizontal wavy lines on the pottery reproduce the wavy surface of the water, and the oblique lines quiet water. The representations of aquatic birds on the vessels, arranged as a frieze of swimming birds on the edge of the vessel, have direct reference to "religion, cosmogenic concepts of local tribes originating mainly from a water cult and the spring awakening of nature."[81]

I think that Yefimenko is right. The symbolism of the Neolithic pottery probably has a common base with Kalevala's myth about the egg from which the universe originated. To this also may be traced the dualistic myths about the creation of the bird world characteristic of both Slavic and Finnic folklore.

So far, to the east of the Urals, such portrayals of aquatic birds on Neolithic pottery and in petroglyphs have not been noted. It is all the more interesting that representations of ducks, so alien to the Cis-Baykal, have been found precisely where indications of ancient cultural ties with the Urals and Karelia had been discovered, i.e., on the banks of the Yenisey, near Krasnoyarsk.[82]

There is still another fact which we should consider. I refer to the traits of resemblance between bone artifacts from the western slopes of Uralian and middle Yeniseyan Neolithic sites. At the Ust-Sobakino site a bone arrowhead with a broadened point was found which duplicated the form of biconate heads so characteristic of the northern European part of the Soviet Union, and the spindle-shaped heads of the Urals. It is the easternmost of all such arrowheads found thus far.[83]

It is very interesting that A. Elenev, while excavating in the Biryusino caves, found seven fragments of composite points, which are of the same type as the Ural points.[84] They are long, narrow shafts, flatly oval in cross-section, with a rounded oval stem and deeply cut grooves for the flat stone side-blades. Points of such type are not known in the Cis-Baykal area if those found in 1937 in the lower reaches of the Angara on Kamennyy Island are not considered. However, the latter site is very near the Yenisey, and, additionally, the composite points found there differ from the rest of the Angara artifacts of such type.[85]

Some findings indicating connections of the Yenisey area with the neighboring region of Central Asia as far as the Aral Sea should also be noted. Thus, for instance, the dunes of Bazaikha also yielded a flint arrowhead with a side notch of the same appearance, size, and mode of preparation as Kelteminar arrowheads. Like the latter it was prepared from a lamellar blade only slightly altered by retouch along the edges. Its stem is long, occupying two-thirds of the overall length, it has a side notch, and the point is correspondingly short and wide.[86]

The provenience and route which arrowheads of this type took to reach the middle Yenisey must be judged by the presence of analogous finds on the eastern slopes of the Urals. There are cultural elements common to both which indicate this. An arrowhead, similar in form and manufacture, found by I. Ya. Slovtsov in the lower layer of the First Andreyevo site (on Andreyevo Lake near Tyumen),[87] confirms the investigations of V. N. Chernetsov. Another arrowhead was found in the Urals [proper], having been noted among the finds from the site near Palkino village.[88]

It is difficult to judge the age of all these finds (including the arrowhead from

Bazaikha). To be sure, none are older than the Kelteminar habitation site, Dzhanbas-kala no. 4 in Khorezmia [Chorasmia], and, consequently, may be dated to the developed ["flowering"] Neolithic. If this is so, these finds easily agree in characteristic traits with the ceramics from the ancient sites of the middle Yenisey, western Siberia, and the Urals, which in turn brings them closer to the Kelteminar sites. As described earlier, the original design consisting of subdivisions of small areas within broad horizontal bands belongs here. The specific wavy or continuous design found often in combination with comb impressions and curvilinear zigzagging, which is made by "rocking" a paddle-stamp, also belongs to this complex.[89]

The data related above are witness to very old cultural and historical ties between the population along the Yenisey river and other peoples and are therefore of great historical interest. If we consider them from the widest historical point of view, we may obtain a number of important results and arrive at definite conclusions both in regard to the history of the Yenisey tribes and to such large and complicated problems as are posed by the questions of the original provenience of the Finno-Ugric tribes and of the appearance of Samoyed tribes in the Sayans and in the Yenisey valley.

The first of such results indicate that during the Serovo period (the end of the 4th to the 3rd millennium B.C.), the tribes of the Yenisey area were not only in proximity, but also in close cultural ties with the population of the Cis-Baykal. At this time there probably existed not only a cultural, but also an ethnic community of the populations in these regions of Siberia. However, even during this relatively early stage of the Neolithic of forested Siberia, there are distinct indications of ties of Yenisey tribes with both the relatively near and the more remote regions to the west. To this time belong the pottery and sculptured artifacts in some way or another similar to ones typical of the Volga-Oka region, Karelia, and even the Baltic area. In agreement are also the bone arrowheads of the Shigir type, [stone] lamellar arrowheads with a side notch, and the early pottery with a continuous wavy design like that found on the Poludeno pottery of the Urals and the Kelteminar of the Aral Sea.

These western connections continue into the period that follows, that is, the 3rd and the 2nd millennia B.C., during which there begin to appear remains first similar to those of the Kitoy [period of the Baykal area] and later to the Glazkovo (the Afontova Gora burials, those near Perevoznaya village, the lower stratum of the Ust-Sobakino site, and others).

It must be emphasized in this connection that the considerable agreement between remains from the middle Yenisey and Cis-Baykal during the Late Neolithic does not imply that the cultural influence traveled in one direction only, i.e., from east to west. The excavations of the Angara Archaeological Expedition at Bratsk in the lower reaches of the Angara at Monastyrskiy Kamen [Monastery Rock] yielded a rich ceramic inventory datable to the Late Serovo and Kitoy periods. It included fragments of round-bottomed vessels completely covered with horizontal bands of caterpillar stamp impressions. From this it follows that during the Kitoy period or even towards the end of the Serovo, a separate cultural stream from the Yenisey to the east reached localities in the region of the Oka river near present-day Bratsk and perhaps farther upriver along the Angara and even to the Lena over the Angara-Lena interfluve. Thus, for instance, a sherd, similar in type, was found in a cultural layer within the area of the Verkholensk Neolithic burial mound.

Some of the changes in the ornamentation of pottery which occur during the Glazkovo period of the Cis-Baykal are not accidental. Up to the end of the Kitoy period only the upper part of the vessel was ornamented. During the Glazkovo, above Bratsk (on the Angara) the appearance of continuous, banded ornamentation covering the vessel from rim to bottom was first noted. But even then such ornamentation was rare in the Cis-Baykal. In care of execution and neatness it is usually inferior to that known from the Bratsk and Krasnoyarsk regions.

It is particularly interesting that on a vessel from a Shivera burial of the Glazkovo period a design of this type is executed very simply and is not stamped but incised with lines and dots. This gives the impression that the Glazkovo potter consciously sought to copy by the most simple means the already existing foreign design obtained from western neighbors.[90]

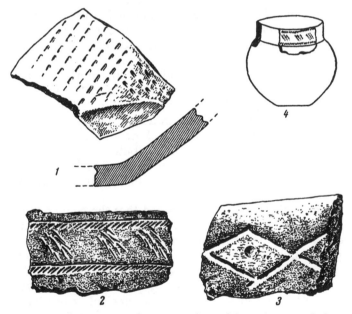

FIGURE 14. Items 1–3, fragments of vessel from the Ust-Sobakino site; 4, reconstructed vessel from the Ust-Sobakino site. (Items 1–3 are about ⅔ natural size.)

With the passage of time, the population of the middle Yenisey changed its economy of hunting and fishing to one of cattle breeding or even agriculture. Such a transition was entirely natural after [prolonged] contact of the forest tribes with the cattle breeders and agriculturists of the steppes. In sequence these were tribes of the Afanasevo, Andronovo, and Karasuk periods. This transition is well illustrated by the contents of the Ust-Sobakino site at Krasnoyarsk. In its upper layers the pottery partially resembles the Andronovo type, but most of it is closer to Karasuk pottery. There are several examples of typically Karasuk pottery in general shape, i.e., spherical vessels with vertical, narrow necks. The ornamentation of several of the vessels is close to the Karasuk. It consists of a horizontal band of rhomboids joined to each other.[91]

The Karasuk-like vessel found at Ust-Sobakino together with a copper ring is apparently of this group. This could be a case of an intrusive burial into a chronologically earlier layer of the site (Fig. 14, items 1–4). Sherds with their surfaces covered with stamp impressions are also found here. These consist of intersecting lines which form a "checkerboard" grid. These impressions are rather large, deepened squares, divided by elevated borders.[92] With these were found fragments of coarse pottery, distinguished by their friability and inclusion of [animal] hairs, which [after firing] leave outlines on the surface of the vessel.[93] It is interesting that these features are characteristic also of pottery from the Late Neolithic sites of Yakutia and also of its Bronze Age. In connection with this may I point out that the findings of physical anthropology agree on the penetration of taiga-dwelling Mongoloids into the steppes of southern Siberia during the Late Bronze Age.

With the Karasuk-type pottery of the third cultural layer of the site (Ust-Sobakino) were uncovered bones of domesticated sheep and swine (?) and, in the second layer, of cows and possibly horses. In this connection V. G. Kartsov wrote: "The economic basis of the (Krasnoyarsk—author) culture gradually changed and from a purely hunting one it became one of cattle breeding, attending to both large- and small-horned cattle although never losing its ties to hunting."[94]

[In this site] there is also pottery similar in appearance to that of the Agar period. It appears also in other regions of Siberia to the west of the Yenisey, including the Ob river region. Consequently, already at this time there are new traits discoverable in the pottery of the [middle] Yenisey peoples which bring it even closer to that of the Andronovo-Karasuk peoples of the forests and steppes of western Siberia. These traits could be attributed entirely to the influence of the Andronovo and Karasuk tribes on their forest neighbors if it were not for a number of specific details of design, strange to the steppe culture. Such is, for example, the net with an octagonal mesh.

Based on the foregoing, it may be proposed that the common cultural traits of middle Yeniseyans and western Siberians (including the population of the east Uralian slopes) at the end of the Neolithic and into the Bronze Age are explainable not only by cultural contact and penetration of particular elements or the adoption of patterns of ornamentation and pottery manufacturing techniques, that is, not only by "diffusion" alone, but by penetration from west to east of a group of tribes. Now, we may identify these ancient tribes with their proper names, and also connect them to particular present-day ethnic groups. It seems to me that here we have principally that ethnic group whose descendants formed the basic layer or substratum of the later Ugrian tribes of the eastern Urals, that is, the Khanty and Mansi, and also the Hungarians. The very early appearance on the middle Yenisey [and] in the foothills of the Sayan mountains of these Uralian tribes from the west, an arrival traceable in the archaeological materials, is also attested to by the presence, in later times, of tribal groups speaking Samoyedic languages.

It was already established in the 18th century and the beginning of the 19th that in the Yenisey valley and in the Sayan-Altay uplands there had lived tribes who spoke languages native to the Samoyeds; these had later become extinct or were Turkicized or, in part, Russified. The distinguished linguistic and ethnographer of the 19th century, A. M. Castren, assembled clear proofs of this view, developed it, and supported it theoretically.

In the light of these accumulated facts it has become clear that the Samoyedic tribes—dispersed in the 18th and 19th centuries over a very large territory—had at one time occupied the entire territory as one ethnic unit.

But where are we to look for the common land of these Samoyedic tribes? Is it to the east, in the Sayans, or to the west, near the Urals? And, which of them remained in the native land and which left it? Did the present-day Nentsi-Samoyeds of the European tundra come to the west or did the ancestors of the Sayan Samoyedic tribes come from the west to the east, to the upper reaches of the Yenisey? And, finally, when did the division of the ancestors of these Samoyedic tribes into two territorially separate parts take place?

Having discovered Samoyedic tribes in the Sayan-Altay uplands, Castren advanced a theory, in his works dealing with the entire epoch, that it was far to the east, in the depths of Asia, that the common, original fatherland of not only the Samoyeds was to be found, but (logically) also that of all Finno-Ugric tribes, since he had also established at that time the linguistic unity of Samoyedic and Finno-Ugric peoples.[95]

However, by the same measure it would seem logical to argue that the reverse could have taken place, that is, that the original home of the Finno-Ugric tribes is to be found not in the east but somewhere in the west, and that they did not come from the Yenisey to the Urals and farther, to the shores of the Baltic Sea, but that the order was the reverse.

In any case, this does not contradict the finding in the east of Finno-Ugric toponymy. Having come from the west, these early Finno-Ugrians naturally would likely have left traces of their sojourn. Nevertheless, toponymy does not, and cannot, offer a solution to the question posed. Consequently, in order to establish their original location and the actual direction of their migrations it is necessary to resort to other sources.

In the absence of written data, there are archaeological remains supplemented by the findings of linguistics, physical anthropology, and other disciplines. It is with these in mind that Soviet investigators have dealt with the material, basing themselves on the results of the new archaeological studies in eastern Europe, in the Urals, and the neighboring regions of central Asia.

A. Ya. Bryusov, for instance, proposes that the ancient tribes of the forest zone of the European part of the Soviet Union who left the Neolithic cultures with "comb-and-pit-marked" pottery were the remote ancestors of the later Finnish tribes.[96] According to the views of S. P. Tolstov, at the time of the Neolithic Kelteminar culture around the Aral Sea (4th to 3rd millennium B.C.), Finno-Ugric tribes dwelt there.[97] In considering this, however, we must realize that the Aral region was not the center but only a periphery of the ancient Neolithic culture of the north, although one which by its geographical position was connected in many respects with the cultures of the southern regions of central Asia and also with those of Iran, Afghanistan, and India. The Kelteminar peoples found here in the 4th to 3rd millennium B.C. were in contact with two diverse groups of central and southern Asiatic tribes. One group comprised the farmers of the oases possessing decorated pottery of the Anau type; the other, those southern[most] hunting and fishing peoples who roamed the vast expanse of the Caspian steppes and wilderness.

Still earlier, before the division into hoe-using farmers and hunters-and-fishers (who later became steppe cattle-breeders), there existed here an Epipaleolithic, in many respects uniform culture of hunters and gatherers with characteristic Capsian or Aurignacian techniques of manufacturing stone implements and miniature geometric artifacts in the shape of segments, triangles, and trapezoids. It is very probable that the Neolithic descendants of these "Capsian" (in culture) tribes, who spread at one time from the Syr-Darya to southern India and Australia, spoke (as S. P. Tolstov believes) Dravidian and Munda languages. As for their northern neighbors, they were probably the ancestors of the Finno-Ugrians. These Finno-Ugric tribes of the Neolithic period apparently were the originators of the Uralian and western Siberian cultures which are so closely related to each other. In a more exact sense, these early tribes of the Aral region, the Urals, and western Siberia do not represent the Finno-Ugrians or Finns but specifically the Ugrian ancestors of the later Uralian tribes, that is the ancestors of the Khanty, Mansi, Hungarians, and Samoyeds.

The neighboring, very similar—but at the same time in many respects different —culture of the Volga-Oka and Karelian Neolithic, with its comb-and-pit-marked ceramics, most likely represented the Finnic tribes, as A. Ya. Bryusov also believes.

These are views which have their origins in the results of archaeological investigations of the forest zone of European Russia and of the closest territories of central Asia. Nowadays, we may assert that they find support also in the east, in the archaeological remains of the very region which Castren selected as the motherland of the Finno-Ugrians. Further, we maintain that here, during this remote time, there existed an altogether different Isakovo-Serovo culture, a different cultural and ethnic world. Today we know this culture fairly well. Its territory began with the banks of the Yenisey and possibly included the Barabinsk steppe and what is now the Kuznetsk basin. From here it spread to the east, beyond the Yenisey to the shores of Lake Baykal and farther, to the upper reaches of the Amur. Its influence is discernible even in the earliest Neolithic of Mongolia and northeastern China, which, because of misunderstanding, is often called "microlithic." The descendants of its bearers include the Altayan (Turkic-Mongolian and Tungus) tribes of Siberia and central Asia.

Within the interrelations of these two large cultural regions of Asia are no doubt hidden interrelations of two large, initially different, groups of ancient populations. To the east lived, in terms of physical traits, Mongoloid tribes; to the west, tribes of initially Europoid type. The process of mixing of the Europoid from the west and the Mongoloids from the east began probably long before the Neolithic, still in the Upper Paleolithic, in that remote time when there existed on the shores of the Angara sites of the Malta and Buret type with an east European culture. It could have happened even then that in the zone of contact between the Urals and the Yenisey there arose a mixed population and began to form the so-called Lappoid or Uralian physical type which agrees in significant measure with that of the speakers of the Uralic and Finno-Ugric languages.

It is difficult to project in detail how this process of imposition, in time and territorial scale, took place. But it is possible to say that it actually took place because of the archaeological facts mentioned *supra* and also because the data of physical anthropology confirm the presence of Europoid traits in the Mongoloid Neolithic population of the Cis-Baykal already in a very early period. One of the more important events of this interaction and interpenetration of Europoids from the west and Mongoloids from the east was the diffusion to the west, into the Yenisey valley, of Cis-Baykalian tribes. This occurred at the end of the Serovo period, if not earlier. This is the period during which Serovo and Kitoy cultural traits appear in the middle Yenisey.

Still more important in its consequences was the counterdiffusion to the east, to the Yenisey, and possibly farther to the Sayans, of inhabitants of the Ural region and western Siberia who carried with them the elements of culture noted earlier. They first penetrated into the region of present-day Krasnoyarsk and to the lower reaches of the Angara in the direction of present-day Bratsk during the Neolithic, then [again] during the Bronze Age.

As a result of the penetration of western tribes to the Yenisey, possibly beginning during the Paleolithic, there also appeared the ancestors of Samoyedic tribes, who later, after reindeer breeding arose, again spread to the northwest into more suitable and richer pasturing places.

The place of origin of the Finno-Ugric tribes and in particular the Samoyeds was not in the Sayan mountains or on the Yenisey. The most ancient land of all of them was the forests and forested steppes of both the eastern and western slopes of the Urals and also the forest regions of eastern Europe adjacent to the Urals.[98]

## *Notes and References*

1. G. Merhart, Neuere Literatur über die Steinzeit Siberiens, *Wiener Prähistorische Zeitschrift*, 11, fasc. 2, 1924; *idem*, Siberien: Neolithikum, *Reallexikon der Vorgeschichte*, vol. 12, 1925; *idem, Bronzezeit am Ienissei*, Wien, 1926.

2. B. E. Petri, *Dalekoye proshloye Pribaykalya, Nauchno-populyarnyy ocherk* (The remote past of the Cis-Baykal. Popular scientific essay), Irkutsk, 1928; *idem, Sibirskiy neolit* (The Siberian Neolithic), Irkutsk, 1926; A. P. Okladnikov, Neolit i bronzovyy vek Pribaykalya (The Neolithic and Bronze ages of the Cis-Baykal), pts. 1 and 2, *MIA*, no. 18, 1950, p. 121.

3. V. G. Kartsov, *Materialy k arkheologii Krasnoyarskogo rayona* (Materials for the archaeology of the Krasnoyarsk region), Gos. muzey Priyeniseyskogo kraya, Opisaniye kollektsiy i materialov muzeya, Otdel arkheologicheskiy, Krasnoyarsk, 1929, p. 1.

4. *Ibid.*, p. 26.

5. *Ibid.*

6. S. V. Kiselev, *Drevnyaya istoriya yuzhnoy Sibiri* (The early history of southern Siberia), Moscow, 1951, pp. 20–21.

7. Okladnikov, Neolit i bronzovyy vek Pribaykalya, p. 244.

8. I. V. Savenkov, *Kamennyy vek v Minusinskom kraye* (The Stone Age in Minusinsk *kray*), Moscow, 1897.

9. *Ibid.*, pp. 21–22, Plate 23, Fig. 1.

10. *Ibid.*, p. 23, Fig. 153.

11. Krasnoyarsk Museum, Collection 124–I.

12. Krasnoyarsk Museum, Collection 187–56 (Ladeyki).

13. Muzey antropologii i etnografii Akademii nauk SSSR, Section of Archaeology, Collection 4083–3 (Ladeyki).

14. Savenkov, *Kamennyy vek v Minusinskom kraye*, p. 20, Fig. 121.

15. E.g., from the acquisitions of the Krasnoyarsk Museum: Collections 110–437; 194–31; 124–3, 6 (Bazaikha); 274–4; 16 (Ladeyki); see Savenkov, *Kamennyy vek v Minusinskom kraye*, p. 37, Figs. 165, 166, 167; Kartsov, *Materialy k arkheologii . . .*, Plate 2, Figs. 20, 25.

16. Savenkov, *Kamennyy vek v Minusinskom kraye*, p. 42, Fig. 181; see Kartsov, *Materialy k arkheologii . . .*, Plate 2, Fig. 29.

17. Krasnoyarsk Museum, Collection 194–26, 29 (Bazaikha); see Kartsov, *Materialy k arkheologii . . .*, Plate 2, Fig. 40.

18. Krasnoyarsk Museum, Collection 122, 111–116.

19. Krasnoyarsk Museum, Collection from Bazaikha, no. 148/1–VII.

20. For instance, from the acquisitions of the State Museum of Physical Anthropology attached to Moscow State University, Collection 16–33, 144 (Ladeyki, collections of G. P. Sosnovskiy), and from Krasnoyarsk Museum, Collection 11–1 (Bazaikha).

21. Krasnoyarsk Museum, Collection 11–1 (Bazaikha). See, for instance, the fragment illustrated by Kartsov (*Materialy k arkheologii . . .*, Plate 5, Figs. 1, 4), although it is not very representative.

22. *Reallexikon der Vorgeschichte*, vol. 12, Plate 9, i; Collection 59–75, Krasnoyarsk Museum.

23. Okladnikov, Neolit i bronzovyy vek Pribaykalya, p. 211, Fig. 56.

24. *Ibid.*, p. 207, Fig. 49.

25. Krasnoyarsk Museum, Collection 110–243 (from below the mouth of the Sobakina river); Collection 179–67 (Sobakina river, Trench II, no. 111–3); Collection 122–108 (Nyasha); Collection 194–68 (Bazaikha).

26. These objects are in the Museum of Physical Anthropology and Ethnography of the Academy of Sciences of the U.S.S.R., Collection no. 1259–9.

27. An adze of this type was found in the vicinity of the village Tes on the Tuba river. Here may be classified the artifacts from the vicinity of Shunera village (see Savenkov, *Kamennyy vek v Minusinskom kraye*, p. 19, Fig. 150; p. 22, Plate 23, Fig. 2).

28. The curving of the blades (cutting edge) of such artifacts may be attributed to the necessity of frequent sharpening after prolonged use. See Okladnikov, Neolit i bronzovyy vek Pribaykalya, Fig. 40.

29. G. F. Debets, *Paleoantropologiya SSSR* (Paleo-anthropology of the U.S.S.R.), Moscow–Leningrad, 1948, pp 62–63.

30. See Kiselev, *Drevnyaya istoriya yuzhnoy Sibiri*, p. 20.

31. E. P. Rygdylon, Zametki o karasukskikh pamyatnikakh iz okrestnostey Krasnoyarska (Remarks on Karasuk remains in the vicinity of Krasnoyarsk), *KSIIMK*, no. 60, 1955, pp. 129–134. Schematic sketches of the bronze artifacts accompany the article.

32. Krasnoyarsk Museum, Collection 210/296.

33. *Ibid.*, Collection 108–19; 123–12/27. See Okladnikov, Neolit i bronzovyy vek Pribaykalya, Fig. 107, p. 361.

34. N. K. Auerbakh, Pervyy period arkheologicheskoy deyatelnosti I. T. Savenkova. Materialy k biografii (The early period of I. T. Savenkov's archaeological activities. Sources for his biography), *Ezhegodnik Gosudarstvennogo muzeya im. N. M. Martyanova v gorode Minusinske*, vol. 6, no. 2, 1928, p. 176.

35. N. Bortvin, Iz oblasti drevney sibirskoy keramiky (From the region of early Siberian ceramics), *ZORSA*, vol. 11, 1915, p. 176, Fig. 1.

36. Kartsov, *Materialy k arkheologii* . . . , Plate 5, Figs. 10, 12, 22. Krasnoyarsk Museum, Collection 122–182 (Nyasha).

37. Kartsov, *op. cit.*, Plate 5, Fig. 14. State Museum of Physical Anthropology attached to Moscow State University, Collection 16–43 (Ladeyki, collections of G. P. Sosnovskiy).

38. Krasnoyarsk Museum, Collection 178–11, 74 (Esaulskoye, collections of V. G. Kartsov, 1929); Collection 129–151, 181; Collection 84–26, 106, 212, 316 (from locality between the Sobakina river and Monastyr, on the left bank of the Yenisey); Collection 110–293 (mouth of the Sobakina river, talus).

39. V. I. Gromov, Iz polevikh arkheologicheskikh nablyudeniy na Yenisey letom 1933 (Selected surface finds made during the summer of 1933 on the Yenisey), *Problemy istorii dokapitalicheskikh obshchestv*, 1934, no. 2; A. P. Okladnikov, Neoliticheskiye pogrebeniya na Afontovoy Gore (Neolithic burials on Afontova Gora [Mountain]), *KSIIMK*, no. 25, 1949, Fig. 1, items 3, 6, 8.

40. Kiselev, *Drevnyaya istoriya yuzhnoy Sibiri*, pp. 65–66.

41. Krasnoyarsk Museum, Collection 79–1.

42. V. N. Chernetsov, Ocherk etnogeneza obskikh yugrov (An essay on the ethnogenesis of the Ob Ugrians), *KSIIMK*, no. 9, 1941, p. 19; cf. V. N. Chernetsov, Drevnyaya istoriya Nizhnego Priobya (Early history of the lower Ob region), *MIA*, no. 35, 1953, Plate 2, item 5.

43. Chernetsov, Drevnyaya istoriya Nizhnego Priobya, Figs. 2, 4.

44. N. N. Gurina, Poseleniya epokhi neolita na Onezhskom ozere (Neolithic sites on Lake Onega), *MIA*, no. 20, 1951, p. 99.

45. B. S. Zhukov, K voprosu o stratigrafii i kulture neoliticheskoy stoyanki bliz s. Lyalova (Contribution to the stratigraphy and culture of the Neolithic site near Lyalovo village), *Russkiy antropologicheskiy zhurnal*, vol. 16, nos. 1–2, Moscow, 1926; A. A.

Spitsin and V. I. Kamenskiy, Raskopki bliz g. Balakhny (Excavations near the town of Balakhna), *ZORSA*, vol. 5, 1903.

46. For a critique of the statements of G. Merhart, see Kiselev, *Drevnyaya istoriya yuzhnoy Sibiri*, p. 17; cf. Okladnikov, Neolit i bronzovyy vek Pribaykalya, p. 121.

47. D. N. Eding, Reznaya skulptura Urala. Iz istorii zvernogo stilya (The carved sculpture of the Urals. Remarks on the history of the animal style), *Trudy GIM*, no. 10, Moscow, 1940, Figs. 21, 22; State Museum of Physical Anthropology attached to Moscow State University, Collection 16–1 (Ladeyki); Museum of Physical Anthropology and Ethnography of the Academy of Sciences of the U.S.S.R., Collection 281–110 (Ladeyki).

48. N. A. Prokoshev, K voprosu o neoliticheskikh pamyatnikakh Kamskogo Priuralya (The problem of the Neolithic remains of the Kama region in the Urals), *MIA*, no. 1, 1940, Plate 6, Fig. 13.

49. Eding, Reznaya skulptura Urala, Fig. 21.

50. J. Ailio, Fragen der russischen Steinzeit, *SMYA*, 1922, p. 32, Fig. 9*b*.

51. Krasnoyarsk Museum, Collection 110–269; 194–68 (Bazaikha); 122–140 Nyasha); MAE, Collection 284–34/33. Cf., for instance, the distribution of the oval, caterpillar-comb impressions in A. Ya. Bryusov, *Ocherki po istorii plemen yevropeyskoy chasti SSSR v neoliticheskuyu epokhu* (Essays on the history of tribes in the European part of the U.S.S.R. during the Neolithic epoch), Moscow, 1957, Fig. 17, item 11 (Lyalovo), Fig. 14, items 13, 14 (Balakhna site), Fig. 24, item 26 (Kubenino); Gurina, Poseleniya epokhi neolita . . . , p. 101, Fig. 16 (Ceramics from the Marie site in Finland); M. E. Foss, Drevneyshaya istoriya severa yevropeyskoy chasti SSSR (The early history of the North of the European part of the U.S.S.R.), *MIA*, no. 29, 1952, Fig. 38, items 3, 4 (Kubenino), Fig. 77, items 2, 5 (The Pechora culture, Yarey-ty, Kolva), Fig. 78, item 2 (Kolva), Fig. 79, item 6 (Suna), Fig. 80, items 1, 3, 4 (Lyalovo site, Lake Krugloye), Fig. 86, items 6, 7 (Konea, Lommi).

52. MAE, Collection no. 276–14; Eding, Reznaya skulptura Urala, Fig. 70.

53. Krasnoyarsk Museum, Collection nos. 194–39 (Bazaikha), 110–251, and 122–19.

54. P. A. Dmitriyev, Shigirskaya kultura na vostochnom sklone Urala (The Shigir culture on the eastern slopes of the Urals), *MIA*, no. 21, 1951, p. 70, Fig. 7, items 6, 7; E. M. Bers, Arkheologicheskaya karta g. Sverdlovska i okrestnostey (The archaeological map of Sverdlovsk and vicinity), *MIA*, no. 21, 1951, Fig. 3, item 11; Eding, Reznaya skulptura Urala, p. 26, Fig. 15.

55. Krasnoyarsk Museum, Collections 194–38, 110–249. State Museum of Physical Anthropology, Collection 48–8 (collections of G. P. Sosnovskiy).

56. Dmitriyev, Shigirskaya kultura na vostochnom sklone Urala, p. 73.

57. V. M. Raushenbakh, Keramika Shigirskoy kultury (The ceramics of the Shigir culture), *KSIIMK*, no. 43, 1952, p. 57; A. Ya. Bryusov, *Ocherki po istorii plemen* . . . , pp. 150–153, 157.

58. O. N. Bader, Kamennyy vek na Urale (The Stone Age in the Urals), *Pervoye uralskoye arkheologicheskoye soveshchaniye, Doklady nauchnykh konferentsiy*, nos. 1–4, Izdaniye Molotovskogo Gosudarstvenogo universiteta imeni A. M. Gorkogo, Molotov, 1948, p. 16; *idem*, Neoliticheskaya stoyanka na r. Poludenke bliz N. Tagila (The Neolithic site on the Poludenka river near Nizhniy Tagil), *ibid.*, p. 40; *idem*, Arkheologicheskiye issledovaniya na Urale v 1946 g. (Archaeological investigations in the Urals in 1946), *KSIIMK*, no. 20, 1948; *idem*, Novye raskopki bliz Tagila v 1944 g. Predvaritelnoye soobshcheniye (New Excavations near Tagil in 1944. Preliminary report), *KSIIMK*, no. 16, 1947.

59. Bers, Arkheologicheskaya karta . . . , Fig. 1, no. 47 (Palkino), no. 48 (vicinity of Sverdlovsk). Cf. the survey of the distribution of such lobed axes published by Chernetsov in his, Drevnyaya istoriya Nizhnego Priobya, p. 10, Plate 2; the case of the relationship of the western Siberian Neolithic to that of the Urals and Cis-Baykal is convincingly argued in the latter publication.

60. Krasnoyarsk Museum, Collection 179–104.

61. N. A. Prokoshev, K voprosu o neoliticheskikh . . . , p. 37, Fig. 18, item 19; Fig. 19.

110   A. P. Okladnikov

62. V. N. Chernetsov, Rezultaty arkheologicheskoy razvedki v Omskoy oblasti—Raboty Severo-barabinskoy ekspeditsii, 1945 (Results of the archaeological survey in Omsk oblast—Fieldwork of the North Barabinsk Expedition, 1945), KSIIMK, no. 26, 1947, p. 82, Figs. 33–47; I. A. Slovtsov, O nakhodkakh kamennogo perioda bliz g. Tyumeni v 1883 g. (On Stone Age finds near the town of Tyumen in 1883), Zapiski Zapadnosibirskogo otdela RGO, bk. 7, no. 1, 1885, Fig. 95.

63. Krasnoyarsk Museum, Collection 110–304.

64. Chernetsov, Ocherk etnogeneza obskikh yugrov, Figs. 2, 10, 13; idem, Drevnyaya istoriya Nizhnego Priobya, Plate 17, Fig. 4.

65. S. I. Rudenko, Graficheskoye iskustvo ostyakov i vogulov (The grapic arts of the Ostyaks and Voguls), Materialy po etnografii, vol. 4, no. 2, Leningrad, 1929, Plate 11, Fig. 1.

66. M. Ebert, Die baltischen Provinzen, Kurland, Livland, Estland, 1913, Praehistorische Zeitschrift, no. 5, 1913, pp. 518–519, Plate 23-a; Ailio, Fragen der russischen Steinzeit; H. Moora, Die Steinzeit Estlands, Tartu, 1932, Fig. 13.

67. Foss, Drevneyshaya istoriya severa yevropeyskoy chasti SSSR, Fig. 102, items 1, 2.

68. Bryusov, Ocherki po istorii plemen . . . , p. 156, Fig. 42, item 3.

69. V. I. Ravdonikas, Naskalnye izobrazheniya Onezhskogo ozera i Belogo morya (Petroglyphs of Lake Onega and the White Sea). Part 1: Naskalnye izobrazheniya Onezhskogo ozera (Petroglyphs of Lake Onega), Trudy Instituta etnografii Akademii nauk SSSR, vol. 9, archaeological series, no. 1, Moscow–Leningrad, 1936, Plate 8, Fig. 24; Plate 16, Fig. 150; Plate 20, Fig. 50; Plate 22, Fig. 9; Plate 33, Figs. 11, 17. Idem, Part 2: Naskalnye izobrazheniya Belogo morya (Petroglyphs of the White Sea), Trudy Instituta etnografii Akademii nauk SSSR, vol. 10, archaeological series, no. 1, Moscow–Leningrad, 1938, Plate 4, Figs. 28–30, 37–40, 74–79; Plate 1, Fig. 169.

70. S. N. Zamyatnin, Miniaturnye kremenye skulptury v neolite severovostochnoy Yevropy (Miniature flint sculptures from the Neolithic of northeastern Europe), Sovetskaya arkheologiya, 1948, no. 10, pp. 103–104, Figs. 2, 3.

71. MAE, Collection 282–130. Cf. Reallexikon der Vorgeschichte, vol. 12.

72. Foss, Drevneyshaya istoriya severa yevropeyskoy chasti SSSR, pp. 198–199, Fig. 102, items 9, 10.

73. Ibid., p. 198. It is interesting that in the Cis-Baykal, together with cliff "writings," there are also profiled anthropomorphic representations in half-sitting pose. These are to be related to the events that occurred here during the Late Bronze Age. Up to that time, during the Glazkovo period, only straight-lined representations of people, exaggerated (in height), were found in the art of the forest tribes of the Cis-Baykal.

74. N. M. Yadrintsev, Otchet o poyezdke v Vostochnyyu Sibir v 1886 g. dlya obozreniya mestnykh muzeyev i arkheologicheskikh rabot (An account of a journey to eastern Siberia in 1886 for the examination of local museums and archaeological works), ZRAO, n.s., vol. 3, 1888, p. 14.

75. Auerbakh, Pervyy period arkheologicheskoy deyatelnosti I. T. Savenkova, p. 176.

76. Krasnoyarsk Museum, Collection 185–21; see Gromov, Iz polevikh arkheologicheskikh nablyudeniy na Yenisey . . . , p. 98.

77. P. Yefimenko, Do pitannya pro dzherela kulturi pizdnoy bronzi na territorii Volgo-Kamya (On the question of the origin of the Late Bronze Age culture in the Volga-Kama region), Arkheologiya, 2, Kiev, 1948; Eding, Reznaya skulptura Urala.

78. Zamyatnin, Miniaturnye kremenye skulptury . . . , p. 108, Fig. 5.

79. Ibid., Figs. 5, 9; A. Ya. Bryusov, Svaynoye poseleniye no reke Modlone (Pile dwellings on the Modlona river), MIA, no. 20, 1951, Fig. 13, item 1.

80. N. N. Gurina, Eneoliticheskiye poseleniya u Poventsa Medvezhegorskogo rayona (Eneolithic habitation sites at Povenets of the Medvezhegorsk rayon), Arkheologicheskiy sbornik, Published by the Scientific Cultural Institute of the Karelo-Finnish S.S.R., Petrozavodsk, 1947, p. 71.

81. Yefimenko, Do pitannya pro dzherela kulturi pizdnoy bronzi . . . , p. 43; cf. p. 28.

82. In 1946 I had the occasion to examine petroglyphs on the Oka river near the

village of Bratskaya Kada, which I think are Neolithic. Here also, there were representations of the swan or goose, analogous to the Karelian ones.

83. Krasnoyarsk Museum, Collection no. 84/231.

84. *Ibid.*, Collection no. 59–60; V. Ya. Tomachev, Drevnosti Vostochnogo Urala (The antiquities of the eastern Urals), *Materialy po pervobytnoy i istorichesko-bytovoy arkheologii Zauralskoy chasti Permskoy gub.*, no. 1, St. Petersburg, 1912, Plate 2, Figs. 7–10; Dmitriyev, Shigirskaya kultura . . . , Figs. 3, 16, 17.

85. A. P. Okladnikov, Masterskaya kamennogo veka na o. Kamennom-Kezhemskom (A Stone Age workshop on Lake Kamenno-Kezhemsk), *Uchenye zapiski LGU*, historical-scientific series, no. 13, 1949, Plate 2, Figs. 1, 3, 6.

86. Krasnoyarsk Museum, Collection 107–6, III/4. cf. S. P. Tolstov, Khorozemskaya ekspeditsiya 1939 g. (The Khorezm Expedition of 1939), *KSIIMK*, no. 6, 1940, p. 70, Fig. 14.

87. Chernetsov, Drevnyaya istoriya . . . , p. 26.

88. P. A. Dmitriyev, Kultura naseleniya Srednego Zauralya v epochu bronzi (The culture of the population of the central Trans-Uralian region during the Bronze Age), *MIA*, no. 21, 1951, p. 11, Fig. 1, item 4.

89. S. P. Tolstov, Drevnosti Verkhnego Khorezma (The antiquities of Upper Khorezm), *VDI*, 1951, no. 1, p. 158; Chernetsov, Drevnyaya istoriya . . . , pp. 30–31.

90. Okladnikov, Neolit i bronzovyy vek Pribaykalya, pt. 3, *MIA*, no. 43, 1955.

91. Krasnoyarsk Museum, Collection 179–127, 110–263, 110–265.

92. Krasnoyarsk Museum, Collection 179, 142, 152.

93. *Ibid.*

94. Kartsov, *Materialy k arkheologii Krasnoyarskogo rayona*, pp. 22–23.

95. For a concise but complete review of the question of the Sayan-Altay mountains as the place of origin of the Finno-Ugrians, see the monograph of E. Molnar, The origin and early history of the Hungarian people, in *Studia Historica Academiae scientiarum hungaricae*, no. 13, Budapest, 1955, pp. 34–43. [In Hungarian.]

96. A. Ya. Bryusov, Zaseleniye severa yevropeyskoy chasti SSSR v neoliticheskuyu epochu (The peopling of the northern European part of the U.S.S.R. during the Neolithic epoch), *Tezisy dokladov i vystupleniy sotrudnikov IIMK podgotovlennykh k soveshchaniyu po metodologii etnogenicheskikh issledovaniy*, Moscow, 1951, p. 23.

97. S. P. Tolstov, *Po sledam drevnekhorezmiyskoy tsivilizatsii* (Traces of the early Khorezm civilization), Moscow, 1948, pp. 72–74.

98. I shall not dwell here on other considerably earlier hypotheses regarding the Samoyedic peoples, particularly those which attempt to tie them to the Yukagirs (Oduls), and will only remark that the present analysis of these problems seems more probable, and less open to doubt.

A. P. OKLADNIKOV

# THE SHILKA CAVE

## REMAINS OF AN ANCIENT CULTURE OF THE
## UPPER AMUR RIVER*

FROM THE ARCHAEOLOGICAL point of view, the upper Amur basin, or more exactly the Shilka river valley (Fig. 1), is one of the most important regions of our Far East.

Administratively, the Shilka valley is part of Chita *oblast* and joins the Amur district proper with the taiga regions of eastern Siberia (Fig. 2). It is bounded on the south by the northernmost regions of Mongolia and northeastern China (Tungpei), and on the north by the taiga of Yakutia. This geographic situation left a definite stamp on the ancient culture and history of the upper Amur.

FIGURE 1. Bank of the Shilka river. The "Columns."

Thus, study of the antiquities of the Shilka river basin help to explain the cultural and historical affiliations of the most ancient peoples of this portion of the Asian continent. Further, study of the Shilka antiquities should show the development of the culture and history of the local population since most ancient times.

*Translated from *Materialy i issledovaniya po arkheologii* SSSR, no. 86, 1960, pp. 9–71.

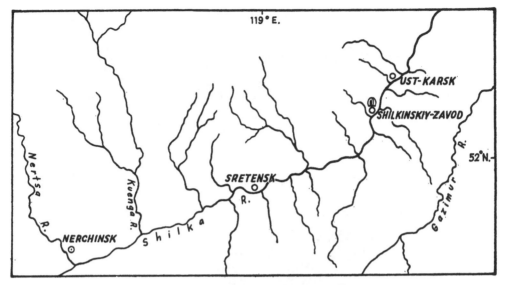

FIGURE 2. Map of the Shilka river valley.

Archaeological remains of this region are of particular interest because representatives of the Altaic linguistic family, peoples of different ethnic groups such as the Tungus-Manchu, Mongols, Turks, and also Paleo-Asiatics, inhabited this region or its immediate vicinity, and may even have originated here some time in the distant past.

By studying the antiquities of the upper Amur and Shilka river basins, we may gain a more complete and profound understanding of the general outlines of the most ancient cultural history and ethnic relationships of this part of Asia, which was the cradle of many peoples—the arena in which important historical events took place, whose significance reached far beyond its boundaries.

However, thus far the archaeological relics of the upper Amur have remained almost unknown. Literary works contain only fragmentary information on the antiquities of the upper and middle reaches of the Amur. The only special publication which mentions the Neolithic remains of this region is one by V. B. Tolmachev, which presents a brief description of collections in the Chita Museum found in two Neolithic burials. Both graves were discovered by chance, the first near Sretensk and the second, apparently, somewhere in Chita *oblast*.[1] The archaeological literature also mentions remains found in another Neolithic grave in Chita *oblast*, at Kokuy station.[2] And that is all!

Therefore, the Shilka cave (Figs. 3 and 4), one of the most ancient and richest archaeological sites in this territory, is of particular interest.

The Shilka cave is situated on the bank of the Shilka river near the village of Shilkinskiy Zavod [factory, works]. The cave is remarkable in that it contained not only numerous and characteristic artifacts of Neolithic man but also skeletal remains and numerous faunal materials—shells of river mollusks and bones of birds, mammals, and fish. Traces of habitation by ancient man were first discovered in the spring of 1952 by a local inhabitant, the topographer I. P. Shabalin, who happened onto the skull and other human bones in the cave and then found stone and bone implements associated with them.[3] Shabalin removed the objects and the skull from the site and entrusted them to N. Kozhin, Inspector

**SHILKA CAVE**

FIGURE 3. The Shilka cave, general view (cave indicated by the arrow).

FIGURE 4. The Shilka cave, entrance.

for the Preservation of Antiquities dispatched for this purpose from Chita. He, in turn, delivered them to the Chita Museum. These finds were reported in April 1952 in the newspaper *Zabaykalskiy Rabochiy* [The Trans-Baykal Worker], where it was stated that

An inhabitant of the village Shilkinskiy Zavod, Shabalin, discovered a cave in the side of a hill near the bank of the Shilka river. Excavations to a depth of 4 m revealed remains of an ancient campsite, with disturbed human skeletons, bones of various animals a stone ax, a flint saw, arrows, bone and flint knives, fish spears, stone and bone arrowheads, beads, and so on. In all, about 100 implements and objects used by the ancient inhabitants of the Trans-Baykal were found.

In June 1952, L. A. Yevtyukhova informed me of these finds and presented me with data obtained from the Administration for the Preservation of Antiquities. Then Ye. I. Krupnov sent me a list of the finds, apparently compiled by N. Kozhin, Inspector for the Chita Oblast [Regional] Executive Committee.

This report states that "the cave, which is situated 18 km from the regional center of Ust-Karsk in a cliff near the village of Shilko-Zavod [i.e., Shilkinskiy Zavod], consists of two chambers (the objects were found in the first chamber). The entrance to the second chamber was blocked by earth and it was not investigated." The finds were described as follows:

DESCRIPTION OF THE FINDS:

1. Human skull with lower jawbone
2. Pelvic bones—5
3. Stones—4
4. Clavicles—2
5. Radii—1
6. Ribs—22
7. Small animal and fish bones
8. Various bones
9. Human vertebrae—17
10. Bone artifacts—3 (unclassified)
11. Adze
12. Stone ax—1
13. Scraper—1
14. Small saws
15. Ornament—bracelet
16. Human sternum
17. Arrowheads and needles
18. Fragments of river shells
19. Composite stone and bone spear head
20. [Stone] side blades
21. Stones
22. Harpoons—11
23. Bone hooks—2
24. Bone daggers—2
25. Bone flakers—2
26. Arrowheads—21
27. Pottery fragments—6
28. Ornament—1 ring
29. Remains of headgear
30. Bone knife
31. Fragments of worked bone
32. Awls made of elk fibulae—2

These reports made it clear that the Shilka finds were worthy of serious study and should be studied on the spot. In this connection, it was also decided that, in so far as possible, the other archaeological remains of the region should be surveyed. Consequently, in September 1952, on behalf of the Institute of Ethnography and the Institute of the History of Material Culture of the Academy of Sciences, I traveled to Chita and Sretensk and then by boat from Sretensk (i.e., from Kokuy station) to Shilkinskiy Zavod and started excavations in the cave.

I continued the excavations in the Shilka cave in 1954. Additional bone and stone artifacts were found and their exact stratigraphic position established.[4] The fieldwork of 1954 was conducted in accordance with the plan of the Far Eastern Archaeological Expedition.*

The history of the discovery and investigation of the Shilka cave is briefly as follows.

The cave had a bad reputation among the local inhabitants, who would not enter it, probably because of the human bones. Local stories have it that in the early days it was warm in the mine shafts on the promontory where the Shilka cave is located and around neighboring Mt. Polosatik and that ore-mining dwarfs lived there. Undoubtedly, these are echoes of the tales popular in miner's folklore about gnomes, reputed to be the guardians of ore treasures.

Now let me describe this unusual site and the objects discovered in the cave.

The cave is located on the left bank of the Shilka river in a picturesque, wooded area, 2 km upstream from Shilkinskiy Zavod. This village has been well known since the 18th century as one of the largest mining centers of the Trans-Baykal region. Ore has long been mined here.

The cave is concealed in a rock outcropping on a low promontory (about 30 m high) between the Shilka river and a small stream, the Chalbucha, which

*[Attached to the Institute of the History of Material Culture, Academy of Sciences of the U.S.S.R.]

empties into it. Limestone outcroppings are found on the promontory. Farther downslope there is a talus train consisting of rubble and crumbled blocks of limestone. The slopes of the promontory are particularly steep towards the Chalbucha. The promontory has a flat summit, which represents the height of an ancient Shilka terrace. On it was a field and a cemetery, and traces of old mine pits. A careful search of the field revealed objects of Neolithic aspect. As mentioned previously, low but picturesque limestone cliffs jut out towards the Shilka river, along a slope dissected by small ruts and crevasses. The cave is situated in one of these crevasses. This small grotto opens towards the south in a niche 2.3 m high and 2.5 m wide; the grotto is 4 m deep (Fig. 5). At the entrance the cave is trapezoidal in outline. Then it gradually narrows and deepens, becoming a fissure. The floor of the cave was level and covered with dark earth mixed with bones, sherds, and various artifacts of ancient man. There was a fairly large block of limestone at the entrance; evidently it had fallen from the roof long ago.

The habitable area of the cave is approximately 10 m². One may stand and move about in the cave without stooping, and the cave is light and dry. Therefore, it is not surprising that ancient men noticed it, and even if they did not use it as a permanent dwelling-place at least they stopped there on hunting trips.

Cultural remains were found even before the entrance to the cave, on the fairly steep slope. Many small fragments of animal and fish bones, pottery sherds, and various stone and bone artifacts were found between the stones strewn along the slope and in the friable black soil around them. The artifacts consisted of bone harpoons, a large fragment of a flat, light-green nephrite ring, and a flint arrowhead of superb workmanship. Shell beads and one stone Daurian-type arrowhead were also found there. No doubt, all these objects came from the cave or the level area in front of the cave entrance, whence they were washed out and slid downslope. Quite probably some of them were thrown there by persons who had dug in the cave in search of treasure.

In the cave itself, the first finds were made immediately at the entrance, in the silt-like friable earth which covered the floor. These were bone fragments and several potsherds. Deeper, there was friable black earth containing stones which had fallen from the roof. This layer was approximately 10 cm thick. Towards the interior of the cave, this layer increased to a thickness of 20–30 cm. It was underlain by yellowish-brown clay, with a particularly large number of stones. The upper part of this layer contained [char]coal, lamellar blades, bones, and sherds of clay vessels, like those mentioned above. The bone fragments found in this layer were darker than those found above them. A spot of red ochre was uncovered in this layer at the cave entrance and small shell beads were found about it. A second ochre spot, even more sharply defined, was discovered near the southeast wall of the cave, 1 m from the first. It is probable that the human skull discovered by Shabalin lay originally at the second spot and the pelvic part of the skeleton rested above the first accumulation of ochre. Thus, the skeleton must have lain across the very entrance to the cave.

A human phalanx and calcaneus were found in the central part of the cave, near the large block of limestone. They apparently belong to the same skeleton as the [above-mentioned] skull. A bone arrowhead with a cleft base was found near the skull. Fragments of a smooth-walled black vessel and a vessel with fine textile impressions on the outer surface were found there. One fragment had a stamped ornamentation. Two horn [antler?] points and one flint arrowhead of the Daurian type were found at the same spot. These finds were associated with a

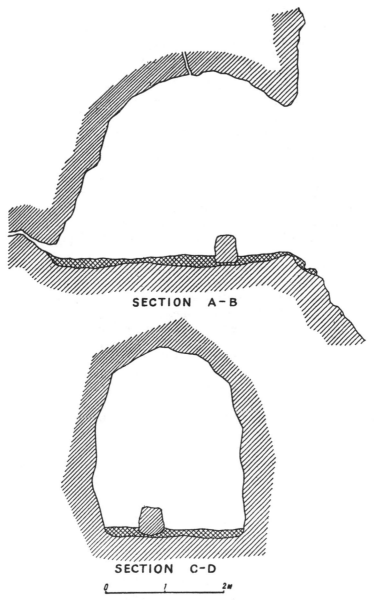

SECTION A–B

SECTION C–D

FIGURE 5. Sectional views of the cave.

fairly well defined burnt layer, or rather the hearth stain, no more than 1 m in diameter, near the skull.

An accumulation of pottery fragments, including ornamental ones, was found near the northwest wall of the cave. A miniature chalcedony arrowhead and a human astragalus and patella were also found there. Fragments of a vessel with wool [hair] impressions on both sides and a broad, flat, triangular superbly retouched arrowhead were found near the southeast wall at a depth of 30 cm at the base of the black humus layer. Fragments of a vessel with textile impressions on the outer surface were found near the arrowhead. Here, at a depth of 20 cm were a mother-of-pearl pendant with two drilled perforations, a chalcedony punch, a fragment of a bone arrowhead with a cleft base, a blank for an arrowhead, and a fragment of vessel with textile impression. An ornamental antler (of a roe deer) found with these objects is particularly interesting. All these objects had been placed in a pocket-like depression in the cave floor.

Many bone fragments and several whole bones were found in the dark layer at a depth of 30 cm in the forward part of the cave, in a pocket-like depression near the east wall. Two scrapers and microblades were also found there. A particularly large amount of fish bones was found near the west wall.

Many bone fragments and animal teeth were discovered in the northern section, deep within the cave, when the upper dark layer of soil had been removed. Sherds with a textile impression, chips, scrapers, and lamellar blades were found. Animal and fish bones and teeth, chipped pebbles, lamellar blades, an ornamented fragment of antler, and bits of charcoal were also discovered at a depth of 20–30 cm. Among the objects found in the northern part of the cave near the west wall were animal bones, a bone knife with side blade[s], scrapers, lamellar blades, and pottery.

A hearth layer was discovered deep within the cave, beneath a 25-cm-thick layer of earth; unbroken animal bones were found in addition to a flint blank for a core and microblades. One human neck vertebra was found nearby. Still farther in the cave, at the point where it becomes a fissure, fish and animal bones and fragments of pottery with textile impressions were discovered close by the wall, at a depth of 25 cm. The cave roof is no more than 1 m high at this point and one must crouch even to sit. More than likely, this area was used for sleeping.

Thus, the Shilka cave was a place where Neolithic man dwelt for a long period of time. Here he slept, made campfires, prepared food, and made bone and stone tools with all, or nearly all, technical means commonly used by Neolithic man. These techniques will be clarified later in the text when the artifacts found in the cave are described. Judging by the abundance of hunting equipment and by the fishing implements (harpoons, fishhooks, sinkers), hunting and fishing, and, in part, gathering, must have formed the economic basis for the cave dwellers. The fish, bird, and animal bones, and the fragments of freshwater mollusks found in the cultural layer of the cave are further evidence of this.

Not only did people live in the cave, but at one time it was used as a burial place as well. One or perhaps more than one inhabitant of the cave was buried here. This burial site was disturbed, probably several times, as indicated by scattering of the human bones about the cave. Possibly these are bones of two skeletons rather than one.

According to Shabalin, a stone ax (really a polished adze of black flinty slate) was protruding from the surface of the ground when he dug up the burial site to which the skull belonged. The skull, its cranium turned upward, lay, facing southeast, "as if on a table," near the ax. The sternum was also lying nearby, as was a bone knife with side blades.

Small beads, in a heap, were found near the limestone rock about 1 m from the skull. Vertebrae of the skeleton were found on the slope near the cave. A heap of flint arrowheads lay there too. Shabalin did not remember seeing any harpoons. Apparently Kozhin had removed them. They probably had lain at the point near the cave entrance where we discovered the ochre spots and the small shell beads (Fig. 6).

Later the cave was used by Iron Age people, who left several characteristic sherds on the surface of the cave floor, including the rims of a vessel and an iron

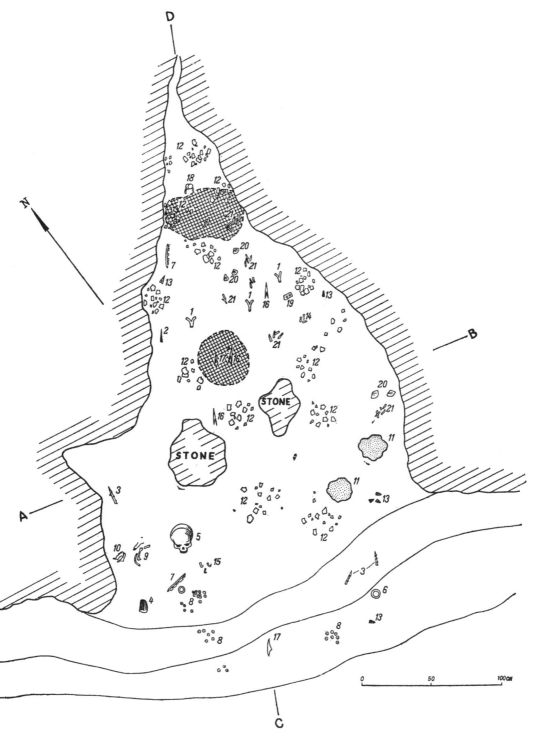

FIGURE 6. Diagram of the cave and objects found in it: 1, carved antler; 2, needle-like point; 3, harpoon; 4, stone ax; 5, human skull; 6, nephrite ring; 7, knife with side blades; 8, mother-of-pearl beads; 9, human bones; 10, human lower jaw; 11, ochre spot; 12, potsherds; 13, flint arrowheads; 14, phalanges, human calcaneus; 15, fishhooks; 16, arrowhead with cleft base; 17, Daurian-type point; 18, human vertebra; 19, mother-of-pearl pendant; 20, scrapers; 21, lamellar blades.

arrowhead. However, the Iron Age people did not remain in the cave for long, or at any rate they did not use it as a more or less permanent dwelling, as had Neolithic man.

The materials belonging to Neolithic man are numerous and characteristic, viz., animal, fish, and bird bones, shells, human bones, stone and bone implements, freshwater shell artifacts, and fragments of clay vessels.

For the description of these finds, we have divided them into groups commonly used in archaeological publications: (1) stone artifacts, (2) bone artifacts, (3) ornaments and art objects, (4) ceramics.

## Description of the Finds

### STONE ARTIFACTS

The stone artifacts found in the cave were the raw material for the products resulting from the manufacture of stone implements—pebbles used as raw material, blanks, chips, and lamellar blades—in addition to broken and intact artifacts, arrowheads, and spear heads, knives, scrapers, punches, and other items.

1. *Cores.* Five cores were found (Fig. 7, items 1, 2, 5, 18, 20). All of them had been worked, used to the limit, and all are miniature in dimensions, especially two of them. The first core (cat. no. 451) is 2.4 cm long and has a striking platform 1.8 cm wide. The material is translucent, yellowish chalcedony. It was prepared from a pebble, the original surface of which remains on one side. The shape of the core is almost conical. Its striking surface was carefully trimmed with transverse, flattening flaking.

The second core (cat. no. 512) is of elongated conical form and is made of close-grained light-yellow flinty rock. It is 2.8 cm long, with a striking platform only 0.7 cm wide. Flint blades could no longer be removed from this core, so it was discarded.

The third core is of the same type as the second, consisting of excellent semi-transparent chalcedony. It is 3.1 cm long, with a striking platform 0.4 cm wide, and has the shape of a polyhedral prism.

The fourth core is 2.6 cm long with a striking platform 0.4 cm wide. It is conical and is made of whitish translucent chalcedony. One side of this core was trimmed by transverse retouching and the other sides present the usual narrow facets of longitudinal flaking.

The fifth core was prepared from a beautiful yellowish-red chacedony pebble with dark concentric bands. The core was split by a longitudinal blow or flaking. There are two flake scars of the burin type at one end of the core, which comes to a thick point resembling the working point of an end burin. The core is 2.4 cm long.

2. *Chips.* About 10 chips, all small, were found.

3. *Pebble blanks.* Ten small pebbles of yellowish and white chalcedony were found in the cultural layer of the cave. These pebbles were intended as the material and blanks for preparing stone artifacts (Fig. 7, item 19). The largest pebble is 6 cm long. They were split in half or partially trimmed by several lateral blows. One pebble, 4 cm long, is of opaque yellowish flint covered with a glossy dark-yellow crust and had been worked at both ends. Evidently it was a blank intended for some small tool.

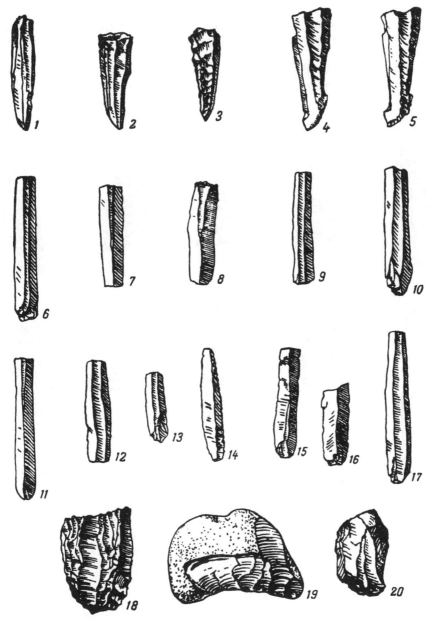

FIGURE 7. Stone implements (cores, blades, chipped pebble).

Another pebble of the same type apparently was intended for a fairly large scraper or even a core scraper. It was trimmed from both sides by broad, flat flaking. One edge is thick; the opposite forms a sharp convex working edge. The material is reddish-yellow translucent chalcedony.

4. *Lamellar blades* (Fig. 7, items 6–17). Eighty-nine blades were found in the cave. They consist of translucent white and yellow chalcedony, often patinated throughout, and also of good-quality flint, dark-gray, light-gray, yellowish- and reddish-banded; green jasper-like slate was also used. For the most part, the

blades are narrow and regularly faceted, with three or, less often, two faces on the dorsal side. They are small. The largest are about 4 cm long and about 0.5 cm wide, but usually they are much smaller, including tiny specimens as small as 3 mm in width.

5. *Stone arrowheads* (Figs. 8 and 9). In all, 29 stone arrowheads were found. They are all of chalcedony and flinty rock of white, red, light-gray, and black

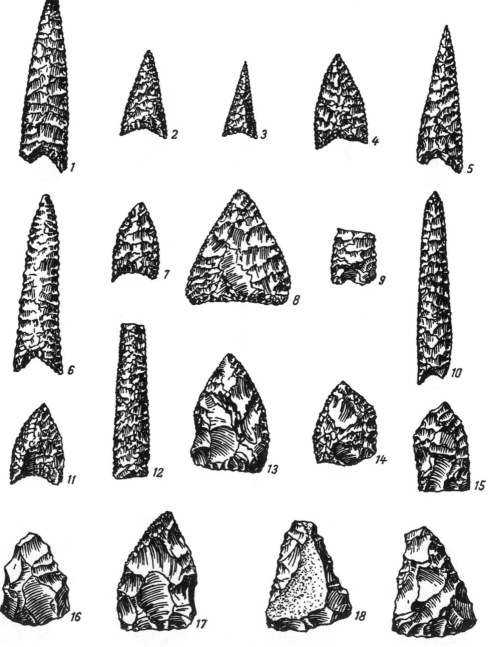

FIGURE 8. Stonework: arrowheads.

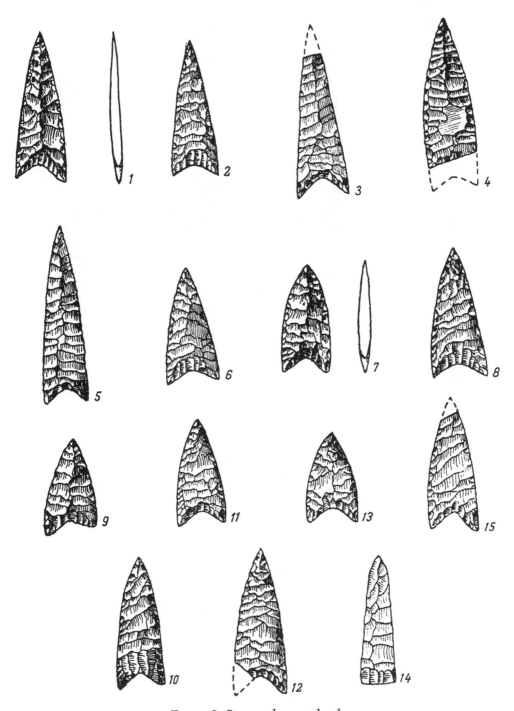

FIGURE 9. Stonework: arrowheads.

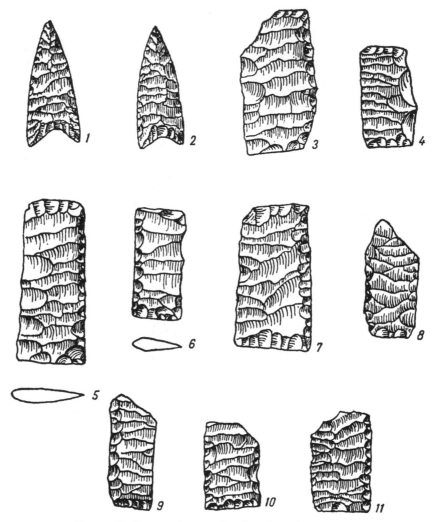

FIGURE 10. Stonework: arrowheads, side and end blades.

colors. In shape and working technique they are ordinary Neolithic heads, bilaterally retouched, with both surfaces worked thoroughly by fine pressure retouching (Figs. 8, 9, 10). They may be subdivided conditionally into four categories.

The first category includes the very broad and short specimens with slightly convex sides and a concave base. There are three such heads (Fig. 8, items 4, 7, 11). They all have a concave base of characteristic asymmetrical shape, i.e., one barb is somewhat longer than the other. These arrowheads are 2.3 cm, 2.4 cm, 3.0 cm, and 2.5 cm long respectively.

The second category comprises 11 arrowheads. The only difference between this category and the first is that the heads are longer and larger. These arrowheads are 4.0 cm, 3.5 cm, 3.3 cm, 3.6 cm, 3.7 cm, 2.8 cm, 3.4 cm, 3.0 cm, 2.4 cm, 3.0 cm, and 2.5 cm long respectively.

The third category consists of narrow and long arrowheads of elongated triangular shape with straight or slightly convex sides (Fig. 8, items 1, 6, 10, 12).

Their bases are also concave and have one asymmetrical barb. These arrowheads were retouched with special care and artistry. They are true works of art of the Stone Age craftsmen. There are six such arrowheads, 4.4 cm, 4.3 cm, 4.7 cm, 5.1 cm, 4.2 cm, and 3.7 cm long, respectively. One of them, broken at both ends, is 4 cm long; originally it must have measured 5 cm in length and about 0.8 cm in width. The minimum width of these arrowheads at the broadest place, the base, is 0.8 cm; the maximum width is 1.3 cm.

The fourth category includes three small arrowheads shaped like isosceles triangles with straight sides and an asymmetrical concave base. They differ from the arrowheads described above in that they are of miniature size and are shorter, being 1.8 cm, 2.1 cm, and 1.7 cm long, respectively. Special mention must be made of the smallest of these three arrowheads, which is made of transparent yellowish-orange chalcedony. The subtle and painstaking nature of the workmanship and the elegance of form of this miniature arrowhead are striking.

The bilaterally retouched arrowheads with a straight base (Fig. 8, items 13–15) may be classified in a special, fifth category. Five such arrowheads were discovered. One of them (cat. no. 176) is made of transparent chalcedony. It is 2 cm long, and 2 cm wide at the base. The second arrowhead (cat. no. 180) is made of the same material but its flaking had not been completed. It is 2.2 cm long and 1.6 cm wide. The third specimen is of yellowish-white flinty material and is 2.6 cm long and 1.3 cm wide. It differs from the other two in that its sides are straight and parallel to each other at the base and then taper sharply. The fourth arrowhead is an unfinished product. It consists of transparent chalcedony with yellowish spots and is 2.2 cm long and 1.8 cm wide. The fifth specimen with a straight base (cat. no. 12) also belongs to the category of blanks or unfinished items. It is shaped like an isosceles triangle. One side is worked completely with broad primary flaking, giving a clear picture of the techniques used for the initial shaping of the arrowheads. All the flake scars run from the edges towards the center. Part of the original smooth surface of the flat stone is preserved on the opposite side, although the edges of the object were worked with primary flaking. This item is 2.8 cm long and 2.3 cm wide and is of flinty shale.

One arrowhead (cat. no. 597), made of transparent chalcedony (Fig. 8, item 8) and superbly formed by pressure retouching, merits a special description. It is flat and thin and is the widest of those found in the cave. Both of its wide surfaces were made smooth by flat faceting. The edges of this arrowhead were worked with equally skillful and painstaking secondary retouching. This item is 3.2 cm long, with a basal width of 2.8 cm, and is shaped like an isosceles triangle.

6. *Fragments of large [projectile] points or knives.* In addition to the arrowheads mentioned, there are four fragments of blanks and unfinished [projectile] points retouched bilaterally. All of these fragments are from quite large artifacts. They include two fragments which must have come from leaf-shaped blades of "men's" knives, or dart points. One fragment, worked with flat, oblique retouching on one side only, is probably part of a unilaterally convex knife. These artifacts were fashioned from white chalcedony and also opaque, yellowish-white or white flint with a reddish line.

7. *Blank for a dart head* (Fig. 11, item 2). This blank is a massive blade, triangular in cross section, of light-gray flinty slate detached from the corner of a flat stone. The faces of the blade were rough-worked with crude preliminary retouching; the edges show secondary retouching. This artifact is 7.3 cm long and 1.6 cm wide.

8. *Points of the Daurian type.* Three points may be classified as Daurian, a

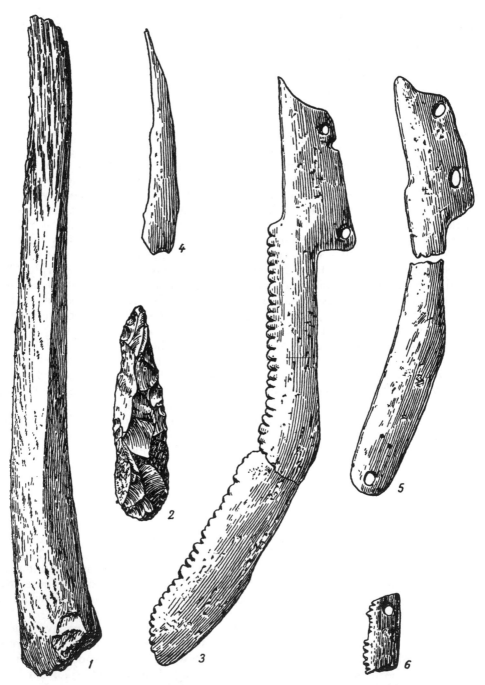

FIGURE 11. Artifacts of antler, bone, stone, and boar tusk.

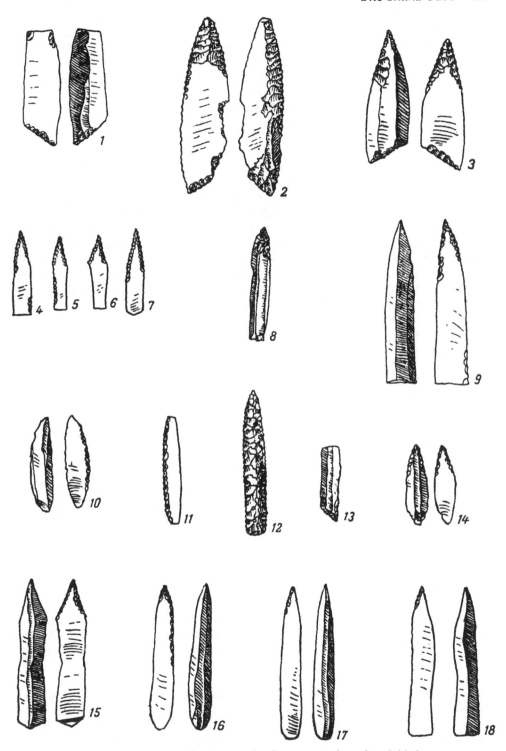

FIGURE 12. Stone artifacts: arrowheads, points, side and end blades.

form characteristic of the upper Amur basin and neighboring regions. Two of these artifacts have been preserved intact, but the upper end of the third is broken. The first point (cat. no. 309; Fig. 12, item 2) was found together with a piece of rock glued to it with a lime crust. It is 4.7 cm long and 1.2 cm wide. It was made from a massive flake with two ridges on the dorsal side. One long edge of the artifact, on the dorsal side, is retouched. The retouching is quite steep, painstaking, and its facets are exactly parallel and regular. On the ventral side, only the ends are retouched. The upper end is worked to a fine point. The opposite end is beveled with bilateral retouching and it is also pointed, but obliquely rather than symmetrically. The cutting edges of the artifact, in contrast to the ends, are not retouched; one of them has a notch, evidently of chance origin. This point is made of yellowish-red, banded flinty slate.

The second Daurian point (cat. no. 60; Fig. 12, item 3) is made of a thin and comparatively broad flake of yellowish-gray flinty slate. There are three parallel ridges on the dorsal side. As in the case of the first point, only the edges are retouched, with most of the flake unworked. Only the dorsal side of the base is actually retouched, since only the convexity of the bulb of percussion is evened out on the ventral side. The end of the point is also worked from both sides, but thoroughly only on the ventral side; just one edge is retouched on the dorsal side and that very negligently. The rest of the flake was not worked. As in the case of the first artifact, the upper end of this flake has a symmetrical point and the lower end was cut obliquely with bilateral retouching and formed into an oblique point. This artifact is 3.5 cm long and 1.1 cm wide.

The third artifact of this type (cat. no. 8; Fig. 12, item 1) was fashioned from a comparatively narrow flake with two ridges on its dorsal side. The material is reddish opaque flint. Only the base of the artifact remains; it is straight, but beveled at an angle to the long axis of the blade, as were the two artifacts described above. The working is characteristic, with fine marginal retouching from the dorsal side only. This fragment is 3 cm long.

9. *Punches* (or arrowheads?). These are small points made from lamellar blades with one end retouched and formed into a point (Fig. 12, items 4–9, 13–18), and are very similar in form and workmanship to the three points described above. The retouching is confined to the lower surface of the blade, and to the very tip of the blade at that. Sometimes the retouching extends to the long edges, but only in part. The five largest artifacts of this type are 4.0 cm, 4.1 cm, 3.8 cm, 4.2 cm, and 3.7 cm long, respectively. Four artifacts of the same type, whose lower ends are broken off, are 1.9 cm, 2.1 cm, 2.2 cm, and 1.9 cm long, respectively.

One artifact, 1.9 cm long, is preserved intact. It was made from a miniature flake of translucent chalcedony with a clearly visible bulb of percussion at its base. The upper end was worked with careful, very fine retouching and made into a point (Fig. 12, item 14). The artistry and delicacy of the workmanship are striking, as is the technique of forming the working points of all the other artifacts of this group.

One point, which was broken in half, is noteworthy because its sharp end was obliquely truncated by very fine retouching. However, it may be that this is not the point itself, but the beveled tang of a Daurian (Shilka) type of point.

Punches of this kind, prepared from narrow, regularly facted lamellar blades, one end of which was worked with very fine retouching on the ventral side and formed into a point, have been found also in the Trans-Baykal. They have been found, for example, among the artifacts picked up from the surface of drifting sands along the bay of Ulan-Khada together with other Neolithic artifacts. They also appear in the inventory listed for Ulan-Khada layers IX and VIII.[5]

10. *A punch-awl.* The collection contains one massive punch of peculiar shape, made of light-gray flint. It is nearly rhombic in cross-section (Fig. 12, item 12) and is worked on both sides with fine, careful retouching. It is 3.8 cm long. Punches of this type are common in the grave complexes of the Glazkovo period in the Cis-Baykal region.

11. *Scrapers.* Scrapers, which comprise one of the most common and numerous finds at Neolithic sites, are relatively scarce in the Shilka cave but show considerable variety of form (Fig. 13). Eleven scrapers and two scraper blanks were found. Eight of the scrapers were made from chips, not flakes. One scraper was made from a long pebble split lengthwise (Fig. 13, item 14), another from a large, thick flake, and a third from a flake knapped off a split nodule of yellow flint. In addition, a core split lengthwise was used as the material for one of the scrapers.

Three miniature scrapers (cat. nos. 524, 189, and 526) were found (Fig. 13, item 12). The first is 1.9 cm long, the second 1.8 cm, and the third 1.2 cm. All these scrapers are convex and have a more or less symmetrical working edge. The working edge of one scraper (cat. no. 437) is asymmetrically convex, arcuate. All the scrapers are retouched from one side only, opposite the bulb of percussion. Both sides of one scraper (cat. no. 523) were trimmed by circular retouching.

Two of the scrapers should be mentioned separately. The first (cat. no. 522) was made from a large, flat chip of light-gray flinty limestone (Fig. 13, item 11). One of its cutting edges is convex, the other (the opposite edge) concave. Merging, these cutting edges form a curved, beak-like point. This artifact is 6.2 cm long and 3.7 cm wide. It can hardly be called an ordinary scraper. More likely it was used as a knife. If it actually was a knife, its nearest analogue would be the peculiar Neolithic knives of the Far East, which are reminiscent in their general outlines of Mousterian triangular points. In fact, the Far Eastern knives of this type, found at the famous Neolithic settlement Suchu I [on the Amur river], at Tetyukhe* and at Gladkaya I, have been called Mousterian. Such knives are not found in the Cis-Baykal region.

The other artifact, the largest of this group, was made from a massive and broad lamella of dark-gray flinty slate (Fig. 13, item 1). The broad end terminates in a bilaterally retouched, convex cutting edge. Both long edges of the artifact are also retouched. The narrow end served as a handle, from which the bulb of percussion had been removed by flat retouching from the ventral side. It is interesting to note that this retouching has a more recent "look" than the surface to which the retouching was applied. Thus, this artifact may have been picked up by the inhabitants of the cave and used afresh. It is 8.3 cm long and its working end is 3.2 cm wide. The working end is highly polished and smooth from long usage.

12. *Almond-shaped flint knives.* The only item of this type in the entire collection is an almond-shaped blade with bilateral retouching (cat. no. 495; Fig. 13, item 6). The retouching on the blade is careful, but fairly rough. One of the long edges is more convex than the other, which is almost straight. The convex edge was used as the actual cutting edge. This is indicated by the fine retouching used to form it, while the opposite edge does not have secondary retouching. Further,

*[In the Maritime Province of the Soviet Far East.]

FIGURE 13. Stonework: scrapers.

the convex edge, especially on one side, has a characteristic oily luster and is polished from long usage. This blade is 7.2 cm long and 3 cm wide.

13. *Fragment of a knife.* The knife was made of white translucent chalcedony, but only part of it, the sharp end, has been preserved. Apparently the overall aspect of the knife was that of an asymmetrical leaf. Its two sides were trimmed differently. On one side, the cutting edge of the knife was retouched only along the edge, in a narrow band, but on the other side it was completely retouched, first with broad, even flaking across the implement, then with fine marginal retouching along the cutting edge. This method of retouching flint knives has been noted in similar objects of the Kitoy period in the Cis-Baykal region and also in Yakutia.

14. *Microblades* (Fig. 12, items 10, 11). These are two thin and narrow blades of excellent translucent chalcedony and could have been used as end and side blades. They are 2.7 cm and 2.4 cm long, respectively. The convex edge of one of them is very finely retouched from the ventral side. The other has fine retouching only along the long, straight blade. Most likely these were miniature blades for needle-shaped points of the insert[able] type, which will be described later in the section on bone artifacts.

15. *Side and end blades.* The collection contains 10 bilaterally retouched flint insertable blades (Fig. 10, items 3, 11; Fig. 14, item 3). They are made of white, translucent, highly patinated flint. They all are elongated rectangles, usually tapering more or less towards one end. Like their analogues from the Cis-Baykal region, these blades, which were inserted into a narrow groove of a bone or wooden handle, have distinctively shaped, broad lateral surfaces. Along one edge they have broad facets with sharply outlined depressions corresponding to the convexity of the bulb of percussion. There is no fine, secondary retouching along this edge. The cutting edge of the blade is coarsely serrated. The opposite side is thicker, worked with finer flaking, has careful secondary retouching along the cutting edge, and its serrations are finer and more regular. The blades are 4.3 cm, 3.8 cm, 3.9 cm, 2.7 cm, 3.0 cm, 2.7 cm, 3.0 cm, 3.2 cm, 2.6 cm, and 2.4 cm long, respectively. The widest blade is 2.1 cm wide, the narrowest, 1.2 cm. Most likely, all these blades are parts of one or two implements with insertable blades whose bone "housings" were found in the cave.

16. *Flint saw* (Fig. 14, item 2). One of the flint chips was fashioned into a saw-like implement by careful retouching. The chip is triangular in outline. One of its long edges is convex and was formed by dulling retouching. The opposite edge has four clearly defined sharp teeth separated by gaps. These teeth, carefully shaped by fine retouching, are similar to the teeth of a metal saw. This artifact is 2 cm long and 1.7 cm wide.

17. *Stone adze.* The large polished stone implements are represented by one specimen with very characteristic contours. It is a heavy, massive adze, made of black flinty slate (Fig. 15). It is quadrangular in cross-section, nearly square. One of its broad surfaces (the dorsum) is almost straight, only slightly convex. The opposite, ventral side, is strongly convex. The special preparation of the butt is a characteristic feature of this adze; it is unusually shouldered on the two broad sides. Thus, the butt is only two-thirds as wide as the middle part of the implement. The facets of the flaking, by which the surface was formed before polishing, are clearly visible on the butt and sides of the implement. The polishing was done carefully. One of the sides has traces of what was the original pebble surface. This side differs from its opposite in being concave. The adze is 14.7 cm long and has a maximal thickness of 4.8 cm.

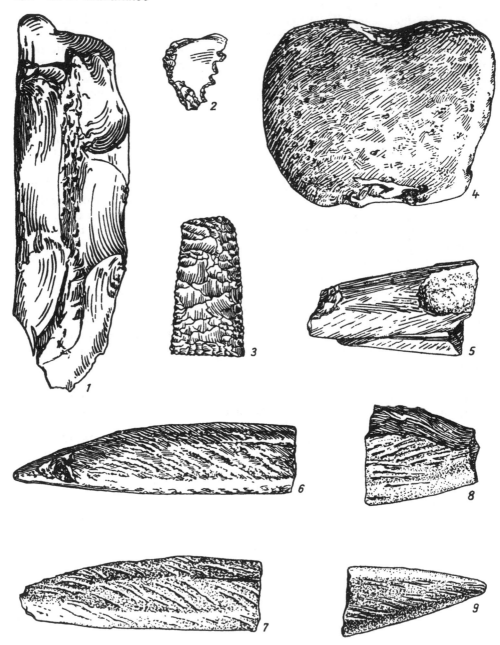

FIGURE 14. Stone and bone tools.

18. *Fragments of an implement made of limestone or slate.* Several fragments of an implement of pinkish limestone or slate in the form of a massive shaft pointed at both ends and plano-oval in cross-section (Fig. 14, items 6–9) have been preserved. The surface of this implement is covered with striations which run along the long axis and somewhat obliquely to it. The purpose of this object and its [initial] size are not known.

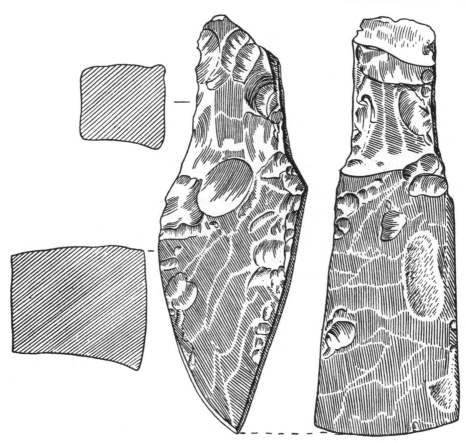

FIGURE 15. Stone adze.

19. *Sinker.* One sinker was found (Fig. 14, item 4). It is 7 cm long and 5.5 cm wide, formed from a flat, well-rounded river pebble with two small notches chipped in its sides [for securing the line].

### BONE AND ANTLER ARTIFACTS

Many artifacts of bone were found in the cave. Most of them were made carefully and artistically and a considerable number of them are preserved intact. However, many are broken, and there are obviously unfinished objects and blanks.

Harpoon heads are the most numerous of the bone objects, arrowheads come next, then various points and other objects. These artifacts were made of bone and antler.

1. *Bone blank.* Among the bone fragments of large animals there are those which could have served as blanks for making bone implements. One such object is a broad, flat fragment of a tubular bone (probably of an elk) split lengthwise (Fig. 14, item 1). There are wide facets on the inner side of this bone, traces of transverse flaking, a sort of retouching by which this piece of bone was to have been given its specific shape. However, the work was left unfinished.[6]

2. *Antler spear head* (dagger?). One fairly large, massive object of elk or maral antler may be classified as a spear head (cat. no. 153; Fig. 16, item 4). It is a thick piece, plano-oval in cross section, terminating in a clearly defined narrow haft. This spear head was worked only in the rough and thus traces of the technique are evident.

A piece of deer antler detached from the base by deep, longitudinal incisions was used as the raw material. Traces of these incisions can be seen distinctly on the inner, spongy part of the artifact. Next, the surfaces of these grooves were scraped out carefully with the cutting edge of a flint instrument. Judging by the characteristic stepped incisions, the same instrument was used to smooth out the irregularities of the surface of the tine. However, the work was left unfinished. The artifact is 18.6 cm long.

3. *Bone handle for a unilaterally bladed knife* (cat. no. 605; Fig. 17, item 3). This handle was made from the tubular bone of a large animal, probably an elk. The bone was split lengthwise. The handle is curved: thus the cutting side, into which the stone blades were placed, is arcuate-convex and the opposite side, which served as the back, is concave. A deep groove for insertion of the side blades was cut along almost the entire length of the convex side of the handle. The groove is 0.4 cm wide, 0.6 cm deep, and 15.6 cm long. The handle is 31 cm long. The entire surface of this artifact was carefully polished, but later became roughened in places by plant roots.

4. *Bone shaft for a bilaterally bladed spear head* (cat. no. 606; Fig. 17, item 1). This shaft is 27 cm long. It is narrow, its maximum width being 1.8 cm, and it is flattened-oval in cross-section. It has narrow grooves 0.2 cm to 0.3 cm deep along both sides, beginning at the base and continuing to the point. The point has a characteristic shape, reminiscent of a duck's bill. Undoubtedly, the blades placed in the grooves of this shaft were narrow and thin; most likely they were ordinary lamellar blades only lightly retouched along the cutting edge. The surface of the shaft was polished carefully. There are long longitudinal striations intersected by oblique striations on the tang of the shaft. Probably all these striations were made intentionally to better secure it to the [wooden] shaft. The shaft was made from an elk or deer tubular bone split longitudinally.

5. *Bone handle of a unilaterally bladed knife* (Fig. 17, item 2). The surface of the handle is smoothly polished. It is of a flattened-oval cross-section. Along one of its sides there is a longitudinal groove which runs practically the whole length of the handle. A flint side-blade with bifacial retouching was set firmly into the groove. This handle is 24 cm long and 2 cm wide.

6. *Bone knife* (cat. no. 152; Fig. 11, item 1). This knife was made from a deer or elk tubular bone split lengthwise. The lower end of the bone, which has traces of longitudinal sawing or incising and which served as the handle, is broken off. The rest of the bone was carefully sharpened and polished so that one edge served as the back of the knife and the other as its cutting edge. The cutting edge is slightly convex and sharpened on one side. The knife is 13 cm long.

7. *Dagger-like implement of elk bone* (Fig. 18, item 1). This implement, similar in shape to a bilaterally bladed dagger, was made from the shank bone of an elk. The bone was split lengthwise. It has longitudinal grooves on both of its broad sides. One of these grooves corresponds to the inner cavity of the tubular bone; the other groove is characteristic of the outer surface of the shank bone of an elk. The working end of the implement looks like a massive convex cutting edge. Identical implements have been found in the Neolithic (Kitoy) graves of

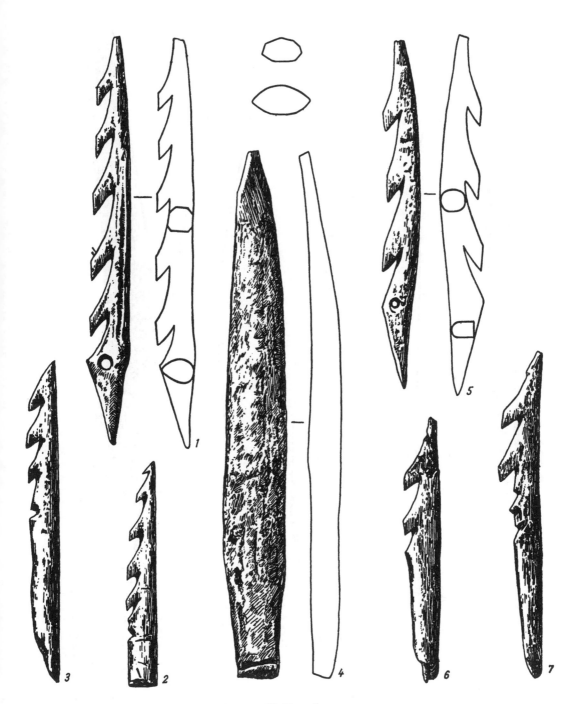

FIGURE 16. Bone harpoons.

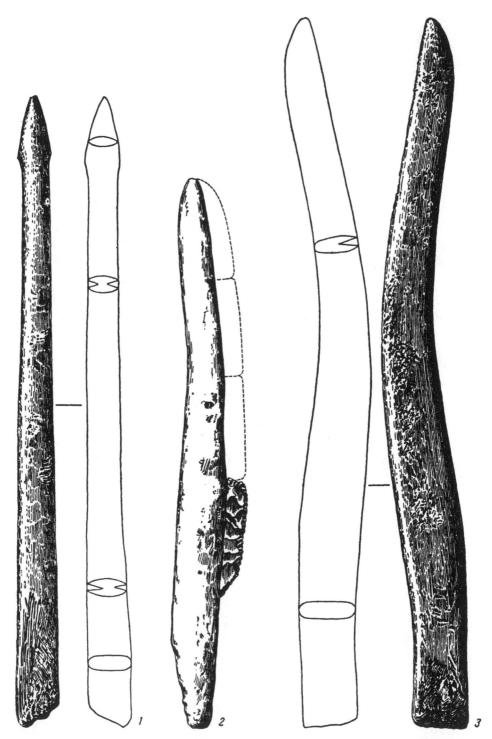

FIGURE 17. Bone shafts for side blades.

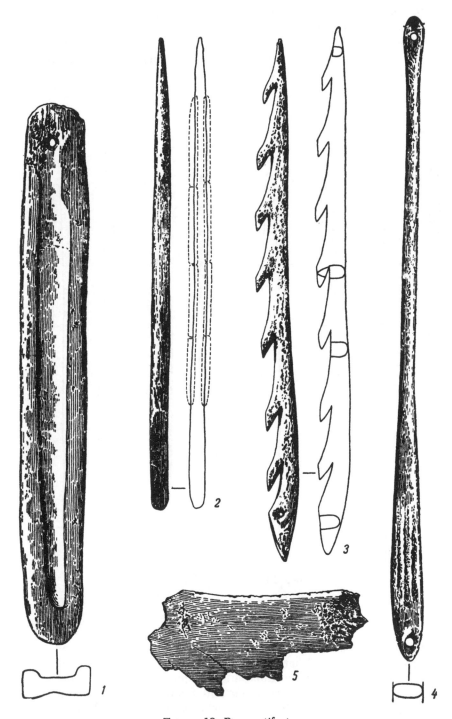

FIGURE 18. Bone artifacts.

the Cis-Baykal region and in burial mounds of the Glazkovo period.[7] The object is 23 cm long and 3 cm wide.

8. *Needle-shaped bone arrowheads.* The distinctive bone points (Fig. 19, items 1–6) occupy a prominent place among the bone artifacts. All, with one exception, are broken. Only small fragments remain of many of them, including fragments of tips and tangs. Judging by these fragments and restored points, all the arrowheads of this type were identical in shape. They were narrow,

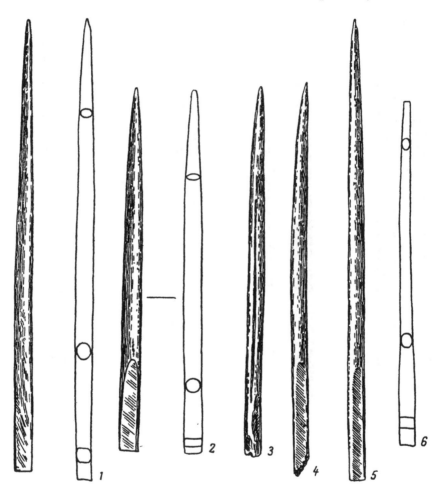

FIGURE 19. Bone arrowheads.

needle-shaped shafts pointed at one end. In all cases the tip is flat, thin, and sharp. In cross-section their upper part is flattened, the middle part oval, and lower end rounded near the tang. Their tangs are wedge-like, cut off symmetrically on two sides in broad flat faces. The entire surface of these artifacts was carefully polished, and fine longitudinal and oblique striations are clearly visible on the tang. These are traces of scraping and smoothing of the surface with a sharp implement, evidently of stone, not metal. One arrowhead (cat. no. 618) of this type is 12.1 cm long; other arrowheads restored from the fragments are 12.2 cm, 13.0 cm, 15.3 cm, and 15.0 cm long, respectively.

9. *Needle-shaped bone shafts for side-bladed arrowheads.* Three such shafts were found; two of these were preserved intact, but only the tip of the third remains. The first (cat. no. 616; Fig. 18, item 2) is narrow, long, and straight; it is oval in cross-section. The tang of the arrowhead was formed by an oblique cut on one of the broad sides. The surface of the cut is smooth and covered with striations. There are shallow transverse incisions on the other side, the convex side. Judging by these incisions, the head was fastened to the shaft as follows: the end of the [wooden] shaft was beveled laterally, the flat side of the correspondingly beveled tang of the [bone] shaft was placed against it, and the shaft and tang lashed together. To keep the binding from slipping, the transverse incisions had been made on the convex side of the tang. Narrow grooves were cut along almost the entire length of the shaft. Judging by the narrowness and shallowness of these grooves, unretouched blades or blades only slightly notched along the edge were fitted into them. The bone shaft is 17.1 cm long.

The second bone shaft of this type is considerably narrower and longer than the first. It is 20 cm long (despite the fact that it had bent) and of a flattened-oval cross-section. In this case, too, the tang has a beveled, flat face. A distinctive feature of this artifact is the carefully incised longitudinal lines. Perhaps they were meant to be ornamental.

10. *Bone points with cleft bases.* One almost intact speciment of this type was found in addition to fragments of tangs of others (Fig. 20). The complete point is 16.4 cm long (Fig. 20, item 1). Its tip is slightly wider and flatter than its middle part and has a longitudinal rib. In cross-section the tip is a flattened rhomb, the middle part round; the cleft extends two-thirds of the total length of the object.

11. *Fishhook point.* Special mention should be made of a small needle-like point, 6 cm long. It has a flattened-oval cross-section and its point is not hooked. Its lower end was [obliquely] cut into two faces. Apparently it was not used as an arrowhead, but as part of a composite fishhook, and was attached to a shank [to form] the hook.

12. *Fishhooks.* Three fishhooks were found in the Shilka cave. All are of bone and differ in shape. Therefore, they are described separately. Two of the hooks are composite.

The first composite hook (cat. no. 607; Fig. 18, item 4) has an unusually long shank and is unusual in shape. The upper part of the shank is very narrow, but it widens towards the tip to provide for an eye. The eye is circular and small. The shank widens gradually and terminates in a flat, oval area, in which there is a fairly large perforation for the insertion of a curved barb. A characteristic feature of this opening is the oblique notch adjacent to it, conforming to the curved barb. In addition, on the lateral edges of the shank there are clearly incised grooves undoubtedly intended for the lashing that firmly secured both parts of the hook together. The shank is slightly curved. The barb is missing, but can be easily pictured. Further, one bone barb found in the cave fits quite well. Antler was used for the shank. This artifact is 27.5 cm long.

The second composite hook (cat. no. 655; Fig. 21, item 6) also consists of two standard parts, the shank and the barb. The shank is short, its chordal length being only 6.7 cm; it is massive and curved. The lower part of the shank is straight, but the upper half curves almost at a right angle. The upper part of the shank is narrow, the lower part wider. The lower part has two perforations which are joined on the outer side of the shank by a longitudinal groove for a fishline. These perforations are broad on the outer side of the shank and narrow

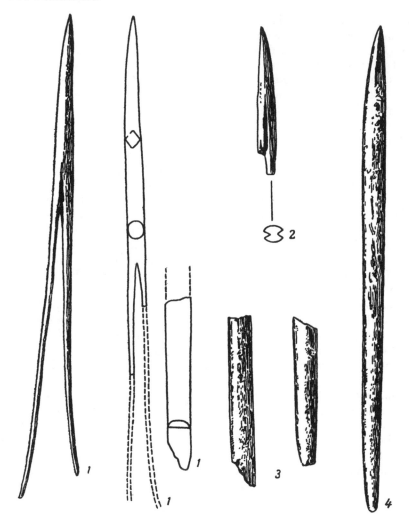

FIGURE 20. Bone arrowheads.

on the inner side. The lower part of the shank has a perforation for the insertion of a curved barb. The barb is 2.6 cm long.

In contrast to the two hooks described above, the third hook was cut from one piece of bone (cat. no. 656; Fig. 21, item 4). This hook is narrow, elongated, and bent into an acute arc. Its upper end bends abruptly inward and is sharp. There are small crosscuts on the upper end to hold it to the line more firmly. The hook's point is barbed. It is 3.7 cm long.

One special feature of the two hooks described above is that their upper end is bent at [nearly] a right angle and is pointed forward, so that it looks like the letter "C." No analogues of these hooks have been found in Siberia. Surprising as it may seem, similar objects have been found far to the west in the Baltic area and the European part of the Soviet Union and in the equally distant east, in Japan. Hooks of this type were found in the burial grounds and settlements of the Early Metal Age, on the Kola peninsula[8] and in neighboring regions of

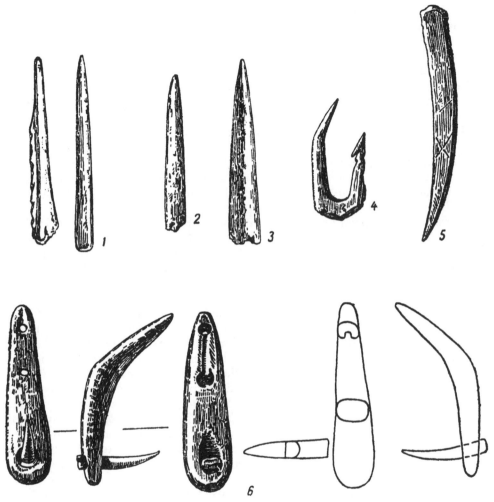

FIGURE 21. Bone fishhooks and barbs.

Scandinavia, where they are associated with the so-called Arctic Neolithic culture and that of the Early Iron Age.[9]

Many such hooks with a bent upper part were discovered in the Early Iron Age graves at Kjelmöy in Varanger Fjord (eastern Finnmark in Norway). Evidently these are graves of the ancestors of the Lapps. Solberg reports that an old Lapp took one such hook from among the things found at Kjelmöy and said that it was used for catching flounder. This hook reminded the old man of iron hooks with a bent shank which the Lapps from Pasvik used in the middle of the 19th century to catch flounder and whose shape indicates that they were patterned after more ancient bone or antler hooks.[10] In Japan such hooks were discovered in the excavations of the famous Neolithic settlement in the Korekawa peat bog described by Oyama.[11] These analogues are so geographically remote that they must be regarded as a convergent development. In any event, in contrast to a number of examples cited earlier, they cannot serve as a basis for concluding that this type of hook was evolved on the Shilka as a result of cultural ties and impetus from the outside.

13. *Harpoon heads*. Sixteen harpoon heads including fragments were found in the Shilka cave. This forms the largest group of bone artifacts found there. Most of the harpoon heads were made from fractured long bones of deer or elk, split lengthwise. Some of them, however, to judge by the state of preservation and the nature of the bone structure, were made of deer or elk antler. The antler harpoons have the following distinctive feature: one side, usually convex, is smooth with characteristic longitudinal grooves, while the other side preserves traces of the original porous inner surface. All of the harpoon heads were polished more or less carefully. On the basis of shape, they can be divided into two main groups.

The first group (the most numerous in our collection) contains 13 specimens. These harpoons have barbs on one side. This group, in turn, may be subdivided into two subgroups: (*a*) those with a line hole in the base, and (*b*) those without a line hole.

There are six harpoon heads with a line hole (Fig. 16, items 1, 5; Fig. 18, item 3; Fig. 22, items 2, 6; Fig. 23, item 1). The first of these (cat. no. 645) is the largest of the 16 harpoon heads found. It is 22.3 cm long, and has seven barbs equally spaced at about 2-cm intervals. The barbs have a characteristic shape, similar to that of Magdalenian harpoons; they are slightly bent near the point and of a beak-like appearance. The shaft of the harpoon is thin, slightly flattened on the sides, and of an oval cross-section. The tang was sharpened by an oblique cut. The line hole is in the broadest part of the tang and was drilled from both sides. The second harpoon (cat. no. 648) has three barbs. Its point is broken off. Its shaft is slightly curved. This harpoon is 12.3 cm long. The third harpoon (cat. no. 647) has four barbs but, judging by the notch at the end, originally it must have had five barbs. The end is broken. This harpoon is 14.2 cm long. The fourth harpoon (cat. no. 649) had six barbs, but its point, with the sixth barb, was cut off or broken off. This artifact is 14.3 cm long. The fifth harpoon (cat. no. 646) has four barbs. It is interesting that there is a lateral notch on the tang below the perforation, indicating that one more barb had been intended there. This harpoon is 14.5 cm long. The sixth harpoon (cat. no. 143) is the smallest of the group with a line hole. It has two barbs and is 9.2 cm long. Only the lower part of the seventh harpoon of this group remains; this fragment has one large barb and is 7.4 cm long.

There are five unilaterally barbed harpoon heads with no line hole (Fig. 16, items 2, 3, 6, 7; Fig. 22, item 3). Two of these are the smallest in the whole collection and have a straight shaft. Cuts were made from the side, across the shaft, and then the bone was snapped in two. One of these harpoons (cat. no. 652) is miniature in size, 5.3 cm long. It has only two barbs. There is a small notch on the side of the shaft. The second harpoon (cat. no. 99) has five barbs. It is 8 cm long. There is a shallow, transverse notch on the side of the haft.

The three other harpoons in this group are larger. In all of these, the haft is obliquely cut on one side; the haft of one of them is obliquely cut and polished on the opposite side as well. The first harpoon (cat. no. 159) has two barbs and the remainder of a third, the lowest, which had been cut off intentionally. The second harpoon (cat. no. 90) has two barbs and its point is broken. The third (cat. no. 650) has three barbs and there is a shallow transverse cut below the bottom barb. These harpoons are 11.6 cm, 9.0 cm, and 11.2 cm long, respectively. In addition, there is a recently fractured fragment of the upper part of a unilaterallly barbed harpoon (cat. no. 144).

Still another type of harpoon head consisting of three specimens were found

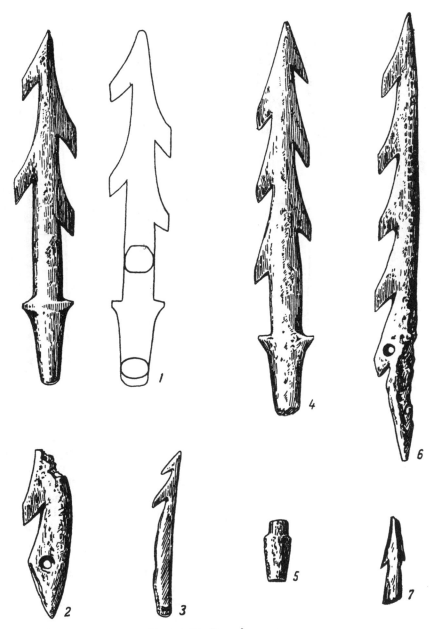

FIGURE 22. Bone harpoons.

(Fig. 22, items 1, 4). These are bilaterally barbed harpoon heads with the haft separated from the shaft by a shoulder. Two harpoon heads of this type have been preserved intact. In both cases, the barbs are not arranged in pairs, but alternate. One of the heads (cat. no. 664) has three barbs, the other (cat. no. 653) has six. These harpoons are massive and oval in cross-section. Their hafts are cut off smoothly at the base. The first harpoon is 12.4 cm long, the second, 11.4 cm. Only part of the third specimen of this type remains; the upper end has been cut off; only one barb remains, the other having been chipped off.

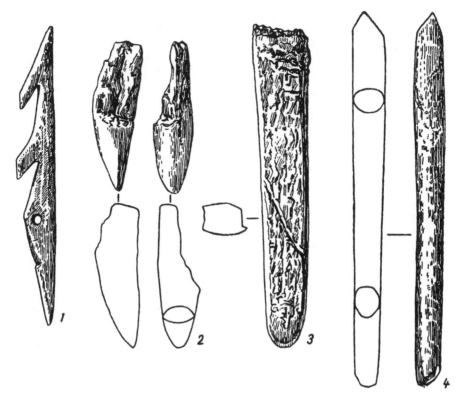

FIGURE 23. Artifacts of bone and antler.

The different shapes of the hafts indicate that the harpoon heads were fastened differently and were used for different purposes. The harpoons with laterally perforated hafts were fastened loosely to the wooden shaft, but could slip easily from the socket, being fastened only by a line passed through the line hole. Harpoons with shoulders were secured in the socket in a similar manner, except that the line was wound around the shank of the head above the shoulders rather than through a line hole. Undoubtedly, the heads without shoulders and line hole were set tightly into the end of a wooden shaft, which was split specifically for this purpose. Consequently, these were not actually harpoon heads, but dart points which could be hurled at the quarry by means of a throwing board or simply by hand.

14. *Small bone points.* Two such artifacts were found (Fig. 21, items 1, 2). One of them (cat. no. 207) is rhombic in cross-section, 5.7 cm long, and has a carefully polished surface. The second point was made from a fragment of the tubular bone of an animal; only the sharp tip is polished. This tip was incised from both sides and then broken off. The point is circular in cross-section and the side of the point bears the trace of a longitudinal incision, by which the point was separated from the bone. The artifact is 5.6 cm long.

15. *Bone polishers.* There are two bone polishers in the collection. The first (cat. no. 95) was made from a fragment of an animal rib split lengthwise. The end of this fragment was sharpened carefully into a narrow oval point. On both surfaces of this implement there are longitudinal striations inclined slightly in relation to the long axis of the artifact. Evidently these are traces of wear. The

second polisher is a flat, massive shaft cut from elk antler (cat. no. 205; Fig. 23, item 3). The sides of the shaft bear traces of longitudinal grooves, made by sawing or cutting, which pass almost through the entire thickness of the shaft. The spongy side of the implement has clear traces of smoothing by flaking of the broad, planing type. The upper end of the shaft bears traces of transverse chopping of the antler after the sawing. The lower end was made into a broad, blunt point. This end shows clear traces of wear in the form of lines running parallel to the long axis of the implement. This polisher is 10 cm long.

16. *Fragment of a bone fish lure* (?). This fragment (cat. no. 93; Fig. 20, item 2) is so small that we cannot be sure of its purpose. The hole drilled in its side, similar to those usually found in fish lures of bone and stone, leads us to classify it as a fish lure. The flattened-oval shape of the remaining end of the artifact and the slightly perceptible linear cuts which form a reticulate pattern (perhaps intended to resemble fish scales) also indicate a similarity to the afore-mentioned fish lures. This type of scale pattern is found on stone images of fish found in the Cis-Baykal region.

17. *Knife made from the shoulder blade of an animal* (Fig. 18, item 5). This is a cutting instrument in the form of a knife with one cutting edge, made from a fragment of a thin, broad blade, probably part of a deer or elk scapula. This knife has a slightly concave cutting edge, sharpened on both sides, and is 9.5 cm long.

18. *Fragment of an elk scapula with traces of cutting* (Fig. 24, item 1). This fragment has a deep incision on one side and a shallow one on the other. The groove of the deep incision is broad and triangular in cross-section. One of its edges is smooth; the other has an abrupt undercut below. The distinct longitudinal lines on the cutting surface indicate that the bone was cut with great effort and slowly with some kind of a stone blade.

19. *Antler and bone chisels.* Deer (or elk) bone and antler were used extensively for chisel-like tools. Three such implements were found, all different in size and shape and type of working end (Fig. 25). Thus, we may assume that the purposes of these tools—generally for woodworking, slotting—were different in detail.

The first artifact in this group (cat. no. 610; Fig. 25, item 2) is a small chisel plano-oval in cross-section, with a symmetrically sharpened convex-oval cutting edge. There is a longitudinal groove along one edge of the implement or, to be exact, this groove consists of three separate but merging grooves which were scraped out, one must assume, with a stone implement. The surface texture of the grooves provides evidence that the work could have been done only with a stone tool. It indicates that great effort must have been required to overcome this strong, hard material: it was not cut, but literally scraped out. We know that the tool was used as a chisel and not an adze, because of its butt. First, a slab of antler was cut out of a tine by incising deep crosscuts and breaking off the required piece. The fracture surface was not in its initial state, but was reshaped by many downward blows as is done with modern metal chisels.

The second chisel (cat. no. 609; Fig. 26, item 1) was not made of antler but of the longitudinally split tubular bone of a large animal, most likely an elk. Actually, this is an unfinished artifact and had never been used. Its working end was fashioned with crosscuts and was not sharpened. One can see clearly how this piece of bone was cut and then broken in two to remove the unwanted end. The artifact is 14.5 cm long.

The third chisel (Fig. 25, item 1) differs sharply from the other two by its

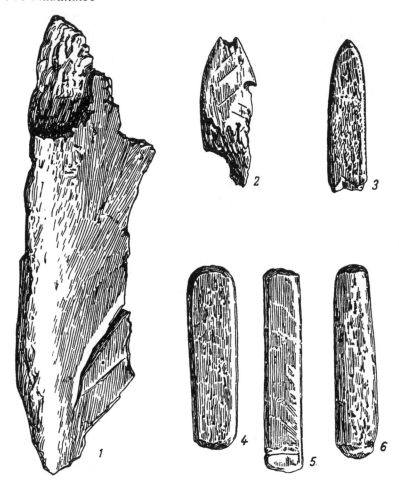

FIGURE 24. Bone artifacts.

massiveness and shape. It was made from a longitudinally split deer tine, more specifically, from the lower part [near] the fork of the antler. Traces of sawing are clearly visible on one of the edges, where parts of the grooves produced by sawing or cutting are preserved. The working end of the chisel is a broad oval point, flat on one side and convex on the other. The shape of the handle, the back part, is a characteristic feature of this chisel; it forms a broad lateral projection, a kind of paddle, which is part of the expanded base, the rosette, or fork of the antler (the latter is less probable). The surface of the handle is flattened and smoothed and one side bears traces of a broad longitudinal [roughened] area where a fracture had occurred. Judging from this, powerful vertical blows must have been administered to the handle when the chisel was used, most likely with a wooden hammer or mallet. The chisel is 9.8 cm long and 3.2 cm wide.

20. *Bone stamp for the ornamentation of pottery* (Fig. 23, item 4). This is an excellently preserved artifact, in the form of a large bone shaft, round in cross-section. The working end has a triangular point formed by two flat lateral faces. The possibility that this point was used as a stamp for ornamenting clay vessels is substantiated by the ornamentation on sherds from a clay vessel found in the

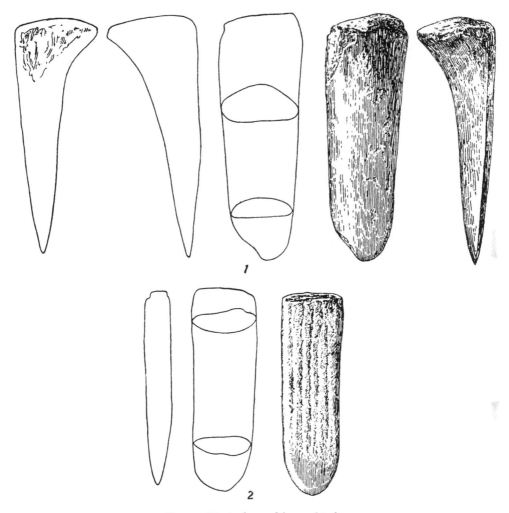

FIGURE 25. Antler and bone chisels.

cave; on these, impressions of a similar, although somewhat narrower, point can be seen. This implement is 12.1 cm long.

21. *Flakers.* Among the bone artifacts found in the Shilka cave there are three peculiar objects whose purpose was not easy to determine at first. These artifacts (cat. nos. 612–614; Fig. 24, items 4–6) are short, small, and flat bars made of deer antler, judging by the characteristic porous surface of one side. The surface of these artifacts was carefully smoothed and even polished. The shape of the ends of these small bars is a distinguishing feature. They have a small convex platform transverse to the long axis of the bar. The platform has a ridge in the middle formed by fashioning inclined on either side of it. Careful examination of the surface of these under a magnifying glass shows that they are covered with tiny scars or scratches running across the area and overlapping each other. These are clear traces of scars left when the end of the shaft was pressed against some sharp and hard object. This object probably was stone worked by man, most probably flint or chalcedony. Thus, all three bone bars are flakers, used

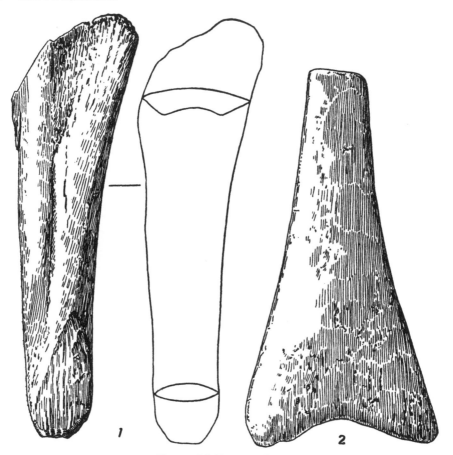

FIGURE 26. Bone artifacts.

for making stone implements by retouch. This is confirmed by the characteristic longitudinal flake scars 0.5 cm, 1.1 cm, and 1.2 cm long at the ends of all three bars. The only explanation of these scars is that sometimes the end of the flaker was pressed so hard against the object being worked that the edge of the bar could not withstand the pressure and a small longitudinal chip would break off. On one of these bars, the area of such a scar was marked off below by a deep crosscut. It might be surmised that this cut was made in order to remove the damaged end of the bar so that the working end could be reshaped and the implement "reactivated." The first bar is 6.4 cm long, 1 cm wide, and 0.4 cm thick; the second is 6.2 cm long, 1 cm wide, and 0.5 cm thick; the third is 5.5 cm long, 1.3 cm wide, and 0.6 cm thick.

It should be added that all three flakers had two working ends. Perhaps one end was used first and then, when it became dull and useless, the other end was used. Also, all three flakers are similar in size and shape. Flakers of this type had not been encountered previously in the archaeology of Siberia and the Far East. The bone and antler flakers most similar to these were found in Glazkovo graves in the Cis-Baykal region. However, there are significant differences between the Shilka and Glazkovo flakers. The latter are at least twice as long as the Shilka specimens and they are round in cross-section rather than flat. Thus,

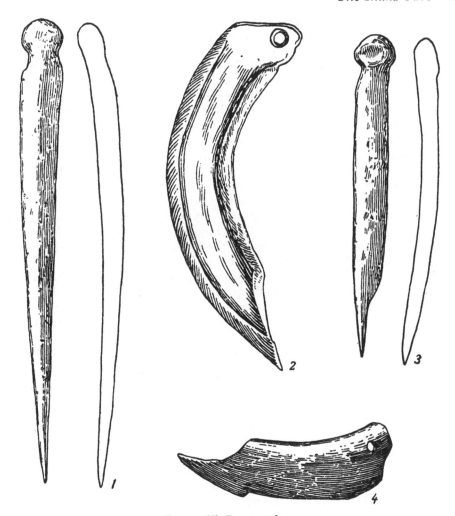

FIGURE 27. Bone artifacts.

the working ends of the Glazkovo flaker differ correspondingly: they do not have the double-sloped elongated working platforms of the Shilka samples, but [the work is done with] the blunted conical ends.

22. *Fragment of an antler pick* (Fig. 23, item 2). This fragment has the massive point characteristic of antler picks or the heads of digging sticks. The pick was curved, with one side convex, the other concave. The point was smoothed through use and covered with characteristic longitudinal lines. Perhaps one other pointed antler fragment in the Shilka collection was part of a pick.

23. *Small bone shovel* (Fig. 26, item 2). The small shovel discovered among the bone objects in the cave has a narrow handle and a broad lower end, terminating in a [very] wide triangular indentation. This end of the implement was smoothed and polished, as was the rest of the surface. It is 13 cm long and 6.8 cm wide at the lower end. Its purpose is unknown.

24. *Points [awls?] made from elk fibulae* (Fig. 27, items 1, 3). Three of the bone points consist of elk fibulae used in their natural state. The articular head of these bones served as a convenient handle. The sharp ends of the bones were

polished. Two of these points have been preserved intact; three others are fragmentary. One point is in a particularly good state of preservation; the tip of the point was not only polished, but was also modified by a lateral longitudinal cut so that the long fossa of the bone was exposed. The points which have been preserved intact are 16 cm and 11 cm long, respectively. Points of this type are found extensively in the Neolithic graves of the Cis-Baykal region.

25. *Comb-like bone artifacts* (Fig. 11, items 3, 5, 6). The first of these artifacts reminds one of a flat curved knife or sickle with numerous, well-defined small teeth along the concave arc of the implement. The handle is subrectangular with a sharp spur at the end and two perforations for attachment. It is separated from the blade by a sharp step. The chordal length of this artifact is 19.3 cm. Only the tip of the second object of this type (cat. no. 92) has been preserved; apparently is was the upper end of the implement, with a perforation. There are teeth along one of its edges; the opposite edge is thick. This group contains one more object, preserved intact. This third object is not dentate, but it does have a third perforation, in the end opposite the handle, in addition to the two perforations in the handle. This artifact is 12.5 cm long.

Most likely these three objects were used as tools for working fibrous plants, e.g., for disentangling or combing fibers prior to making thread from them. This assumption is well supported by the presence of textile impressions on potsherds found in the cave. However, we do not have conclusive evidence that this was the function of these strange objects.

Their closest analogues are the bone artifacts found in the excavations at Ipiutak, Alaska, where similar objects appear in the inventory of typical Ipiutak burials together with hunting equipment, harpoons, and arrowheads. In Ipiutak, these objects are listed among those of "unknown use."[12]

The comb-like objects from Ipiutak differ from those of the Shilka cave in that they have larger teeth and are hook-shaped. Further, the Shilka artifacts are very much like the antler "combs" found in the Early Iron Age graves at Kjelmöy in Varanger Fjord. They have the same short, blunt teeth along one edge. Solberg, who reported these objects, doubts that they could have been used to comb hair, but thinks that such objects, in the form of long unilaterally dentate blades, were used in the manufacture of clay vessels. According to Solberg, this assumption is supported by the clearly defined comb-stamped punctate ornamentation of vessels found at Kjelmöy.[13]

ORNAMENTS AND ART OBJECTS

Ornaments made of shells (sea and river) and white nephrite were found in the cave. They include rings of white nephrite, a rectangular mother-of-pearl ornament, and small shell beads. There are also two ornamented pieces of antler.

FIGURE 28. Mother-of-pearl ornaments.

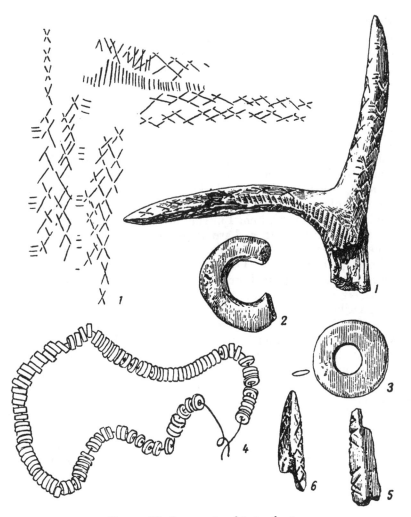

FIGURE 29. Ornaments, objects of art.

1. *Mother-of-pearl ornament.* This ornament, a rectangular thick plaque with slightly rounded corners (cat. no. 651; Fig. 28), was made from the valve of an edentate river mollusk. There are short transverse marks along the edges of the smooth inner side of the plaque, and there are two perforations, through which thread was passed to fasten the ornament to clothing. This object is 5.9 cm long and 3.5 cm wide.

2. *Nephrite rings.* Two nephrite rings were found among the ornaments in the cave (Fig. 29, items 2, 3). Both rings are flat and small. The first is not complete, about one-third of it being missing. It is irregular in outline; the inside diameter is one-and-a-half times wider than it is high. This ring was made from a white, translucent platelet of light-greenish nephrite. It is 3.3 cm in diameter.

The second ring has been preserved intact. The width of the hole is almost equal to the thickness of the rim. It was made from light-green nephrite with light yellowish spots. It is 2.2 cm in diameter.

3. *Beads* (Fig. 29, item 4). The beads are small, miniature, flat, regularly shaped rings and appear to be segments of a cylinder. The lateral planes of the circles are completely even and flat. The perforations of the beads are regular and cylindrical. Thus, a cylindrical shaft must have been made first, then polished, and a cylindrical perforation drilled through it; next the shaft must have been sawed into short segments, rings, each having its "ready-made" perforation. The average bead is 4 mm in diameter and 1.0 cm to 1.5 cm thick. However, some of the beads are only half that size. The beads are yellowish-white on the outside and some of them have a slightly reddish tint, probably traces of painting with red ochre. Evidently the beads seem to be made of sea-shells, but perhaps they were made from a compact, hard white stone. One hundred and third-eight such beads were found in one heap.

4. *Ornamented pieces of antler.* A forked piece of the upper part of a roe deer (?) antler, completely covered with a design, was found among artifacts of antler (cat. no. 660; Fig. 29, item 1). The ornamentation consists of short linear incisions made carefully with a sharp implement. The broken end of one tine, also ornamented, was found separately (cat. no. 479). The design pattern conforms to the shape of the antler. Oblique crosses were made on the narrow, sharp faces. These crosses merge to form a kind of a rhombic network or a herring-bone and zigzag pattern. On one tine these ornamental bands are separated by a band of parallel, slightly oblique incisions. On the other tine, the ornamental bands on both sides of the antler are separated by "clusters" of alternating and regularly spaced short parallel lines. There are three lines to the cluster, except for one row at the tip of the antler where there are two lines. There is no evidence that this ornamented antler was used as a tool.

Perhaps one more antler fragment ornamented in exactly the same fashion, i.e., with intersecting slanting incised lines, belongs to this artifact. Still another ornamented antler fragment was found. Only half of its design has been preserved because the fragment was split lengthwise. It consists of crosses, but they are larger and more deeply incised than in the first example (Fig. 29, item 5).

5. *Boar-tusk artifacts* (Fig. 11, item 4; Fig. 27, items 2, 4; Fig. 30). Seven artifacts of boar tusk were found in the cave. They were fashioned from tusks split (sawed?) lengthwise. In all cases, the outer, enamel surface of the tusk was left unworked. Actually, the inner side was not worked either and only the edges of the incision were trimmed and polished. One of these artifacts (cat. no. 658) is an exception. In this case, the inner surface bears traces of careful trimming and polishing. The upper end of each of these artifacts forms a curved thick point formed by oblique notching. One artifact was made from a narrow splinter of a tusk formed into a narrow curved point. It should be noted that there are very fine but distinct linear incisions on one side (the concave side) of the object, including lines forming an oblique cross. Holes were drilled through the broad end (the handle) of five artifacts. Three have one hole, and two of them have two holes. These holes were drilled from the inner side (i.e. the side opposite the enamel) of the artifacts. These artifacts have chordal lengths of 11.5 cm, 7.4 cm, 7.3 cm (with the end of the point broken off), 9.2 cm, 4.2 cm, and 7.2 cm, respectively.

At first glance one might take these objects for ornaments similar to those of the Cis-Baykal of the Neolithic and Early Bronze Ages, i.e., of the Glazkovo and Shivera periods. This seems to be supported by the drilled holes and the fact that these artifacts were made of split boar tusks. As we know, ornaments of the Neolithic epoch in the Cis-Baykal region, since the Serovo period and especially

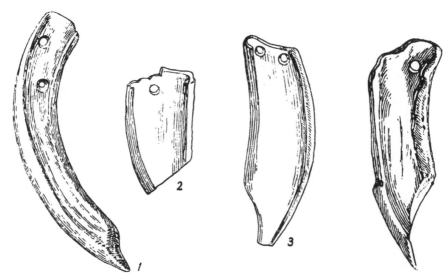

FIGURE 30. Artifacts of boar tusks.

the Kitoy, were made of split, not whole, tusks. However, the carefully made and painstakingly polished points of the boar-tusk objects found in the Shilka cave leave no doubt that these were used as tools and not ornaments. This still does not explain their purpose, however. Perhaps they were used as knives of a sort for some special work (where a convex cutting edge was required), or perhaps they were used as awls or small gravers.

POTTERY

Several dozen fragments of clay vessels were found in the cultural layer of the cave (Figs. 31 and 32). Judging by these, the finds in the cave represent six Neolithic vessels.

Apparently most of the fragments belonged to two quite large vessels made of clay with a large admixture of very fine sand and quartz or feldspar specially ground for the purpose. Small angular fragments of this material may be clearly seen in the fractures of the sherds. This was the standard admixture used for the Neolithic pottery of the Cis-Baykal region and eastern Siberia in general. The walls of these vessels are relatively thin and brittle and appear brownish-red when fractured. They have an average thickness of 0.4–0.5 cm. Both vessels had slightly everted rims, smoothly edged. The walls were convex, and the bottoms seem to have been round. The inner surface of the vessels was smooth, while the outside was covered with characteristic ribbed, elongated [tiny] protuberances and depressions corresponding to the texture of a fairly coarse fabric, probably made (judging by its coarseness) from plant fibers (Fig. 31, items 3–9). Plasticine impressions of the sherds show clearly that a fabric must have been used. Only a narrow band along the very rim edge of the vessel was free of these textile impressions. The ornamentation on three of the sherds was stamped over the textile impressions. It consists of a horizontal straight bands formed by indentations of a stamp whose flat working end was shaped into a triangular point. The bone object described above as a stamp for ornamenting

Figure 31. Pottery.

vessels is exactly of this nature (Fig. 23, item 4), and might very well have been used to decorate the surfaces of the pottery described here. To produce this ornamentation, the stamp could have been moved steadily at a certain depth in the clay, producing a characteristic groove with stepped ribbing on its bottom (Fig. 32, item 9). This is the typical technique of ornamenting clay vessels in the Cis-Baykal. The ornamental grooves were so deep that they show on the inner side of the thin sherds as longitudinal ridges separated correspondingly by longitudinal depressions. The ornamental bands, running in parallel rows about ten to the group, were not continuous and thus did not gird the vessel completely. They were separated by non-ornamented spaces about 2 cm wide (Fig. 32, item 10). In this particular case, the ornamental band was not at the rim, but below it, apparently on the body of the vessel. The ornamented sherds also have repair holes, drilled in pairs and used to hold cracked vessels together (Fig. 32, item 10).

The second vessel differed from the first in that its walls were only about 3 mm thick. The outer surface of the vessel was covered completely with textile impressions, but much finer ones, and the texture as a whole was also finer and more delicate than that on the first vessel. The rim ornamentation consists of two parallel rows of short vertical, slightly inclined impressions, made by a stamp whose working end was somewhat curved. Between these two horizontal rows of ornamental impressions, there is a band of symmetrically placed holes (Fig. 32, items 1, 6).

The third vessel was essentially the same as the one described above, except in the details of its rim shape and ornamentation. The rim was thickened and everted, with a clearly defined ridge on the outer side. The ornamentation consisted of two parallel rows of oblique stamped impressions, as above. However, these impressions were made with a straight and wide stamping paddle instead of a curved one (Fig. 32, items 2, 4, 5). A similar ornamental band also was placed along the inside edge of the neck (Fig. 32, item 3).

The fourth vessel differed from the rest in that both its inner and outer surfaces were smooth, but, at the same time, they were covered completely with very fine hair impressions, interwoven and crossed and often in clusters (Fig. 31, items 1, 2). The upper edge of the rim was decorated with paddle-stamp impressions, while the outside of the vessel was decorated with broad, extensive bulges pressed out from the inside. Farther down on the outside of the vessel, 2 cm from the rim, there was a narrow ornamental band of two parallel incised lines (Fig. 31, item 2).

Three other small fragments of clay vessels should be mentioned separately. One of them, 4 mm thick, has on its outer surface quite distinct square impressions arranged in a checkered pattern. This checkerboard, pseudo-textile ornamentation is well known in the Soviet Far East and in northeastern Asia. Here, they are fairly large, with meshes about 0.4–0.5 cm wide. The second sherd is even thinner, being 1.5–2.0 mm thick (Fig. 31, item 4). Its entire outer surface is covered with distinct and deep impressions of a textile nature. Apparently, these are impressions of a fabric of coarse, tightly twisted threads set parallel to each other. The third fragment is noteworthy in that its outer surface shows clear impressions of thin, but well-twisted threads crisscrossed (not parallel as in all the other sherds) so that they form a kind of rhombic net (Fig. 31, item 10). This type of textile impression can be explained most easily if we assume that a wooden mallet (or paddle) wound or wrapped with crisscrossed threads was used in the manufacture of the vessel. Blows struck with such an object upon

FIGURE 32. Pottery.

the wet clay of an already prepared vessel would leave this kind of impression on its outer surface.

Three fragments of Iron Age vessels were found in the surficial deposits of the cave, including one vessel rim. These fragments are quite thick and compact and have massive appliqué strips pressed onto their outer surfaces.

## Dating of the Finds

As mentioned above, the finds in the Shilka cave consist of two chronological groups. Some of the finds, numerically few, belong to the Iron Age, that is, to the 1st millennium A.D. The bones of a horse and horned cattle found in the Shilka cave should be dated also to the Iron Age.

The overwhelming majority of the finds are Neolithic. All of them were discovered in the thin cultural layer, chiefly in its upper part, in the dark earth with a high humus content. They are all in the same state of preservation. Therefore, without further ado, one may regard them as belonging to the same period, viz., to the "Neolithic in general."

However, upon more detailed analysis one finds that the problem of dating this main group of Shilka articles is not so simple, because the Shilka finds, when compared with the better-known Neolithic remains from the Cis-Baykal region, show an unexpected association of objects of different age which is difficult to explain in terms of the periodization for the Cis-Baykal Neolithic.

The stone and bone objects include a considerable number that are similar to the most common and widely diffused in the Cis-Baykal region typical of the Serovo and Kitoy periods. First among these are the side- and end-blade implements characteristics of the Cis-Baykal.

The Shilka cave contained three types of combined stone and bone artifacts characteristic of the Cis-Baykal Neolithic, viz., (1) unilaterally bladed, almost straight knives with closely joined, bifacially retouched flint blades inserted into grooves, (2) large knives, also unilaterally bladed, but recurved in a characteristic manner, with a cutting edge consisting of flint side-blades on the convex side and a concave side serving as the handle of the implement, (3) bilaterally bladed spear heads, of a flattened-oval cross-section, with a characteristic duck-bill point. Such objects have been found in Cis-Baykal burial sites chiefly of the Isakovo and Serovo periods, and occasionally of the Kitoy period.[14] They are not present in Glazkovo sites, however. Hence, they represent elements of a fairly early stage of the Cis-Baykal Neolithic, not later than the end of the 3rd millennium B.C. This is also supported by the single large polished implement found in the Shilka cave—the large shouldered adze. In Yakutia such adzes belong to the fully developed Neolithic.[15] In the Cis-Baykal region they were found in the graves of the Verkholensk burial ground, which is also dated no later than the 3rd millennium B.C., and is probably even earlier.[16]

One antler chisel with a characteristic handle having a side flange is particularly interesting for establishing a connection between Shilka and the Neolithic of the Cis-Baykal. Chisels of this type are one of the most characteristic objects in the inventory of the early graves of the Verkholensk burial ground.

The stone arrowheads found in Shilka cave are also quite archaic. In many cases the basal notch has a specific angular shape and one barb is considerably longer and larger than the other. Thus, the base of the arrowhead is asymmetrical; in this respect they duplicate the earliest Cis-Baykal arrowheads, those of the Isakovo, Serovo, and Kitoy periods which preceded the Glazkovo, the earliest stage of the Bronze Age.[17]

Most of the harpoons found in the Shilka cave may be compared with the Cis-Baykal harpoons of the Kitoy period. Their similarity includes such a characteristic feature as the small perforated lateral flange in the lower part of the harpoon for attaching the harpoon to the shaft. However, there is one difference, the Kitoy harpoons have barbs on both sides, while the Shilka ones have barbs on one side only.[18]

As for the ornaments, the small shell beads lacking the lustrous mother-of-pearl surface are archaic in their general appearance. Beads of this type have been found in a number of Serovo graves on the Angara, but not in Glazkovo ones. In the latter there are relatively large flat beads of riverine or marine shells with the lustrous surface preserved.[19]

Of particular interest in this connection are the fragments of vessels whose outer surface is covered with impressions of fine fabric rather than net-impressions. Impressions of this type first appear in the Cis-Baykal in the Kitoy period, as indicated, for instance, by the excavations on the Angara above Irkutsk in 1951–1955, at a habitation site near the village of Patrony and on Sosnovyy Island. On the Angara, this type of pottery is usually accompanied by the ancient Cis-Baykal pottery which has characteristic net-impressions on the outer surfaces of the vessels, but which differs sharply from it in general appearance. It is also significant that Shilka pottery shows these textile impressions together with ornamentation of the Cis-Baykal technique and in its compositional pattern. The ornamentation was applied with a stamp and was arranged in parallel rows of horizontal grooving. Ornamentation of this type is present on the vessels of the Serovo and Kitoy periods. Along with these objects, however, others were found in the Shilka cave which in the Cis-Baykal region can be dated no earlier than the Glazkovo period. First, there are the greenish-white nephrite rings, which did not appear in the Cis-Baykal until the Glazkovo period.[20] The bone artifacts of specific form, the needle-like arrowheads with a cleft base, also belong to the Glazkovo period of the Cis-Baykal.[21] The flint arrowheads with a straight base should also be grouped with them.[22] There were no such arrowheads in the Cis-Baykal region before the Glazkovo period. However, they are most common and widespread artifacts in the Glazkovo burial complexes.

Earlier we have described a massive dagger-like implement with a broad, oval blade made from a longitudinally split shank bone of an elk. Dagger-like implements of this type are found extensively in the Neolithic graves of the Cis-Baykal region, beginning with the Kitoy period. They are particularly characteristic of the Glazkovo sites on the Angara, Lena, and Selenga rivers.[23]

One of the three bone fishhooks found on the Shilka, namely, the largest of them, is also worthy of special attention. It is comparable with the Glazkovo bone fishhooks of the Cis-Baykal region. These Cis-Baykalian hooks also have a long shank with a curved point at the end. In their details, the Glazkovo hooks differ from the Shilka hook in that they have no perforation at the upper end and their shank is simpler, but in general these hooks are very much alike in construction and form.[24]

Among the harpoon heads recovered from the Shilka cave, there are some which are similar to those with wide lateral flanges near the tang found in the classical Glazkovo burials on the Angara.[25]

One could explain the combination of objects of different general aspect by assuming that the Shilka cave contained two groups of objects of different age, viz., early, Isakovo-Serovo or Kitoy, and late, Glazkovo, that is, items from the 4th to the 2nd millennium B.C. For example, the group of objects found in the burial may be considered of different age than the objects found in the cultural layer of the cave. It seems unlikely that the cave would have been used simultaneously as a burial ground and as a dwelling place. It seems likely that people dwelt in the cave first and later abandoned it after using it as a burial place. However, the reverse could have been true; the cave could have been used as a burial place first and then, much later, after this had been completely forgotten, it could again have been used as a dwelling place. In this case, the grave should have been either of the Glazkovo (i.e., late) period or of early, Neolithic age.

Unfortunately, the grave had been disturbed even before Shabalin visited the cave. He was therefore able to remove only part of the scattered human skeleton.

Thus, we cannot be sure what objects were associated with the skeleton and, consequently, we cannot be certain of its age, of its chronological relation to the other objects of Neolithic aspect found in the cave. However, artifacts such as the large composite knives and spear heads, fishhooks, and others are found in such a good state of preservation (i.e., intact) only in graves. The splendid group of bone artifacts (and the polished adze of the Verkholensk-Vilyuy type) which Shabalin found and removed from the cave is of the same array and typological character as objects typical of the Cis-Baykal Neolithic burial inventories. The shell beads of the archaic Serovo type must also be classified with this group. They are found in such large quantities only in graves and not in habitation sites, which is quite natural and obvious.

The presence of ochre in the grave also supports the conclusion that the burial belongs to the Serovo period. Red ochre in small amounts is usually found not only in the Kitoy graves, but also in the earlier Neolithic burial sites of the Cis-Baykal region. Thus, the Neolithic grave in the Shilka cave may be considered contemporary with the graves of the Serovo stage in the Cis-Baykal region, i.e., not later than the first half of the 3rd millennium B.C.

If this is true, the other objects, including two nephrite rings, one of which is broken, and the needle-like arrowheads with a cleft base, should be dated to a later period, represented by the Glazkovo in the Cis-Baykal. Consequently, people began to live in the Shilka cave more or less continuously during the Glazkovo period, leaving behind a culture-yielding layer and evidently not suspecting that the skeleton of a previous inhabitant of the cave lay beneath them.

The general appearance of the pottery and sherds, i.e., the bulk of the material which is most important for determining the cultural aspect and the mode of life in Neolithic settlements, and also for dating them, points to a much later, Glazkovo dating for the majority of the finds.

This material does not contain any examples of net-impressed pottery, which is the most archaic type of the Cis-Baykal and neighboring regions. It is found in the earliest Neolithic sites not only in the Cis-Baykal and in Yakutia, but also in neighboring Mongolia, including the Gobi desert and the region of the Great Wall of China.[26]

Unfortunately, the interesting surface finds of pottery from habitation sites in Tungpei [Manchuria] assembled and published by Licent were not well illustrated [reproduced] and one cannot always draw reliable conclusions from them. However, among the sherds from the well-known Lin-hsia site in the Shara-Muren valley there are some from vessels which are similar to the Isakovo and Serovo pottery of the Cis-Baykal region and which are almost identical with those from Baindzak (Gobi) and the Great Wall of China. According to the description, these are (ovoid) vessels whose walls are covered on the outside with a "wafer" impression in the form of a honeycomb and which have large perforations along the edges of the rim.[27] One might assume that these highly interesting vessels, buried among the great mass of surface-collected Lin-hsia material of different ages and types, are the most ancient.

This means that pottery of this type might very well have appeared in the Shilka valley, which occupies an intermediate position between the [above-mentioned] regions of Asia, and specifically in the Shilka cave, provided the cultural layer of the Shilka cave is of the same age as the Isakovo and Serovo or Kitoy sites in the Cis-Baykal region.

The only argument against the deduction that the burial complex of the Shilka cave is not of the same age as its cultural layer is the presence, among the harpoons taken by Shabalin, of the two harpoons which are very similar to Glazkovo ones or even identical with them. These harpoons are in the same [perfect] state of preservation as the archaic combined bone and stone implements.

It is also possible that the objects of different aspects [periodizations] found on the Shilka (Glazkovo, on the one hand, and Isakovo, Serovo, and Kitoy, on the other) were used simultaneously during the Glazkovo period, i.e., at the end of the 2nd millennium B.C. Something similar has occurred in other regions adjacent to the Cis-Baykal, for instance, on the lower Angara below Bratsk, and in Yakutia. For example, an Early Bronze Age grave was found near Pokrovsk on the Lena, 86 km upriver from Yakutsk. This grave was close in time to the Glazkovo sites in the Cis-Baykal region. It contained a composite bone and stone spear point of an archaic aspect.[28] The Early Bronze Age site in the village of Siktyakh on the lower Lena is even more interesting in this respect. An asymmetrically triangular knife of an ancient form for the Cis-Baykal was found there.[29] A similar knife was also found among the objects associated with the skeleton of an Early Bronze Age man at Bugachan, in the Zhigansk region.[30]

Thus, more than likely, the ancient burial in the Shilka cave belonged to the earlier, Serovo or Kitoy period and the cultural layer belonged to a later, Glazkovo period. This question cannot be answered definitively until further investigations are made and new materials are obtained for a more accurate dating of the Neolithic remains of the Shilka river valley.

In general, however, and for the convenience of presentation, we shall define the culture of the early inhabitants of the cave, those who had left the bulk of the finds, as generally Neolithic, keeping in mind that at least part of the material belongs to a period contemporaneous with the Glazkovo Eneolithic of the Cis-Baykal, i.e., the initial stage of the Bronze Age.

## A Description of the Life and Economy of the Neolithic Population of the Shilka River Valley, based on Materials from the Shilka Cave

The excavations in the Shilka cave have provided significant material for the characterization of the culture of its most ancient inhabitants, the people of the Neolithic (and Glazkovo) periods. These people had developed great abilities in working stone and bone. Their stone blades, points, side and end blades, as well as their bone implements and weapons are on a level with the highest attainments of the Neolithic peoples of eastern Siberia;* they are as good as the Serovo artifacts of the Cis-Baykal.

On the whole, the culture of the Neolithic tribes of the Shilka valley was tied intimately with the sources of their subsistence, hunting and fishing. As inhabitants of taiga country, in the valley of one of the largest rivers of Asia, it was

*[I.e., present-day administrative eastern Siberia is meant here—the territory between the Yenisey river and the Pacific drainage.—Editor.]

natural that the Neolithic inhabitants of the Shilka cave should have been primarily hunters and fishermen. The whole complex of stone and bone artifacts found in the cave shows that fishing and hunting were their main occupations. Hunting equipment is represented by stone and bone arrowheads, various types of daggers, and knives which could be used for splitting hides, for dressing game, and for various domestic chores.

As already mentioned, 29 common stone arrowheads were found, 3 of the Daurian type, and about 9 bone arrowheads. There was one stone dart head, one common bone spear head, and one composite bone and stone spear head. In addition, 1 simple and 2 composite knives and daggers were found. According to N. K. Vereshchagin, the cave contained bones of various animals, apparently game animals:

1. Wolf (bones of 1 wolf)
2. Domestic dog (10 or 11 animals)
3. Fox or steppe fox (3 animals)
4. Lynx (1 animal)
5. Sable or marten (2 or 3 animals)
6. Polecat
7. Weasel or ermine
8. Flying squirel (3 animals)
8. Flying squirrel (3 animals)
10. Far Eastern gray rat (1 animal)
11. Alpine pika [mouse hare] (1 animal)
12. Snow hare (6 animals)
13. Mouse
14. Barabinsk hamster
15. Marmot (4 animals)
16. Musk deer (2 animals)
17. Maral
18. Roe deer
19. Elk
20. Wild boar
21. Mountain sheep (?)
22. Birds[31]
23. Fish[32]

As can be seen from this list, the inhabitants of the Shilka cave hunted ungulates, beasts of prey, and rodents.

They fished as well as hunted. They fished with improved tackle, employing hooks of highly original design and also nets, as can be surmised from the stone sinker found in the cave. Hunting for large fish with harpoons and implements of the fish-spear type was highly important. Sixteen harpoons, 3 fishhooks, 1 sinker, and 1 fish lure similar to the bone lures of the Cis-Baykal were found in the Shilka cave.

No evidence of agriculture was found. The fragment of a pick indicates that the inhabitants practiced gathering.

The abundance of dogs attracts one's attention. Dogs are the most numerous of the animals listed (10 or 11 animals). This is interesting because the Amur tribes of the recent past, especially the Gilyaks, raised dogs in particular. L. I. Shrenk wrote: "The Gilyaks, more than any other of the aboriginal tribes of the Amur, travel by dog[-pulled sleds] and therefore this animal occupies a much more important place in their lives and economy than in the lives and economy of other inhabitants of the country. The dog is not only a work animal but provides the most common and favorite fur for their clothing and its meat is used as food."[33]

At funerals, for example, the Gilyaks offered dogs as sacrifices to the spirits. In describing the use of dogs for food, Shrenk notes features of a special cult attitude towards the dog: "The place in the yurt between the rear wall and the hearth, where a bear head was displayed during the bear festivals, was set off by two poles and between them on the floor lay a wand with a tuft of thin, curly shavings, the so-called *tsakha*."[34] As Shrenk's words indicate, in this case the dog was afforded the same cult honors as the sacred animal of the Gilyaks, the bear. In honor of the dog, the wand, *tsakha*, was placed in the sacred place, as an indispensable item in the rituals of the Gilyaks and also of the Ainus, where such a wand is called *inau*.

According to Shrenk, the dog was equally important to other Paleo-Asiatics, among whom it was connected with hunting and fishing: "From times immemorial they all were fishermen or had been fishermen and their only domestic animal, the dog, served mainly as their means of transportation."[35]

A similar practice was observed among the Ulchis and Goldis (Nanays). I. A. Lopatin wrote:

The only domestic animal of the Goldis is the dog; they have neither horses nor cattle nor sheep nor goats . . . among the Goldis, the dog is used both as a draft animal and for the hunt. Every Goldi always has many dogs (a minimum of 10). Many Goldis have 15 or even 20 dogs and the rich have as many as 30. . . . It is not surprising that Goldis, who use the dog for transportation, have many dogs, but it is very strange to see so many dogs among the Goldis who live in the outskirts of towns and have horses and who rarely use the dogs as means of transportation. Moreover, in camps in the vicinity of towns, one often finds Goldis who live in abject poverty, yet keep many dogs. Obviously, the dog was so necessary and close to the indigenous Goldi that he could not conceive of any type of life without dogs. One cannot convince a Goldi that it is far better to keep a horse, which would be economically more profitable than [keeping] 10 or 12 dogs.[36]

The Goldis harnessed their dogs to special sledges, using up to 10 dogs. The dogs were fed soup made from dried fish. The attitude of the Goldis towards the dog, according to Lopatin, "is one of affection, one might say, and their attachment to their dogs is very touching. Little wonder that the dog is the subject of many of the legends and tales of the Goldis."[37]

An abundance of dogs was a characteristic feature of a Goldi camp. Lopatin sketches a vivid picture of such a camp: "Whining, howling, and barking can be heard for a great distance, enlivening a camp lost deep in the remote taiga or on a distant [river] island. When a stranger approaches the camp, the dogs raise a deafening howl and offer no little danger to the traveler."[38]

Apparently, the importance of the dog in the lives of the Amur tribes can be attributed to the nature of their economy, in which fishing played a major role.

Judging by the abundance of fish and dog bones in the cultural layer of the Shilka cave, the situation must have been similar in the case of its Neolithic inhabitants. Evidently the occupants of the cave, like the Gilyaks, the Itelmens, and the Ainus, used dog meat for food; otherwise there would not have been so many dog bones in the cave. In this connection, it is interesting to note Shrenk's general conclusion about the use of dog meat as food: "The custom of eating dog meat exists only among those northern peoples whose sole domestic animal is the dog, which serves them as a work animal."[39]

Despite the small size of the cave, it was a place where stone and bone were worked into tools. This is indicated by the presence of animal bones with traces of cutting and by the discovery of lamellar blades, pebble blanks and cores, broken bone implements, and blanks for bone artifacts. The cave seems to have been used mostly as a temporary dwelling by small hunting parties. It is very interesting that the cave apparently was used first as a burial place. In this connection, it should be noted that in neighboring eastern Siberia, during the Neolithic, caves were very rarely used as both dwelling and burial places. In this respect the Shilka cave is unique.

## Cultural and Ethnic Ties of the Neolithic Peoples of the Shilka Valley with other Regions

While describing the materials found in the cave, on several occasions we pointed out certain analogues to objects used by Neolithic peoples of other regions and with this we were able to determine the purpose or the nature of the various objects. The same applies to the problem of dating the finds. As we have seen, analogues of this type indicate that the Shilka material may be classified as Serovo-Kitoy and Glazkovo, and thus that it belongs to the 3rd and 2nd millennia B.C.

Apart from the matter of their dating, the Neolithic finds from the Shilka cave, taken as a whole, are very interesting in another aspect: they shed light on the cultural ties and relationship between the ancient peoples of this and other regions. The accompanying table presents the general relationship of these analogues.

Let us treat them in more detail, subdividing them territorially in the following order: Cis-Baykal, Trans-Baykal, Yakutia, Mongolia, Tungpei (Tsitsihar, Ch'ih-fêng, and Lin-hsia, the Yangshao culture), China proper, and the Soviet Far East (the lower Amur and Primore [Maritime Province]). It will become evident from what follows why the Neolithic sites of the lower Amur and Primore are examined last instead of first.

We have already discussed the relationship between the Shilka and Cis-Baykal Neolithic periods and have seen that the general aspect and inventory of stone and bone artifacts from these two regions coincide in many respects. Analogues have also been found among the pottery.

The examples of art—the ornamented objects found in the Shilka cave—may be added to the list of common cultural elements. The ornamented pieces of antler from the cave have elements that are characteristic of the earliest examples of Neolithic ornaments from the Cis-Baykal region, namely, short parallel incisions arranged in clusters of three.[40] Ornamentation in the form of crosses, including crosses arranged in parallel rows and contiguous so that they form rhombic patterns, is also well known in the Cis-Baykal among the ornaments of the Kitoy and Glazkovo periods. An anthropomorphic image in M. P. Ovchinnikov's collections in Glazkovo,* which may be classified in a preliminary way as of the Kitoy period, has a rhombic and chevron design.[41] Patterns of crosses and transverse lines identical with those on the objects found in the Shilka cave are also present on an anthropomorphic image from the burial ground at Anosovo on the Angara.[42]

The economy and pattern of life of the inhabitants of the Shilka valley and the Cis-Baykal region were essentially the same during the Neolithic period. The ancient inhabitants of the Shilka cave were the same type of hunters and fishermen as their contemporaries on the Angara and Lena rivers, except that fishing perhaps played a greater role on the Shilka than in the Cis-Baykal.

As already mentioned, the role of the dog is a special feature of the life of the Shilka people. True, the dog also occupied a prominent place in the way of life and the cult of the Neolithic tribes of the Cis-Baykal region. This is

*[The locality not the period.—Editor.]

CORRELATIONS IN THE INVENTORY OF THE SHILKA CAVE AND OTHER NEOLITHIC SITES

| Item | Shilka cave | Trans-Baykal | Cis-Baykal | Yakutia | Mongolia | China Tsitsihar | China Lin-hsia | China Yangshao culture | Lower Amur and Primore |
|---|---|---|---|---|---|---|---|---|---|
| 1. Prismatic cores | + | + | + | + | + | + | + | Exceptionally rare | Exceptionally rare |
| 2. Lamellar blades | ++ | ++ | ++ | ++ | ++ | ++ | ++ | Rare | Rare |
| 3. Bilaterally retouched arrowheads | ++ | ++ | ++ | ++ | ++ | ++ | ++ | No flint arrowheads; slate arrowheads occur | + |
| 4. Daurian-type arrowheads | + | + | Exceptionally rare | – | + | + | + |  | – |
| 5. Punches made from lamellar blades | + | + | + | + | + | + | + |  | Rare |
| 6. Flint scrapers | +++ | ++ | +++ | +++++ | ++++ | ++ | ++ | – | Rare |
| 7. Almond-shaped knives | +++ | – | +++ | +++++ | ++++ | – | – | – | + |
| 8. Bilaterally retouched side-blade | +++ | + | ++ | +++ | ++ | – | – | – | – |
| 9. Shouldered (stepped) adze | +++ | + | Rare | + | + | Only in extreme southern China |  | – | Rare |
| 10. Mousterian-type knife | ++ | – |  | + | – |  |  |  | + |
| 11. Combined spear heads and knives | + | Rare | + | + | Side blades occur | ++ | – | Rare | + Side blades occur |
| 12. Needle-like bone points | ++ | – | – | – | – | – | – | + | – |
| 13. Bone heads with cleft base | + | – | + | + | – | – | – | – | – |
| 14. Unperforated harpoon heads with oblique haft | + | + | + | – | – | + | – | – | – |
| 15. Harpoon heads with side perforation | + | – | + | – | – | + | – | – | – |
| 16. Harpoon heads with a side projection at the tang | ++ | – | ++ | – | – | – | – | – | – |
| 17. Bone fish-shaped lures | +++ | + | +++ | + | – | – | – | – | – |
| 18. Nephrite rings | +++ | + | ++ | + | – | – | + | ++ | Rare |
| 19. Shell beads | +++ | – | + | + | – | – | – | ++ | – |
| 20. Ornamented antler | +++ | + | + | + | – | – | – | ++ | – |
| 21. Dagger-like points of split elk bone | ++ | + | + | + | – | – | – | + | – |
| 22. Boar-tusk artifacts (tools) | ++ | – | – | + | – | – | – | ++ | – |
| 23. Rectangular plaques of mother-of-pearl | ++ | ++ | – | – | + | – | – | – | – |
| 24. Round-bottomed vessels | ++ | ++ | + | + | + | – | + | – | Flat-bottomed vessels |
| 25. Ornamentation applied with direct stamping and with stab-and-drag method | + | + | + | – | + | – | + | – | – |

*Note:* The plus sign indicates the presence of the item.

evidenced by ritual burials of dogs in the Lenkovka and Kalashnikova ravines [Lenkovka Pad and Kalashnikova Pad] on the Angara, and also in the lower Neolithic layer at the mouth of Berezovka creek, which empties into the Selenga near Ulan-Ude. However, on the Angara and Selenga dogs were used primarily for the hunt, while on the Shilka they were probably used as draft animals and for food, as among the Gilyaks. On the whole, there were more common than diverse features in the ways of life of the ancient Shilka and Cis-Baykal peoples. Thus, the cultural similarities embrace all of the most essential features, not just individual cultural factors. This similarity may be explained by convergence, that is, by a natural coincidence of general cultural traits due to similar economic structures and living conditions. The Neolithic tribes of the Shilka valley and their contemporaries on the Lena or the Angara lived under the same taiga conditions and their mainstays were hunting and fishing supplemented by collecting edible wild plants.

Moreover, the specific aspects of the stone and bone implements as well as ornamentation indicate that there must have been yet another reason for the similarity of these two ancient cultures, namely, cultural and ethnic ties between the two peoples. During the Neolithic and Bronze ages these regions must have been occupied by tribes having a similar culture and history.

As was mentioned earlier, the analysis of the materials from the Shilka cave uncovered analogues with another Neolithic culture from a region geographically closer and more akin to the Cis-Baykal Neolithic, namely, the Neolithic of Yakutia on the middle and lower Lena. The primary evidence of this association is the shouldered adze discovered unexpectedly in the Shilka cave. Details of pottery ornamentation such as the widely used textile impressions and the custom of decorating the vessels with just a narrow band along the rim also point to a similarity between the Neolithic of Shilka and Yakutia. The sherd with the pseudo-textile or check-stamped ornamentation and especially the fragments of a vessel with impressions of wool or hair on its inner and outer surfaces found in the Shilka cave are even more characteristic of the Neolithic of Yakutia. The hair impressions are especially characteristic of the Late Yakutian Neolithic and also of the later, vestigial Neolithic cultures of the Arctic, which survived until the 17th and 18th centuries A.D.[43]

The similarity between rock paintings in the Lena valley and those of the upper Amur and Shilka rivers is particularly interesting in this respect. These latter were studied by the Far Eastern Archaeological Expedition of the Institute of the History of Material Culture in the summer of 1954. The upper Amur rock paintings, like those of the Lena, were made with red ochre and their principal subjects were likewise the elk and elk hunting. The Lena and upper Amur paintings are also similar in style, in the specific manner of representing human and animal bodies. Besides, both the Lena and the upper Amur paintings differ in the same respect from the rock paintings left by the Bronze Age cattle breeders of the steppes of the Trans-Baykal. This similarity indicates a kinship between the ancient cultures of the forest tribes of the upper Amur, the neighboring tribes of the northern taiga, and the tribes still farther north in the forested tundra and the tundra proper of remote northeastern Asia.

Further, the Neolithic culture represented by the finds in the Shilka cave is quite similar to the Neolithic of the forest regions of the Trans-Baykal. The entire microlithic inventory of the Shilka cave closely resembles, in materials and shapes, the stone artifacts found in the dune campsites [blowouts] of the Trans-

Baykal. The basic materials used for these small stone implements from the Shilka cave and all the steppe and forested steppe regions of the Trans-Baykal were white translucent chalcedony and greenish jasper-like rocks. The artifacts of the Trans-Baykal also include many small lamellar blades with superb faceting, and scrapers similar in shape to those of the Shilka, including miniature discoid scrapers. Punches made from lamellar blades are numerous. Bilaterally retouched arrowheads, similar to the Shilka types, are common in the Trans-Baykal; these include arrowheads with an asymmetrical, notched angular base. The Shilka cave also yielded points of the Daurian type, which appear in two variants in the Trans-Baykal region: (a) those with a side notch, and (b) those shaped like a willow leaf.[44]

Pottery with textile impressions and a stamped pattern similar to that found in the Shilka cave is frequent in many Neolithic sites of the Trans-Baykal region. Similar sherds with ornamentations consisting of straight, short indentations are found—for instance, the surface finds of sites along the shores of Lake Irgen near Beklemishevo and elsewhere.

The mother-of-pearl ornament described above is also typical. It consists of a thick, nearly rectangular plaque. Plate-like ornaments of this type, cut from thick shells, probably river shells, were found in the Neolithic grave discovered by me in the Khamnigan region of the Chikoy river basin in the Buryat Autonomous Republic not far from the Mongolian border.

Daurian points and rectangular plate-like ornaments of river shells are not found in the Cis-Baykal region or are an exception there. For example, only one Daurian point was found in a Serovo cliff grave on the Lena near Vorobyevo. Not a single rectangular mother-of-pearl plate-like ornament has been found in the Cis-Baykal. Thus, the Shilka Neolithic is even more closely related to the Trans-Baykal Neolithic than to the Yakut and Cis-Baykal Neolithic. The connection between the Neolithic of the Shilka valley and the Trans-Baykal region becomes evident when one examines the pottery materials found in the Shilka cave. Shilka pottery is quite similar to that of the ancient Trans-Baykal.

The Shilka finds are also quite similar to the Neolithic remains of northern Tungpei, which borders on the Trans-Baykal. For example, unilaterally barbed bone harpoons similar to the Shilka specimens, including a harpoon with a perforated side flange, were found at Ang-ang-hsi (near Tsitsihar), where small flint artifacts similar to the Shilka specimens in composition and technique of manufacture, including Daurian points, were also found.[45]

The question of the relationship between the Neolithic culture of the Shilka valley and the Neolithic cultures of Mongolia and the neighboring regions of the Far East, the Amur region, and the Primore is especially interesting.

The Mongolian Neolithic first became known through the work of P. K. Kozlov's expedition (the finds in the T'ola region), the American expedition of Andrews to the Gobi Desert,[46] and the Japanese archaeologists who explored the region of the Great Wall of China.[47] Extensive data were also collected by the Sino-Swedish expedition of Sven Hedin in Inner Mongolia along the Great Wall.[48]

S. V. Kiselev's Mongolian Historical and Ethnographic Expedition of 1949, sponsored by the Academy of Sciences and the Science Committee of the Mongolian People's Republic, had a special team which studied the Mongolian Stone Age. This group discovered many traces of the Stone Age in the Orkhon valley near Erdeni-Dzu, in the Kerulen and Khalka [Halhaiin] river basins and in the

Gobi at Dalan-Dzadagad [43° 30′ N.; 104° 30′ E.]. New observations were made in the well-known Baindzak region of the Gobi (Shabarak-usu).

From these investigations we can get a general picture of the Mongolian Neolithic and its relation to the Neolithic cultures of the neighboring territories. Neolithic campsites have been discovered from Altan-Bulag [50° 18′ N., 106° 30′ E.] and Ulan-Bator [Ulaan Baatar] in the north to the Great Wall in the south and from Aksu [Wensu—40° 15′ N., 80° 30′ E.] in the west to Jehol in the east. They have been found not only along the rivers and in the grassy steppes, but in the middle of the arid sand and stone wastelands known as the Gobi.

Traces of agriculture and gathering, as evidenced by finds of querns and grinders, are a distinctive feature of the Mongolian Neolithic. Nothing similar has been found in the Cis-Baykal and Trans-Baykal regions or in the Shilka valley. Other stone artifacts discovered in Mongolia have also distinctive, quite unique features.

No large stone implements of the taiga type (like the Isakovo and Serovo slate adzes) have been found in Mongolia and the occasional large polished artifacts of the Cis-Baykal type found here are imports from the Cis-Baykal. Of this nature is the characteristically flat, green-nephrite adze of the Serovo or, more likely, Kitoy period found in Inner Mongolia and reported by Maringer.[49]

However, stone cutting tools of a distinctive, uniform, morphologically stable type were used extensively in Mongolia during the Neolithic period. These are bilaterally convex, flaked or polished, often relatively thin almond-shaped axes with a sharp, narrow butt and a broadened lower end with a convex cutting edge. These axes are oval or lenticular in cross-section.[50]

Axes of this type have been found in the Soviet Primore and in northern China. In China they are associated with Yangshao sites. Such axes have also been found in the Kurile Islands and in Japan and represent the earliest type of ax found in that "island realm" of East Asia.[51] These axes are also characteristic of the earliest stages of the Neolithic on the Amur, near Khabarovsk. The most primitive axes of this type have been found in the Neolithic settlements of Japan together with fragments of vessels with pointed bottoms in the settlements of the Inaridai type, which are regarded as the earliest Neolithic settlements of Japan.[52]

The large, elongated, tongue-shaped scrapers (or, more likely, knives) with a distinctive narrow handle and a broad convex working end are another component of the unique stone inventory of the Neolithic settlements of Mongolia. Such objects have not been found in the Cis-Baykal and Trans-Baykal regions, but are quite reminiscent of the Primore knives, such as those found at Tetyukhe Bay.[53]

The peculiar triangular implements, which are completely retouched on all sides and are triangular in cross-section with a base formed into a grooved-concave cutting edge, are also very interesting. Maringer calls them chisels or adzes. Similar artifacts have been found in one of the most ancient Amur settlements near Osipovka in the vicinity of Khabarovsk.

Beads made of drilled pieces of dinosaur and ostrich eggshells, and pieces of such shells with a fine, engraved geometric pattern, are a distinctive feature of the Mongolian Neolithic.[54]

However, the Mongolian and the Cis-Baykalian Neolithic have much in common. Similar traits are found in the dwelling and settlement patterns, and a real likeness is revealed by the hearths. In fact, the hearth arrangement is identical. Further, the stone inventories of the two regions have many features in common. The similarity can be seen primarily in the numerous prismatic cores and the blades detached from them, the side and end blades, the large punches [perforators], and the arrowheads, often made of the same type of material [as the Cis-Baykal heads], viz., translucent white and reddish-yellow chalcedony and various types of flints. The miniature, often discoidal scrapers and the scrapers of other shapes are identical.

The similarity of the Mongolian Neolithic to the Cis-Baykal and Trans-Baykal Neolithic is even more apparent in the pottery. Since it was made of a plastic substance, different Neolithic groups tended to work it in a manner of their own. The vessels with pointed bottoms, traces of net impressions, the ornamentation applied with a stamp, and the comb-impressed dot-patterns, the dominance of zonal patterns [on the vessels], and other characteristic features form a strong bond between the Mongolian Neolithic remains and those of eastern Siberia.

The Daurian arrowheads represent a specific link between the Mongolian Neolithic culture and that of the Trans-Baykal and upper Amur (Shilka valley) regions. Let us review their distribution. These arrowheads are as characteristic of the ancient settlements near Hailar [Hailaerh] and Tsitsihar as they are of the Shilka cave. A good specimen of these artifacts has been published, for instance, by Ye. I. Titov from his collections obtained from blowouts near Hailar.[55] Several such points may be discerned in the photographs of Neolithic objects published by A. S. Lukashkin and collected by him at Tsitsihar.[56] The arrowheads collected by M. Ya. and V. F. Volkov[57] were also collected at Hailar and Tsitsihar. I found similar points in 1949 on the banks of the Khalka [Halhaiin] river, which empties into Lake Buir-Nor [Buyr-Nuur] in the Mongolian People's Republic.

Arrowheads of this type were discovered in the Amur valley between Sretensk and Blagoveshchensk at a number of sites other than the Shilka cave. They were found by S. M. Shirokogorov and N. L. Godatti, and by S. M. Sergeyev in the Sretensk area.[58]

Several points of this type, selected from Sven Hedin's collections from Mongolia, were published by Maringer, who considers them perforators.[59] Maringer differentiates between perforators and drills by stating that perforators are retouched on one side only, drills on both. He also distinguishes a group of awls with a thicker and broader tip. Such objects were also found in the collections from Neolithic sites in the Shara-Muren basin, including those from Lin-hsia.

The discovery of blade-type arrowheads (flaked-off prismatic cores) in northern Japan was somewhat unexpected. Arrowheads of this type have now been found on Hokkaido and Honshu. Usually they have a small basal notch, but sometimes the base is almost straight. Both the notch and the long edges of these arrowheads from northern Japan were formed by fine retouching on the ventral side of the blade only, not the dorsal. Blade chips with a beveled edge and the same kind of fine retouching were found in northern Japan together with the blade-type arrowheads. Medium-sized gravers were also found there. Japanese investigators, for instance, M. Yoshizaki, hold that the "blade culture" is older and of different origin than the Developed Neolithic culture of Japan, that is, the Jomon. Yoshizaki claims it to be derived from the northern regions of the Asian mainland, specifically from the Amur basin, where, according to him, blade arrowheads were found by O. Naosige. In the light of the data given above, his conclusions seem quite probable and actually confirm the existence of ancient cultural ties between the people of northern Japan and the peoples of the Amur basin and Tungpei, and also of Mongolia.[60]

It should be added that stone artifacts similar in form to the perforator points described are found not only in the Trans-Baykal region, but also in other Neolithic sites on the shores of Lake Baykal and Olkhon Island.[61] The presence of such objects in Ulan-Khada and neighboring settlements on the shores of Lake Baykal is natural, since this is a "border zone" between the Cis-Baykal taiga and the Trans-Baykal forested steppes. At times, such objects could penetrate from here farther into the forest regions of eastern Siberia, as indicated by the presence of Daurian arrowheads in a cliff grave of the Serovo period near Vorobyevo village on the Lena river.

In analyzing these artifacts, we may divide them into a number of typological variants or groups as follows.

The first group contains points of the simplest type, lamellar blades in which the end opposite the bulb of percussion is worked with marginal retouching on the ventral side and thus is made into a point. Such points are very widely distributed, from the Great Wall to Lake Baykal.

The second group consists of points with a side notch. These have been found only in the Trans-Baykal region, where they were first described by G. F. Debets. The most ancient arrowheads found in the Khinskaya and Chastyye ravines [Khinskaya Pad and Chastyye Pad] are very similar to these in form, although they differ in details.

The third group contains willow-leaf points with narrowing retouching on the lower part. Such points have been found in the Trans-Baykal region and on the middle Amur, near Blagoveshchensk.

The fourth group consists of points whose lower end also has been worked. However, it was not rounded, but formed into a tang with a straight base by retouching. Points of this type are found in collections from the Trans-Baykal and middle Amur regions.

The fifth group is comprised of points of the type found in the Shilka cave, those with the beveled tang. They may be called Shilka points, inasmuch as they have been found only in the Shilka cave and have not been identified elsewhere.

Despite their diversity, these objects have common archaic features of form and technique. All of them are made from lamellar blades and are retouched only in part, along the edges or the tips. The retouching on the working edges is fine and marginal, while on the tips it is flat and smoothing. Often the retouching is of the opposite type [on both sides], or only one side of the blade is worked.

The shape of these blades points and the retouching of only part of the cutting edge and tip indicate a definite relationship with the ancient, Mesolithic flint-working technique and with characteristic Mesolithic points. Therefore, in type and technique, these points should not be compared with the ordinary Neolithic arrowheads which are retouched bilaterally, but rather with the Mesolithic arrowheads of the West and with the heads from Khinskaya ravine which preceded those of the developed Neolithic of the Cis-Baykal region.

It must also be kept in mind that certain other artifacts always accompany these arrowheads, such as the miniature perforators of flint and chalcedony, formed with the same type of mraginal retouching from narrow, regular lamellar blades, and miniature scrapers, including those of regular discoidal form. The miniature dimensions of these artifacts, which may be attributed to the small size of the chalcedony pebbles available, is one of their characteristic features. Another characteristic feature is the type of workmanship, often a specific, steep Mesolithic retouching which dulled, as it were, the edges of the implement.

However, in stating these points of similarity between the objects found in the Neolithic sites of the Upper Amur, Mongolia, and northern China and those found in the Mesolithic sites of Europe and Asia, it must be kept in mind that the similarity is indubitable but very general. For example, the Daurian points, especially their Shilka variant, are quite distinctive. They do not agree in their details with the Swideran points of eastern Europe, the Khin arrowheads of the Cis-Baykal, the Kelteminar arrowheads of central Asia (the Dzhebel [Jebel] cave, the site of Dzhanbas-Kala 4 in Khorezm [Chorasmia]) or earlier points with a dulled edge, such as the points found in the Kaylyu cave in Turkmenia or the site of Khodzhi-Gor in the Isfara river valley, among which one may see direct vestigial outgrowth of the ancient Gravettian tradition. Moreover, the very conditions under which these objects were found in Mongolia and the Trans-Baykal region leave no doubt that they belong to a considerably later period than do the Mesolithic objects of central Asia, the Cis-Baykal, and Europe.

As we have noted, in the Shilka cave these points were found together with typical Neolithic artifacts and pottery. In Tungpei, at Hailar and Tsitsihar, they were found in blowouts together with Late Neolithic flint arrowheads, scrapers, and rectangular, bilaterally retouched side-blades. Polished stone adzes were also found there. Thus, they must be regarded merely as a late and unique reflection of ancient, Early Neolithic or Mesolithic traditions.

Further, for a solidly based history of the ethnological and cultural relations of the Neolithic tribes of Asia it is important that these traits exist in specific territories—the Trans-Baykal steppes, Mongolia, and the upper reaches of the Amur. They form one of the most characteristic general features of the Neolithic culture of these regions. It should be emphasized that the archaeological literature frequently mentions the microlithic nature of the Mongolian Neolithic. For example, even Nelson, on the basis of a typological analysis of all the stone implements found in Baindzak, attempted to establish a chronological and typological similarity between the Mongolian finds and the Mesolithic Azilian artifacts of western Europe. In his opinion, these similarities appear in the mallets (hammers), which are predominantly round, in the multifaceted angular and spherical cores, the drills [borers], the retouched chips, the small Azilian-type end-scrapers, and the round scrapers or scrapers with two points. The discoidal beads made of ostrich and dinosaur eggshells and of mollusk shells also belong here. All stages of manufacture are represented, from crudely chipped lamellar blades to drilled and even ornamented beads. This led to the hypothesis that the Late Stone Age culture of Mongolia was nicrolithic in general, i.e., that the whole sequence of microlithic cultures of the Old World began in Mongolia and that tribes arriving from Mongolia initiated the Azilian-Tardenoisian culture of western Europe and Africa.

The above conclusions seemed confirmed to a certain extent by the first impression of the surface finds from the Neolithic sites of Mongolia. Compared with the neighboring sites in the Cis-Baykal region and the Soviet Far East, the Mongolian sites actually have relatively few large stone implements, polished axes, chisels, and adzes. Further, any investigator observing the Neolithic collections from Mongolia is struck by the abundance of small stone artifacts, often of quite minute size: arrowheads, miniature scrapers, lamellar blades, and various points of the awl and punch type. The graceful, often geometrically regular outlines, the high degree of perfection of the manufacturing technique, together with the high quality of the materials, viz., semiprecious stones (agate, jasper, and chalcedony) with a rich gamut of colors, brilliance, and refraction of light on the facets, almost give these miniature objects the appearance of jewelry.

Nevertheless, despite the presence of small stone artifacts in the Gobi, my investigations, in 1949, of the well-known Baindzak site where Nelson discovered two cultures, the Azilian and Neolithic, led me to unexpected conclusions. Actually, there are two stratigraphically distinct cultures at Baindzak, but the upper layer yielded material which must be dated as Early Bronze Age, while the lower yielded pottery analogues to the net-impressed Isakovo-Serovo pottery of the Cis-Baykal.

Neither the scarcity of large polished implements nor the abundance of small flint artifacts can be considered proof of the existence of a microlithic culture in Mongolia. The rarity of large polished implements in Mongolia may be explained by the essentially nomadic life led by the steppe tribes and by the lack of heavily forested areas. Thus, there was no need for large chopping tools, i.e., specialized instruments for working wood. Moreover, it cannot be said that no large stone artifacts were found here, since axes, pestles, querns, and other large objects have been found, albeit few of them.

The abundance of very small stone artifacts, in turn, may be explained by the type of raw material available in the Gobi. As is known, there are sandy plains strewn with beautiful chalcedony pebbles near the basaltic mantles of Mongolia. These pebbles were picked up by the Neolithic peoples at their very campsites. The abundance of this material explains the general aspect of the entire stone inventory of the Gobi finds, just as the abundance of nephrite and flinty slate determined the Cis-Baykal inventory, and the abundance of chalky slate and boulder slate the inventory of the sites in the northern part of European Russia. In general, however, there is nothing specifically microlithic about the inventory of the Mongolian Neolithic sites, if the term "microlithic" is taken in the established Western sense. There are no small flint artifacts of geometrically regular outlines, shaped like trapezoids, symmetrical and asymmetrical triangles, segments, and the like.

All these objects and the corresponding flint-working technique, including the dulling of the edges of the blades, have a specific origin and are associated with the special cultural and ethnic traditions of the Capsian culture of the South. They are characteristic of a specific time and specific territory. They appeared in southern Europe and Asia and also in Africa already at the end of the Paleolithic and sometimes remained until the Late Neolithic and Early Bronze Age as vestiges of an ancient technique.

Thus, the term "microlithic culture" or "microlithic Neolithic" cannot be applied to the Mongolian Neolithic as it can only lead to error and confusion. This does not mean that some technological features inherited from the Pre-Neolithic, i.e., Mesolithic past, could not have been used in the stone-working techniques of the Mongolian Neolithic. In fact, the Daurian points described above are of this very type and are a characteristic component of the material found in the Shilka cave; moreover, they link the culture of the Shilka cave dwellers with the cultures of the Neolithic peoples of the Trans-Baykal region and Mongolia.

In passing from Mongolia to the neighboring Tungpei and Inner Mongolia, one must begin with the nearest regions of Tungpei, i.e., with the northeastern provinces of China. The group of Neolithic sites near Tsitsihar (Ang-ang-hsi) have provided the richest finds and are described in detail in this volume by V. Ye. Larichev.* At Ang-ang-hsi, as in the Shilka cave, prismatic cores and regularly faceted lamellar blades are the most abundant items in the inventory; next come the Daurian points made from such blades, and the bilaterally retouched side- and end-blades.[62] The triangular arrowheads with a concave base, and sometimes with asymmetrical barbs, and the scrapers are identical with the Shilka artifacts in shape and manufacturing technique.[63] Of the bone implements the unilaterally barbed harpoon heads, including those with the perforated tang, the composite stone and bone daggers, and the small points are similar to those found in the Shilka cave.[64] Small stone implements, similar to the Shilka specimens in shape and manufacturing technique, were found in Neolithic sites in the Hailar region.[65]

Chinese and other archaeologists classify all these sites as belonging to the so-called Large Microlithic culture of Mongolia and Tungpei on the basis of the small retouched flint artifacts found in them, as opposed to the large polished slate implements of the other cultures of northern China, the Yangshao and Lungshan.

Features of the so-called Microlithic culture of Mongolia can be traced also farther south, where there was a zone of direct contact between the most ancient agricultural civilization of the Chinese Neolithic and the northern cultures contemporary with it. Fragments of vessels covered with painted ornaments characteristic of the Yangshao culture have been discovered at various places on the edge of the Gobi, in Inner Mongolia.[66] In Ch'ih-fêng ("first prehistoric culture") there are traces of a Neolithic settlement whose inhabitants, like the people of the Yangshao culture, tilled the soil, employing not only stone hoes but heavy polished tools, apparently [points of] ploughshares.[67] Querns with grindstones and pestles were used to mill flour. The same kind of querns and stone ploughshare points have been found in other places in Inner Mongolia as well.[68] The pottery of Ch'ih-fêng graphically indicates who taught the inhabitants of Neolithic Mongolia the art of agriculture; vessels in the shape of high amphora-like jars with side handles and vessels painted in a style similar to that of the Yangshao culture were found here.[69]

Further, the "first prehistoric culture" of Ch'ih-fêng contained small flint implements of northern types, unknown to the Yangshao culture. Here the principal material for the manufacture of knives, awls, and arrowheads was soft stone, slate, and bone, not flint. Evenly faceted lamellar blades detached from prismatic cores, as well as perforators and side blades fashioned from such lamellae, have been found in Ch'ih-fêng; flint scrapers and large, broad laurel-leaf knives similar to the Isakovo-Serovo knives of the Cis-Baykal region have also been found here. Significant among the small stone artifacts are bilaterally retouched arrowheads, triangular in outline and with a notched base. In Ch'ih-fêng, these arrowheads, like those from the oldest Neolithic sites in the Cis-Baykal, usually have asymmetrical barbs.[70] The Chinese archaeologist Liang Ssŭ-yung has reported also a crude, short and broad arrowhead similar to the Daurian heads. One edge is retouched from the dorsal side. The point was formed by retouching along the edge from the ventral side, and the tang was beveled in the same manner as in the arrowheads from the Shilka cave.[71]

The material from the Shilka cave contains objects which show somewhat unexpected correspondence with the cultures of China proper. The first such

*[See pp. 181–231.]

correspondence is found in the ornamented forked piece of deer antler. Among the Neolithic objects found in China and published in the album *The most important objects found in excavations in the five provinces Shensi, Kiangsu, Jehol, Anhwei, and Shansi* (Peking, 1958),\* there is an ornamented forked antler fragment exactly like the one found in the Shilka cave. True, the ornamentation differs in the details, having spiral markings instead of the straight-line geometric pattern of the Shilka fragment, but the general similarity of the two is unquestionable, and a more analogous piece has not been found anywhere.

The boar-tusk implements found in the Shilka cave constitute a second analogue with Chinese materials. Andersson noted "knives or razors" and "thin slivers pared off a boar's tusk" in the remains of the Yangshao culture.[72] A third coincidence is represented by the long, needle-like arrow points from the Shilka cave which have no immediate parallels in the Neolithic complexes of the North. However, there are similar and even identical points at a number of sites in China, including the classical settlement of Yangshao-ts'un. They have the same circular cross-section and sometimes they have the same type of tang, bilaterally cut to form a wedge, as do the Shilka specimens.[73]

The final example of coincident traits between the Neolithic of China and the Shilka cave is a special case, viz., the polished, shouldered adze. As already noted, it may be compared directly with Yakut [Neolithic] artifacts of the same type; however, in shape, it is also very much like artifacts found in southern China. The Chinese archaeologist Lin Hui-hsiang devoted a special article to these implements and to the problem of their origin. He writes that shouldered axes of this type are found extensively in southeastern China, in Fukien, Chenkiang, Kwangtung, Kiangsu, and Kiangsi. They do not appear at all in northwestern China and but rarely in northeastern China. In turn, stone tools of the types characteristic of the Yangshao culture prevail in the North and are very rare in southeastern China. Shouldered tools of the adze type are found in the Phillippine Islands, Sulawesi [Celebes] in Indonesia, and in Polynesia. Such implements are so abundant in the Philippines that they have been called Philippine mattocks [hoes] or hatchets. In Polynesia, tools of this type were used widely to make dugout canoes and for other work at the time the first Europeans came there; they were one of the principal tools used in the everyday life of the Polynesians. Lin Hui-hsiang believes that shouldered axes or adzes originated in southeastern China, where they were associated with the local Neolithic culture. This culture also produced pottery with stamped geometric designs, shouldered mattocks or adzes, and stepped axes. The ancient Malay people, the Yüeh, were the carriers of this Neolithic culture. In migrating, the Malay tribes carried such axes to the islands of the South China Sea, to Taiwan, the Philippines, Sulawesi, and northern Borneo, and thence to Polynesia. Carried northward, these axes reached Anhwei, Kiangsu, and Shantung.[74]

It would be tempting to assume that the shouldered adze found in the Shilka cave and the artifacts of this type found in Yakutia on the northern Lena originated in China. The finds in the Verkholensk burial ground, dated to early Serovo, indicate a rather different origin, namely, that the Siberian forms of the stepped or shouldered adzes appeared and developed independently (convergently) in the Lena basin.

Now let us compare the Neolithic finds in the Shilka cave with those of the Soviet Far East. In doing so, let me state, a priori, that the Shilka materials have very few cultural elements which could be considered specific and characteristic

\*[Title given in Russian by Okladnikov. Chinese original not traced.—Translator.]

of the most ancient cultures of our [Soviet] Far East. Artifacts such as the bilaterally retouched flint arrowheads with a notched base or bilaterally retouched rectangular side-blades are common to most Neolithic cultures of a considerable part of eastern and northern Asia. These elements are inherent in the Neolithic of all of eastern Siberia, Mongolia, Tungpei [Manchuria], and even Korea and Japan. The only Shilka object which might be regarded as a specific stone implement of the Far East, i.e., as typical of the Amur basin and the Primore, is the artifact resembling the knives of the Far East, which [in turn] are similar in shape to the Mousterian points of Europe and central Asia.

The art objects, which are so important for determining cultural history, show clearly the difference between the Shilka Neolithic and the Neolithic cultures of the Soviet Far East, on the one hand, and the similarity of the Shilka Neolithic and the Neolithic of the Cis-Baykal, on the other. We have already seen that the Shilka ornamentation is typically Cis-Baykalian, having nothing in common with the specific, curvilinear ribbon ornamentation of the Amur and Primore with its spirals, fillets and, later, meanders. Analysis of the pottery shows the same thing. In shape, surface-working, technique of manufacture, ornamentation, in a word, in all respects, the pottery from the Shilka cave differs basically and in principle from that of the ancient Far East. The vessels of the Far East are flat-bottomed. The Shilka sherds do not show any evidence of flat bottoms; in fact, judging by their general appearance, the vessels must have had round bottoms. Often the surface of the Far Eastern vessels is burnished, but the Shilka specimens are rough, decorated with textile impressions. The motifs and the composition of the ornamentation of the Far Eastern Neolithic pottery, as well as the means of applying it, are entirely different.

True, some relation to the Amur pottery may be detected in the check-stamped (pseudo-textile) ornamentation. The Shilka cave contained one sherd with this type of ornamentation; however, this feature is also similar to that of the Neolithic pottery of Yakutia. This lends support to the idea that the check-stamped ornamentation of Yakutia was introduced from the Far East through the upper and middle reaches of the Amur at the close of the Neolithic period.

In summary, it should be stated that the Neolithic culture, traces of which were discovered in the Shilka cave, was not primitive and simple, but was unexpectedly complex. Its complexity was due, in considerable measure, to its ties with the Neolithic cultures of other regions of Asia.

The Shilka cave was a kind of focus, where cultural elements of various neighboring regions and cultures of the remote past converged.

We have sketched some common features linking the Neolithic of the Shilka river valley with that of Mongolia, Tungpei, and even China, i.e. with the Yangshao culture. However, the strongest cultural, and in this case ethnic, ties between the Shilka Neolithic and that of the neighboring steppe and forested-steppe regions of the Trans-Baykal and Cis-Baykal are readily apparent. There is no doubt, too, that the Neolithic tribes of the upper Amur were in close and lively contact with the peoples north of the Amur in what is now Yakutia, on the middle Lena, and perhaps even with those farther to the northeast, towards Bering Strait. Of course, this was due in considerable measure to the geographic position of the Shilka cave, which neighbors on the Cis-Baykal and Trans-Baykal regions and Yakutia.

The geographic position of the Shilka cave is not merely a matter of latitude and longitude, but one of position on one of the great rivers of the Asian continent, the Amur, which connects eastern Siberia with the Far East. For ages rivers have linked together tribes living along their banks, facilitated the spread of ethnic groups, and contributed to the exchange of cultural elements, leveling the differences among cultures.

Thus, it might have been assumed that the most natural direction of cultural and ethnic affiliations would have been determined by the course followed by the river, i.e. towards the Pacific. It might have been expected that direct evidence of close relationships to the cultures of the Far East and Mongolia would be revealed in the Neolithic remains of the Shilka valley. However, observations have revealed a different picture. Here the river did not unite, but rather it separated.

The relationship between the Shilka and Mongolian Neolithic cultures followed the same pattern. In its most important specific features, the Mongolian Neolithic is similar to the Neolithic of the Soviet Far East. Examples are the almond-shaped axes with an oval cross-section, the triangular retouched implements with the grooved cutting edge, the knives with a narrow handle and a semicircular working end. As we have seen, the Mongolian Neolithic is much more remote from the Shilka Neolithic than the Shilka Neolithic is from the Cis-Baykalian, Trans-Baykalian, and Yakutian Neolithic.

There must be some definite and weighty reason for this. It must be concluded that this situation resulted from a complex historical situation of long standing and from ethnic kinship of the cultures.

At that time, the road of contact probably led from Shilka directly north to eastern Siberia, and eastward, into the Trans-Baykal, and also into Yakutia. The tribes and cultures of these regions were akin to those of the Shilka valley, while farther south, down the Amur and Ussuri rivers, the cultural and ethnic world was strange and alien, if not hostile, to the inhabitants of the Shilka valley and the upper Amur. This idea is not contradicted by the unexpected similarity of certain cultural elements of the Shilka cave dwellers and those of the Yangshao agricultural society of northern China. Such individual features might have been brought in through the Mongolian tribes.

It must be assumed that river fishermen and tillers of the soil, long associated with the ancient agricultural culture of China, lived for millennia along the middle and lower Amur, in the Ussuri basin, and around Lake Khanka, as well as in what is now Tungpei, and developed rich and unique cultures. First, there were the ancient Su-shêng, I-lu, Wuchi, and Moho; then the P'o-hai and Jurchen. Anthropological data indicate that representatives of the southern Mongolian races, the "Far Eastern" races according to modern antropological nomenclature,[75] lived here in antiquity as well as later. This group also includes the inhabitants of ancient Tungpei and Liaotung, who were closely akin to the peoples of the Yangshao culture, who created the great agricultural civilization of the Yellow river valley and were the Neolithic ancestors of the Chinese.[76]

Evidently the language of these inhabitants of Primore and the Amur country was Paleo-Asiatic at first, and later the Tungus-Manchu languages began to be used.

The entire culture of the Shilka cave people and particularly their art, their ornaments, suggest the large ethno-cultural group to which these people belonged. The ornaments, namely, the shell beads and the nephrite rings, were identical with those of the inhabitants of eastern Siberia in the Neolithic and Early Bronze Ages, the people of the Serovo, Kitoy, and Glazkovo cultures.

It is a matter of great importance that a complete human skull was found in the Shilka cave. The state of preservation of the Shilka skull, i.e. its degree of fossilization, and the conditions under which it was found in the cave (along with burial objects) indicate that it is Neolithic or, at the latest, of the Glazkovo period. Judging by the skull features, the man from the Shilka cave belonged to the northern group, to the Baykal physical type.[77] The present northern Tungus tribes, the Evens and Evenkis and also the Yukagirs, belong to this group.

Thus, the Shilka finds provide new and significant data which shed new light on important and major problems of the ancient ethnic history of northern Asia, such as the problem of the origin of the Tungus peoples and their culture. However, this problem, as a whole, goes far beyond the limits of this special study, whose purpose is to provide information on that remarkable archaeological site in the Shilka river valley, the Shilka cave.

## Notes and References

1. V. P. Tolmachev, Predmety "kostyanogo veka" iz Vostochnoy Sibiri ("Bone age" objects from eastern Siberia), Soobshcheniya Instituta istorii materialnoy kultury, vol. 2, 1929, pp. 334–338; idem, Otchet o sostoyanii i deyatelnosti Chitinskogo otdeleniya Priamurskogo otdela RGO za 1898–1900 gg. (Progress report on the activities of the Chita Division of the Amur Department of the Russian Geographical Society from 1898 to 1900), Russkoye geograficheskoye obshchestvo, Otchet za 1902 g., no. 11, 1903.

2. G. Merhart, Neolithikum, Reallexikon der Vorgeschichte, vol. 12, 1926, p. 69; A. P. Okladnikov, Neolit i bronzovyy vek Pribaykalya: istoriko-arkheologicheskoye issledovaniye (The Neolithic and Bronze ages of the Cis-Baykal region: A historical and archaeological study), pts. 1 and 2, MIA, no. 18, 1950, pp. 61, 160.

3. Concerning the skull found in the Shilka cave, see the paper by M. G. Levin, Drevniy cherep s reki Shilki (An ancient skull from the Shilka river), Kratkiye soobshcheniya Instituta etnografii, no. 18, 1953, pp. 69–75; cf. also A. P. Okladnikov, Neolit i bronzovyy vek Pribaykalya, chast 3: Glazkovskoye vremya (The Neolithic and Bronze ages of the Cis-Baykal region, pt. 3: The Glazkovo period), MIA, no. 43, 1955, pp. 8–10.

4. I wish to take this opportunity to express my special thanks to M. I. Rizhskiy, lecturer at the Chita Pedagogical Institute, for his valuable assistance in this research and to G. A. Tsyrelnikov of Shilkinskiy Zavod for his hospitality and cooperation.

5. B. E. Petri, Neoliticheskiye nakhodki na beregu Baykala: predvaritelnoye soobshcheniye o raskopke stoyanki "Ulan-Khada" (Neolithic finds on the shore of Lake Baykal: Preliminary report on the excavation of the "Ulan-Khada" site), Sbornik Muzeya antropologii i etnografii, vol. 3, 1916, pp. 118–119 and Plate 12, nos. 4–6; Okladnikov, Neolit i bronzovyy vek Pribaykalya, pts. 1–2, pp. 100–103.

6. Compare similar bone-working techniques by retouch flaking in the Neolithic of Yakutia: cf. A. P. Okladnikov, Lenskiye drevnosti (Lena antiquities), no. 3, Moscow, 1950, Plate 31, no. 5 and Plate 37, no. 6.

7. Okladnikov, Neolit i bronzovyy vek Pribaykalya, pts. 1–2, p. 389, Fig. 120 (object on the far right); pt. 3, p. 109, Fig. 41.

8. N. N. Gurina, Pamyatniki epokhi rannego metalla na severnom poberezhye Kolskogo poluostrova (Early Metal Age sites on the northern coast of the Kola peninsula), MIA, no. 39, 1953, p. 382, Fig. 25, nos. 1, 6, 8; idem, Osnovnyye etapy drevneyshey istorii Kolskogo poluostrova po dannym arkheologii (Principal stages of the earliest history of the Kola peninsula on the basis of archaeological data), Uchenyye zapiski Leningradskogo gos. universiteta, no. 115, 1950, Fakultet narodov Severa, no. 1, Plate 3, no. 6.

9. G. Gjessing, Yngre stenalder i Nord-Norge (The Late Stone Age in northern Norway), *Instituttet for sammenlignende kulturforskning, ser. B, Skrifter*, vol. 39, 1942, p. 207, Figs. 164–165 (from Rauhellaren, Traena, Nordland, Selnes, Balsfjord, Troms).

10. E. Krause, *Vorgeschichtliche Fischereigeräte und neuere Vergleichsstücke: eine vergleichende Studie als Beitrag zur Geschichte des Fischereiwesens*, Berlin, 1904, p. 90 and Plate 9, Fig. 375 (fishhook from a burial at Kjölmo in Norwegian Lapland); B. L. Bogayevskiy, Tekhnika pervobytnokommunisticheskogo obshchestva (Technology of a primitive communist society), *Istoriya tekhniki* (History of technology), vol. 1, pt. 1, Moscow–Leningrad, 1936, p. 209, Fig. 124, no. 6; O. Solberg, Die Eisenzeitfunde aus Ostfinmarkin, *Videnskabs-Selskabets skrifter, Historisk-filosofisk klasse*, no. 7, 1909.

11. K. Oyama, Korekawa-Funde vom Korekawa, einer charakteristischen steinzeitlichen Station vom Kame-ga-Oka der Nord-Ost-Jomon Kultur, *Shizengaku zasshi*, vol. 2, no. 4, Tokyo, 1934.

12. H. Larsen and F. Rainey, Ipiutak and the Arctic whale hunting culture, *Anthropological papers of the American Museum of Natural History*, vol. 42, 1948, Plate 75, nos. 14, 15.

13. O. Solberg, Die Eisenzeitfunde aus Ostfinmarkin.

14. Cf. Okladnikov, Neolit i bronzovyy vek Pribaykalya, pts. 1–2, pp. 180–181, Fig. 30 right, p. 183, Fig. 31 right (Isakovo stage), and p. 235, Fig. 69; *idem*, Novyye neoliticheskiye nakhodki na Angare (New Neolithic finds on the Angara river), Sovetskaya arkheologiya, vol. 16, 1952, p. 325, Fig. 4.

15. A. P. Okladnikov, *Istoricheskiy put narodov Yakutii* (The history of the peoples of Yakutia), Yakutsk, 1943, p. 29, Fig. 4, no. 1; *idem*, Yakutiya do prisoyedineniya k Russkomu gosudarstvu (Yakutia before its annexation to the Russian State), *Istoriya Yakutskoy ASSR* (History of Yakut A.S.S.R.), vol. 1, Moscow–Leningrad, 1955, p. 78, Fig. 18, no. 1.

16. *Ibid.*; cf. also A. P. Okladnikov, Arkheologicheskiye raskopki na Angare i za Baykalom (Archaeological excavations on the Angara and beyond the Baykal), *KSIIMK*, no. 51, 1953, pp. 16–19.

17. Okladnikov, Neolit i bronzovyy vek Pribaykalya, pts. 1–2, p. 179, Fig. 29, p. 230, Fig. 68, and p. 363.

18. *Ibid.*, p. 367, Fig. 111.

19. *Ibid.*, p. 272, p. 274, Fig. 85, and p. 275, Fig. 86; pt. 3, pp. 275–276 and Appendix 3, pp. 368–369.

20. *Ibid.*, pt. 3, pp. 174–185.

21. *Ibid.*, pt. 3, pp. 64–65, Fig. 19.

22. *Ibid.*, pt. 3, p. 63, Fig. 18, nos. 1–7 and 20–26.

23. *Ibid.*, pts. 1–2, p. 389, Fig. 120; pt. 3, p. 108, Fig. 41, nos. 1–3.

24. *Ibid.*, pt. 3, p. 95, Fig. 34, no. 1.

25. *Ibid.*, pt. 3, pp. 78–79, Fig. 23.

26. A. P. Okladnikov, Novyye dannyye po drevneyshey istorii Vnutrenney Mongolii (New data on the ancient history of Inner Mongolia), *Vestnik drevney istorii*, 1951, no. 4, p. 174; [author ?], Inner Mongolia and the region of the Great Wall, *Archaeologia Orientalis*, ser. B, vol. 1, 1935, Plate 15, Fig. 1.

27. E. Licent, Les collections néolithiques du Músee Hoang Ho et Pai Ho de Tientsin, *Publications du Músee Hoang Ho Pai Ho de Tientsin*, no. 14, 1932, Plate 70, nos. 10–11, Plate 71, no. 4.

28. Okladnikov, Lenskiye drevnosti, no. 3, p. 17.

29. *Ibid.*, no. 2, Yakutsk, 1946, Plate 13, Fig. 1.

30. *Ibid.*, no. 2, Plate 11, Fig. 11.

31. Not identified.

32. Not identified.

33. L. I. Shrenk, *Ob inorodtsakh Amurskogo kraya* (Aborigines of the Amur region), vol. 2, St. Petersburg, 1899, p. 175.

34. *Ibid.*, p. 124.

35. *Ibid.*

178   A. P. Okladnikov

36. I. A. Lopatin, Goldy amurskiye, ussuriyskiye i sungariyskiye (The Amur, Ussuri, and Sungari Goldis), *Zapiski Obshchestva izucheniya Amurskogo kraya Vladivostokskogo otdeleniya Priamurskogo otdela Russkogo geograficheskogo obshchestva,* vol. 17, 1922, p. 120.

37. *Ibid.,* pp. 120–121.

38. *Ibid.,* p. 122.

39. Shrenk, *Ob inorodtsakh Amurskogo kraya,* vol. 2, p. 125.

40. Okladnikov, Neolit i bronzovyy vek Pribaykalya, pts. 1–2, p. 277, Fig. 88.

41. *Ibid.,* pts. 1–2, p. 394, Fig. 123.

42. *Ibid.,* pt. 3, p. 296, Fig. 148, nos. 1–2.

43. Okladnikov, Yakutiya do prisoyedineniya . . . , vol. 1, pp. 129, 171–172, 174.

44. Okladnikov, Neolit i bronzovyy vek Pribaykalya, pts. 1–2, pp. 158 and 160; G. F. Debets, Opyt vydeleniya kulturnykh kompleksov v neolite Pribaykalya (An attempt at isolating the cultural complexes of the Cis-Baykal Neolithic), *Izvestiya Assotsiatsii nauchno-issledovatelskikh institutov pri fiziko-matematicheskom fakultete Moskovskogo gos. universiteta,* vol. 3, 1930, no. 2-A, pp. 151–169.

45. P. Teilhard de Chardin and Pei Wên-chung, Le néolithique de la Chine, *Institut de géobiologie,* no. 10, Peking, 1944, Fig. 21. [For more detail see V. Ye. Larichev, Neoliticheskiye pamyatniki basseyna verkhnego Amura (Neolithic sites of the upper Amur basin), *MIA,* no. 86, 1960, pp. 81–126. This article has been translated and published on pp. 181–231 of this book.]

46. R. C. Andrews, *On the trail of ancient man: A narrative of the field work of the Central Asiatic Expeditions,* London, 1926; Russian edition published by P. P. Soykin, Leningrad, 1929.

47. N. C. Nelson, Prehistoric archaeology of the Gobi Desert, *American Novitates,* no. 222, New York, 1926, pp. 10–16; *idem,* The dune dwellers of the Gobi, *Natural History,* vol. 26, 1926, no. 3, pp. 246–251; *idem,* Notes on the archaeology of the Gobi, *American Anthropologist,* vol. 28, 1926, no. 1, pp. 305–308; *idem,* Notes on cultural relations between Asia and America, *American Antiquity,* vol. 2, 1937, no. 4, pp. 267–308; *idem,* Archaeology of Mongolia, *Compte rendu du Congrès International des sciences anthropologiques et ethnologiques,* Copenhagen, 1939, pp. 259–262; *Archaeologia Orientalis,* ser. B, vol. 1, 1935.

48. J. Maringer, Contribution to the prehistory of Mongolia, *Reports from the scientific expedition to the north-western provinces of China under the leadership of Dr. Sven Hedin; the Sino-Swedish Expedition,* Publication 34, pt. 7, Archaeology, no. 7, Stockholm, 1950.

49. *Ibid.,* Plate 36, no. 7.

50. *Ibid.,* pp. 188–193 and Plates 34–35.

51. *Ibid.,* Plate 36, no. 7; J. G. Andersson, Researches into the prehistory of the Chinese, *Bulletin of the Museum of Far Eastern Antiquities,* no. 15, Stockholm, 1943, Plate 8, Fig. 2, Plates 12, 13, 64; J. Schnell, Prehistoric finds from the Island World of the Far East, now preserved in the Museum of Far Eastern Antiquities, *ibid.,* no. 4, 1932.

52. G. Groot, *The prehistory of Japan,* New York, 1951, Plates 1, 4.

53. Maringer, Contribution to the prehistory of Mongolia, Plate 21, no. 5, Plate 30, no. 15, Plate 33, no. 2.

54. *Ibid.,* p. 143, Fig. 41 and p. 195.

55. Ye. I. Titov and V. Ya. Tolmachev, Ostatki neoliticheskoy kultury bliz Khayloara po dannym razvedok 1928 g. (Remains of a neolithic culture near Hailar resulting from survey data of 1928), *Istoriko-etnograficheskaya sektsiya Obshchestva izucheniya Manchzhurskogo kraya,* ser. A, no. 30, Harbin, 1928, Plate 3, Fig. 9.

56. A. S. Lukashin, New data on the Neolithic culture of Northern Manchuria, *The China Journal,* vol. 15, Shanghai, Oct. 1931, no. 4, pp. 198–199.

57. Cf. Larichev, Neoliticheskiye pamyatniki. . . .

58. Okladnikov, Neolit i bronzovyy vek Pribaykalya, pts. 1–2, p. 160.

59. Maringer, Contribution to the prehistory of Mongolia, p. 178.

60. A. P. Okladnikov and V. N. Goreglyad, Novyye dannyye o drevneyshey kulture kamennogo veka na severe Yaponii (New data on the most ancient Stone Age culture in northern Japan), *Sovetskaya arkheologiya*, 1958, no. 3, pp. 246–250.

61. Cf. also Petri, Neoliticheskiye nakhodki . . . , Plate 12, nos. 4–6. Petri calls them awls. Such objects were found primarily in the ninth layer of Ulan-Khada. (Cf. Okladnikov, Neolit i bronzovyy vek Pribaykalya, pts. 1–2, p. 101.)

62. Cf. Liang Ssŭ-yung, Doistoricheskoye mestonakhozhdeniye Anantsi (The prehistoric site Ang-ang-hsi), Peking, 1932, p. 12–16, 21.

63. *Ibid.*, pp. 18, 20.

64. *Ibid.*, pp. 28, 31, 33–34.

65. Titov and Tolmachev, Ostatki neoliticheskoy kultury.

66. Maringer, Contribution to the prehistory of Mongolia, Plate 40, nos. 1–6. According to Maringer (pp. 199–200), most of the painted sherds were found in settlements in the region of Gurnai, but finds were also distributed over the eastern and central parts of Inner Mongolia and extended even to the north, but are lacking in the Black Gobi. They are not found at all in Outer Mongolia. There are specimens similar to the early, Yangshao stage, as well as to the later period, the third stage (according to Andersson), the Ma-Chang; cf. also J. Maringer, Vorposten chinesischer Kolonisation in der Mongolei der Steinzeit, *Studies on Inner Asia*, Tokyo, 1955, pp. 1–19.

67. K. Hamada and S. Mizuno, Prehistoric sites at Hung-Shan-Hou, Ch'ih-feng in the Province of Jehol, Manchukuo, *Archaeologia Orientalis*, ser. A, vol. 6, Tokyo-Kyoto, 1938.

68. Licent, Les collections néolithiques . . . ; Teilhard de Chardin and Pei Wên-chung, Le néolithique de la Chine; E. Licent and P. Teilhard de Chardin, Note sur deux instruments agricoles du néolithique de Chine, *Anthropologie*, vol. 35, 1925, nos. 1–2, pp. 63–74.

69. Hamada and Mizuno, Prehistoric sites at Hung-Shan-Hou. . . .

70. *Ibid.*, Fig. 30, Plates 27, 40.

71. Lü-Tsun-Ê, Arkheologicheskoye obsledovaniye na gore Khunshan, Chifen, Vnutrennyaya Mongoliya (Archaeological investigation on Mt. Hungshang, Ch'ih-fêng, Inner Mongolia), *K'ao-ku hsüeh-pao*, no. 3, Peking, 1958, p. 27, Fig. 5, p. 37, Figs. 1–6.

72. J. G. Andersson, Prehistoric sites in Honan, *Bulletin of the Museum of Far Eastern Antiquities*, no. 19, Stockholm, 1947, p. 66, Plate 76, nos. 1–3.

73. *Ibid.*, Plate 74, nos. 2, 12, 14, 16; cf. also Andersson, Researches into the prehistory of the Chinese, Plate 46, no. 1.

74. Lin Hui-hsiang, The stepped adze: One of the characteristics of the Neolithic culture in the south-eastern region of China, *K'ao-ku hsüch-pao*, no. 3, Peking, 1958, pp. 16–23. It is interesting to note that adzes of this type were found by N. N. Dikov in the Ust-Belaya settlement in the Anadyr basin, Chukchi peninsula.

75. N. N. Cheboksarov, Osnovnyye napravleniya rasovoy differentsiatsii v Vostochnoy Azii (Main trends in racial differentiation in eastern Asia), *Trudy Instituta etnografii*, n. s., vol. 2, 1947, pp. 24–83; *idem*, K voprosu o proiskhozhdenii kitaytsev (On the problem of the origin of the Chinese), *Sovetskaya etnografiya*, 1947, no. 1, pp. 30–70; *idem*, Severnyye kitaytsy i ikh sosedi (The northern Chinese and their neighbors), *Kratkiye soobshcheniya Instituta etnografii*, no. 5, 1949, pp. 64–69; M. G. Levin, Antropologicheskiye tipy Amura i Okhotskogo poberezhya (Anthropological types of the Amur region and the Okhotsk Sea coast), *Kratkiye soobshcheniya Instituta etnografii*, no. 1, 1946, pp. 65–68; *idem*, Antropologicheskiye tipy Sibiri i Dalnego Vostoka: k probleme etnogeneza narodov Severnoy Asii (Anthropological types of Siberia and the Far East: The problem of the ethnic origin of the peoples of northern Asia), *Sovetskaya etnografiya*, 1950, no. 2, pp. 53–64; *idem*, Etnicheskaya antropologiya i problemy etnogeneza narodov Dalnego Vostoka (Ethnic anthropology and problems of the ethnic origin of the peoples of the Far East), *Trudy Instituta etnografii*, n.s., vol. 36, 1958 [Translated into English and published by the University of Toronto Press as: Arctic Institute of North America, *Anthropology of the North: Translations from Russian Sources*, no. 3, "Ethnic origins of the peoples of northeastern Asia"].

76. D. Black, A study of Kansu and Honan aeneolithic skulls and specimens from later Kansu prehistoric sites in comparison with North China and other recent crania, *Geological Survey of China, Palaeontologia Sinica*, ser. D, vol. 6, pt. 1, fasc. 1, Dec. 1928; Kenji Kiyono, Takeo Kanaseki, and Takashi Hirai, Die menschlichen Skelettreste der Steinzeit aus P'i-tzu-wo in Kwantang, Südmandschurei, *Archaeologia Orientalis*, vol. 1, 1929; S. Miyake, T. Yosimi, and M. Nanba, Über die menschlichen Skelettfunde in den Gräbern von Hung-shan-hou bei Ch'in-feng in Dschehol, Mandshukuo, *Archaeologia Orientalis*, ser. A, vol. 6, 1938.

77. M. G. Levin, Drevniy cherep s reki Shilki.

V. Ye. LARICHEV

# NEOLITHIC REMAINS IN THE UPPER
# AMUR BASIN AT ANG-ANG-HSI
# IN TUNGPEI*

THE NEOLITHIC REMAINS known to us in the Tungpei† territory of northeastern China are of great value in clarifying the history of the earliest tribes inhabiting the Far East, and especially in establishing connections between eastern and northern Asia during the Stone Age. Tungpei is situated between two sharply contrasting areas, from a cultural and historical point of view—agricultural China and the world of the taiga-hunters and fishermen of the Cis-Baykal, Trans-Baykal, upper Amur, and Yakutia.

These two cultural-historical areas were not isolated but, on the contrary, have been closely connected since earliest times. The ancient cultures of northern Asia long felt the powerful influence of one of the most important and advanced centers of man's development—ancient China. This influence penetrated to the neighboring areas of Korea, southern Manchuria, and Inner Mogolia, to the steppes of central Asia and the Siberian taiga as far as Lake Baykal, and reached even further, into Yakutia.

In the literature concerning the Siberian Neolithic, connections have been noted between the Kitoy stage of the Cis-Baykal Neolithic on the one hand, and the culture represented by the finds in the cave of Sha-kuo-t'un on the other.[1] There is much in these connections that still remains unexplained, but the very fact that they exist is indisputable, and quite natural if one takes into account the nearness of Siberia to China and Mongolia. There are adequate grounds for postulating a close relationship between the cultures of Siberia and China also during Paleolithic times.[2]

To elucidate these connections and follow the actual channels of their development, one must become familiar with the Neolithic cultures of the regions adjacent to the one where the agricultural—animal husbandry complex of the Yangshao culture flowered 3000 years before our era, i.e., Tungpei.

But the value of ancient archaeological remains in a study of the primitive history of Asia is not limited to their reflection of the transitory connections between Neolithic cultures of northeast Asia. There is another equally interesting possibility—that of establishing the specific cultural characteristics of the Tungpei Neolithic, that is, of defining the culture of an ancient population which covered a wide territory in Neolithic times, by tracing its historical development through the course of the millennia.

Yet, in spite of the significance of the Tungpei Neolithic sites they have been given little space in Soviet archaeological literature, even in that which includes general surveys of the Stone Age in Asiatic countries. To repair this omission,

*Translated from *Materialy i issledovaniya po arkheologii SSSR*, no. 86, 1960, pp. 81–126.
†[Tungpei simply means "northeast" and refers to that region of China known as Manchuria. —Editor.]

at least in part, is the aim of this report which deals with a group of Neolithic sites near Ang-ang-hsi [a small town south of Lung-chiang] (Tsitsihar). These sites are among the most clearly defined and richest Neolithic finds in Tungpei.

A glance at the map reveals at once that the central part of Tungpei consists of low-lying areas adjacent to the main rivers of the region, the Nonni [Nên Kiang], Sungari [Sung-hua], and the Liao. This great Manchurian plain is surrounded on all sides by mountainous areas. On the west the Great Khingan ridge separates it from the Mongolian uplands, with comparatively gradual slopes on the west, and with abrupt, precipitous slopes on the east. On the north and northeast the plain is bordered by the Little Khingan range, which follows the course of the Amur, and on the east by the many ridges of the East Manchurian mountain complex.

The mountain system as a whole forms a gigantic horseshoe open to the south, where the plain borders the Gulf of Liaotung (Fig. 1). In only one other place are the mountains interrupted, and that is in the extreme northeast of Tungpei, where the narrow valley of the lower Sungari thrusts between the spurs of the Little Khingan range and the East Manchurian mountain country. As a whole, therefore, the Manchurian plain is a geographically isolated area. Its character is that of steppe country. Forests are rare and encountered only occasionally along the Nonni and Sungari rivers.

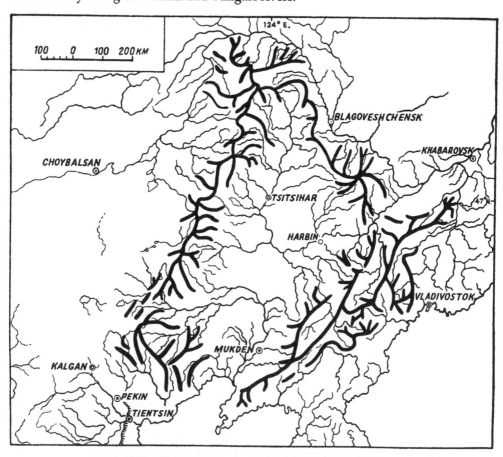

FIGURE 1. The mountains and lowlands of Tungpei.

The Ang-ang-hsi site is located in the driest, most steppe-like part of the Great Manchurian plain. Here the varied steppe grasses and black soil typical of the northeastern plain are replaced by dry wormwood and grass steppe on brown or gray soil. Characteristic of this region are the numerous depressions with interior drainage in which there are salt marshes and salt lakes. Occasionally there are aeolian sands and shifting dunes. In geographic terms this is the northeastern extremity of the region typical of Central Asia. These natural conditions could not help but affect the history of the ancient population of Tungpei and the development of its culture, which was in many ways related to the cultures of the neighboring steppes and deserts of Central Asia.

The Ang-ang-hsi station is located in the valley of the river Nonni, one of the main tributaries of the Sungari. Near Ang-ang-hsi the Nonni in its many windings makes two sharp curves. Opposite the Neolithic sites the river branches into several channels with a series of sandy islands. Quiet and peaceful in spring and autumn, the Nonni is transformed during the heavy summer rains into a broad, violent river which floods a great area and reaches at times almost as far as Ang-ang-hsi. At such periods all the land between the river and the high, second terrace where the site is situated becomes a regular sea with occasional remnant, sandy protuberances on which Neolithic remains have occasionally been found. These protuberances appear to be what is left of the first and second terraces, which have been strongly disturbed by the periodic flooding of the river. The action of the river and the character of the place seem to have changed little since the time when the banks were inhabited by Stone Age people. Possibly the river has shifted slightly to the west, since the old terraces are situated some distance from the present riverbed.

Study of the Neolithic remains near Ang-ang-hsi began more than thirty years ago and is associated with the name of the Harbin archaeologist, A. S. Lukashkin. During an ornithological excursion in the vicinity of Ang-ang-hsi in September, 1928, he discovered, five kilometers to the west of the locality, a great number of chips, finished tools, and pottery sherds in the bottom of a "blowout" in a sand dune.[3]

In 1929 and later, between 1932 and 1939, Lukashkin visited these places repeatedly and discovered six Neolithic sites, among them the only [known] burial ground in northern Tungpei and Mongolia used by the people of that epoch.

In September 1939 the well-known Chinese archaeologist, Liang Ssŭ-yung, who has done so much towards the study of the Stone Age in China, visited Ang-ang-hsi. He excavated a Neolithic burial and obtained rich material, including bone artifacts.[4]

A seventh site was uncovered in 1933 in the vicinity of the village of Hsin-chia-tz'ŭ by Zheleznyakov.[5] The Japanese archaeologists K. Komai and S. Mitsuno visited the Neolithic settlements in the summer of 1933, excavated at site no. 6, and made collections at site no. 1.

In 1937 V. V. Ponosov opened up a site with an ossuary near the Russian cemetery and also a site to the southeast of Elasu village. V. S. Makarov investigated this area in 1941–1942. He uncovered a number of sites to the northwest of Ang-ang-hsi, thus increasing the number of sites to sixteen.

The sites are arranged in two rows (Fig. 2). The first consists of nos. 1–3, 8, and 15; the second, of nos. 4–7, 9–14, and 16. some of them (for example, no. 7) are situated at some distance from the river, and can be associated more with the interior lakes of the region than with the river.

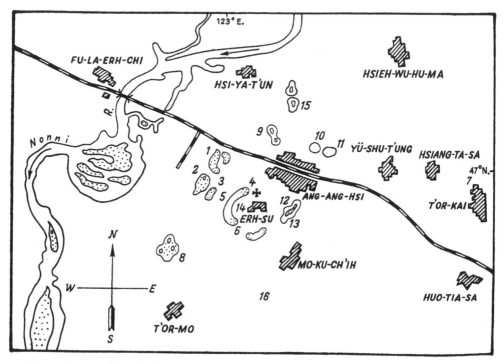

FIGURE 2. Distribution of Neolithic sites in the vicinity of Ang-ang-hsi (Tsitsihar) station.

The first row of settlements obviously corresponds to the first terrace, the second row to the second terrace, of the river Nonni. Liang Ssŭ-yung holds a different view, insisting that settlement no. 2 is associated not with the river terrace, but with the dunes along the shore of a larger lake which was later filled in with sand. Yin Ta also adheres to this theory, and cites the relevant passage from the work of Liang Ssŭ-yung.[6] However, this point of view is not shared by other investigators.

Site no. 1, the first Neolithic settlement to be uncovered in the Ang-ang-hsi area, lies to the west of the station (Fig. 3). It is situated 0.5 km south of the railroad, figuring from the 659th km.[7] It lies on an elongated hill 320 m by 85 m, oriented from southwest to northeast. As a result of plowing of the surrounding area and the terrace remnant, the surface of the hill has been dispersed by wind, leaving an extensive "blowout." At the bottom of the blowout innumerable bone fragments whitened by the sun and bits of pottery, shells, tools, and chips were found. This site is one of the richest in the Ang-ang-hsi region. Suffice it to say that all the stone axes known from Ang-ang-hsi come from here. Food refuse of Neolithic man is also abundant here—shells, bones of animals and fish, especially the sheatfish [Silurus glanis].

Site no. 2 (Fig. 4), a burial ground, is located about 1 km to the southwest of site no. 1. It is also situated on an elongated rise of ground stretching from southwest to northeast, with its summit blown away by the winds. At the northern end of a blowout, Lukashkin discovered several disturbed graves with bone and stone tools and pottery. In the southern half of the blowout were found many ceramic fragments, stone tools and rejects, proving the site to have been a settlement as well as a burial ground.

FIGURE 3. Site no. 1.

FIGURE 4. Site no. 2 (cemetery).

A small knoll some 100–120 m from the southern end of site no. 2 was probably originally a part of the [larger] formation. The summit of the knoll has been eroded by the wind. in the summer of 1932, Lukashkin discovered here the remains of a burial—a skull and two or three cracked and broken ends of bones. There were no objects associated with the burial. At the eastern end of the depression were found fragments of pottery as well as some flint and chalcedony artifacts. Lukashkin named this place site no. 3.

Site no. 4 lies about 3 km to the southwest of Ang-ang-hsi, on the second terrace near Elasu village. The heavy summer rains and spring winds have disturbed a part of the mound-like rise and caused a wide blowout (50 by 15 m) in which Lukashkin discovered several stone-lined fire pits. Nearby could be traced strata of freshwater mollusks and splintered bones. Compared with site no. 1, the stone inventory was small. The fact that the wind and water only slightly eroded the upper layers covering the cultural layer of the settlement explains this. Makarov obtained abundant material from this site upon excavating it in 1941–1942.

Site no. 5 is situated on the same rise as no. 4, about 150–200 m distant. Pottery fragments and many chips and stone tools were collected on the surface of the cultural layer uncovered by the wind. Some of the pottery, according to Lukashkin, belongs to the Metal Age.

Site no. 6 concludes the row of settlements on the terrace remnant. It occupies the southern extremity of the rise, and is located on the shore of a small lake opposite Elasu. Here fish bones were especially numerous—mullet [*Trigla hirundo*], perch, and carp. The conditions differed here from those at the previous sites. The cultural level was damp, owing to the proximity of the water, and this ensured the preservation of the bone artifacts, uncovered during the excavations of the Japanese archaeologists Komai and Mitsuno.

Site no. 7 is east of Ang-ang-hsi, near the village of Hsin-chia-tz'ŭ. It was investigated by Zheleznyakov in 1933, but the material has yet to be published.

Site no. 8 was discovered by V. S. Makarov as were all the remaining sites with the exception of the last. It lies to the southwest of the railroad station on a rounded eminence with its summit blown away by the wind, and presents the appearance of a series of high mounds separated by deep blowouts. Within these were found fragments of arrowheads and numerous knife-like blades.

Site no. 9 was discovered to the northwest of Ang-ang-hsi in what is called the Platov mound area. Sites nos. 10 and 11 are situated opposite this mound on a high promontory of the terrace.

Sites nos. 12 and 13 are located 1.5 km to the south of Ang-ang-hsi, on a small hill with two summits; no. 14 is on the road from this hill to Elasu. Site no. 15 was discovered by Makarov 5 km north of the railroad station on one of the two dunes that rise out of the river plain. No. 16, discovered by V. V. Ponosov, is situated on a promontory of the second terrace. The collections from it have not been published.

It is important to acquire an understanding of the living conditions and natural environment which Neolithic man in Tungpei, and particularly in the Ang-ang-hsi region, must have encountered. For this it is indispensable to review the stratigraphy of the cultural remains of the settlements on the basis of information contained in the works of A. S. Lukashkin and P. Teilhard de Chardin. The stratigraphy of the Neolithic settlements near Ang-ang-hsi is in general analogous to that of Neolithic settlements in the northern part of Jehol and Mongolia. According to Lukashkin, the upper stratum consists of "black earth" up to 2 m thick, which rests on deep, Late Pleistocene sand deposits of the Shara-Osso-gol type. The so-called black earth is a dark-gray, humus-enriched fine sand consisting of round and sharp-edged particles varying in size from 0.5 mm in diameter to those scarcely visible to the naked eye. Mixed with the "black soil" are irregular, friable, many-faceted particles (*zhuravchiki*), which reach a maximum diameter of 20 mm. These particles carry on their surface minute grains of white sand, which Lukashkin thinks indicates podzolization. From this it obviously follows, though Lukashkin does not mention it, that in former times the banks of the river Nonni were, in part, covered by forests. The lower part of the "black earth" has another coloration. Because of this, Lukashkin treats it as a separate stratum, calling it "brown earth." It is greenish-brown in color and consists of fine sand with humus of just the same consistency as the layer above it. This "brown earth" is more friable than the earth above. The particles (*zhuravchiki*) found in the "black earth" layer are present here also, though more rarely. The "brown earth" appears thus to be a thin transitional stratum between the "black earth" and the thick layer of yellow sand lying below it.

In general, the cultural remains are associated with the "black earth." In this deep, 2-m layer were found burials at sites nos. 2 and 3, and fire pits at site no. 4. The gray-brown stratum is almost entirely devoid of cultural remains. It probably was the surface of the terrace when Neolithic man settled there.

The stratigraphy has been studied most thoroughly at site no. 4 (Fig. 5). The top layer is "present-day" soil, gray in color. Below it is the black buried soil, full of cultural remains of the Neolithic period. This stratum attains 40–50 cm in thickness. Below this lies a mixed layer, brown in color, with scant cultural remains; those that do occur are associated primarily with the part of the stratum closest to the main cultural level. The thickness of this brown layer is about the same, 40–50 cm. Below it lies yellow sand not less than 10 m thick.

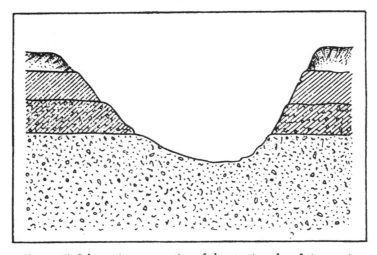

FIGURE 5. Schematic cross-section of the stratigraphy of site no. 4.

This careful analysis of the stratigraphy of the Neolithic settlements near Ang-ang-hsi is most valuable, since the published material from Neolithic sites of southern Manchuria and of eastern and western Mongolia is, for the most part, made up of surface collections from disturbed sand terraces. For this reason there has been a growing opinion, expressed in the literature on the subject, that the Neolithic remains in this large area of Asia with its deserts or dry steppes are associated with dunes and represent simply "dune settlements." Teilhard de Chardin and Pei Wên-chung in their most recent summarizing work on the Chinese Neolithic call the Neolithic of Manchuria and Mongolia the "microlithic dune Neolithic."[8] The same opinion is held by Lukashkin, Yin Ta, Liang Ssŭ-yung, and others. However, this name leads to a misunderstanding about the conditions under which Neolithic man lived here. It can give the impression that he lived under conditions similar to those existing today, and that his dwellings were situated on moving sand dunes. Actually it must be recognized that conditions have changed radically since the time in question. In some cases the reason for the change has been man himself. E. Licent, who has travelled a great deal in southern Mongolia, describes how the character of the landscape was literally transformed in a few years by the introduction of agriculture. Without trees and with ground under cultivation, the climate became significantly drier and hotter, erosion of the soil was soon under way, and presently in place of fertile plains there appeared sandy dunes filling in the furrows of the plough. Deserts appeared and grew right before people's eyes.[9]

The stratigraphic conditions under which the cultural remains were found in the Neolithic settlements at Ang-ang-hsi station convincingly show that man settled here not on dunes but on the thick soil covering of the alluvial sands on the terraces of the Nonni. Evidently the original soil cover was a brownish layer composed of gray-brown humus-enriched material. The process of soil formation was not at an end, but was even intensified during the Neolithic period. This is supported by the fact that the thickest stratigraphic layer from Lukashkin's cut is the humus-enriched buried stratum with the cultural remains. The soil-forming process has evidently continued up to very recent times, as indicated by the fact that the top soil cover, along with the lower levels, has just now entered a cycle of active erosion. The ploughing and the strong drying spring winds and the harsh continentality of the climate brought about the rapid destruction of the layers lying above the yellow sand. In place of the steppe and the forested steppe with their abundant spring grasses, there appeared the lifeless desert with its shifting dunes.

A. P. Okladnikov observed approximately the same stratigraphy in 1949 in the Baindzak area (Shabarak-usu) of Mongolia during his work with the Mongolian Historical-Ethnographic Expedition.[10] He had the good fortune to observe a uniquely important stratigraphic cut with two culture-bearing levels associated with clearly defined geologic and climatic stages of the Quaternary of Mongolia. This cut was produced by natural causes and formed the steep wall of a blow-out. The layer underlying all others consisted of dense, whitish clay. Actually it formed the bottom of the depression. Above this lay 2 m of reddish-brown loam, sharply lifferentiated from the stratum above it, a light-gray buried soil about a meter in thickness. In the loam were found cultural remains of the early Neolithic, small flint scrapers of the Mongolian type, a fire pit, and ceramics of the Serovo type. The culture-bearing layer was covered by a fine-grained, friable, dark sand about 0.5 m in thickness, and above it were 3 m of compact sand. On top of this was a second culture-bearing level, in turn topped by flaky, yellow sand. In the sand were found remains of the closing Neolithic, very probably proximate in time to the Yangshao culture or to the early Shang-Yin culture. At least in this layer one begins to encounter isolated sherds of the Bronze Age period.

Thus in Mongolia, as in Manchuria, the Neolithic period was evidently associated not with a period of erosion, but on the contrary with an active period of soil formation. This indicates a milder and rainier climate for the time. Forests grew in the river valleys, and the depressions, which are so plentifully scattered throughout Mongolia and Manchuria and which are now dry and have interior drainage, were at that time full of water. Concerning the type of climate, the remains of fauna at the Ang-ang-hsi settlements are also interesting. According to the data of Liang Ssŭ-yung and Lukashkin, there were bones of deer, boar, gazelle, roe deer, fox, wolf, hare, dog, birds, fish, and frogs. The fauna thus reflects the varied character of the landscape in which Neolithic man lived and obtained his food. Along with the types of fauna characteristic of the steppes are those usually associated with wooded, swampy, and even mountainous terrain.

Such, in general terms, were the conditions when there arose many Neolithic settlements on the banks of the large river **Nonni.**

Investigations of the Neolithic settlements in the Ang-ang-hsi area have resulted in the accumulation over the past fifteen years of a large quantity of unusual material which provides us with a general characterization of the Neolithic culture of northern Tungpei. If formerly such characterization was significantly more difficult for Soviet archaeologists because no Stone Age materials from northeast China were available to them, the task is now facilitated by printed publications as well as the collection of stone artifacts assembled from the Ang-ang-hsi area. The collection described in the present article was made by M. Ya. and V. F. Volkov, who turned it over to A. N. Formozov.[11]

In quantity and variety this material is in no way inferior to any published earlier. These finds give us a new approach to a series of cultural-historical problems connected with the ancient cultures of the Far East. The Volkovs collected the material with the greatest of care. They attached to the collection annotated charts showing the location of the stone tools. As a result it was possible to identify the origin of the collection as sites nos. 1 and 4, the richest of the Ang-ang-hsi settlements.

Below is given a detailed description of the stone artifacts of the Volkov collection. For the sake of completeness we shall include materials published earlier. This is necessary not only because a new evaluation of the published materials is now possible in the light of the thorough study of the Cis-Baykal Neolithic, but also because these publications have, for the most part, become collectors' items or are inaccessible to investigators.

The principal raw materials used in the manufacture of the tools are siliceous rocks of the varied and rich color range typical of Mongolia and Manchuria. Each of these siliceous rocks is distinguished by its own color spectrum, with varied colors speckled throughout it. The chalcedony is usually white, varying in coloration from almost entirely transparent to milky-white with yellowish streaks. Rarely, artifacts of yellow, dark yellow, or reddish chalcedony are found. The rich variety of coloration in this stone is increased by numerous shadings from one color to another and by colored speckling. A jasper-like flint of greenish coloring was also used. Such flint seems to be typical of Manchuria and Mongolia and is found nowhere else. In the collection there are also flakes of opaque white flint. Many of the artifacts are of yellowish-white, light or dark gray siliceous slate. Rarely we see a red-brown siliceous slate. The material used less frequently than any other is black slate, which is called lidite* in the Trans-Baykal. In this collection small implements are definitely more numerous than large ones.

Among the small artifacts are various types of cores (Fig. 6, items 1–3, 6, 8).

1. Unilateral, conical cores. Core no. 1 was used to the limit and was only discarded after the flakes began to break up. It is interesting in that it has on the side a well-preserved longitudinal ridge with remains of transverse preparatory chipping [undercutting], and with fine supplementary retouching along the edge of the ridge (Fig. 6, item 2). Originally the nucleus was a trihedral core-blank with three facets cut by transverse chipping; one remains. Such trihedral blanks with transverse retouching are widely distributed in eastern Siberia.[12] One of the three sides of this core is the flat, smooth surface of the original pebble. The core has a striking platform 1.5 cm by 0.7 cm carefully formed by small, transverse chipping. The length of the core is 3.7 cm; the material, reddish-brown translucent chalcedony (carnelian).

*[Basanite, a variety of flinty jasper.—Editor.]

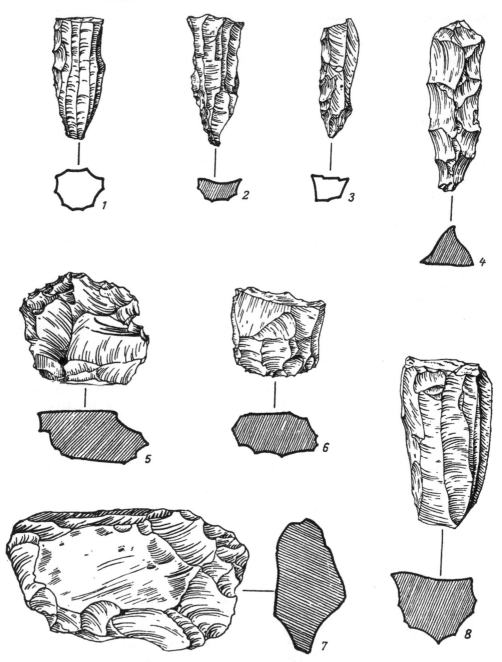

FIGURE 6. Stone artifacts from Ang-ang-hsi.

Core no. 2 is the largest in the collection (Fig. 6, item 8). It was prepared from a crudely flaked pebble of greenish jasper. One of the facets retains the original, natural surface of the nodule. The core has a well-defined platform shaped by transverse blows with traces of secondary retouching along the edge. The platform measures 2.4 cm by 2 cm, and the length of the core is 4.5 cm.

Core no. 3 is round, cylindrical at the top and slightly narrowing at the end opposite the striking platform. One side retains the untouched pebble surface. The platform is oval, prepared by the usual chipping, and measures 0.9 cm by 1.2 cm. The length is 2.7 cm; the material, light-gray jasper-like slate.

Cores nos. 4 and 5 were prepared from the small, flat pebbles of flinty slate, light yellow in color, which are often found in the Neolithic sites along the Shilka and Amur. No. 4 is especially interesting. One longitudinal ridge is well preserved, slightly retouched with small transverse strokes. Because of the natural form of the pebble, the sharp end of the core is not strictly conical, but lengthened and flattened. There are dents along ridges on two sides of the core, apparently traces of retouching with [beginning] fractures. It may be projected that this end of the core was resting on something hard at the time a flake was being detached, and when strong pressure was exerted on the platform, scales flew off, creating something like retouching by indirect [unplanned] percussion. This phenomenon did not occur in other cases, no doubt owing to the fact that the material was harder than the soft, yellow, flinty slate of the core in question. Its length is 2.5 cm, the dimensions of the platform 1.3 cm by 0.6 cm.

Core no. 6 is similar to no. 4, just described. It also has traces of the original pebble surface; the sharp, lower end is elongated, flattened, and seemingly sharpened by retouching with indirect percussion with fracturing. Its length is 2.3 cm. The platform is nearly round and formed by fine, shallow, transverse retouching.

Core no. 7 is roughly prismatic, prepared from a flat pebble of light-yellow flinty slate, and with the same trait mentioned above, that is, its ridges were formed by indirect percussion (Fig. 6, item 3).

2. Conical cores. There are two examples of this type of core in the collection, one complete, the other split. A transversely flaked edge has been preserved in nos. 8 and 9 (Fig. 6, item 1). It was made by flaking from both sides, without secondary retouching along the edge. The platform has been carefully hewed by transverse flaking and secondarily corrected on one side; its size is 1 cm by 1.4 cm. The length of the core is 3.4 cm; the material is gray flinty slate.

Core no. 10 was prepared from a piece of dark-green jasper-like flint. It is conical, with a concave platform, shaped along the edge by secondary retouching. Its length is 2.5 cm.

3. A celt-like variant of a core-scraper. Core no. 11 is flat as a result of the original form of the pebble, which shows in several places (Fig. 6, item 6). The platform was formed by a series of slipshod transverse blows. The end of the core opposite the striking platform has been changed by retouching into a scraper-like blade; hence its name "core-scraper." The retouching on one of the flat sides of the blade is fine and carefully done "creeping" retouch; however, on the opposite, convex side it is carelessly done, with fracturing similar to that of the striking platform. The length of the core is 2.2 cm, and the size of the platform 1.2 cm by 2.4 cm. Apparently the artifact published by Makarov, which he calls an ax (Fig. 11, item 1), belongs to this same type of core, to judge by the drawing.

4. Fragment of a core of even cylindrical shape. A blow struck at the edge of the platform was apparently too strong and a sizable part of the core was broken away. On the surface of the fracture in the part that remains, a bulb of percussion and ripples from the force of the blow may be clearly seen. Apparently the toolmaker wished to separate a flake from one of the sides of the core, but the line of fracture moved from the side near which the blow was struck across to

the opposite side, which obviously could not have been his intention. The core was spoiled. Its length is 4 cm; the material is a white, opaque flint.

Illustrations of several fine cores were published by Lukashkin and Makarov. They are all miniatures, cylindrical and conical in shape, with numerous grooves left where flakes have been taken off (Fig. 11, items 7–10; Fig. 13, items 6, 7; Fig. 22, item 2).

In concluding the section on cores, we shall describe the side flakes from the cores (Fig. 6, item 4). They are triangular in cross-section and have a smooth, slightly concave surface on the inner side. The two remaining sides were formed by a series of transverse flakings which produced an uneven, wavy edge with a sawtooth pattern. This in no way differs from that of the triangular core-blank which was discussed earlier under the description of conical and unilateral conical cores. Of interest are two flakes from small cores which have a fine, shallow retouching on the inner side. The edge, formed by the retouching, is even on one side and dentated on the other. The length of the flakes is 2 cm and 4.7 cm respectively. Two such core flakes were published by Lukashkin with the caption "lamelles with sharpened ends" (Fig. 14, item 7).

The next fairly large group in the collection is made up of knife-like blades with and without retouching. The blades without retouching may be divided into two groups: three-faceted and two-faceted. There are 34 examples of the former, from 1 cm to 3.5 cm long and 0.3–1.5 cm wide. The materials are flinty slate, jasper, chalcedony, and flint. More than half of these blades show signs of use because of the minute chipping of the fine cutting edge. A number of them, particularly the so-called profile blades,* were no doubt used as inset blades [side blades]; others were put to use simply in the form in which they were struck off (Fig. 7, item 11). There are 28 examples of two-faceted blades. Their dimensions are the same as those of the three-faceted, and some of them show signs of use.

The retouched blades may be divided into several groups.

1. Blades with one edge retouched (six specimens). As a rule the retouching has been done on the convex side. The retouched edges of some blades have become polished through use. They are almost all regularly rectangular in outline and of nearly uniform length from 1.8 cm to 2.1 cm. These also were apparently used as insets [side blades].

2. Blades with one retouched edge (Fig. 7, item 8). The fine, shallow retouching has been done on the dorsal part. The length of one blade is 3 cm; of the other, 2.4 cm.

3. Blades with retouching on opposite sides. One of these blades, no. 1, is three-faceted and massive; it had been struck from a prismatic core, and intentionally broken off at one end. The retouching on the convex side is deeper than that on the spine. The length of the blade is 2.1 cm, the width 1.8 cm, the thickness 0.5 cm. It is made of a dark, flinty slate.

Blade no. 2 is three-sided and very thin. The retouching on the back is deep and polished; that on the underside is shallow. The length of the blade is 2.9 cm, the width 0.9 cm, and the thickness 0.2 cm.

Blade no. 3 is three-faceted and slender. The retouching on the back is deeper than on the face, as in blade no. 2. However, the retouching covers only half of one of the sides. The length of this blade is 2.2 cm, the width 0.8 cm, and the thickness 0.2 cm.

---

*[Secheniya in the original.]

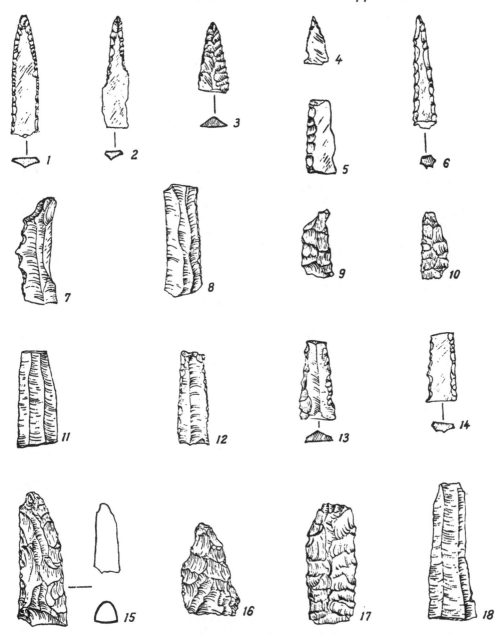

FIGURE 7. Stone artifacts from Ang-ang-hsi.

Blades nos. 4 and 5 have the same characteristics: the retouching of their long lateral faces is deeper on the back than on the underside (Fig. 7, item 12). The length of blade no. 4 is exactly 2.5 cm and of blade no. 5, 2 cm; the width in both is 0.9 cm.

There is one additional blade in the collection, and two flakes made from core-like pebbles whose edges are worked over with the characteristic dentated retouching (Fig. 7, item 7). These served as a kind of a saw blade. The edge of the core flake (judging by the published illustration) is well polished (Fig. 6, item 4).

There are also a great number of rectilinear, knife-like blades in the collections of other investigators (Fig. 11, items 11, 12, 14, 15; Fig. 13, item 12). The dorsum of some of the blades has been carefully worked over with deep retouching. These blades were apparently used as knives (Fig. 11, items 2, 13, 19; Fig. 13, items 5, 11).

There are two flakers in the collection (Fig. 6, item 5; Fig. 9, item 4). The first was made from a flat, disk-shaped pebble. With the exception of the part opposite the cutting edge, the edges are worked with broad, transverse retouching. The cutting edge is treated with fine, longitudinal secondary retouching, and is strongly indented. It has the appearance of a semicircle with a dentated edge. The upper platform of the flaker retains the original pebble surface. It is flat, and of elongated oval appearance in outline. The length of the flaker is 3 cm and the width 2.8 cm. The material is light-gray flinty slate.

A. S. Lukashkin has published an artifact that is apparently of the same type. He calls it a "chipped ax."[*] Judging from the illustration, this tool, like our flaker, has a semilunar blade, secondarily retouched. Its surface also has been worked over with the same broad, transverse chipping. The dorsum of the tool is flat. In some cases such flakers undoubtedly served as large, scraper-like tools or crude knives (Fig. 10, items 1, 11).

The second flaker is significantly larger (Fig. 6, item 7). It was prepared from an elongated, flat pebble of dark-gray flinty slate. A considerable part of the pebble's original surface has not been retouched. A broad, semilunar blade has been formed by chipping. The working edge of the flaker is badly damaged in its center and on one side. The length of the flaker is 6.5 cm and the width 3.5 cm.

Scrapers make up a significant part of the collection because of their number as well as the variety of types represented (Fig. 8). They were for the most part prepared from large, lamellar chips and, more rarely, from knife-like blades. Several different types may be distinguished among them.

1. Scrapers with a semilunar working edge produced by deep, fine retouching. One of the sides was fashioned with blunting retouching. Scraper no. 1 was prepared from a large flake of dark-green jasper, obviously struck from a prismatic core. On its dorsum are traces of flakes struck off previously. The thickened end opposite the working edge has the marks of secondary retouching customary to the striking platforms of cores. The length of the scraper is 2.3 cm and the width of the working edge 1.1 cm.

Scraper no. 2 was prepared from a thick flake struck from a pebble. Part of its dorsum preserves the original surface. The blade is massive, semicircular, formed by steep retouching which continues around to one of the sides. The other side, like that of scraper no. 1, is not retouched. The length is 1.2 cm, the width of the working edge 1.3 cm. It is possible that the retouched side-edge was also used as a working blade. A good portion of it is polished [through use].

2. Scrapers with a straight working edge and two retouched side-edges. Scraper no. 3 was prepared from a flake struck from a core (Fig. 8, item 4). The working edge of the scraper was shaped by steep retouching. Almost the entire dorsum was carefully retouched. It is interesting that the face has also been worked and shows signs of at least three fractures. Two of them were produced

*[*Obbityi topor* in the original.]

FIGURE 8. Scrapers from Ang-ang-hsi.

symmetrically on the two sides. The length of the scraper is 1.3 cm and the width of the working edge is 1.3 cm. The material is dark-green jasper-like flint.

Scraper no. 4 was prepared from a chip struck from a pebble-core (Fig. 8, item 2). It is triangular in shape, with the usual steep retouching forming the working edge and less steep retouching on the side edges. The scraper's dimensions are 1.7 cm by 1.5 cm.

Scraper no. 5 is made of beautiful, dark-green jasper. It s narrow and long, with steep, finely retouched lateral edges. The blade is straight, worked by deep retouching. The length is 2.2 cm, and the width of the working edge, 1.3 cm.

The remaining types of scrapers made from lamellar chips are represented by single specimens.

Scraper no. 6 was prepared from a broad, flat flake struck from a core (Fig. 8, item 8; see also Fig. 8, item 7). It has a rounded working edge formed by the usual steep retouching. The side edges of the scraper are not retouched. On the face there is fine undercutting near the bulb of percussion, which a sharp blow seems to have rendered too large. The length of the scraper is 1.8 cm, and the width of the working edge, 2.2 cm. The material is a dark flinty slate.

Scraper no. 7 is the largest in the collection (Fig. 9, item 8). The flake from which it was prepared is unusually long and wide for the industry at Ang-ang-hsi. The dorsum does not show supplementary retouching. It bears the traces of earlier irregular and rough chipping. The high bulb of percussion was struck out with a strong, slipshod blow. Contrary to the usual, the working edge of the scraper does not occupy the entire forward edge of the flake but only half of it; it continues, however, around part of the long lateral edge. The length of the blade is 5.5 cm, the width 3.5 cm. The material is dark-brown flinty slate.

Scraper no. 8 is the most nearly perfect in form and workmanship (Fig. 8, item 12). The dorsum is completely worked over with fine retouching. One side of the face shows the same accurate retouching. This serves to emphasize the general asymmetry of the scraper. The working edge is almost straight, slanting to one side. This scraper is triangular in form and lenticular in cross-section. The length is 2.1 cm, the working edge measuring 1.7 cm.

The scrapers made from chips may also be divided into several typological groups.

1. Scrapers with a working edge coming to a point (three specimens). Scraper no. 9 was prepared from the flake struck off a pebble of reddish flinty slate. The dorsum retains the pebble surface and only the lateral edges have been modified by two long fractures. Steep retouching covers their entire length. The sides of the scraper are bowed rather than straight and come together at an obtuse angle forming a pointed working edge, which is retouched even more steeply. The original retouching of the working edge has been smoothed with use. The lower edge of the working blade was later repaired with fine, sharpening retouching done from the face, and covering the original retouching. The length of the scraper is 2.2 cm, and the width of the working edge 0.9 cm.

Scraper no. 10 was prepared from a flake of light-green flinty slate. It is triangular in outline, with the side edges smoothly meeting to make a pointed blade. In contrast to no. 9, it has only one retouched side, worked from the dorsum. The second side is without retouch. The end opposite the working edge was flattened with longitudinal chipping, which helped cut away the bulb of percussion. The working edge is much smoothed with use—the traces of retouching are barely visible. The length of the scraper is 1.9 cm, and the width of the working edge, 0.9 cm.

Scraper no. 11 has a pointed blade formed by the smooth joining of the two semicircular, retouched lateral edges (Fig. 8, item 5). The retouching is steep. The length of the scraper is 2.2 cm, and the width of the working edge, 0.8 cm. The material is jasper-like flint.

2. Scrapers with semicircular working edges (three specimens). Scraper no. 12 was made from a solid flake struck off a pebble (Fig. 8, item 11). Its lower surface [face] is flat and level. One side is not worked from the dorsum, but the other is; it constitutes the working edge formed by regular, although unevenly spaced chipping. Finer supplementary retouching was done along the edge—with some fracturing. The length of the scraper is 2 cm and the width 2.5 cm.

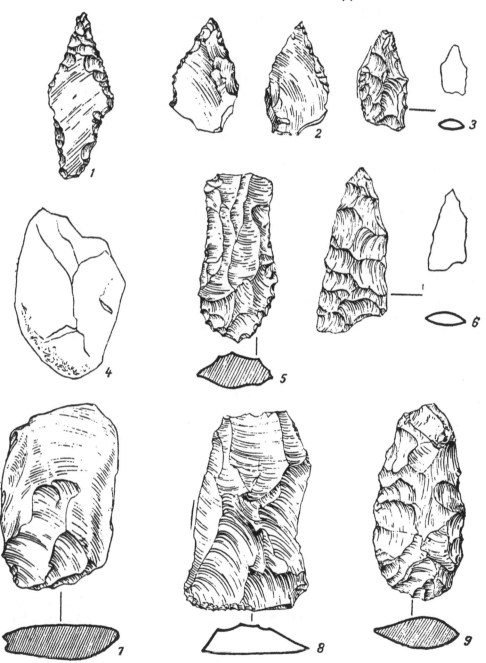

FIGURE 9. Stone artifacts from Ang-ang-hsi.

Scraper no. 13 has a broad working edge (Fig. 8, item 9; see also Fig. 8, item 13, 14). It was formed by careful, shallow retouching. The length of the working edge is 2 cm; the width of the scraper is 1.5 cm. The material is jasper-like flint, dark green in color.

Scraper no. 14 is minute (Fig. 8, item 1). The dorsum retains traces of flakes struck off earlier. It is apparently made of a flake from a "microcore." The lateral working edge of the scraper was made with fine, steep retouching. The other long lateral edge has not been retouched. The bulb of percussion on the face has been lightly undercut. The length of the scraper and of the working edge is 1 cm and the width is 0.7 cm. The material is flinty slate of a light yellow color.

The remaining types of scrapers, prepared from chips and large flakes struck off pebbles, are represented by single specimens. Scraper no. 15 is made from a thin, wide flake struck from a chalcedony pebble (Fig. 8, item 15). It has a broad working edge and a narrow hilt. The heavier side-edge is partially retouched, the other edge not at all. The retouching of the working edge is flat. The scraper is 2.8 cm long and 2.2 cm wide.

Scraper no. 16 is round, graceful, formed by steep retouching. It was prepared from a small pebble of light yellow flint which had been broken in half. The semilunar blade tapers smoothly into a narrow handle which has in part retained the pebble surface. The dorsum of the scraper had been flattened before the retouching of the working edge and handle. The working edge shows traces of secondary retouching. The length and breadth of the scraper are the same— 1.6 cm.

Scraper no. 17 was prepared from a large fragment of a pebble of light-yellow flinty slate. A large part of the dorsum retains the original surface of the pebble. Careful, low retouching forms the pointed working edge of the scraper. The side edges are roughly chipped. This chipping occurs closer to the side opposite the working edge, and forms the narrow handle of the scraper. The length is 2.6 cm; the width, 2.7 cm.

Scraper no. 18 was made from a small flake struck from a chalcedony pebble. The blade is semilunar, formed by fine retouching. One lateral edge is covered with fine facets of retouching. Traces of this work can clearly be seen on the blade. A large part of the back still retains the original surface of the pebble. The length of the scraper is 1.3 cm and the width of the working edge is 1.4 cm.

Scraper no. 19 is of the grooved type. It is made from an irregularly shaped flake in which the master was able to make use of a natural hollow (Fig. 8, item 6). The shallow retouching was done from the face of the flake. The material is green jasper-like flint. Scrapers of this type were apparently used in the final working of arrow shafts.

A. S. Lukashkin and V. S. Makarov have published illustrations of some scrapers of significance. Among these are scrapers made from straight, knife-like blades (Fig. 10, item 2; Fig. 11, item 20; Fig. 13, items 4 and 8) and from large, elongated flakes, sometimes completely retouched from the dorsum (Fig. 10, items 3, 10, 12; Fig. 11, items 16 and 21). One scraper was prepared from a small, round chip (Fig. 10, item 8).

In our collection, the large, retouched knives are of great significance; one of these was prepared from a solid, elongated pebble and the second from a large flake struck from a pebble.

Knife no. 1 is elongated and rounded, rhomboid in cross-section (Fig. 9, item 9). In some places it retains the surface of the pebble. The rounded end of the knife was formed with special care and here the retouching is the finest and the shallowest; this end has been secondarily retouched along the edge. The rest of the blade has a sawtooth edge, with the teeth slanting in different directions like those of a crosscut metal saw. This apparently increased the cutting power of the tool. The handle is narrowed. The dorsum is worked with cruder chipping than the edge. The length of the knife is 5.5 cm and the width, 2.4 cm. The material is gray flinty slate.

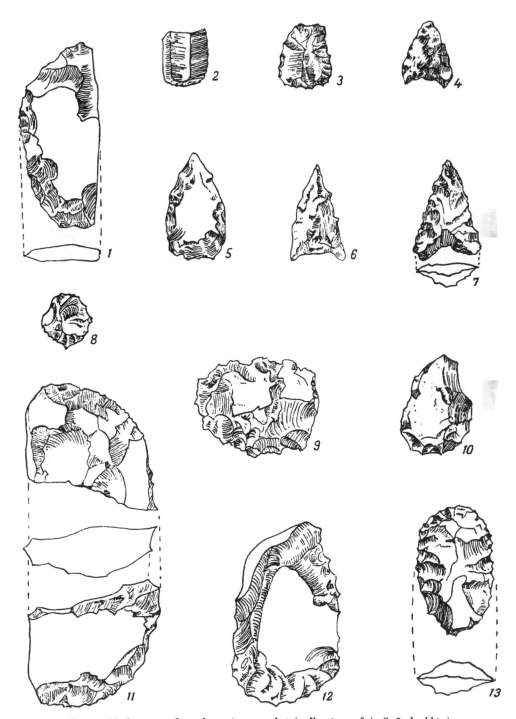

FIGURE 10. Stone artifacts from Ang-ang-hsi (collections of A. S. Lukashkin).

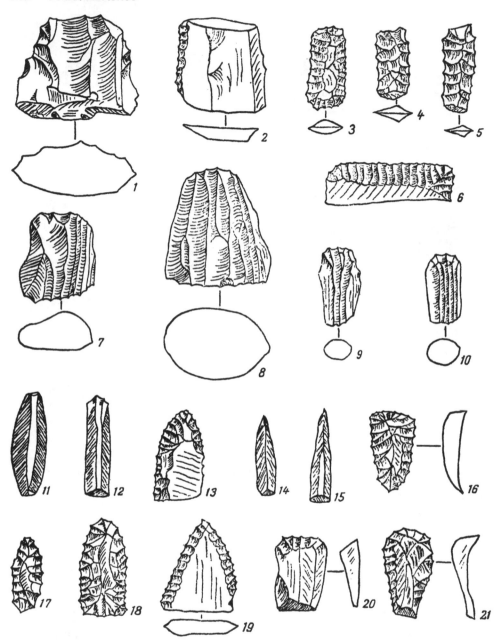

FIGURE 11. Stone artifacts from Ang-ang-hsi (collections of V. S. Makarov).

Knife no. 2 is oval in cross-section (Fig. 9, item 5). The handle occupies nearly half the length, and is peculiarly made. One side of it was formed by striking two large chips off the face and by steep dulling retouch along the edge. On the dorsum the retouch was transverse. The working edge of the knife is rounded, [and was sharpened] with small shallow retouch. The length of the knife is 4.5 cm, the width 2.2 cm, and the material is flinty slate.

In the collection there is a fragment of a triangular-shaped knife made of a slender, flat flake struck from a pebble. On one side the knife has retained the pebble surface; on the other the surface is uneven owing to the ripples resulting from the blow; it has been finely retouched along one edge. The peculiar characteristic of the knife is its deep, opposite retouching.

Among the materials published by Lukashkin there is a knife of the type described above. As a matter of fact he does not usually call such objects knives, but more often spear points.[13] P. Teilhard de Chardin makes the same error when classifying his inventory and is usually followed in this by Lukashkin and Makarov.[14] Most of these knives have a semilunar working edge, a straight or slightly arched dorsum, and a narrow handle. In planar aspect the knives are triangular. One end is thicker than the other. It is very likely that they were hafted to bone or wood. The broad surfaces are covered almost entirely with narrow, parallel facets of retouching (Fig. 10, item 13; Fig. 15, items 1–3, 6). One particularly significant knife illustrated by Makarov (Fig. 15, item 4) is of another type. Its blade is triangular, broader than the handle, from which it is clearly separated by small shoulders.

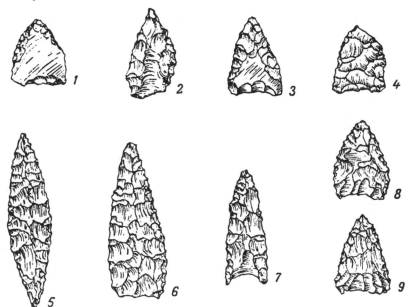

FIGURE 12. Arrowheads from Ang-ang-hsi.

Among the material collected by Lukashkin there are two types of knives not represented in our collection. One of these types is prepared from a flat piece of flinty slate. Such knives might be called "plate" knives (Fig. 14, item 11). They have a broad lower part which, gradually narrowing, forms a point at the top. The end of the point and one edge of the knife are carefully retouched from both sides; the remainder is not retouched. The three "plate" knives found by Lukashkin come from site no. 2 (the cemetery), and one was discovered during the excavation of burial no. 1. It is possible that the knife is part of the grave inventory. Lukashkin published an illustration of one such knife which had been preserved in entirety. It is 92 mm long, the width at the base is 27 mm, and the thickness 5 mm.[15]

Knives of another aspect are also typologically quite distinct (Fig. 13, item 9; Fig. 17, item 13; Fig. 22, item 4). They have a handle broad in comparison with the blade, and the blade [edge] is somewhat curved rather than straight. The greater part of the flat surfaces is covered with careful retouching. These tools are described as spear points by Teilhard de Chardin and Lukashkin, although with the reservation that they really might belong to the so-called rectangular knives. This, however, is not exact. The "rectangular knives" are nothing but two-sided, retouched inset blades. According to Teilhard de Chardin one knife was found during the excavation of burial no. 1.[16]

Knife-like blades which, in a number of cases, served as inset blades are described above. In our collection there are also inset blades, beautifully worked on both sides with pressure flaking, set in bone or wood and used as knives or points. Two inset blades show especially fine workmanship. No. 1 is triangular in cross-section, rectangular in planar outline, closely covered with flat, leveling retouching (Fig. 7, item 17). The length of the inset blade is 3.3 cm, the width is 1.5 cm, and the material is light-brown flinty slate. It was made from a flat, elongated pebble; traces of the pebble surface remain here and there between the marks of retouching.

No. 2 is regularly rectangular in shape. The edges of the artifact are worked with steep retouching. The retouching of the long sides is symmetrical [i.e. worked from both surfaces]. As in the knife-like blades with symmetrical retouching described above, so on this inset blade the retouching on the dorsum is steeper than that on the face, and has a high shiny polish. The inset blade is prepared from a flake struck from a prismatic core. It measures 2.5 cm by 0.8 cm.

Related to the side [inset] blades is the group of end [inset] blades; some are made from "plates" and some from chips (Fig. 7, items 9, 10, 16; Fig. 9, items 3, 6). They have a triangular outline, almost a straight base, and a tapered point. The working edge is heavier than the opposite side, which is sharpened and has a hollow, a depression, near the top. This hollow is repeated in three specimens and obviously has something to do with the method of securing the end blade to the bone or wood socket. Three of the four specimens are of chalcedony.

The "small, right-angled knives" or "pointed knives" of Teilhard de Chardin and Lukashkin are nothing but side and end inset-blades.[17] All of these inset blades are made of transparent chalcedony and are covered on both sides with fine shallow retouching. Their length is from 26–27 mm, width 8–12 mm, and maximum thickness 4 mm. These inset blades were found, as were the knives described above, in site no. 2, during the excavation of burial no. 1 (Fig. 11, items 3–5 and 17; Fig. 13, item 3; Fig. 17, items 6 and 10–12).

Punches may be divided into two groups on the basis of the method of manufacture: (1) those made from flat [lamellar] chips and (2) those made from regular flakes or pieces struck off pebbles or cores.

The first group consists of three punches having a specific shape, with small shoulders. Typical of these is the most representative punch, no. 1 (Fig. 8, item 10; see also Fig. 8, item 3). Retouching with long, narrow facets completely covers one side of the artifact, and clearly defines the transition between the broad handle and narrow point. On the shoulders this retouching is finer and shallower. On the opposite side the retouching is more casual and only covers the point. The handle of the punch is slightly straightened on the sides, showing traces almost everywhere of earlier gross flaking. The tool is not complete, the point having been broken off. The length of the preserved part is 2.5 cm, and the width of the point 0.7 cm. The point is rhomboidal in cross-section. The material is green jasper-like flint.

Teilhard de Chardin and Lukashkin also have published illustrations of two such punches with diminutive shoulders, made from transparent white chalcedony[18] (Fig. 13, items 10, 13). They describe them as drills or augers. The exact purpose of these instruments is hard to establish. It is possible that in some cases these were neither punches nor drills but simply perforators, that is an all-purpose tool of the drill type. This is upheld by the fact that the punches from the collection of the Volkovs show traces of smoothing and polishing from action, which could have been produced only if the working edge were not at the end, but on the sides of the tool. They were not used to make perforations but only to widen openings which had been made with another [type of] instrument.[19]

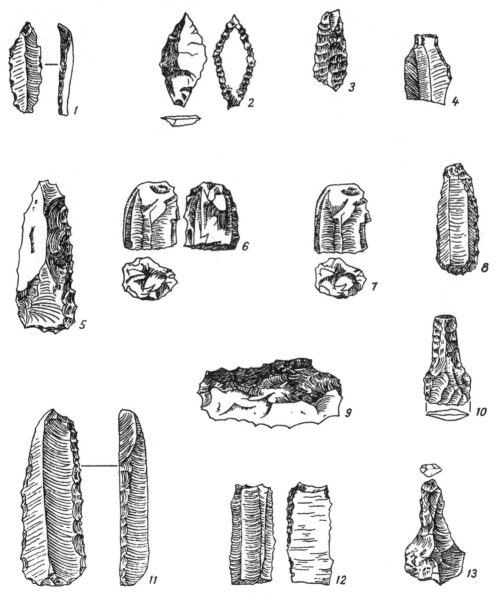

FIGURE 13. Stone artifacts from Ang-ang-hsi (collections of A. S. Lukashkin).

Punch no. 2 was made of a two-faceted plate-like flake. One side has been partly worked over with fine retouching. As with punch no. 1, the point has been retouched with special care. The transition between the broad handle and the narrow (partly missing) point is clearly defined. The length of the punch is 2.6 cm, the length of the handle 2 cm, and the width of the working part 1.2 cm.

The classification of a third artifact is more difficult although it also may be grouped among punches with small shoulders. It was made from a small but thick two-faceted flake of light-green flinty slate. The shoulder is clearly defined on one side. The retouching which shapes the point is steep, blunt, and executed from the dorsum. The handle, short in comparison with the point, is almost without retouching. The length of the punch is 1.7 cm, the length of the handle 0.5 cm, its width 0.8 cm, and the width of the point is 0.3 cm. The tip of the point is missing. This type of punch has been published by Lukashkin (Fig. 14, items 3, 4, and 8). Its most distinctive characteristic is the steep, symmetrical retouching which shapes the point from both dorsum and the face. The end of the point is usually worked with special care.

There are several variants in the second group of punches:

1. Punches of the Mongolian-Siberian type. These are prepared from straight two- or three-faceted flakes struck from prismatic or conical cores. Their second characteristic trait is the steep, blunt retouching, done from the face. There are two punches of this type in the collection.

Punch no. 1 was made from a straight, three-faceted flake, the lower end of which had been broken off (Fig. 7, item 1). The thin portions of the two side-facets have been half-removed by steep, almost vertical retouching. Thus, only the thick central portion of the flake is left. The punch thus became almost oval in cross-section. The retouched sides gradually become constricted towards the end of the punch, forming the point. The length of the preserved part of the punch is 3 cm, and the width at the base is 0.6 cm. The material is light-brown flinty slate. The same characteristics are found in punch no. 2 (Fig. 7, item 6).

Only the handle of punch no. 3 is left, the point having been broken away (Fig. 7, item 14). The steep retouching here also was done from the face. A considerable portion of the sharp side-facets has been removed and because of this the handle has an oval cross-section. The length of the preserved part is 1.8 cm; the width, 0.8 cm. The material is the same light-brown flinty slate.

Punch no. 4 is complete. The point, formed by steep retouching from two sides, is rhomboid in cross-section. The handle has been worked with obtuse retouching from one side only, in this case the dorsum. The length of the punch is 3 cm and of the point, 1.8 cm.

Lukashkin has published illustrations of a number of punches with deep, blunt retouching dulling a considerable part of the [originally] sharp edges of the flake. Some have the points retouched from the face (Fig. 14, items 2, 5), others from the dorsum (Fig. 14, item 6). There is a particularly notable punch prepared from a long, massive trihedral blade (Fig. 14, item 13). The edges and the point have been worked with deep retouching from both face and dorsum. The end of the punch is shaped by flat, planing retouch.

2. A punch shaped like a point of the Daurian type. The very tip is retouched. The sheen of the material is also a trait particular to the Daurian type of arrowhead.

3. Massive, trihedral, punch-like points retouched on three sides. There are two points of this type. One point, prepared from a lateral flake of a core, has been worked with special care. The material is white, transparent chalcedony.

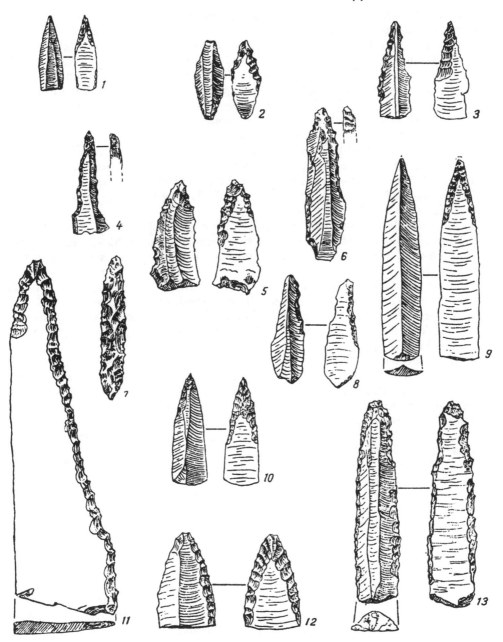

FIGURE 14. Stone artifacts from Ang-ang-hsi (collections of A. S. Lukashkin).

4. Punch-like points made from lateral flakes of a core. On the first point only one of the three facets was retouched. The second is interesting in that the retouching which forms its working edge is symmetrical (Fig. 7, item 15). The length of the first point is 3.5 cm; of the second, 3.7 cm.

5. A punch-like point prepared from a thick, angular chip with symmetrical retouching forming the working edge (Fig. 9, item 2). The presence in the

collection of such a point, reminiscent in outline of triangular Mousterian points, has great significance as the distribution of such points is rather restricted.

There are two less-remarkable objects which belong in the category just described, but do not fit into any of the above-mentioned groups. Punch no. 5 was made from a chalcedony chip and has a sawtooth edge. Punch no. 6 is only partially preserved. The usual steep retouching covers it from the face. The upper part of the flake has been supplementarily planed off. The master, however, was not content with this. He also retouched both sides from the dorsum. Such a manner of working is rather unusual.

Lukashkin also published an illustration of a beak-like punch (Fig. 13, item 1). It was made from a two-faceted flake, whose curved end has been worked into a point. The lateral edges of the punch were partially retouched.

The arrowheads are worked with great skill. Among them the following types can be distinguished:

1. Arrowheads of elongated triangular shape (two specimens; Fig. 12, item 7). They are rhombohedral in cross-section, have a concave base, and are completely covered with fine retouching. The length of one point is 3.8 cm; of the second, 2.7 cm.

2. Short and broad arrowheads, triangular in outline (four specimens; Fig. 12, items 4, 8, 9). They have broad concave bases, slightly asymmetrical.

3. An arrowhead of elongated, rhomboidal shape, rhomboidal in cross-section (Fig. 12, item 5). It is covered with fine, careful retouching. The length of the arrowhead is 4.2 cm.

4. An arrowhead with a tang (Fig. 9, item 1). It was prepared from a broad, heavy flake. It has a triangular point separated from the tang by shoulders. The retouching covers only the edges except at the point, which is retouched from both sides. The length of the point is 4.5 cm; of the tang, 1.9 cm.

There are several additional pieces in the collection which are either arrowheads or spearheads but, owing to their fragmentary state, it is impossible to say anything definite about them.

There is also a tool prepared from a carefully retouched, curved flake. It is probably a cutting instrument.

5. Arrowheads made from triangular chips (Fig. 12, item 1). One of these is thin and flat; the second, thick, the thickest part being at the point. The base of the arrowheads is slightly concave.

6. Arrowheads with a straight base (two specimens; Fig. 12, item 2). These arrowheads have a broad lower portion and a sharply narrowing, asymmetrical tip. Almost the entire surface is covered with fine retouching. The length of one arrowhead is 1.3 cm; of the other, 2.2 cm.

Lukashkin and Makarov published illustrations of several arrowheads with concave, slightly asymmetrical bases. They are all triangular in shape, with broad bases. Most of them are almost entirely worked over with fine retouching (Fig. 10, items 4, 6, 7; Fig. 15, item 7). Three arrowheads are retouched only on the face (Fig. 15, items 5, 9, 10). There are also arrowheads with straight, round, and triangular bases (Fig. 10, item 5; Fig. 13, item 2; Fig. 15, items 8, 11). One rhomboidal arrowhead has been retouched on the face along the edge but the rest of the surface remains unworked.

7. A fragment of an arrowhead of Daurian type (Fig. 7, item 3). It was prepared from a straight, trihedral flake. On its face, the point was fashioned along the edge with fine, planing retouch. Teilhard de Chardin describes a series of such arrowheads collected by Lukashkin as "long arrowheads."[20] Those tools, he

writes, are of an extremely precise and consistent type and consist of a long, straight blade, triangular in cross-section, with the edge skilfully retouched with "oblique" retouching. A considerable part of the blade has not been worked. The length of the Daurian arrowheads varies from 18 mm to 50 mm. On some of the short Daurian arrowheads or on fragments of long blades there is sometimes, besides the smoothing retouching from the face which forms the point, also retouching along one edge from the dorsum. It is possible that some of these so-called Daurian heads were used as punches (Fig. 14, items 1, 9, 10, 12; Fig. 15, item 12). Makarov published illustrations of two long, narrow arrowheads. In contrast to the Daurian points, they are retouched overall (Fig. 15, items 13, 14).

Small artifacts, however, do not exhaust the varied list of stone tools characteristic of the settlements near Ang-ang-hsi. Although they predominate in our collection, as they do in the materials from other sites in the northern regions of Tungpei and Inner Mongolia, larger tools are equally characteristic. Such are, for example, the polished chopping tools—axes or adzes.

One of the remarkable objects in the collection is a miniature adze, prepared from an elongated pebble of dark flinty slate (Fig. 9, item 7). The adze bulges unilaterally in cross-section. The lower part is smoothed to a beautiful polish. The sharpened blade is accurately defined. The width of the blade edge is irregular: at one side-facet it is about 0.8 cm, and near the other, 2 cm. The evenly arched dorsum preserves the pebble surface at the butt-end. The lower half of the adze and the blade are just as carefully polished [on the face] as on the dorsum. The side facets are cut and rounded. In this way the transition from the arched to the flat face takes place smoothly and gradually, which completes the general elegance and classic look of the adze. Its measurements, like those of all the known axes from Ang-ang-hsi, are small: the length is 5 cm, the width 3.5 cm, and the thickness in the center 1 cm. The thickness decreases at the blade and butt to 0.5 cm and 0.7 cm, respectively. The measurements of the axes described by Lukashkin are as follows: breadth to 2 cm, thickness to 0.2 cm, and length to 10 cm.

The remaining nine axes known from Ang-ang-hsi (site no. 1) from other publications were also made from small pebbles of argillaceous slate. In cross-section these tools are either round or unilaterally bulging. The butt in some specimens is somewhat narrower in comparison with the blade (Fig. 16, items 1, 2; Fig. 21, item 2; Fig. 22, item 6).[21] An ax lenticular or rather flat-oval in cross-section was found by Liang Ssŭ-yung in burial no. 4 (Fig. 16, item 4). It is small and flat with a beautifully sharpened blade. It has been ground off more on its dorsum than on its face. The remaining surface has been well polished. This is the most outstanding ax in the collection. As a whole all the axes represent a unified, typologically consistent group.

In addition to the adzes made from pebbles there are in the Ang-ang-hsi settlements "microcelts" (as Lukashkin calls them) made from nephrite of a milky-green color. These small tools are almost square in outline and have two keenly sharpened edges. From repeated grindings both these have been worn "to nothing." The length of these "microcelts" is 30 mm, the width 20 mm, and the thickness 8 mm. One was found by Lukashkin, a second by K. Komai. Unfortunately Lukashkin does not illustrate a cross-section of the "microcelts" but merely states that they are "flat in form." To be sure, being rectangular in cross-section, they may differ typologically from the local axes.

There are stone disks among the artifacts, and these have been illustrated in Lukashkin's book (Fig. 20, item 10). There are none in our collection. Lukashkin

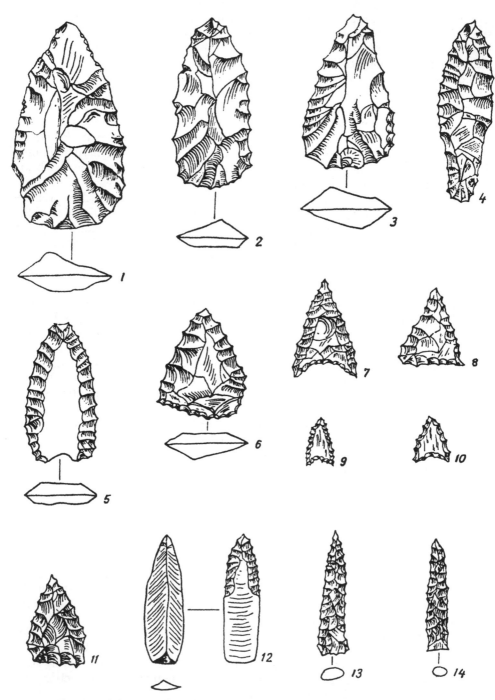

FIGURE 15. Stone artifacts from Ang-ang-hsi (collections of V. S. Makarov).

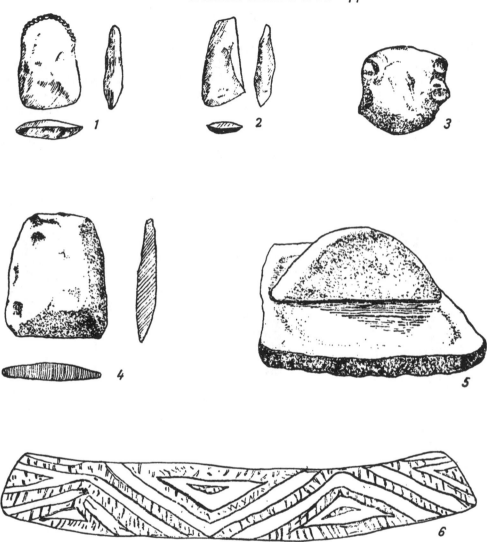

FIGURE 16. Stone tools and vessel ornamentation from Ang-ang-hsi.

uncovered three such artifacts at site no. 1, and a fourth specimen was found at the same place by Komai. These disks are made of soft argillaceous slate, and are of irregular rounded outline. Their diameter is not over 8.5 cm, and they are from 4 to 6 mm thick. One of them was flaked; the others were polished and worn along the edges.

Lukashkin made another important find near Ang-ang-hsi, a grinding stone of a quern. It consists of a cylinder broken in half, three-faceted in cross-section, with the lower surface polished smooth from long use. The grinding stone is made from tufa or tufaceous sandstone. Its length is 16 cm, width 6.1 cm, and height 4.2 cm. Lukashkin is of the opinion that before it was broken its length must have been about 25 cm. V. V. Ponosov has published an illustration of a complete quern (together with the grinding stone).[22] The lower part of the quern has the appearance of a large, four-cornered slab with the surface ground to a bowl shape. The grinder is semilunar in form with a high [arched] back (Fig. 16, item 5). Finally, there is a stone tool, the last, a small, flat pebble with little dents on both sides; it served as a weight (Fig. 16, item 3).

Among the ornaments in the collection there is a fragment of unusual hexagonal bead of greenish stone. Only one half of it has been preserved. The bead was broken while it was being made, when the openings, which were being drilled from both sides, had almost met. The opening on one side is shorter and broader than on the other. The walls of the openings are highly polished from the drilling. All the facets of the head which have been preserved are finely polished and are of a beautiful, velvet-green color. The length of the bead is 2.5 cm; the width of the preserved half, 0.7 cm.

Among the ornaments mentioned by Teilhard de Chardin and Lukashkin, we note the following: (1) a small fragmented ring, made from the shell of a fresh-

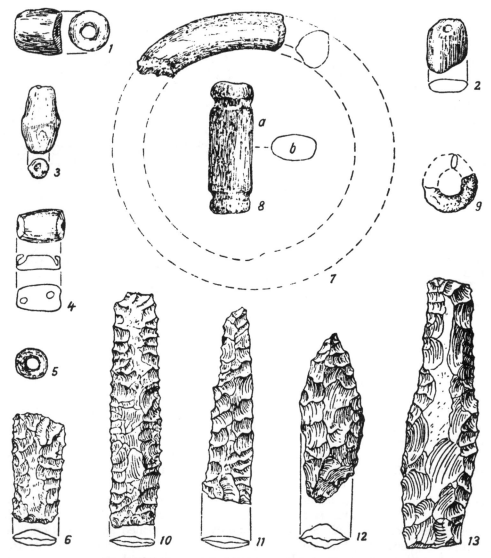

FIGURE 17. Ornaments and stone tools from Ang-ang-hsi.

water mollusk[23] (Fig. 17, items 9, 5); (2) a many-faceted, roughly polished bead in the form of two hexagonal pyramids joined at their bases, made of cornelian (Fig. 17, item 3); (3) an elongated bead (amulet), rounded above and flat below, of amazonite, with two openings drilled at an angle at each end of the bead (Fig. 17, item 4); (4) a pendant of amazonite made from an irregular, flat, polished pebble (Fig. 17, item 2); (5) a large, elongated, cylindrical bead, rounded, made of muddy-white soapstone (steatite).

Among other artifacts, the fragment of a carefully polished marble ring, semi-oval in cross-section (Fig. 17, item 7), is particularly worthy of note.

FIGURE 18. Schematic plan of burial excavated by Liang Ssŭ-yung in Ang-ang-hsi.

Along with the numerous Neolithic sites in the Ang-ang-hsi region, there was uncovered a burial mound unique for Mongolia and Tungpei. It was of the same period and had an unparalleled bone inventory. A particularly rich burial was excavated by Liang Ssŭ-yung. It contained, in addition to the ax described above, a pot and about 12 bone artifacts (Fig. 18; Fig. 19, items 1–6; Fig. 20, items 9, 11; Fig. 22, item 9). This burial, though it was better preserved and more complete, like those excavated by Lukashkin, seems to have been plundered by robbers in antiquity. This is evident from the displacement of the leg bones, the absence of the arm bones, and the disturbed nature of the inventory—all of the articles were tossed beyond the feet by the robbers. They lay in disorder and, for the most part, in pieces. Apparently only the pot, near the left shoulder of the skeleton, was in its original position. The body was placed with its head pointing north. Harpoons were the most numerous of the bone artifacts, and may be divided typologically into several varieties. Three harpoons have the same specific type of spur. The latter are made by two cuts produced at an angle to each other. The spurs are short and hardly protrude beyond the surface of the shaft. They are placed on one side of the shaft, and do not occupy the whole length of it, but only a small part below the point, in the form of two or three teeth. The pointed base—the long lower portion of one of the harpoons—is rounded and pointed at the end. The opposite end, the point, is finer, not so heavy, and carefully sharpened on the very tip (Fig. 19, item 1). The third harpoon differs from the other two in being curved rather than straight.

The second variety is represented by the preserved lower part of a harpoon head. This harpoon is oval in cross-section. Two preserved spurs are very different from those described above. They protrude beyond the surface of the shaft "like bird claws" (Fig. 19, item 3), and are far apart from one another. The spurs of this harpoon are characteristically bent towards the base, recalling the characteristic Magdalenian harpoons of western Europe. The end of the hafting base is triangular, broad and heavy, with a round hole through one side. The tip of the base is pointed. A harpoon fragment found by Lukashkin in 1929 in the burial ground at the spot where grave no. 2 was later dug may also belong to this variety (Fig. 22, item 3). On this harpoon there were four remaining spurs, broader and less curved than those on the preceding specimen. It was made from a long piece of leg bone of a large mammal. The shaft is carefully rounded and polished. The length of the preserved part is 9 cm and the thickness, 0.9 cm.

A third variety is represented by a harpoon which has no parallel in either the Cis-Baykal or the Trans-Baykal. It is somewhat L-shaped, with the transition between the point and the base clearly marked by a ridge. A little above the ridge, nearer to the point, is a rectangular projection with an oval perforation. The base of this projection is heavier than the upper part of it, which had been thinned. The perforated projection was used for attaching a line. The location of it about the middle instead the end is also rather unusual. The end of the harpoon is pointed, triangular in shape, with a triangular projection on one side, which is smoothed and thin, the opposite side being rounded and heavy. Perhaps it represents one of the varieties of toggle harpoon heads (Fig. 19, item 5).

There was also found a fragment of a knife [handle] made for inset blades, manufactured, according to the words of Liang Ssŭ-yung, "from a flat long bone of a large animal" (Fig. 19, item 2). The [natural] groove of the bone can be seen on the broad surface. Narrow slots for the insertion of side blades have been sawed in the two sides. The slot does not reach the end, the remaining part serving as the handle. The end is flattened and rounded. The handle is separated from the blade by small shoulders. At a level with the shoulders there is a longitudinal, oval opening through the center of the shaft, fitted into the natural groove of the bone. The end of the knife has been broken off.

Among the objects of undetermined use is a bone slab with one perforation at one end and two at the other. The slab has a peculiar shape, like a figure eight made of two adjacent rhomboids whose ends have been cut off on a straight line (Fig. 19, item 6). The upper rhomboid with one perforation is the longer, broader, and more geometrically regular of the two, while the lower one has the two openings fairly widely spaced, and is slightly wider at the very end. It is difficult to say anything definite about the possible use to which the bone slab may have been put. Most likely it was a piece of fishing gear. The purpose of a

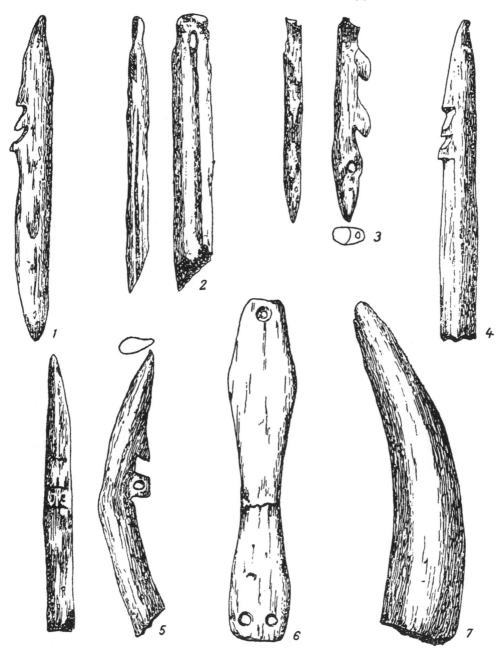

FIGURE 19. Stone tools (from burial excavated by Liang Ssǔ-yung) and antler tools.

bone slab with three round openings along its edge (Fig. 20, item 11) is also uncertain.

An awl or point (Fig. 20, item 9) was made from a flat, metacarpal bone. Its point is narrow, long, and "highly polished as though from long use." The handle, to the contrary, is broad, short, with a small round hole through it. There are transverse grooves about the rounded end of it, two of which go all the way

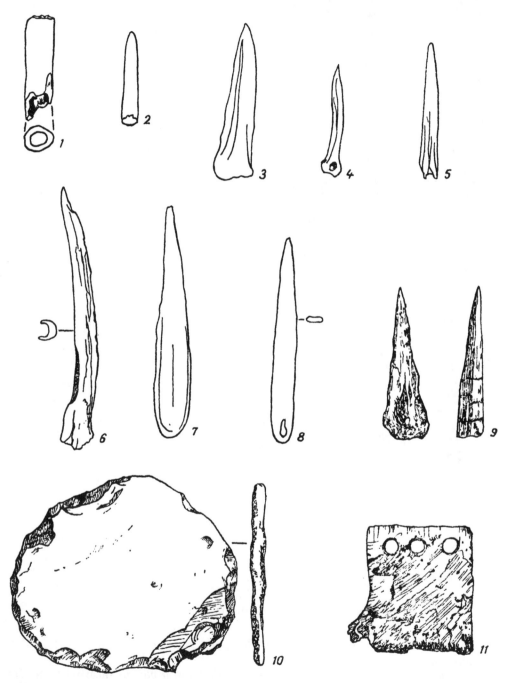

FIGURE 20. Bone artifacts and flat scraper from Ang-ang-hsi.

around and two just halfway. Of objects not belonging to the inventory of the burial, Liang Ssŭ-yung published an illustration of a worked antler [point] (Fig. 19, item 7).

Among the other bone tools we should mention an awl broken in half, 7 cm in length, with a polished surface. It was found by Lukashkin in the area of the burial ground. He also found a number of broad awls or punches made from metacarpal or metatarsal bones of the roe deer (Fig. 20, items 2–6). There are also several needles with eyes, and sections of the tubular bones of the roe deer which apparently served as needlecases (Fig. 20, item 1). The Japanese archaeologist Komai found such a case at site no. 6 with a needle placed in it.

Lukashkin describes two additional bone artifacts. One of them is a flat slab of bone about 11 cm long and 1.2 cm wide (Fig. 20, item 8). At one end it comes to a point; at the other it has an elongated opening. Lukashkin is of the opinion that this tool is a shuttle for net-making. The second artifact he calls a spoon-like object (Fig. 20, item 7). It is 12 cm long, and is broad at one end and narrow and pointed at the other. The broad surface apparently has a depression which caused Lukashkin to call it "spoon-like." He makes mention of an armor plate in his first report on the collection from the vicinity of Ang-ang-hsi; the plate had been found in the region of burial no. 1.

There are some pottery sherds in our collection. They are undistinguished and it is impossible to say anything definite about their periodization. Therefore all descriptions and conclusions about ceramics will be based on the data of Liang Ssŭ-yung, P. Teilhard de Chardin, and A. S. Lukashkin.

Pottery was collected from the surface in the blowouts, and in part it was extracted from the cultural level. Most of the material consists of fragments of pots. They are not uniform in character and date from different periods. They may be divided into two groups, according to the technique of manufacture. One group of pots was made by hand, and the second with the help of the potter's wheel (during the historic period). The sherds which show a comparatively poor firing belong to the older ceramics. They are made of black or rosy-black clay, either pure or mixed with ground shells or with large grains of sand and quartzite. Some sherds are ornamented with projecting knobs with incisions on them, or fingernail impressions or incisions interspersed with appliqué knobs (Fig. 22, item 1).

Four whole pots obtained in the excavation of burials nos. 1, 3, and 4 also belong to the older ceramics. Lukashkin uncovered two flat-bottomed pots from burial no. 1. The first pot has straight sides and a scarcely perceptible rim (Fig. 21, item 4; Fig. 22, item 8). It has a distinctive detail unusual among the pots from eastern Siberia—a spout for pouring the liquid. Almost the entire outside surface is covered with a pseudo-textile imprint, not effaced by subsequent reworking and therefore clearly visible. The diameter of the pot at the base is 9.8 cm, the diameter of the mouth is 14.7 cm, the thickness of the bottom is 1 cm, and of the walls 0.6 cm.

The second pot is jug-like in shape. It is broadened at the base, and smoothly narrows towards the top forming the neck and a slightly flared rim (Fig. 21, item 1; Fig. 22, item 10). There is no textile imprint on this pot—the surface is smooth and free of ornamentation. The diameter of the base is 7.2 cm, the maximum diameter of the lower part s 13.6 cm, and the diameter of the mouth is 8.5 cm. The thickness of the walls is 0.7 cm.

In burial no. 3 fragments of a cup with narrow base and wide mouth (Fig. 21, item 5; Fig. 22, item 7) were found. The diameter of the bottom of the cup is 9 cm and of the rim 42 cm; the height is about 23 cm. The outer walls are smooth and only along the upper edge is there a simple design consisting of a double row of nail-shaped indentations, interrupted by two appliqué knobs.

A fourth pot, spherical in shape, was found by Liang Ssŭ-yung in burial no. 4 (Fig. 16, item 6; Fig. 22, item 9). It has a clearly defined neck with straight walls. Below the shoulders which separate the neck from the body, there is a broad band of distinctive ornamentation encircling it completely. This clearly

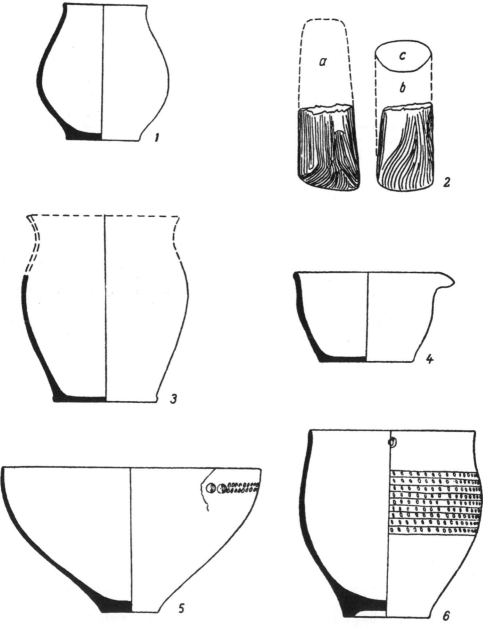

FIGURE 21. Stone ax and ceramic vessels from Ang-ang-hsi.

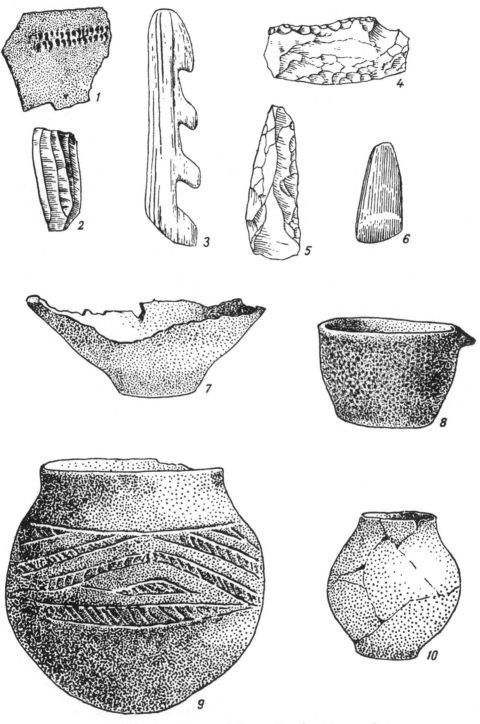

FIGURE 22. Stone, bone, and clay artifacts from Ang-ang-hsi.

separates the upper from the lower, unornamented part of the vessel. The decoration is confined between two incised parallel lines which encircle the pot. The central, unifying decorative element is a broad, zigzag strip made of two incised parallel lines. Between them are drawn vertical and slanting incisions. In each section formed by the zigzag isosceles triangles are drawn with the same incised [parallel] lines. The space between the lines which make up the sides of the triangles is filled with slanting or vertical lines. The upper row of triangles is turned with the tops down, the lower row with the tops up. Some of the triangles have an additional semilunar or flat triangular figure drawn inside them, also filled with slanting lines. The net result is a well-thought-out and organized band of decoration.

Lukashkin gives illustrations of two additional, reconstructed pots (Fig. 21, items 3, 6). The latter one, according to him, was also found in burial no. 4. But as there is no corroboration from Liang Ssŭ-yung or Yin Ta, we shall limit ourselves here to showing an illustration of the reconstruction without giving a detailed description. Attention is called to the incisions forming a banded type of decoration, with the [horizontal] lines enclosing short hatching, and to the fact that the pot has a small false bottom.

Lukashkin also uncovered sherds from pots of the *li* type which belong to the Bronze Age. The rest of the sherds with corded decoration can be dated to the Iron Age. They are thick-walled and have a considerable admixture of sand.

Such are the basic traits of the material culture of ancient tribes who lived along the banks of the Nonni river during the Stone Age. There is much in this culture which is characteristic and significant—the arrowheads with their specific planing and creeping [fish-scale] flaking at the point, shouldered punches, and punches retouched with a steep, almost blunting flaking which removes a considerable part of the edge of the blade. The bone harpoons are no less original. They are closest to the Trans-Baykal harpoons (Sretensk, Shilka cave), and are distinct from those of the Cis-Baykal. All these traits unite the inventory of Tungpei with the remains of material culture in the Trans-Baykal and the eastern areas of Inner and Outer Mongolia.[24]

First of all it is important to establish the area of ancient cultures to which the Tungpei Neolithic complex belongs. The literature devoted to this subject contains contradictory or insufficiently formulated deductions. All investigators are agreed on one point: the Neolithic culture of Tungpei and Inner Mongolia is quite distinct from the synchronous Neolithic cultures of the wooded plateau (Yangshao) and the Great North Chinese Plain (Lungshan [Ch'êng-tzŭ-yai]). The sharpest distinction between this culture and the more southern cultures lies in the former's mastery of chipping, flaking, and fracturing techniques in the preparation of stone tools—techniques altogether unknown in the south. The nature of the raw materials used dictated the preponderance of miniature stone tools at the Neolithic sites in the northern regions of Tungpei and in Central Asia. For this reason Teilhard de Chardin ascribes the Neolithic sites near Ang-ang-hsi to a microlithic culture.[25]

The most characteristic element in the material culture of the tribes of northern Tunpei (and that which places them immediately in the group of northern peoples) is the wide use of conical and prismatic cores and the removal from them of straight, knife-like flakes. The cores as well as the flakes struck from them are found with the greatest frequency in the Neolithic sites at Ang-ang-hsi. Teilhard de Chardin writes that the "micro-cores" of Ang-ang-hsi are the most characteristic trait of every Neolithic site in the Manchurian-Mongolian region. Such cores are found all through Manchuria and the Gobi, as well as in the northern part of Shansi.[26] He also writes that the cores from Ang-ang-hsi are distinct from the specimens collected by Licent in Lin-hsia. According to him the cores from Lin-hsia consist of "complete and perfect cores," while the "micro-cores" from Ang-ang-hsi have a less regular shape (some parts of them have not been worked). The blades struck from Ang-ang-hsi cores are also less regular and not as elegant. Lukashkin reiterates the same. V. S. Makarov, however, published illustrations of two straight cylindrical cores with even facets which came from site no. 5.[27] At Lin-hsia, moreover, irregular cores like those at Ang-ang-hsi are also found. This becomes convincing when the illustrations in Licent's mono-graph, where the material has been published in full, are studied.[28] Here, side by side with the "perfect cores," are cores resembling those of Ang-ang-hsi. On the whole, the cores from Ang-ang-hsi and the similar ones from Lin-hsia, known in the two variations conical and unilaterally conical, belong to a variety of materials widely distributed through neighboring Mongolia.[29] They are also widely distri-buted in the region of Lake Baykal,[30] in the Trans-Baykal, on the Shilka, and on the upper Amur.

Together with this type of core and the straight, knife-like blades struck from it, side flakes were found at Ang-ang-hsi. These three-faceted, curved flakes with the characteristic transverse chippings on one or two of the facets which meet at an angle are always encountered where conical or prismatic cores are found. Before knapping blades from the core, the master struck one or two sides off the trihedral blank core which he had already prepared, leaving a third side, which, according to A. N. Rogachev, enabled him to secure a grip on the core. It is probable that this method of preparing and working the cores was universal here. The flakes struck from the blank in order to prepare a conical core constitute the side flakes of that core. They are found in great numbers in the Neolithic sites of Inner Mongolia, the Cis-Baykal, and the Trans-Baykal.[31] The presence in north-ern Tungpei of prismatic and conical cores together with the knife-like blades struck from them has great value in characterizing the culture as a whole, as well as its place among the Neolithic cultures of northeastern Asia. This type of core testifies to the relationship between the Neolithic of northern Tungpei and the cultures of Mongolia, the Cis-Baykal, and the Trans-Baykal; it also defines northern Tungpei as a specific cultural area in relationship not only to the cul-tures of Yang-shao and Lungshan, but also to cultures of the nearest neighboring areas, southern Tungpei, Korea, Japan, the adjacent maritime region [Primore], the lower Amur and Sakhalin.

Even the first investigators of Far Eastern history, after comprehensive study, were able to survey all the cultural variations of this area and to come to the important conclusion about profound differences between the southern and southeastern regions and the northern and northwestern ones. R. Torii, a promi-nent historian and ethnographer at the beginning of this century, after having made several long trips through Korea and Manchuria, first separated the terri-tory into two great culture areas: "South-Manchurian" and "Mongolian." A pre-ponderance of polished stone tools was characteristic of the first of these, which was situated in the basin of the river Liao; chipped tools were rarely encountered here.[32] The same conclusions are arrived at by an analysis of the material col-lected by Licent. The number of conical and prismatic cores with straight-faceted, knife-like flakes struck from them, which are found in such profusion in the territory of Inner Mongolia, sharply decreases in the direction of Kalgan.

Farther to the east and south they disappear altogether, being replaced by irregularly shaped flakes.[33] I. Yawata defines two separate culture areas on the same grounds. The material collected by the First Scientific Expedition to Manchuria was published by him in two volumes, one of them devoted to the northern region, the other to the southern.[34] Liang Ssŭ-yung came to the same conclusions as a result of extensive work in the north.[35] He writes about the existence of a single Neolithic culture which extended from Sinkiang in the west to Tungpei in the east, and about the influence of Siberian culture on it. Like his predecessors, Liang Ssŭ-yung defines two culture areas for Tungpei and Jehol: (a) the regions to the north of the rivers Hsi-liao [Shara] and Sungari where flaked tools predominate; (b) the region of the river Liao, where polished tools predominate. The boundary between them is obviously along the river Liao.

As a result of extensive archaeological investigations by Chinese scientists in Kirin province it has been possible to specify precisely the most eastern as well as the southern boundary of the culture characterized by conical and prismatic cores with the straight, knife-like flakes struck from them. The latter boundary approximately follows the river Sungari, occasionally crossing onto its right bank. Remains of the so-called "microlithic" culture have been found near the district towns of Ch'ang-ch'un, Te-hui [Chang-chia-wan], Sung-hua-chien, Kan-nan, Lang-ya-pa, and Shuang-liao.[36] They are absent to the east of Kirin in the basins of the rivers Tuman [T'u-men], Matan, and Yalu.[37] This is not a matter of chance which could be explained by the lack of study of the region. The cores and knife-like blades are also absent in northern Korea,[38] on Liaotung peninsula[39] and in Soviet Primore [Maritime Province].[40] Anybody with even a rudimentary knowledge of the Neolithic of the Cis-Baykal, Mongolia, and Trans-Baykal, where conical cores and knife-like flakes are commonly found at the sites, cannot but notice that the cores in this [eastern] area and the technique of working them are altogether different from in the regions to the west of Korea and the Primore. Here the striking did not produce knife-like flakes, but irregular, elongated chips. Of course, in the hundreds or thousands of chips it is not impossible that a chip closely resembling a knife-like blade might be found. But this form would appear to be incidental and not one by which the material as a whole can be characterized. In places where conical cores and prismatic blades really exist, it does not take special efforts or particular analyses of the material to discover them.

Thus, the southern regions of Tungpei together with the Liaotung peninsula, Korea, the Primore, the lower Amur basin, Sakhalin, Japan, and possibly Kamchatka comprised, in Neolithic times, an extensive cultural area where one of the characteristic traits was the absence of conical cores and straight, knife-like flakes. Hence, the area where chipped stone artifacts predominate may be thought of as a culture zone quite distinct from those of Central and South China, where, in general, chipped tools are absent, and from Siberia and Central Asia, where the technique of striking knife-like flakes from conical cores prevails. As may be deduced from the map of the distribution of general cultural "blocks" in Asia, the Neolithic of northern Tungpei with that of Inner Mongolia constitutes the most eastern outpost of Siberian-Mongolian culture (Fig. 23). In this lies its great historical and cultural significance; it is the connecting link between the cultures of Siberia and Mongolia to the northwest and the agricultural cultures of China to the south.

The northern elements in the Neolithic culture of the river Nonni are not limited to the cores and knife-like flakes which reflect its common origin with cultures north and west of it. Among its stone tools there are some so significant

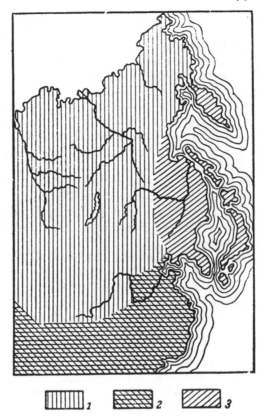

FIGURE 23. Distribution of prismatic cores and
knife-like blades: 1, area with both prismatic
cores and knife-like blades; 2, area in which
flaked artifacts are not found; 3, area in which
flaked artifacts are found, but prismatic cores
and knife-like flakes are absent.

and so clearly typed that we may speak not only of affinity with the Cis-Baykal
culture, but of its genuine nearness. Such are, first of all, the knives. Two knives
with curved blades and a broad handle from the collections of Lukashkin are
exact copies of Serovo knives which are found repeatedly in the accurately dated
burial complex of the Cis-Baykal (Fig. 13, item 9; Fig. 17, item 13; Fig. 22, item
4). The lamellar knives, triangular in shape and with characteristic retouching
along the edge and an unusually formed point, belong to a later period (Fig. 14,
item 11). Such knives are well known in the Cis-Baykal from graves of the
Kitoy period.[41] The rectangular and triangular inset blades are more difficult
to date but are important for establishing connections. Inset blades are an
indispensable article in the Neolithic sites of Tungpei, Eastern Siberia, and
Mongolia. Inset blades of this type are very characteristic of the Cis-Baykal
Neolithic. They are encountered through the entire Neolithic period and disappear only during the Glazkovo period with the appearance of metal. The Ang-
ang-hsi inset blades resemble most closely those from burials of the Serovo and
Kitoy periods.[42] These artifacts are also found in Mongolia,[43] on the Shilka and
upper Amur. It is interesting that no artifacts of this type have been found in

western Siberia. The arrowheads and scrapers from Ang-ang-hsi also have a Siberian look. All the types of scrapers described earlier were also found in the Neolithic sites in Mongolia, Sinkiang, Trans-Baykal, Shilka, upper Amur, and the Cis-Baykal. But like the prismatic cores they are not to be found in China proper.

The round, flat disk, chipped along the edge, found by Lukashkin has also its counterparts in neighboring cultures. The precise purpose of these stone disks had not yet been established, but it is clear that they must have played an important role in the Neolithic economy of Eastern Siberia and the Far East since the area of their distribution is extensive. Such disks were encountered in early Glazkovo burials in the Cis-Baykal and many were found near Lake Irgen in the Trans-Baykal. They are also known from settlements of the Yangshao culture and in southern Tungpei. J. G. Andersson found such a disk in the cave of Sha-kuo-t'un. Disks of a larger size were reported by Lukashkin from the Neolithic settlements at Cape Laoteshan near Port Arthur [Lüshun]. Similar round disks made of sandstone were found in the sixth stratum of the Mesolithic cave Dzhebel [Jebel] in Turkmenia.

The [short] triangular, lance-like knives of which V. S. Makarov has published illustrations are often encountered in Mongolia as well as in the Neolithic sites along the Shilka and the Amur and in the Cis-Baykal.

The axes from Ang-ang-hsi are similar to the Kitoy ones in their oval cross-section, but the general configuration and some details of treatment have a local, Tungpei character. The opinion has been expressed in print that the stone axes from the Neolithic sites at Ang-ang-hsi had no definite form, but the shape was determined in most cases by the natural configuration of the pebble. Polishing during the preparation of the axes was practiced only to a limited degree, according to this opinion, and basically only to sharpen the blade. The remaining surface was left unpolished or was only polished in what would be considered a very primitive manner—"with disorder, shallow, irregular strokes." It is impossible to agree with this opinion. The description of an adze from the collection of M. Ya. and V. F. Volkov testified to the fully developed and consummate technique of preparing polished hewing tools.

The axes from Ang-ang-hsi, as Teilhard de Chardin has already noted, are quite distinct from those of the Yangshao culture. They can be distinguished not only by the material used (slate pebbles rather than hard igneous rock), but above all typologically, which is the most important, and certainly not by any primitive method of polishing. Axes of the type prevalent at Ang-ang-hsi are not to be found in the Hwang Ho [Yellow river] valley, while, at the same time, axes, rectangular in cross-section, which are characteristic of Yangshao are absent from northern Tungpei.

Of course, the explanation is not connected with the type and level of economic development, since the Neolithic tribes of the Amur, Cis-Baykal, and Trans-Baykal related to those of Tungpei possessed at the same time large, heavy axes and adzes. Nor did the lack of appropriate material play any important part, as is generally said to be the case. The riddle may be explained on the basis of the environment in which the Neolithic tribes of Tungpei lived. This was basically a forested steppe with numerous lakes around which the hunting and fishing economy usually centered. The Neolithic population of Tungpei was composed of nomadic hunters, fishers, and gatherers whose economy was organized rather differently from that of the neighboring Primore, Amur, and Cis-Baykal which were covered with taiga. It was not necessary for them to make heavy dugouts or to construct wooden fences and pens for battue hunting.

In the Mongolian collections of F. Bergman there are no axes of the Tungpei type. The axes of Mongolia basically are chipped axes or, in the late stage of the Neolithic, fully polished. Axes that are partially polished have also traces of primary flaking. There are examples with dot flaking on the handle and a polished blade. Characteristic of Mongolia are axes with an outline that is rectangular or slightly tapered towards the butt, and with the broad surfaces almost entirely covered with long facets of diagonal retouching. To put it briefly, the axes and adzes of Mongolia are characterized in the majority of cases by preliminary and careful preparation of the blank with flaking, chipping, and dot [fish-scale] retouching.[44]

At the same time the axes of the Ang-ang-hsi region do not constitute a narrow local group. Licent discovered the same type of axe in southern Tungpei and in eastern Mongolia (Yufangt'ou, Pat'ou-yinchieh),[45] while J. G. Andersson found them in the Sha-kuo-t'un cave not far from the mouth of the Liao river, although the latter display certain individual peculiarities along with their general resemblance to the Ang-ang-hsi type.[46]

The large collection of stone axes gathered by Andersson and his Chinese assistants in the Chinese-Manchurian-Mongolian border territories (the provinces of Chihli, Chahar, and Jehol) is composed of a group of tools quite distinct from the axes of China proper. Andersson assigns to a separate group, which he calls the "northern type of ax," those which are characterized by an oval cross-section, tapered butt, and semicircular working edge.[47] In their basic traits, the Ang-ang-hsi axes belong to this group. The "northern axes" of Andersson appear to be identical with the Kitoy and especially the Glazkovo axes of the Cis-Baykal and Trans-Baykal.

The "microcelts" of nephrite noted by Lukashkin are especially interesting, first of all because of the very fact that tools of nephrite were found. This is undeniably an import, since nephrite is known from only two regions in Asia: Sinkiang and the Cis-Baykal, and possibly also the basin of the Vilyuy river. The Sinkiang nephrite occurs in dull, grayish tones. The white and milky-green nephrite came, for the most part, from the Cis-Baykal.[48] Through exchange it reached the banks of the Nonni, where they knew the value of tools made from such a strong and rare material. Not infrequently the "microcelts" were "ground to nothing" by repeated sharpening.

The bone inventory from the settlements about Ang-ang-hsi has a definitely northern look. It includes bone awls, needlecases of hollow bones with needles, and also a spoon [all of which] resemble similar implements from Kitoy and Glazkovo burials in the Cis-Baykal. One of the harpoons with a projection near its base, found in a burial by Liang Ssŭ-yung, recalls to a degree harpoons of the Kitoy period of the Cis-Baykal.

Northern connections are also indicated in the ornaments, namely the shell beads. Small ornaments of marble were found in Kitoy burials and these are similar to the little disks of shell found by Lukashkin. But even more similar are the small mother-of-pearl rings and disks from Glazkovo graves.[49] A. P. Okladnikov thinks that they were made from thick seashells rather than from the thin local river or lake shells. These seashells, and in some cases the beads themselves, were apparently obtained by the Cis-Baykal tribes from the Far East through trade with the Neolithic peoples living along the Shilka, Amur, and in Tungpei. The most numerous artifacts found by Andersson during the excavation of the cave of Sha-kuo-t'un were the small, flat, shell beads of just this type.

The fragment of a marble ring may be considered proof of close cultural ties between the Neolithic tribes of Tungpei, China proper, and the Cis-Baykal. Nephrite, marble, and clay rings are the ornaments most frequently encountered in the graves of the Kitoy and Glazkovo periods of the Cis-Baykal[50] and of the Yangshao and Yin periods of China.

Beads with the openings drilled at an angle are also occasionally found in the north. In a burial of the Glazkovo period in Yakutia, a shell with such openings was found, together with small beads of the Glazkovo type. However, there are more numerous analogies with such beads in the south and west. Andersson has found an interesting collection of beads and pendants in the Chu-chia-chai burial ground. Among these a group of shell beads which represent a curious imitation of the cowrie shell stands out. They have a flat undersurface, a bulging, rounded upper surface, and two openings which are not drilled along the axis of the bead, but at an angle.[51] The large stone beads from Sha-kuo-t'un cave have just such openings drilled at an angle into one of the flat surfaces.[52] Andersson describes them as buttons. One should also mention the bone beads from the Mariupol cemetery. M. Makarenko has already pointed out the resemblance between the beads from Mariupol, with their characteristic openings drilled at an angle, and the stone beads from the cave of Sha-kuo-t'un.[53]

Beads with holes drilled at an angle are not the only evidence of distant western connections for the Neolithic tribes of the river Nonni. The pebble flakers from Ang-ang-hsi confirm that the analogy with the Mariupol beads is not accidental. In spite of the variety of their exterior form—which is dictated to a large extent by the form of the pebble—they represent a distinct type of tool with consistent, recurring characteristics. Flakers of this type do not occur in the Neolithic of eastern Siberia. They are also unknown in the Mongolian Neolithic. The flat, disk-like flakers with semicircular working edge are only encountered in the Mesolithic cave of Dzhebel [Jebel] in Turkmenia,[54] and also in the early Tripolye settlement of Luka-Vrublevetskaya,[55] although in the latter case they are less flattened and are nearer to the spherical.

To sum up what has been said above, we may conclude that the majority of basic stone tools, bone artifacts, and ornaments have an undoubted Siberian-Mongolian appearance.

The Neolithic tribes of northern Tungpei resemble the tribes of the Cis-Baykal in their way of life, the conditions under which they live, and their economy. The stone and bone inventory from the settlements and burials proves that their basic occupations were hunting and fishing. The greatest quantity of the food refuse at the sites consists, according to Lukashkin, of fish bones, river mollusks, and wild animal bones. The basic occupations were reflected in the burial inventory, through which relatives of the deceased carefully provided him for his life in the next world. Just as in the Cis-Baykal, they buried the dead in pits dug in the earth, in a supine, extended position; beside him were placed bone harpoons, simple and inset knives, fish spears, bow and arrowheads, and so forth—typical, basic equipment of a hunter and fisherman. These traits can be appreciated if we compare them with burials of the farmers of the Yangshao culture, for instance, with such a clear-cut site as the Pan-shan burial ground. Here the deceased is usually accompanied not by weapons but by many richly decorated clay vessels, ornaments, and only rarely by stone axes of adzes. A comparison of these two types of burial clearly emphasizes two ways of life, two traditions, two types of culture sharply differing from each other.

However, in the inventory there are also implements which occur in the southern and eastern culture areas. In the collection of the Volkovs there are characteristic points of a Mousterian type. The basic distribution area for such points includes the Amur, Sakhalin, Soviet Primore, and northern Korea.[56] A knife with handle illustrated by V. S. Makarov (Fig. 15, item 4) also has a Primore character. Such knives were especially numerous in the settlement of Tetyukhe.* The characteristic dentated or sawtooth retouching (Fig. 7, item 7) is also evidence of ties with the east and south. This is the first evidence of ties between northern Tungpei and the southeast. The second piece of evidence of such ties is the presence in the sites of Ang-ang-hsi of "table-like" querns and grinders. Querns and grinders are not found in the Neolithic sites of eastern Siberia. A distinguishing feature of the Neolithic settlements of southern Tungpei, Mongolia, Korea, and our Soviet Primore is the frequent presence in them of boat-shaped or flat, "table-like" stone querns with grinders, mortars, and pestles. This feature was remarked upon by the first investigators of the Neolithic of Tungpei, Licent and Teilhard de Chardin. Querns and grinders are characteristic of the broad culture "belt" which stretches from the northwest to the northeast and includes the areas of the agricultural Neolithic of China, the areas of the Yangshao and Lungshan cultures. This belt is not limited to the regions immediately bordering the loess plateau of inner China, but reaches far to the north, as supported by the fact that Lukashkin found grinders, and Ponosov querns with grinders, in the Ang-ang-hsi area.

Grinders of the type called "semilunar" are the most archaic of those known from Tungpei and the maritime regions of East Asia. In a much later period there appear grinders of another type. They are more compact, but chiefly they have a broader grinding surface. Often the grinders are larger than the querns themselves, and then their central part is worn down, forming a ledge which separates the thicker, unworn ends from the flat, even grinding surface.[57] In the Soviet Primore we find the same picture; in the earlier sites of the Tetyukhe type are found archaic querns with semilunar grinders, while in the Shell Mound period the querns are not "plate-like," but boat-shaped, with heavy, compact grinders having a broad working surface.[58]

Lukashkin's finds of pestles and fragments of grinders of querns indicate a great deal: the Neolithic population of Ang-ang-hsi did not limit itself to the proceeds of the hunt, but was employed in extensive gathering, and activity foreign to the forest hunters and fishers of the Cis-Baykal and Yakutia. But there is one step between gathering and agriculture. And this step was taken by the Neolithic population of Tungpei, particularly in the southern reaches of Lin-hsia, Chu-chia-chai, and Hung-shan-hou [Tzŭ-ch'uan] where the important role was that of the Yangshao culture. But this subject should occupy a study of its own in which cultural remains from a far wider area would be analyzed and not only those from the region of the river Nonni.

Here it is important to emphasize the revolutionizing influence of the southeastern cultures and especially that of agricultural Neolithic China on the culture of northern Tungpei. This influence prepared a real transition in the economy as a result of which the culture of northern Tungpei was established as a specific culture area among northern cultures. Important facts unveil an authentic picture of the formation of such a culture area. Agriculture and other elements of the southeastern cultures did not spread to northern Tungpei simply as a result of contact between ancient tribes. The matter was a great deal more complicated. One must speak here of the penetration of the representatives of both cultures themselves into the territory of their neighbors, and of the resulting creation of

*[A coastal settlement ca. 200 miles northeast of Vladivostok.—Editor.]

a particular, hybrid culture which combined elements of both. This is the only possible explanation for the surprising and unusual fact that in the culture of northern Tungpei we find together the Siberian-Mongolian flint inventory, with the ceramics which are characteristic of the southern regions of Soviet Primore, Korea, and Liaotung. None of the types of vessels found in the burials near Ang-ang-hsi are encountered in Mongolia, the Cis-Baykal, Trans-Baykal, or the upper reaches of the Amur or Shilka. The deep, flat-bottomed cup with clear textile impressions and a spout for pouring has analogues in the maritime cultures. The spout is a trait characteristic of the eastern areas. Another cup with a narrow base and wide mouth, decorated along the rim with parallel rows of elongated impressions, also is reminiscent of cups from the southern Primore and northern Korea. The spherical, jug-like pot with a smooth surface and low, slightly flared rim has exact analogues in the ceramics from the settlement of Tan-to-tz'ŭ on Liaotung peninsula, excavated by Japanese archaeologists.

It would be difficult not to overvalue the meaning of the vessel found by Liang Ssŭ-yung in burial no. 4. In form it resembles that just described; it has a spherical body and almost vertical, low collar-rim. But most important is the decoration, which leaves no doubt about its southeastern connections. The broad decorative band encircling the pot, consisting of a pattern of prolonged zigzagging triangles, is well known from the Neolithic settlements of the southern Primore (Gladkaya I), North Korea (Nonpkhori), and the later stages of the Jomon culture of Japan. Everything is in agreement, beginning with the character of the composition (triangles alternately pointing up and down between the zigzagging bands), the technique of execution (incised double lines with slanting transverse hatching), and ending with such a characteristic detail as the incised semilunar figures inside the triangles, also filled in with diagonal hatching.[59] One must not, however, suppose that the tribes of northern Tungpei were only passive [receivers]. On the contrary, excavations of a large Neolithic burial ground near Yen-tz'ŭ in Hsiao-yin-tz'ŭ showed how strong was the return influence of this northern culture. Here are found the typical Cis-Baykal and Trans-Baykal stone and bone arrowheads, stone axes, tools with inset blades, bone daggers, spoons, pins, needles, and so on.[60] Here only the ceramics were local. There were no prismatic cores or knife-like flakes with straight edges. Thus the culture of northern Tungpei was combined as closely with the southeastern cultures as with the northern. In this it differs sharply from the Neolithic of Inner and Outer Mongolia where the Serovo and Glazkovo ceramics of the Cis-Baykal predominate.

All these distinctive features of the stone and ceramic inventory allow one to consider northern Tungpei during the Neolithic as an area culturally as well as ethnically unique. It is apparent that it is here that the earliest representatives of the Manchus lived.

We have not yet at our disposal enough material for a complete study of all aspects of the culture of the people who left the numerous remains in the Ang-ang-hsi area. We can only say that they were a hunting and fishing people who basically spent their lives near the rivers and numerous steppe lakes, which not only abounded in fish but also attracted wild game. Their stone and bone artifacts indicate a typical eastern Siberian culture. Their collection of tools was rich, varied, and highly specialized. At the same time they were also gatherers. Gathering took on greater and greater significance with the passage of time. Mention has already been made of numerous querns, grinders, and pestles for breaking up vegetable foods throughout a wide cultural zone, from the Primore in the east to the Gobi in the west.

The problem of dating the Neolithic sites of Ang-ang-hsi can only be posed against a broader historical background, by including materials from neighboring areas that have been studied more thoroughly—China, the Cis-Baykal, the Soviet Far East, Korea, and Japan. This was exactly the way the problem was first presented by the Chinese investigators Pei Wên-chung, Yin Ta, and Liang Ssǔ-yung. Yin Ta, in a short section devoted to the characteristics of the Neolithic culture north of the Great Chinese Wall, uses the materials of Liang Ssǔ-yung from Ang-ang-hsi as a basis, and writes that the Neolithic culture of Tungpei could have preceded that of Yangshao. But at the same time this does not exclude the possibility that it existed contemporaneously, and it is [also] possible that the Tungpei culture survived substantially longer than the Yangshao, since the outlying provinces were greatly retarded in their development in comparison with China proper, as is evident even now upon examination of an ethnographic map of Chinese frontier country. Yin Ta in this connection poses the question of the stages of tribal development, and presents a general analysis of the stage of cultural development attained by the Neolithic tribes of Tungpei.

Pei Wên-chung approaches the problem from a slightly different angle.[61] In reviewing the stages of development of the Tungpei Neolithic, he separates it into various stages corresponding to the level of technical and economic development. Pei Wên-chung writes in general terms of the relationship between the Neolithic of Tungpei and the Cis-Baykal. Assigning to himself the task of explaining the origins of the Tungpei Neolithic, he avers that at the end of the Paleolithic the Cis-Baykal tribes moved southward because of the onset of a cooler climatic era and reached the vicinity of Talai-Nor [Hu-lun-ch'ih]. He refutes the claim of Japanese archaeologists that the finds in the region of Talai-Nor belong to the Paleolithic and asserts that they are Mesolithic. These finds represent, according to Pei-Wên-chung, the earliest remains of the Tungpei Neolithic. He calls this stage the "Talai period."

The descendants of the "Talai" people then crossed the Khingan range and penetrated to the interior of Tungpei, where, in the Ang-ang-hsi region, are found remains of the next stage, the "Lungchiang period." The people of this stage practiced the same type of hunting and fishing, but they already had invented crude pots and had learned to polish stone. Pei Wên-chung ascribes the Lungchiang period to the early Neolithic, without attempting an exact chronological date. With this move the Neolithic tribes did not come to a standstill. They divided into two branches, one of which moved to the Sungari ([to it are ascribed] the Neolithic finds in the region of Kirin and Yenchi) and the other towards the southwest, as far as Lin-hsia and Inner Mongolia. Here the tribes encountered a culture with painted pottery in the vicinity of the Great Chinese Wall (Tzǔ-ch'uan in the Ch'ih-fêng basin). Pei Wên-chung ascribes this culture to the "Lin-hsia period" and judges it to be chronologically much later than the Lungchiang period. According to his periodization, this would be Middle Neolithic.

The Lin-hsia people, influenced by different geographic conditions, changed from hunting and fishing to agriculture. The remains from the so-called mixed culture, where the Yangshao type of painted pottery is found together with chipped tools (as at Sha-kuo-t'un, Chu-chia-chai, Ching-hai, and other sites), are ascribed to the late stage of the Neolithic. Upon encountering the painted pottery culture, part of the Neolithic Tungpei tribes moved westward, along the Gobi, to Sinkiang.

The mixed culture experienced the strong influence of the earlier Chinese

Neolithic and, later, of the Yin and Chou cultures, and from the other, western side, the Siberian-Scythian culture. According to Pei Wên-chung the mixed culture gradually formed into that of the different peoples of northwestern and northeastern China, peoples who are mentioned in the early Chinese chronicles.

The views that Yin Ta and Pei Wên-chung have presented about the sources of the Tungpei Neolithic and the role of the Neolithic remains of Ang-ang-hsi in the history of the North Chinese Neolithic are extremely interesting and deserve particularly careful scrutiny.

At present, however, we shall limit ourselves to one particular problem, that of the age of the Ang-ang-hsi remains. Yin Ta quite properly admits the possibility that the Tungpei Neolithic and the Yangshao culture may have existed contemporaneously. In his turn Pei Wên-chung asserts, with justification, that the Tungpei Neolithic was genetically related to the Cis-Baykal Neolithic. Taking cultural ties into account, one may confidently approach the question of relative as well as absolute dating of the Neolithic remains at Ang-ang-hsi. On the Cis-Baykal chronological scale these settlements come closest to the Serovo-Kitoy. Isolated Serovo elements, which were noted during our analysis of the stonework, occur in the Cis-Baykal as survivals into the Kitoy period. The stone discoid scrapers which appeared in the Cis-Baykal during Glazkovo times may have been adopted by the Glazkovo people from the Tungpei tribes; hence, their late appearance.

Dating the Neolithic remains in the Ang-ang-hsi area to the Kitoy period does not contradict their approximate dating as contemporaneous with the Yangshao of China proper, and with the culture of Soviet Primore which is best represented by the site of Gladkaya. Thus, the Neolithic sites near Ang-ang-hsi station may be dated within the limits of 2500 B.C. and 1800 B.C.

The wide scope of archaeological work now being carried on by Chinese archaeologists in the Tungpei region should allow us, in the near future, to draw a more complete picture of the traits of a Neolithic culture in one of the central [cultural] areas of northeastern Asia. This work should also lead towards the establishment of a more precise chronology, a division into developmental stages and their characteristics.

## Notes and References

1. M. Makarenko, *Mariyupilskiy mogilnik* (The Mariupol cemetery), Kiev, 1933, pp. 22–23, 47–48, Plates 7, 10, 20, 31, 23; A. P. Okladnikov, Neolit i bronzovyy vek Pribaykalya (The Neolithic and Bronze ages of the Cis-Baykal), Istoriko-arkheologicheskoye issledovaniye, pts. 1 and 2, MIA, no. 18, 1950, pp. 134, 136.

2. N. K. Auerbakh, Paleoliticheskaya stoyanka Afontova gora III (The Paleolithic site Afontova Gora [mountain] III), *Trudy Obshchestva izucheniya Sibiri i yeye proizvoditelnykh sil*, no. 3, Novosibirsk, 1930, p. 41; G. F. Debets, Opyt vydeleniya kulturnykh kompleksov v neolite Pribaykalya (The problem of periodization of the culture complexes of the Cis-Baykal Neolithic), *Izvestiya Assotsiatsii nauchno-issledovatelskikh institutov pri fiziko-matematicheskom fakultete MGU*, vol. III, no. 2-A, 1930, pp. 157–158, 161–164; G. P. Sosnovskiy, (1) Sledy prebyvaniya paleoliticheskogo cheloveka v Zabaykalye (Traces of the presence of Paleolithic man in the Trans-Baykal), *Trudy Komissii po izucheniyu chetvertichnogo perioda Akademii nauk SSSR*, vol. 3, 1933; (2) Paleoliticheskiye stoyanki Severnoy Azii (Paleolithic sites of northern Asia), *Trudy II Mezhdunarodnoy konferentsii Asssotsiatsii po issledovaniyu chetvertichnogo perioda*, vol. 5, 1934; A. P. Okladnikov, (1) *Neolit i bronzovyy vek Pribaykalya*, pp. 141–156; (2) Proshloye Yakutii do prisoyedineniya k Russkomu gosudarstvu (The

history of Yakutia before its union with the Russian state), in: *Istoriya Yakutskoy ASSR*, vol. 1, Moscow-Leningrad, 1955, pp. 56–71; S. N. Zamyatnin, O vozniknovenii lokalnykh razlichiy v kulture paleoliticheskogo perioda (On the origin of local differences in the cultures of the Paleolithic period), *Trudy Instituta ethnografii Akademii nauk SSSR*, n.s., vol. 16; *idem*, Proiskhozhdeniye cheloveka i drevneye rasseleniye chelovechestva (The origin of man and the distribution of mankind in antiquity), 1951 [?]; *Vsemirnaya istoriya*, vol. 1, Moscow, 1955, pp. 78–81.

3. A. S. Lukashkin, New data on the Neolithic culture of northern Manchuria, *The China Journal*, vol. 15, no. 4 (October), Shanghai, 1931. See also: A. S. Lukashkin, New data on the Neolithic culture in northern Manchuria—A report on some excavation work in the valley of the Nonni river near Tsitsihar station on the Chinese East Railway, *Bulletin of the Geological Society of China*, vol. 11, no. 2, Peiping, 1931; *idem*, Issledovaniye neoliticheskikh stoyanok bliz stantsii Tsitsikar (Investigations of the Neolithic sites near Tsitsihar station), *Vestnik Manchzhurii*, no. 3, Harbin, 1934; Yin Ta, Neolithic cultures of China, Peking, 1955, p. 13 (in Chinese); An Chih-min, Microlithic cultures, *K'ao-ku tung-hsin*, no. 2, 1957 (in Chinese).

4. Liang Ssŭ-yung, The prehistoric site Ang-ang-hsi, *Byulleten Nauchno-issledovatelskogo instituta istorii i filologii*, vol. 4, no. 1 (in Chinese). See also: Pei Wên-chung, *Cultures of the Stone Age in China*, Peking, 1954, p. 4, Plate 10; Yin Ta, Neolithic cultures of China, p. 15; An Chih-min, Microlithic cultures; P. Teilhard de Chardin and Pei Wên-chung, Le néolitique de la Chine, *Institut de géobiologie*, no. 10, Peking, 1944, p. 39; Lukashkin, New data . . . , *Bulletin of the Geological Society of China*, vol. 11, no. 2, pp. 180–181: Lukashkin, Issledovaniye neoliticheskikh . . . , p. 21.

5. V. S. Makarov, Novye dannye o neoliticheskoy kulture rayona stantsii Anantsi (New data on the Neolithic culture area of Ang-ang-hsi station), *Zapiski Kharbinskogo obshchestva yestestvoispytateley i etnografov*, no. 8, Harbin, 1950.

6. Yin Ta, Neolithic cultures of China, p. 20. Yin Ta cites the following conclusion of Liang-Ssŭ-yung: "Basing ourselves on geological studies we may conclude that the sand dunes in the region of Ang-ang-hsi were formerly the shore accumulations of a large lake. The remains uncovered by us appear to be the remains of a shore culture."

7. The principal data on each of the sites to be described here are taken from the works devoted to the Neolithic settlements in the region of Ang-ang-hsi station and cited above.

8. Teilhard de Chardin and Pei Wên-chung, Le néolithique de la Chine.

9. E. Licent, Les collections néolithiques du Musée Hoang Ho et Pai Ho de Tientsin, *Publications du Musée Hoang Ho Pai Ho de Tientsin*, no. 14, 1932.

10. Personal communication and diary of A. P. Okladnikov.

11. I take this opportunity to express my gratitude to A. N. and A. A. Fromozov and to A. P. Okladnikov for making possible my publication of this material.

12. Okladnikov, Neolit i bronzovyy vek Pribaykalya, p. 199.

13. Lukashkin, Issledovaniye neoliticheskikh stoyanok . . . , p. 13, Plate 1, item 10.

14. Teilhard de Chardin and Pei Wên-chung, Le néolithique de la Chine, pp. 186–187, Plate 1, items 6, 7.

15. Lukashkin, Issledovaniye neoliticheskikh stoyanok . . . , p. 13, Plate 2, item 6. See also: Teilhard de Chardin and Pei Wên-chung, Le néolithique de la Chine, p. 185, Fig. 1, item c.

16. Lukashkin, Issledovaniye . . . , p. 13, Plate 1, item 11; Plate 2, item 2. See also: Teilhard de Chardin et Pei Wên-chung, Le néolithique de la Chine, p. 187, Fig. 2, item A.

17. Teilhard de Chardin and Pei Wên-chung, Le néolithique de la Chine, p. 187, Fig. 3, items A–D; Plate 2, items 9, 10; Lukashkin, Issledovaniye . . . , pp. 13–15.

18. Teilhard de Chardin and Pei Wên-chung, Le néolithique de la Chine, p. 191, Fig. 7, items A, B. See also Lukashkin: Issledovaniye . . . , p. 17, Plate 3, items 14, 15, and p. 18; *idem*, New data . . . , *The China Journal*, vol. 15, no. 4, Plate 1.

19. Cf. J. G. Andersson, The cave deposit at Sha-kuo-t'un in Fêngtien, *Paleontologia Sinica*, ser. D, vol. 1, Peking, 1923, p. 11, Plate 6, item 4.

20. Teilhard de Chardin et Pei Wên-chung, Le néolithique de la Chine, p. 190, Fig. 5, Plate 2, items 2, 12–14, 17. See also: Lukashkin, Issledovaniye . . . , Plate 2, Fig. 8, items *a, b, c.*

21. Lukashkin, New data . . . , *The China Journal,* vol. 15, no. 4; *idem,* Issledovaniye . . . , Plate 1.

22. V. V. Ponosov, Agriculture and cattle breeding in North Manchuria in the Stone Age period, Reprinted from the *Bulletin of the Institute of Scientific Research, Manchoukuo,* vol. 1, no. 3, pp. 163–166, Fig. 2.

23. Teilhard de Chardin and Pei Wên-chung, Le néolithique de la Chine, p. 185; Lukashkin, New data . . . , *Bulletin of the Geological Society of China,* vol. 11, no. 2, pp. 177–178, Plate 2; see also illustrations in the text; Lukashkin, Issledovaniye . . . , p. 12, Plate 1.

24. Tolmachev, Predmety "kostyanogo veka" iz Vostochnoy Sibiri (Objects from the "bone age" of Eastern Siberia), *Soobshcheniya GAIMK,* no. 2, 1929, pp. 334–338, Fig. 1.

25. The terms "microlithic culture" and "microlithic tools" in their real sense are not applicable to the cultures of Central Asia and northern Tungpei, since not a single tool of the geometric form characteristic of the true microlithic cultures of the West has been found in these places. If other traits are to be added to the characteristics defining the microlithic culture, i.e. the geometric form and specific technique of fashioning, then the realm of microlithic cultures must include eastern and western Siberia, not to mention other more remote areas where prismatic and conical cores and the thin, knife-like flakes stuck from them are found. The use of the term "microlithic culture" when referring to the early cultures of Mongolia and Tungpei should be dropped in order to avoid confusion. The irresponsible use of this term only leads to unfounded generalizations and unaccountable conclusions.

26. Teilhard de Chardin, Some observations on the archaeological material collected by Mr. Lukashkin near Tsitsihar [source?], p. 189, Plate 2.

27. Makarov, Novye dannye o neoliticheskoy kulture rayona stantsii Anantsi, p. 30, Plate 2, nos. 9, 10.

28. Licent, Les collection néolithiques du Musée Hoang Ho et Pai Ho de Tientsin, Pates 29, 30.

29. J. Maringer, *Contribution to the prehistory of Mongolia,* Stockholm, 1950, Plate 26.

30. B. E. Petri, Neoliticheskiye nakhodki na beregu Baykala (Neolithic finds on the shores of Lake Baykal), *Sbornik MAE,* vol. 2, 1916, p. 132, Plate 12.

31. Maringer, *Contribution to the prehistory of Mongolia,* Plate 21, item 1; Plate 23, item 14; Plate 25, items 7, 8; Plate 29, items 9 and 10.

32. R. Torii, Populations préhistorique de la Manchourie méridionale, *Journal of the College of Sciences, Tokyo Imperial University,* vol. 36, art. 8, 1915; R. Torii and K. Torii, Populations primitives de la Mongolie Orientale, *ibid.* vol. 36, art. 4, 1914.

33. Licent, Les collections néolithiques du Musée Hoang Ho et Pai Ho de Tien Tsin. See also Andersson, The cave deposit Sha-kuo-t'un in Fêngtien; K. Hamada, Huug-Shan-Hou, Ch'ih-Fêng, *Archaeologia Orientalis,* ser. A, vol. 6, 1938.

34. I. Yawata, (1) Contribution to the prehistoric archaeology of South Jehol, *Reports of the First Scientific Expedition to Manchoukuo,* sect. 6, pt. 1, Tokyo, 1935, pp. 1–106; (2) Contribution to the prehistoric archaeology of Northern Jehol, *ibid.* sect. 6, pt. 3, Tokyo, 1940.

35. Liang Ssŭ-yung, (1) The prehistoric site Ang-ang-hsi; (2) Neolithic remains from the sites of Ch'ih-fêng, Shuang-ch'eng, Lin-hsia, Sh'a-pu-kan in Jehol province, *Tien'e k'ao-ku pao-k'ao,* vol. 1, Shanghai, 1936, pp. 1–68 [in Chinese].

36. Tung Chu-ch'ên, Three types of Neolithic cultures of Kirin, *K'ao-ku hsüeh-pao,* 1957, no. 3, pp. 31–38 [in Chinese]; Ching Sang-ling, Report on the investigations of prehistoric cultural remains along the river It'ung in Ch'ang-ch'un, *Hsia-mên ta-hsüeh hsüeh-pao,* 1954, no. 1 [in Chinese]; Pang Hêng-chieh, Neolithic settlements in the environs of Ch'ang-ch'un, *K'ao-ku tung-hsin,* 1957, no. 1, pp. 6–10 [in Chinese].

37. Tung Chu-ch'ên, Three types of Neolithic cultures . . . ; V. V. Ponosov, Prehistorical culture of Eastern Manchuria, *Bulletin of the Institute of Scientific Research, Manchoukuo,* vol. 2, no. 3, Hsinking, 1938, pp. 337–346; Fusida, Report on investigations in Hsiao-yin-tz'ŭ, Yen-chi), vol. 1, Ch'ang-ch'un, 1943 [in Japanese]; Watanabe Hiroshi, On the methods of striking off flakes and the so-called stone knives, *Ningtsai,* 11, 1948 (in Japanese). Watanabe included in his report a map of the distribution of knife-like flakes in northern and eastern Asia.

38. Hwan Gi Dok, Stone Age sites and remains in the province of North Hamgyŏng, *Munkhva yusan,* 1957, no. 1, pp. 72–102 (in Korean). The material published by Hwan Gi Dok does not relate to "microliths."

39. Pi-tzŭ-wo, Prehistorical sites by the river Pi-Liu-Ho, South Manchuria, *Archaeologia Orientalis,* vol. 1, 1929;* Mu-yang-chêng, Han and pre-Han sites at the foot of Mount Lao-t'ien in South Manchuria, *Archaeologia Orientalis,* vol. 2, 1931;* Torii, Populations préhistoriques de la Manchourie méridionale.

40. A. P. Okladnikov, Drevneyshiye kultury Primorya v svete issledovaniy 1953–1956 gg. (The early cultures of the Primore in the light of investigations in 1953–1956), *Sbornik statey po istorii Dalnego Vostoka,* Moscow, 1958, p. 63.

41. Okladnikov, Neolit i bronzovyy vek Pribaykalya, p. 361, Figs. 107–108.

42. *Ibid.,* p. 235, Fig. 69; p. 239, Fig. 71; p. 315, Fig. 109; p. 366, Fig. 110.

43. Maringer, *Contribution to the prehistory of Mongolia,* Plate 30, items 1, 2, 4, 5.

44. *Ibid.,* pp. 188–193, 205, Plates 35, 36.

45. Licent, Les collections néolithiques du Musée Hoang Ho et Pai Ho de Tientsin, Plate 38, items 1, 10; Plate 45, items 1, 4.

46. Andersson, The cave deposit at Sha-kuo-t'un in Fêngtien.

47. J. G. Andersson, Researches into the prehistory of the Chinese, *Bulletin of the Museum of Far Eastern Antiquities,* no. 15, Stockholm, 1943, pp. 46–47, Plates 12, 13.

48. A. P. Okladnikov, Neolit i bronzovyy vek Pribaykalya, pt. 3, *MIA,* no. 43, 1955, pp. 174–189.

49. *Ibid.,* pp. 189–200, Figs. 82, 73, 102.

50. *Ibid.,* pts. 1–2, Fig. 122; pt. 3, Figs. 75, 76, 78, 79, Plate 6, Fig. 130.

51. J. G. Andersson, The site of Chu-chia-chai, *Bulletin of the Museum of Far Eastern Antiquities,* no. 17, Stockholm, 1945, pp. 61–62, Plate 27.

52. Andersson, The cave deposit at Sha-kuo-t'un in Fêngtien, Plate 8.

53. Makarenko, *Mariyupilskiy mogilnik,* pp. 47–48, Plates 38, 39.

54. A. P. Okladnikov, Peshchera Dzhebel—pamyatnik drevney kultury Prikaspiyskikh plemen Turkmenii (Dzhebel cave; remains of an early culture of the Cis-Caspian tribes of Turkmenia), *Trudy Yuzhno-Turkmenskoy arkheologicheskoy kompleksnoy ekspeditsii,* vol. 7, Ashkhabad, 1956, p. 26, Fig. 11, items 4, 5.

55. S. P. Bibikov, Poseleniye Luka-Vrublevetskaya (The site of Luka-Vrublevetskaya), *MIA,* no. 38, 1953, p. 293, Plate 1.

56. Okladnikov, Drevneyshiye kultury Primorya . . . , p. 26, Fig. 4.

57. A. P. Okladnikov, Archeologicheskiye issledovaniya v Primorye 1953 (The 1953 archaeological investigations in the Primore), *Soobshcheniya Dalnevostochnogo filiala im. V. L. Komarova Akademii nauk SSSR,* no. 8, Vladivostok, 1955, pp. 3–13; *idem,* Primorya v I tysyacheletii do n. e. Po materialam poseleniy c rakovinnymi kuchami (The Primore in the 1st millennium B.C. Materials from the shell mound settlements), *Sovetskaya arkheologiya,* no. 26, 1956, pp. 54–96.

58. Okladnikov, Drevneyshiye kultury Primorya . . . , p. 50.

59. *Ibid.,* pp. 27–42.

60. Fusida, Report on investigations. . . .

61. In addition to the work of Pei Wên-chung cited earlier, see also: Pei Wên-chung, The problem of the microlithic culture of China, *Yen-chi hsüeh-pao,* no. 33, 1947, pp. 1–7 (in Chinese).

*[Transliterated from the original. The first word in both entries indicates a place-name rather than a name of a person.—Editor.]

V. Ye. LARICHEV

# ANCIENT CULTURES OF NORTHERN CHINA*

IN THE HISTORY of relations between ancient China and surrounding peoples northeastern China (Tungpei) occupies a special place. From the time of the first dynastic chronicles to the late Middle Ages, the attention of Chinese historians is invariably drawn to their northern neighbors—the Tungus-Manchu and the Turkic-Mongolian tribes of Tungpei. Before their eyes the early "barbarian" Su-shêng, I-lu, and Moho tribes became a formidable power which played a considerable role in the historical destiny of China. Not without reason does the Great Wall twist over the mountains along the entire [northern] border of China proper. Since Han times this wall deterred the pressure first of the "barbarian" Tung-hu, Tung-i, and Huns, and later of the P'o-hai, Liao, and finally the Chin states. Since ancient times it had represented the border between two worlds and two cultures—the cattle-breeding nomads of the north and one of the most ancient agricultural civilizations in the world, that of Central China. Historically, these two parts of China are difficult to conceive without each other.

Along its other borders Tungpei is contiguous with the eastern and southeastern regions of the Soviet Union. Through the basins of the Amur and Argun and the steppes and deserts of Mongolia this part of China is connected with the Trans-Baykal and Cis-Baykal regions. To the east and northeast through the Ussuri and Sungari basins, the Suifen valley and Lake Khanka, Tungpei is connected with the Maritime Province [Primore] and northern Korea. From antiquity the tribes of Tungpei and Inner Mongolia have been in very close contact with the population of eastern Siberia and our [Soviet] Far East.

The subject of interrelations among these cultural regions in ancient times is usually confined to the first mention in the chroniclers of Chinese dynasties. The literature for the period of more than one and a half thousand years that preceded the history of the Neolithic and Bronze ages in China is meager. Yet, already during the Neolithic, large local centers became clearly defined in the extensive area of northern China, and during subsequent millennia a distinctive development of culture took place in them, and later out of these cultures one of the greatest civilizations on earth evolved. The relationships among these cultures differed. Their influence can be found far beyond the boundaries of China. In turn, there can be found in them the reciprocal influence of countries far away from China. In summing up the voluminous material accumulated over recent decades, we may separate, within the territory of northern China, three cultural and historical areas, three large groups of Neolithic remains.

The first Neolithic culture occupies the loess regions of northern China. It is known as the Yangshao culture or the Painted Pottery culture.[1]

Just as the landscapes of the loess plateau are inimitable in the rest of China, so are its unique Neolithic remains. This first Neolithic culture turned out to be so surprisingly highly developed that its first discoverer, J. G. Andersson, initially refused to admit to himself its Neolithic aspects.

*Translated from Akademiya nauk SSSR, Sibirskoye otdeleniye. Dalnevostochnyy filial imeni V. L. Komarova, *Trudy*, seriya istoricheskaya, vol. 1, 1959, pp. 75–95.

The Yangshao culture is basically an agricultural one. Its principal settlements are characterized by their large size and rich cultural layer sometimes reaching the thickness of 4 or 5 meters. One such typical settlement has been discovered and has been under excavation by Chinese archaeologists for the past few years in the village of Pan-p'o-tsun in the vicinity of Sian in Shensi (see Fig. 1).

FIGURE 1. Semisubterranean dwelling at Pan-p'o-tsun.

In this settlement both round and rectangular dwellings with walls of baked clay plastered with clay mixed with straw and with a centrally located hearth were found. The roofs were made with rafters set close together and also covered with clay mixed with straw. The rafters were tied together with hemp or grass ropes. The roof rested on upright posts inside the dwelling. Around each such dwelling were several pit-cellars. In one of the pits Chinese archaeologists have found a clay jug with bristle grass [*Setaria viridis*] and in another a heap of grain husks. Andersson also came across the imprints of rice kernels on fragments of pottery. Specific vessels of the *yen* type—steam cookers consisting of two vessels—also are evidence that foods were prepared from rice, bristle grass, and barley. Steam was produced in the lower vessel, and entered the second, upper vessel filled with grain through holes in the bottom.

That agriculture was a basic occupation is evidenced by the stone tools, among which the first and most important place is occupied by the specialized tools for tillage, the narrow, long plough, shouldered ax-hoes, small stone plummets for digging sticks. Often querns, grinders, and slate and clay sickles are also found.

The most striking trait of the stone artifacts of the Yangshao culture, apart from their typological uniqueness, is the prevalence of polished tools. The people of this culture did not know the techniques of flaking and chipping. In manufacturing axes, spear points, and arrowheads, and ploughs, sickles, and punch awls, they used the grinding method. This demonstrates the specific cultural traditions and sources of the Yangshao culture, which are not related to the north where the percussion technique was prevalent but to the south and the eastern maritime regions of China.

The well-developed agriculture of the Yangshao culture was combined with the taming and breeding of domestic animals, particularly of dogs and swine. At the same time hunting and fishing were highly developed. For instance, in the Pan-p'o-tsun settlement there were found large numbers of stone and bone spear points and arrowheads, fish hooks and harpoons fashioned out of bone, and small stone sinkers for nets. The importance of fish and wild animals in the lives of the inhabitants of Pan-p'o-tsun is evidenced by the find of clay vessels with drawings of fish, deer, and other animals, which apparently were subjects of special reverence or of a cult. Possibly, there also existed a primitive type of orcharding because around the hearths and about the dwellings many fruit pits were uncovered.

Spinning, weaving, and pottery-making were additional economic endeavors.

Special kilns were used for firing vessels. East of Pan-p'o-tsun two kinds of kilns were discovered, single-tiered and double-tiered. The first, simpler type consisted of a horizontal tube; at one end of it fuel was burned and the other served as a chamber for the unbaked articles. Such a kiln was less than a [cubic] meter in size. In them, according to Chinese archaeologists, one could fire five or six articles of medium size or two or three articles of large size. The second type of kiln is larger. It consisted of a pit for fuel, a horizontal stone floor, through the openings of which passed the gases [and heat] of combustion, and an upper chamber where the unbaked vessels were placed (Fig. 2).

The abundance and diversity of clay artifacts is unusual, At Pan-p'o-tsun there were between 40 and 50 varieties. Among them the most striking were vessels of the *li*, *tou*, and *ting* types. With the general upsurge of agriculture is coupled the

FIGURE 2. Kilns for the firing of pottery at Pan-p'o-tsun.

development of art. The painted pottery is a striking example of this. The beautiful funeral urns from the burial grounds of this culture, embellished with spiral and geometrical ornaments in the forms of squares, checkerboards, triangles, and arcs, are widely known. This ornamentation was painted on the light-brown background of the vessels with black and purple-red colors. The fine artistic taste of the master craftsman is seen in everything—the color scheme, its relationship to the principal background color and to the shape of the vessel, and in the artistic combination of lines and spirals. Together with the geometrical, "abstract" ornaments (associated in the opinion of some with the cult of the dead and concepts of the otherworld, the sky, the elemental forces of nature, the cult of fertility),[2] animals and occasionally human figures executed in a simple, realistic manner are depicted on some vessels.

The beginnings of sculpture are contemporaneous with the above. Among the finds of Andersson there are two lids in the form of human busts. Details of the human face of one of the lids are done in black—the characteristically narrow, slanting eyes, a broad, flattened nose, narrow lines for the lips, and a thin beard can be distinguished very clearly. The head is shaved and only from its top falls a wavy streak, representing a queue or braid. There is no doubt that it is a portrait. The second figure represents a schematized full-length figure of a man. The legs, arms, and head are well formed.

Besides the settlements there are also some burial grounds that have yielded particularly valuable materials. They allow a reconstruction of the funeral rite and, what is particularly important, the establishment of the physical type of the Stone Age people of China. Besides the well-known burial grounds excavated by Andersson, Chinese archaeologists have of late excavated a number of new burial sites. Of particular interest is the burial ground near the Pan-p'o-tsun settlement. It is separated from the settlement by a ditch. Here, in rectangular pits, adults were buried, most pits containing only one dead. Only two cases of group burials were uncovered—one of two, another of four. The dead were laid out in a well-defined position—head to the west and face up (Fig. 3). In fifteen burials they lay face down. Characteristically there were also "secondary burials," in which the corpses had been moved to another place, after some time. Dead children were put in large, jug-like clay coffins, covered by finely worked cups or pans. They were not buried in the cemetery (where only adults were interred) but close to the dwellings. The majority of such burials were group burials; in all more than 50 jug-like coffins were found.

Among articles placed with the dead, clay vessels with beautiful drawings and varied ornamentation are found most frequently. Not a single burial contained tools.

The large burial ground of Pan-p'o-tsun is an outstanding example of a clan cemetery—all of its graves lie in orderly rows oriented from east to west or north to south.

Research in physical anthropology conducted by Black, based on materials acquired by Andersson from burial grounds, resulted in the extremely important conclusion of the essential a likeness of the people of Yangshao and the present population of China. Thus, the Yangshao people were the Neolithic ancestors of the present-day Chinese.[3]

The second, but not less distinctive, Neolithic culture of northern China is connected with another outstanding physical geographical region—the Great Plain. [To the west] this lowland is sharply defined by the continental ledge of a hilly loess plateau. In the north, numerous mountain ranges separate it from

FIGURE 3. Collective grave in the burial ground near Pan-p'o-tsun.

the lowlands of Sung-Liao [the Manchurian Plain]. The southern regions of China are separated from the North China Plain by the Huai-yang Shan [range], the Yangtse river, and the ancient, much-worn South China Mountains.

In this geographically closed region of lowlands, occupied by the provinces of Honan and Shantung, towards the end of the 1920's, a well-known Chinese archaeologist, Wu Gin-ding, discovered a new Neolithic culture, which he called Lungshan (after the site of the first finds near the town of Lungshan).[4] In one respect it differs strongly from the Yangshao culture. There is no painted pottery, but polished black vessels with eggshell-thin walls predominate. Their ornamentation consists mostly of parallel [appliqué] ribs. Sometimes it consists of incised lines or stamped imprints producing a pseudo-textile pattern. The shapes are no less varied than in the Yangshao culture. There are also such specific types as the *li, tou,* and *ting,* but with their own characteristic profile and general shape.

In size, thickness, and content of the cultural layers the settlements are not inferior to those of the Yangshao culture. Some of them are fortified by walls made of compactly pressed layers of clay mixed with stones. The site of Ch'êng-tz'ŭ-yai, for instance, is surrounded on four sides by walls 12 m thick and up to 5 m high. In the northern districts of Honan the sites contain round, semi-subterranean dwellings with clay stoves and chimneys. Around the dwellings are found numerous storage pits.

On the whole the Lungshan culture is characterized by well-developed agriculture, in which together with the cultivation of cereals cattle breeding was practiced. Besides dogs and swine the Lungshan people began to breed sheep, large horned cattle, and horses. The cattle and horses may have been used for agricultural work. In general, the Lungshan culture gives the impression of being more developed than the Yangshao. It resembles the Yin culture (e.g., in some essential elements in the distinctive clay tamping technique in building walls, fortune telling from bones, clay vessels resembling the Yin bronze vessels).

With these various differences between the Yangshao and Lungshan cultures there are also many similarities. At some Yangshao sites Andersson has singled out black ceramics which are no different from those of Lungshan. Many types of vessels, among them a number of special type, are similar.

The articles of stone—axes, sickles, and arrowheads—do not differ in type from those of Yangshao. They too are made by the grinding technique.

All this testifies not only to close cultural contacts but also to close, apparently congeneric, relationships between the two cultures. However, until extensive and specialized excavations of the Lungshan sites are carried out, the problem of cultural, ethnical, as well as chronological relationships between the two greatest Neolithic cultures of eastern Asia will remain unresolved. Its solution is especially important because of its connection with even deeper historical problems of the origin of the slave-owning state of the Shang or Yin dynasty.

In the summary works dealing with the Stone Age of northern China, questions concerning the well-known and more thoroughly studied Yangshao and Lungshan cultures are analyzed best.[5] However, up to this time there remains in the background a third culture, no less important for the understanding of the history of northeastern Asia—the Neolithic culture of Tungpei. It is also related to a distinctive geographical region—the Sung-Liao lowland [Manchurian Plain], surrounded on all sides by mountains. The Tungpei culture is characterized by the predominance of flaking and pressure retouching techniques used in the manufacture of stone tools, techniques unknown to the southern Neolithic cultures of China. The tools usually have archaic traits or shapes (Fig. 5). Here belong the knives of Mousterian form found in almost all Tungpei sites, the triangular knives and inset-blades, shouldered punches, flat-plated [lamellar] tools with steep, almost blunting retouching, core scrapers of the Gobi type, and arrowheads with asymmetrical barbs. Most of these tools are characterstic of all subarctic cultures and are distributed from Mongolia in the southwest to the Chukchi peninsula in the northeast. This very important point demonstrates that the Neolithic of Tungpei, Mongolia, and eastern Siberia apparently had been preceded in all these regions by the considerably more ancient Mesolithic and perhaps even older Upper Paleolithic cultures. The limited number of stone tools common to all sites of the Tungpei Neolithic includes oval and flat-oval axes and adzes, as well as wedge-shaped gouges with thick butts. Because of the similarities in the stone inventory, the Chinese investigators are correct in including all Neolithic remains of Tungpei and Inner Mongolia in one so-called "microlithic" culture. However, with all their indisputable general similarity they can be divided into four clearly defined local groups.

The first group comprises Neolithic finds of Inner Mongolia south of the Shara-Muren [Hsiliao river]. This is the Neolithic of Jehol and Liaoning provinces. The most important remains of this culture are in the sites of Ch'ih-fêng [Hung-shan], Kao-chia-yin-tz'ŭ, Ta-t'ung, Yin-t'a-kou-ts'un, Wa-p'ên-yao-ts'un, Pao-t'ou [Pao-tow], Sha-kuo-t'un, and in part Lin-hsia. It is characterized primarily by the extensive spreading of painted ceramics, most outstanding examples of which have been found at the site of Hung-shan. Its culture-yielding zone constitutes the core of the Tungpei Neolithic and on the basis of it alone one could establish the distinctness of this culture.

Although the sites of the Hung-shan—Sha-kuo-t'un type exhibit many original traits, the presence of painted ceramics and a similarity in certain shapes of vessels bring them close to the Yangshao Neolithic. Among the various ornamentations that are close to those of Yangshao are patterns of spirals, curved lines, checkered rhombic designs, and slating lines with volutes. There is also a similarity in such details of the vessels as handles and the imprint of matting on the base. Further, these remains are brought closer to the Yangshao by the stone tools (rectangular slate sickles and axes rectangular in cross-section). The ornaments of the Huang type made of clay, shell, or nephrite as well as the cylindrical beads are also identical. At the same time many of the painted ceramics are strikingly distinctive. Among them are the deep cups, a vessel of the *tou* type, and an amphora-like vessel with a low neck unusually elongated in comparison with the known specimens of the Yangshao culture. The patterns with which the majority of the vessels are ornamented represent thoroughly altered and sometimes fundamentally reworked designs of the Yangshao culture. These are flattened voluted triangles, combinations of elongated rhombs, wide horizontal belts, and concentric arcs.

The second leading characteristic of these vessels, removing them even further from the Yangshao, is the presence of high, [both] narrow and wide vessels of a truncated conical shape. They are decorated with a specific incised or combed ornament in the form of parallel horizontal or vertical zigzags. Just as characteristic of this region are jar-shaped vessels, decorated with linear ornamentation forming shaded triangles and a particular kind of netting of intersecting lines. Exceptionally interesting is the presence of polished and flaked tilling tools, the great majority of which are not characteristic or are very seldom found in Yangshao sites. Among these are wide, cleaver-like, shouldered hoes, rectangular or trapezoido-triangular hoes with characteristic grooves for the butt, wide, heavy hoes with a concave [bent] working edge which the Chinese investigators call stone shovels. Widespread also are stone ploughshare points, slate or shale sickles of semilunar shape, querns, pounders, grinders, and mortars.

In general, the combination of painted ceramics and the so-called microlithic artifacts found in the majority of the largest sites of the cultural area is original and unique. Such a combination of the most striking cultural elements of northern and southern areas of eastern Asia came about as a result of the overlapping and [therefore] close contact between two cultures—the hunters and food gatherers of the north and the farmers of the south. This resulted in something basically new and qualitatively sharply different from both. Distinctive reworking of the achievements of a hunting and cattle-breeding culture and an agricultural one led to the rise, in the south of Tungpei, of a new agricultural region with highly developed farming. This is why such an outstanding development as the plough arose among the tribes of Tungpei during the Stone Age. A great upheaval in the economy occurred with the creation of a plough culture, with the use of domesticated animals. The rise of plough agriculture in the Neolithic culture of Tungpei extends in significance far beyond its borders. This development puts the question of the rise of plough agriculture in China and in northeastern Asia in general on a completely different basis. Before these discoveries, the plough was surmised to be a rather late cultural borrowing from the west. Now we know that the plough had appeared in China already during the Stone Age and that its appearance was not connected with borrowing but with a propitious development of extensive local cultural contacts.

The ties of the Jehol and Liaoning Neolithic with that of Lungshan are more

difficult to trace. But that such ties undoubtedly existed is documented by fragments of the gray and dark gray vessels with thin walls; these are a constant
discovery at the sites. The only complete vessel found was a vase with a high
stem[-leg] found at Sha-kuo-t'un, a site characteristic of the Lungshan culture.
In stone tools these contacts are evidenced by axes, rectangular in cross section,
and semilunar sickles.

The Jehol and Liaoning Neolithic and the Neolithic of northern Korea and the
early remains of the southern part of the Maritime Province [Primore] are considerably closer in their cultural and apparently ethnic relationships. In the latter
places we find the characteristic truncated conical and jar-shaped vessels and
cups. These are ornamented with incised vertical and horizontal zigzags. There
are fragments of vessels with stamped arched bows. The stone artifacts include
a great number of ploughshare points found in northern Korea. Cultural contacts
also may be traced farther to the east and northeast, to Japan, the lower Amur,
and northern Sakhalin, through the truncated conical vessels and zigzag type of
ornament.

A completely different aspect of the Tungpei Neolithic is presented by the
outlying districts. Here we find a distinctive trend of a historical development
related to the Stone Age cultures of Mongolia, the Cis-Baykal, Trans-Baykal,
and upper Amur river. This second zone of the Tungpei Neolithic is known from
a series of remains located in the Lin-hsia region. Characteristic for this zone is
the predominance of tools manufactured by the flaking technique, and of sherds
decorated with stamping and net imprints. The latter are fragments of round-
bottomed vessels similar to those widely distributed in the Serovo period of the
Cis-Baykal, in the Trans-Baykal, and in Mongolia. They are known in the literature as Serovo vessels. The stone artifacts strongly resemble those of Mongolia
and the Cis-Baykal. In other words, this is the most eastern outpost of the
Mongolian Neolithic.

The third cultural zone is represented by such unusual remains as those of
Ang-ang-hsi and the sites on the rivers Nonni and Sungari. Characteristic for this
zone are small flint artifacts manufactured by flaking and pressure retouch. The
majority of these stone artifacts are typologically close to the finds from the
sites and burial grounds of the upper Amur and the Trans-Baykal region. To it
belong first of all the inset blades, the Daurian arrowheads, and the lamellar
Kitoy knives. Just as characteristic for this region is a great variety of bone
artifacts, particularly the harpoons which in type are close to the Serovo-Kitoy
harpoons of the Cis-Baykal region. Here, painted ceramics are absent and vessels
with characteristic side-profiles and low, wide cups predominate. This zone
represents the southern outpost of the Trans-Baykal and upper Amur Neolithic.

The fourth cultural province of Tungpei occupies the lower reaches of the Sungari river and the region adjacent to the inland reaches of Soviet Primore. It
has been studied least of all. Here there are very few Stone Age remains. Its
characteristics are best expressed at Po-k'ên-ha-tieh and in the sites near Lake
Chingpo. In much, this province is close to the adjacent inland [area of the]
maritime culture and to the Neolithic of the lower Amur.

Thus the principal trait of the Tungpei Neolithic is its mosaic nature: the
presence of many, sometimes essential elements of neighboring cultures brought
about by its intermediate position between two extensive cultural areas of northeastern Asia—that of the hunters and fishers of eastern Siberia and that of the
agriculturalists of central China. As is now becoming clearly established, the
numerous exchanges between these two cultures flowed across Tungpei. The

tribes of Tungpei were in a sense the transmitters of their cultural achievements. Tungpei was that melting pot where the cultures of east and west, north and south, united and acquired a new form and content. As a result there appeared something unique and original in the form of the Neolithic of Jehol.

In the 2nd millennium B.C. essential changes in the culture, economy, and social structures took place in southeastern Asia and its neighbouring regions to the north. Important political events, which involved the majority of the countries of the Far East, took place. In the territory of central China, there arose, based on the Yangshao and Lungshan cultures, the first slave-owning empire in the east, the Shang-Yin. In the extreme south of the Asian continent the distinctive Bronze Age cultures of the Samrong Sen and Bosson arose in Indo-China. In the maritime regions north of the Huang-Ho there arose other new cultures: in North Korea, the dolmen culture, in the Primore, the Shell Mound culture. With time these changes reached beyond the continent to include far-away Japan, where during 400–300 B.C. there arose the new Yayoi culture. Contemporaneously there arose in the central steppes and semi-arid regions of Tungpei and Mongolia a Bronze Age culture founded on the local Neolithic. This is the "culture of slab-lined graves." The region of its distribution extends far beyond the borders of central China and Tungpei. To the north it reaches Lake Baykal, spreading across the Minusinsk steppes, and the influence of this culture reached eastern Europe during Andronovo-Karasuk times.

The rise of a Bronze Age culture and its subsequent development had a profound influence on the course of historical events not only in Tungpei but also far beyond its borders. The Tungpei culture and that of its neighbors appear to be the basis on which, during the 1st millennium B.C., there arose powerful tribal unions of the steppe barbarians, against which the Chinese of the Chou and Han dynasties were forced to engage in hard struggle. This was a prototype of the future more extensive "barbarian" alliances that have played such a significant role in the cultural and political history not only of Asia, but also of Europe. In this connection the remains of the Tungpei Bronze Age, located close to the Yin-Shang area, are of particular interest.[6] Unusual remains of this culture have been discovered in recent years by Chinese and Japanese archaeologists in the following places: Hung-shan-hou in Ch'ih-fêng; Hsi-t'uan-shan, Sao-ta-kou, Tung-t'uan-shan-tz'ŭ, T'u-chêng-tz'ŭ in Kirin province; Hsiao-kuan-chuang in T'ang-shan; in the Wang-ch'ing district near T'ien-ch'iao-ling village, and at Hsiao-yin-tz'ŭ in Yen-chi. The basic characteristics of the Tungpei slab-lined grave culture are now quite clear. The great burial grounds, which are usually situated on hill-tops, are noteworthy. Graves or stone cists in these burial grounds were set in orderly rows, extending from north to south. The interred was usually positioned with his head to the east. The cist was built of stone slabs set on edge (Fig. 4). The floor was lined with square or rectangular slabs. The top of the cist was covered by large slabs laid in a row across the long side-walls. Sometimes an extension for artifacts was built on one side of the cist. Stone slabs were used for its construction. The deceased was positioned on his back with arms folded on the pelvis or at his side. The legs were slightly bent. Artifacts interred with him were not numerous but signifcant. On either side of the head and sometimes near the waist or feet was placed a vessel. Usually there are some ornaments: beads of pyrophyllite or of shell, malachite, nephrite, agate, steatite, and rarely of bronze (pendants of the Andronovo type and open rings with overlapping ends). Near the waist were found perforated stone axes of the Yin type (Ch'ih-fêng), axes rectangular in cross-section, thick edges trapezoid in cross-section, and

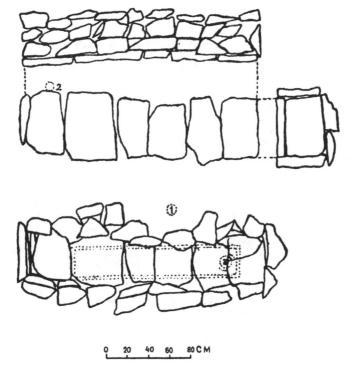

FIGURE 4. Slab-lined grave from T'u-chêng-tz'ŭ.

round chisels with a narrowed blade (e.g., at T'u-chêng-tz'ŭ, T'ang-shan). Arrowheads usually come in heaps of several dozens each. They are made of bone, flint, or slate. At the waist of a skeleton in one of the Hsi-t'uan-shan graves was found a slate sickle unique for its size. In the burial grounds of Kirin province slate knives, daggers, spear points, spindles, and clay net sinkers, are often found. Bronze artifacts are extremely rare; they consist of arrowheads of the Yin or Chou type (Ch'ih-fêng), bronze knives (at T'u-chêng-tz'ŭ and Sao-ta-kou), and celts with a rectangular socket (at Sao-ta-kou and Hsü-chia-t'un).

On the slabs that formed the lid of the grave are found animal bones: the skulls and lower jaws of dogs and sometimes the jaws of sheep, oxen, deer, and pigs. The vessels contained vegetable food. In several graves at Hsi-t'uan-shan in cups and a teapot [-like container] were found millet grains of two types: *Echinochloa villosa* (glutinous millet) and *Setaria lutescens*.

The settlements of the Tungpei slab-lined grave culture are often located with the burial grounds, on hilltops. They are areally extensive, e.g., Sao-ta-kou occupies 52,000 m². The cultural layers are thick. The excavations in Ch'ih-fêng and T'u-chêng-tz'ŭ established that the semisubterranean dwellings were constructed of compacted and well-fired clay walls. The floor was coated with clay. The outside of the dwelling was covered with thick wooden planks. Near the dwellings were also discovered numerous storage pits, slab-lined fire pits and ash pits. The abundance of vessels of the *li*, *ting*, and *tou* types is remarkable. Stone tools are represented by axes, adzes, chisels, slate sickles, and arrowheads (Fig. 6).

Characteristic also are the perforated axes of the Fatyanovo type and the round Yin maces, sometimes irregularly shaped [scalloped] at the edge. Agricultural

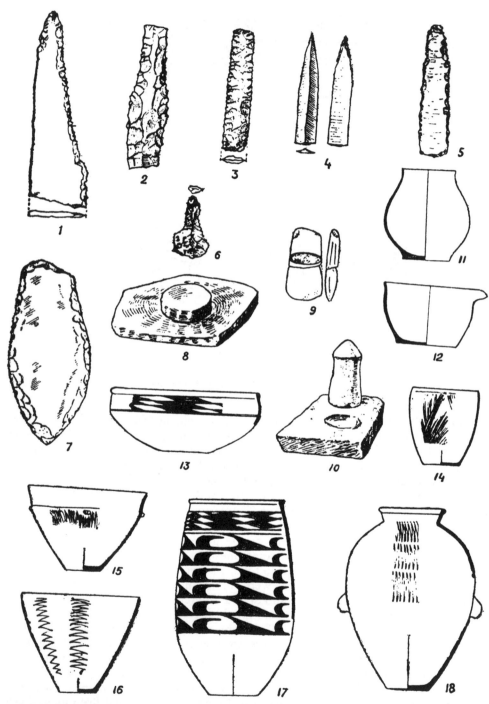

FIGURE 5. Neolithic artifacts from Tungpei sites: 1, knife of the Kitoy type; 2, knife; 3, inset blade; 4, point of the Daurian type; 5, lamella with rough retouching; 6, shouldered awl; 7, ploughshare point; 8 and 10, querns, mortar and pestle; 9, polished axe; 11 and 12, vessels from Ang-ang-hsi and Tsitsihar; 13, 17, and 18, painted ceramics; 14, 15, and 16, ceramics with rubbed and incised zigzag ornamentation (Hung-shan-hou, Ch'ih-fêng).

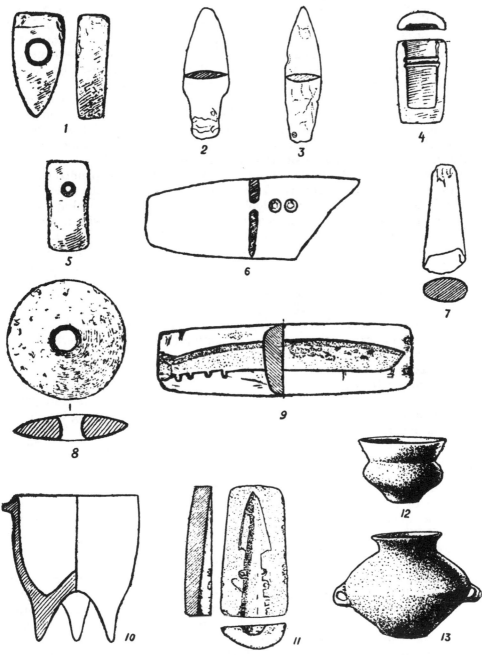

FIGURE 6. Bronze Age artifacts from Tungpei: 1, perforated ax of the Fatyanovo type (Hung-shan-hou); 2 and 3, slate knives (T'ien-ch'iao-ling, Wang-ch'ing district); 4, casting mold for a celt (Hung-shan-hou); 5, ax of the Yin type; 6, slate sickle (T'ien-ch'iao-ling); 7, polished ax (Pai-ts'ao-kou); 8, mace (Hung-shan-hou); 9 and 11, casting molds for knife and spear points (T'ang-shan, Hsiao-kuan-chuang); 10, vessel of *li* type (T'u-cheng-tz'ŭ); 12 and 13, vessels from the Hung-shan-hou site, Ch'ih-fêng.

tools are represented by hoes, querns, grinders, mortars, and polished plough-share points of dainty, leaf-like shape. Bronze articles are not numerous—a sickle resembling the slate sickles, knives, and ornaments (the "khuany"* and Karasuk ternate hasps). Besides these there were found at Ch'ih-fêng and Tang-shan molds for making celts, knives, and spear heads. Tools that were cast in them are in many respects similar to the Yin ones. The fauna of the settlements is represented by cracked and burned bones of domestic animals: pigs, horses, oxen, dogs, and sheep. Bones of deer are occasionally encountered.

The remains of the Tungpei Bronze Age culture spread over an extensive territory from Jehol and Inner Mongolia in the west to North Korea and Soviet Maritime Province in the east. The appearance of this culture did not interrupt the development of local tribes. The roots of this culture are in the Neolithic which preceded it and above all are connected with the complex of outstanding remains which constitute the core of the Tungpei Neolithic (the painted ceramics and the truncated conical vessels). These ties are confirmed by the extensive use of typologically characteristic stone tools. Among them are axes oval and rectan-gular in cross-section, adzes trapezoid in cross-section, chisels with a wide butt and narrow blade, chipped arrowheads with a notched base. For tilling, the same kind of stone plowshares were used, as were shouldered ax-hoes, slate sickles, querns, grinders, and mortars. The ceramics of the Bronze Age reflect a relationship with the preceding culture just as clearly. The people of the slab-lined graves continued to make squat cups, truncated conical vessels, amphora-like vessels with small handles, spheroid vessels with a high neck, and finally, specific jar-shaped vessels with the same characteristics as those of the Neolithic—haphazard comb imprints and appliqué horizontal or vertical rolls. There are also some vessels of the *tou* type.

In the economy, agriculture and cattle breeding still occupy the central place. The agriculture is evidenced in the cultivation of millet the traditional grain of northern China. It is possible that rice was also spottily grown. This is supported by the find of a cup with holes in its bottom in one of the T'u-chêng-tzǔ graves. In China and Korea such cups were used for the steaming of rice. Plants from whose fibers textile fabrics were manufactured (the *lochang* of the Chinese chronicles) were also grown. The fibers were spun with spindles, which were often found in women's graves. The wool of domesticated sheep was also used.

As during the Neolithic, the soil was tilled not only with stone hoes, but mainly with ploughs with stone and apparently bronze [ploughshare] points. Oxen and horses, domesticated already during the Stone Age, were used as draught power. The meat of these animals was also used. Domesticated swine and dogs continued to be bred in great quantities. In addition, sheep appeared in the western districts of Jehol. Among other economic activities fishing played an important role (in the graves are sometimes found clay sinkers, remnants of the fishing nets interred with the dead). Judging by the remains of bones, deer were the main object of the hunt. Thus the economy of the slab-lined grave culture of Tungpei was diversified. It is not surprising, then, that, as in the Neolithic, the sites of the Tungpei Bronze Age culture have all indications of the settled way of life and of prolonged occupation. They are characterized by their extensive area, their rich content in the cultural layer, as well as by the great number of storage pits and semisubterranean dwellings. All this is a continuation of the Neolithic traditions.

*[Transliterated only; meaning of term not clear.—Editor.]

At the same time much new appeared in the culture of the Tungpei tribes. The innovations represented above all the mastery over metal. In the graves and settlements, together with the stone implements, are found bronze celts, spear points and arrowheads, sickles, knives, and ornaments. The local origin of such bronze implements is proved by the discovery of casting forms. [At the same time] new types of tools appeared— perforated axes of the Yin and Chou type, perforated wedge-shaped axes, and flat or ball-shaped mallets with a perforated center. Instead of flint, slate becomes the primary material for the manufacture of tools. Slate arrowheads, knives, daggers, and spear heads became widely distributed. Tilling tools also undergo a noticeable change. The ploughs are usually beautifully polished and have a geometric, regular willow-leaf shape. The shouldered hoes acquire a characteristic, violin-like shape. The rectangular sickles are replaced with semilunar ones of the Yin type. Into the old forms of ceramics new features are incorporated—the profile of the cups becomes more angular, truncated conical vessels become smaller and lose their caracteristic zigzag type of ornament, the neck of the spheroid vessels becomes lower and in some cases disappears altogether. In vessels of the *tou* type the reservoir becomes smaller but the base increases in size and is sometimes larger than the reservoir. There are also new types of ceramics: low cylindrical vessels in the form of a pot and plates with low, vertical walls. But of the greatest importance is the appearance of a great number of specifically Chinese vessels of the *li* and *ting* type. They were absent during the Neolithic, but represent the most distinctive artifacts in Bronze Age Tungpei sites.

The new also includes the uniformity of the slab-lined grave culture across the entire territory of Tungpei. The culture lacks such local differences as were characteristic of the Tungpei Neolithic. Now all local cultures are more or less equalized and overlain by the uniform Bronze Age culture. This common culture spread beyond the territory of Tungpei proper.

With its rise is connected an event of considerable historical significance, the spread of this culture to the north, towards Lake Baykal. This expansion had been facilitated by the preceding trend of historical development: the close cultural and ethnic relations of the tribes of Daurian culture in the Trans-Baykal with those of northern Tungpei (Ang-ang-hsi) during the Neolithic and Early Bronze Age. At this time, rather suddenly there arises the culture of the slab-lined graves, which by its principal ethnic characteristics (clan cemeteries, orientation to the east, *li* vessels in the graves) represents essentially the same culture as in the Trans-Baykal region.[7] All this also indicates great changes in the social structure of the Tungpei and Trans-Baykal tribes. In the remains of their cultures features of large, powerful ethnic and tribal unions may be discerned. Traits of a political alliance, possibly a predecessor of the Hun union of tribes, can be surmised from this. There is no mention of this alliance in Chinese chronicles, but the entire complex and course of events which took place during the Yin and Chou periods are support for this possibility.

Only one group of remains distributed over the Manchurian upland east of Ch'ang-po-shan retains sharply distinguishing features. This is the Yen-chi culture. Here there are no vessels of the *li* and *ting* type; the chipped flint tools continue to be widely used together with the slate artifacts. There are the daggers, arrowheads, pressure flakers, and so on. The pattern of orienting the dead in this culture is the opposite of that observed in the burial grounds of Jehol, Kirin, Liaoning, and the Trans-Baykal region. It represents the mountain tribes of Tungpei cut off from the key centers of the Bronze Age culture.

Together with the ethnic and cultural consolidation [described above] also a new step was made towards cultural contacts with neighboring territories, principally with Yin China.

Many cultural achievements, notably the appearance of metal, came to Tungpei because of the influence of the Yin [Shang] culture. Characteristically, the bronze weapons of Tungpei are copies of Yin models. Spear points, celts, and knives, which were cast in the molds at T'ang-shan, are exact copies of those of Yin. The celts of a later period copy those of the Chou type. The ornaments are bronze artifacts which may be regarded as belonging to the Tungpei proper— rings with overlapping ends, sewn on pendants of small, round, connected plaques, and an Andronovo pendant. Through Tungpei the Yin culture reached Lake Baykal (Fofanovo, Glazkovo) and even distant Yakutia.[8] This explains the unexpected appearance in Yakutia of celts and spear points of the Yin type, with imitations of the characteristic Yin designs.

The Tungpei people built their dwellings in the Yin way "pressing earth into dike-like form." This method survived, according to the chronicles, to the Han period. By then the Chinese themselves had already forgotten the method because it surprises the Han chronicler. It may be surmised that cultural contacts between Tungpei and China go far beyond mere exchange and influence. The appearance in this region of great numbers of vessels of *ting* and *li* type is evidence of infiltration of the Chinese themselves. This infiltration was obviously caused by significance [political] changes in China, resulting first in the appearance of the Shang-Yin state, and then in its destruction by the western Chou. It is known that not the least part in this debacle was played by the Tung-i, "the eastern barbarians" who formed a part of the Chou army. Some of the Yin tribes fled to the north. Some time later the expansion of the Chou dynasty took the same direction. It is in this period, crowded with stormy political events, that the slab-lined grave culture of Tungpei was being born. Its spread north to the Trans-Baykal is evidently connected with the Tung also. The region east of Ch'ang-po-shan was the only one that remained relatively unaffected by these changes. Here the high ranges of the Manchurian upland stopped the Chou [troops]. The Han emperors could not conquer it either.

In the east, the dolmen culture of Korea and particularly the shell mound culture of the [Soviet] Maritime Province are close to Bronze Age Tungpei. Here are the same rectangular semisubterranean dwellings; slate is used widely for the manufacture of tools and chipped stone tools disappear completely. The ceramics are similar except for the vessels of the *li* and *ting* type. Hunting, formerly the basic economy, is supplanted by agriculture and stock breeding.[9]

To the west and northwest the Tungpei culture continued its cultural relations with those of Ordos, Mongolia, western Siberia, and Kazakhastan.[10] The contacts are reflected in the finds of the Andronovo pendant, Karasuk beads from large ternate clasps, knives, and buttons.[11] During the Karasuk age, spheroid vessels resembling those of Tungpei prevailed in the west. Also, in Kazakhstan were found udder-shaped vessels, recalling somewhat those of the *li* type (Dyndybay [Dymboi?]). These contacts are not limited to Kazakhstan, but spread even farther to the west, to the Volga valley and the northern Caucasus. They explain the appearance in Tungpei of the wedge-shaped perforated axes characteristic of the Fatyanovo culture and of ball-shaped maces characteristic of the Catacomb culture of the lower Volga. Other traits connect the Tungpei with the Seima-Turbino culture of the Volga valley: these are the Yin-like bronzes.

In connection with these data concerning the most ancient cultures of northeastern China there arises the important and complex question of the ethnic affiliation of the peoples of ancient Tungpei, whose cultures show distinctly different traits from the time of the Neolithic. This question, of course, deserves special study. However, taking into consideration all the facts, one can relate even now, with a great deal of confidence, the Tungpei Neolithic and Bronze Age tribes to the ancestors of the Tungus-Manchu. This corresponds to ancient traditions according to which the Su-shêng and I-lu were ancestral to the Manchu. The new archaeological materials also show the outstanding role, both cultural and ethnic, played by the Chinese Yangshao and Lungshan cultures during the Neolithic and by the Yin during the Bronze Age.

## Summary

This article was written on the basis of published archaeological data of Soviet and Chinese investigators. The author has established that the Neolithic culture of northern China was most closely connected with the Yangshao Neolithic. Already during the Neolithic the population of Tungpei had begun to till the soil with ploughs. The author emphasizes that during the Neolithic the Tungpei tribes had ties with both the continental [inland] and maritime Neolithic cultures of the lower Amur basin. The rise of a Bronze Age culture in the 1st millennium B.C. seems to have been the basis for the formation of tribal unions of the barbarians of the steppes.

The author proposes that the Tungpei tribes of the Neolithic and Bronze Age epochs were ancestral to the Tungus-Manchu.

## Notes and References

1. J. G. Andersson, An early Chinese culture, *Bulletin of the Geological Survey of China*, no. 5, 1923; *idem*, Preliminary report on archaeological research in Kansu, *Memoires of the Geological Survey of China*, no. 5, Peking, 1925; *idem*, Research into the prehistory of the Chinese, *Bulletin of the Museum of Far East Antiquities*, no. 11, 1943; *idem*, Prehistoric sites in Honan, *Bulletin of the Museum of Far East Antiquities*, no. 19, 1947.

2. J. D. Black, The human skeletal remains from the Sha Kuo T'un cave deposit in comparison with those from Sha Tsun and with recent North China skeletal material, *Paleontologia Sinica*, ser. D, vol. 1, fasc. 3, Peking, 1925.

3. J. G. Andersson, Symbolism in the pre-historic painted ceramics of China, *Bulletin of the Museum of Far East Antiquities*, no. 1, 1929.

4. Li Chi, Liang Ssŭ-yung, Wu Chiang-ting, *et al.*, Ch'êng-t'zŭ-yai, *Archaeologia Sinica*, no. 1, 1934; Yin Huan-chang, Hua-tung hsin-shi-ch'i shih-tai i-chih, Shanghai, 1955.

5. C. W. Bishop, The Neolithic Age in Northern China, *Antiquity*, vol. 7, 1933, pp. 389–404; Andersson, Research into the prehistory of the Chinese.

6. Li Wên-Hsing, Ch'i-lin fu-chin-chih li-chi i-wu, *Li-shih-yü ka'oku*, 1946, no. 1, pp. 21–25; Chia Lan-p'o, Ch'i-lin Hsi-t'uan-shan Ku-mu-chih fa-hsüeh, *K'ao-hsüeh t'ung-pao*, vol. 1, no. 8, 1950, pp. 573–575; K'ang Chia-hsing, Ch'i-lin Chiang-pei T'u-ch'êng-tz'ŭ ku-wên-hua i-chih chi-shih-kuan-mo, *K'ao-ku hsüeh-pao*, 1957, no. 1, pp. 43–52; An Chih-min, Kang-shan shih-kuan-mo chi-ch'iu hsing-kuan-ti i-wu, *K'ao-ku hsüeh-pao*, 1954, no. 7, pp. 77–86; T'ung Chu-chü, Ch'i-lin-ti hsing-shih-ch'i

248    *V. Ye. Larichev*

shih-tai wên-hua, *K'ao-ku t'ung-hsin*, 1955, no. 2, pp. 5–10; T'ung Chu-ch'ên, Ch'i-lin hsing-shih-ch'i wên-hua-ti san-shui-lê-hsing, *K'ao-ku hsüeh-pao*, 1957, no. 3, pp. 31–38; R. Torii, Populations préhistoriques de la Manchourie méridionale, *Journal of the College of Science, Tokyo Imperial University*, vol. 36, art. 8, 1915.

7. S. P. Sosnovskiy, Plitochnye mogily Zabaykalya (Slab-lined graves of the Trans-Baykal), *Trudy otdela istorii pervobytnoy kultury*, Leningrad, 1941, pp. 273–309; A. P. Okladnikov, Arkheologicheskiye issledovaniya v nizovyakh reki Selengi (Archaeological researches in the lower reaches of the Selenga river), *Kratkiye soobshcheniya Instituta istorii materialnoy kultury*, no. 35, 1950, pp. 85–90; *idem*, Arkheologicheskiye issledovaniya v Buryat-Mongolii v 1947 g. (Archaeological researches in Buryat-Mongolia in 1947), *Vestnik drevney istorii*, no. 3, 1948, pp. 155–163; *idem*, O raskopkakh v doline r. Selengi letom 1947 g. (Excavations in the Selenga river valley in the summer of 1947), *Zapiski Buryat-Mongolskogo nauchno-issledovatelskogo instituta kultury*, no. 10, 1950, pp. 61–64; *idem*, Epokha pervobytno-obshchinnogo stroya na territorii Buryat-Mongolii (The epoch of primitive social structure in the territory of Buryat-Mongolia), *Izvestiya Akademii nauk SSSR, seriya istorii i filosofii*, vol. 8, no. 5, 1951, pp. 440–450; *idem*, Zabaykale v I tysyacheletii do n.e. (The Trans-Baykal in the 1st millennium B.C.), *Narody Sibiri*, 1956, pp. 83–92.

8. A. P. Okladnikov, Neolit i bronzovyy vek Pribaykalya (The Neolithic and the Bronze ages of the Cis-Baykal region), pt. 3 (Glazkovo period), *MIA*, no. 43, Moscow-Leningrad, 1955, pp. 166–202; *idem*, *Istoriya Yakutskoy ASSR* (The history of the Yakutsk A.S.S.R.), vol. 1, Moscow–Leningrad, 1955, pp. 170–185.

9. A. P. Okladnikov, Primore v I tysyachiletii do n.e.; po materialam poseleniy s rakovinnymi kuchami (The Primore [Maritime Province] in the 1st millennium B.C.; based on materials from shell middens), *Sovetskaya arkheologiya*, 1956, no. 26; V. Ye. Larichev, Stoyanki kultury rakovykh kuch v rayone bukhty Tetyukhe (Shell midden sites in the region of Tetyukhe Bay), *Sovetskaya arkheologiya*, 1958, no. 1.

10. A. P. Okladnikov, Arkheologiya i izucheniye drevney istorii Zabaykalya. Soveshchaniye po osnovnym voprosam istorii Akademii nauk SSSR, 27 oktyabrya 1952 (Archaeology and the study of the ancient history of the Trans-Baykal region. Academy of Sciences of the U.S.S.R., Conference on the basic questions of history, October 27, 1952), *Tezisy dokladov*, Ulan-Ude, 1952, pp. 21–22.

11. S. V. Kiselev, *Drevnyaya istoriya Yuzhnoy Sibiri* (The early history of southern Siberia), *MIA*, no. 9, 1951; S. A. Teploukhov, Drevniye pogrebeniya Minusinskogo kraya (Ancient burials of Minusinsk *kray*), *Materialy po etnografii*, vol. 3, no. 2, Leningrad, 1927.

G. I. ANDREYEV

# CERTAIN PROBLEMS RELATING TO THE
# SHELL MOUND CULTURE*

UNTIL RECENTLY the "shell mounds" were considered the oldest of the archaeological sites known in the territory of the Maritime Province [Primore].† The work of the Far Eastern Expedition, which started in 1953 and grew in scope under the direction of A. P. Okladnikov, furthered the study of the history of the Maritime Province. Sites (although still not numerous) preceding the shell mound culture and attributed by investigators to the Paleolithic and Neolithic eras were discovered by this expedition.[1] Thus, both the place of this culture among other antiquities of the Maritime Province and its relative and absolute chronologies were determined.

Although in the determination of the absolute age of these cultures a certain clarity and understanding has already been achieved by investigators, this is not quite applicable to the chronological classification of particular groups of sites.

The purpose of this paper is to propose a chronological division of the shell mound culture of the Maritime Province and to trace its historical development.

The problem of dating the shell mounds attracted the attention of the very first investigators. As far back as 1881, M. I. Yankovskiy considered his excavations of shell mounds in the Maritime Province as "the first discovery of kitchen middens of the Stone Age on the Siberian coast."[2] However, he had difficulty in determining their absolute chronology. Yankovskiy considered gathering, fishing, and, to a lesser degree, hunting to be the bases of the economy of the people who had left the shell mounds.

I. S. Polyakov, who surveyed the southern Ussuri region, Sakhalin Island, and Japan in 1881–1882 differentiated the shell mounds of the Maritime Province from those of Sakhalin on the basis of both their content of bones of mammals and their artifacts. He wrote that whereas the presence of flint and masses of chipped "rounded" implements is characteristic of Sakhalin Island, the shell mounds of the Maritime Province contain "ground spears and arrows," which "are very reminiscent of those which only recently were still in use in the Aleutian Islands."[3] V. P. Margaritov, who studied the shell mounds on the coast of Amur Bay somewhat later, attributed them to the period of "complete lack of familiarity with metals,"[4] but thought that they could be dated to the first half of the 1st millenium A.D. He reasoned that the people of the shell mound culture were primarily engaged in gathering, hunting, and fishing, and were familiar with weaving. Somewhat later, F. F. Busse attributed the shell mounds of the Maritime Province to the time of "primitive man," and he believed they continued up to about 700 A.D.[5] V. K. Arsenyev, who uncovered and described a significant number of archaeological sites in the Maritime Province between 1910 and 1920, came quite close, in his published works, to the chronology of Busse.[6]

---

*Translated from *Sovetskaya arkheologiya*, no. 4, 1958, pp. 10–22.

†[Most of the sites mentioned in the initial pages of this article are located within a 40–50-km radius of Vladivostok (43° 10′ N., 131° 55′ E.).—Editor.]

In the 1920's, A. I. Razin studied the shell mounds of the Maritime Province and dated them to the Neolithic. He made the first attempt to classify the pottery of the shell mounds on the basis of its ornamentation. As for the dating, Razin agreed with his predecessors. He thought that these sites existed prior to 600–700 A.D.[7]

N. G. Kharlamov, who visited the Khabarovsk and Vladivostok museums in the late 1920's, after familiarizing himself with the shell mound materials, wrote that he agreed with Razin that they belonged to the Neolithic.[8]

The Amur Expedition, which began its work in 1935 under the direction of A. P. Okladnikov, yielded materials which made it possible to distinguish several stages in the ancient history of the Amur region and, in particular, the Amur Neolithic.[9] On the basis of this work, L. N. Ivanyev in 1951 re-evaluated the archaeological material in the Vladivostok Museum and dated it to the range of the 2nd–1st millennium B.C. to the 14th–15th century A.D.[10] Unfortunately, Ivanyev did not support his dating with an analysis of the material. Also, he erroneously attributed to Okladnikov the term "Amur and Maritime Province Neolithic." The latter was considering only the Amur Neolithic.[11] In order to prove the existence of a single "Amur and Maritime Province" Neolithic, Ivanyev, without foundation attributed vessels with rounded bottoms and spiral-ornamented pottery to the shell mound culture of the Maritime Province.[12]

In 1952, while working on the collections in the Vladivostok Museum, I determined that the sites of the Mariime Province known at present lack round-bottomed vessels and pottery with spiral designs,[13] a point on which the cultures of the Maritime Province and the Amur region of the 2nd–1st millennium B.C. differ substantially. This was also noted by Okladnikov.[14]

Archaeological work, started in 1953 in the Maritime Province under the direction of Oklandnikov, has made it possible to determine the general date of sites of the shell mound type (the end of the 2nd to the beginning of the 1st millennium B.C.). New finds afford the opportunity to date the shell mounds, for the most part, to the 1st millennium B.C.; in particular, this date is supported by finds of green and bluish-green perforated cylindrical beads of jasper as well as a number of jade artifacts, which, in the opinion of G. G. Lemmleyn, are Chinese and which he dates to the middle of the 1st millennium B.C.

Thus, the new data have confirmed the conclusions of Okladnikov, M. V. Vorobyev,[15] and myself.

The study of a considerable number of collections from the shell mounds which are kept in the Vladivostok Museum and in the repositories of the Institute of the History of Material Culture indicated that the material on hand makes it possible to periodize them in two stages. The suggested periodization is as yet preliminary and probably will be clarified and modified in the future. Since the shell mounds are to be considered remains of a single culture, it is natural that the stages are very closely connected. Many common features can be observed in both the chronologically early and late shell mounds. In the pottery, the following are such common features: modeling of vessels by hand, flat bottoms, superior polishing, appliqué coils, ornamentations in the form of incised horizontal lines and various combinations of geometric elements—angles, meanders, and others. Spiral, oval, and voluted designs are absent. The vessels are very frequently slipped with a crimson or light-brown color. Common to the stone inventory is the presence of polished slate knives, arrowheads, dart points, and spear heads. Axes and adzes are usually rectangular in cross-section. Finds of bone implements, among which punches, awls, and needles predominate, are quite frequent.

The first stage can be characterized by shell mounds of the type found on the narrow isthmus that joins Brinner Cape with Yankovskiy peninsula[16] and also by certain sites on the narrow isthmus that joins Besargin Cape with the Muravyev-Amurski peninsula.[17] Here also belongs the shell mound situated alongside the narrow isthmus joining Bolshoy peninsula and Popov Island.[18]

Along with the general features noted above, the pottery of the early stage is characterized by some specific traits. Traces of a black or dark-gray slip are observed on a number of vessels. The profiles of the vessels are smooth; the rims are either straight or slightly everted, or inverted. There are many bowls, plates, and pot-shaped vessels (Figs. 1, 2). The proportion of vessels with decorations in the form of small nodules and non-perforated handles is quite small.

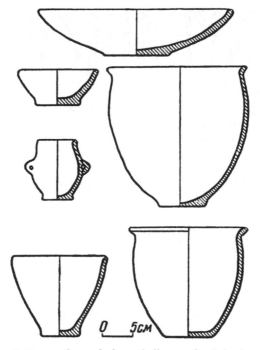

FIGURE 1. Types of vessels from shell mounds of the first stage.

Comparatively short and thick types of axes and adzes are predominant (Fig. 3, items 1, 2). Arrowheads are either elongated, lanceolate with points that are rhomboid in cross-section, and have a slightly flattened base, or they are long and narrow with almost straight parallel sides and hexahedral in cross-section.[19]

Daggers with shaped handles are devoid of crosspieces.[20] There are also knives and daggers which in form resemble Karasuk types.[21] The sites are rich in bone artifacts.

The second stage is represented by sites of the shell mound type situated between the village of Yuzhniy and Cape Sedlovinniy,[22] on Vinogradny Cape, close to the former Popov Fishery,[23] as well as by the sites in the vicinity of the mouth of the Seyfun river near the Chapayev *kolkhoz* [collective farm][24] and on Peschany [Sandy] peninsula.[25]

A number of new features appear in the pottery of the second stage. To begin with, the rim is recurved (Fig. 4, second and third rows) and is frequently

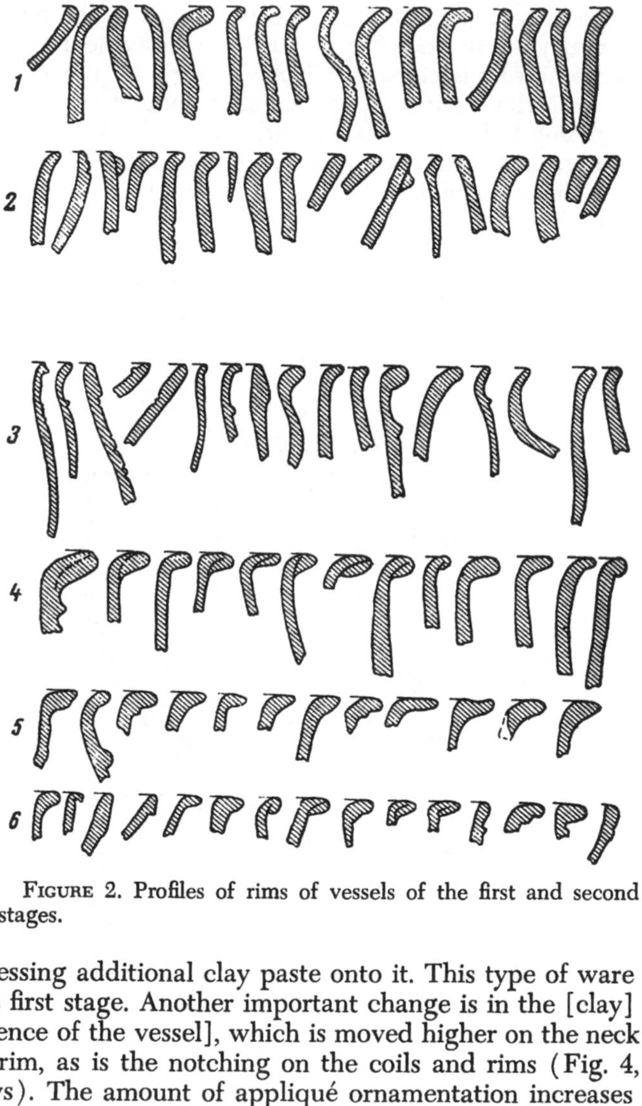

FIGURE 2. Profiles of rims of vessels of the first and second stages.

formed by pressing additional clay paste onto it. This type of ware is completely lacking in the first stage. Another important change is in the [clay] coil [running the circumference of the vessel], which is moved higher on the neck of the vessel, closer to the rim, as is the notching on the coils and rims (Fig. 4, first, second, and third rows). The amount of appliqué ornamentation increases considerably; there are different kinds of nodules and unperforated handles. Also there is a substantial increase in appliqué designs in the form of belts with vertical [incisions] sometimes forming acute angles (Fig. 5, fourth row).[26] Lattice-work and double oblique crosses appear in geometric designs. The practice of applying paint to parts of the surface of the vessels during the first stage is discontinued.

The stone implements also take on new features. Axes and adzes become longer and thinner (Fig. 5, items 1–5). Among the arrowheads, a new type with grooved base appears (Fig. 5, items 16–17). There are slate daggers with crosspieces (Fig. 5, item 8).

Finds of perforated jasper beads (Fig. 5, item 23), and disks and pendants of nephrite, are also new among the ornaments of the late stage.

Overall, the significance of bone implements declines somewhat.

An unusual site on Cape Sedlovinniy, which lacks shell mounds, is very closely related to sites of the second stage. The pottery is multiform and, along with the

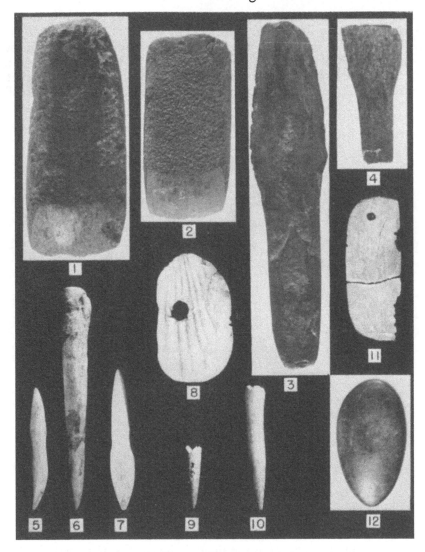

FIGURE 3. Artifacts from shell mounds of the first stage: items 1–4 and 12—stone; items 5–7 and 9–11—bone; item 8—shell.

elements common to both stages, it includes features typical for the second stage only. The daggers with crosspieces (Fig. 6), cylindrical stone beads, a number of axes and adzes, and other objects found here also resemble the finds of the later stage.

It is still impossible to determine whether sites of the Cape Sedlovinniy type are chronologically somewhat later than the sites with shell mounds, or whether both types existed at the end of the late stage when changes in the economy of the people had already begun and gathering had started to lose its former significance.

We shall base our periodization of the shell mounds of the southern Maritime Province on the late stage, since it contains materials that are more easily datable.

It was not by chance that we turned our attention to the presence of such features in the pottery of the late shell mounds as the strongly curved rim and coil, located closer to the rim. In the pottery of the Middle Ages, these features are already definitively shaped, while here we have, so to speak, only the beginning of their development, which then continues in such transitional sites as those in the vicinity of the Artem Government Regional Electric Power Plant.*

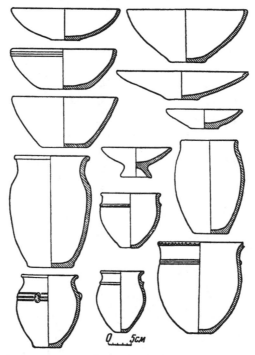

FIGURE 4. Types of vessels from shell mounds of the second stage.

It is no less important to realize that the late shell mounds lack materials pointing to links with Han China. It is difficult to believe that the links between the Maritime Province and China, which can be traced as far back as the first half of the 2nd millennium B.C., were terminated during the Han period when the territory of the latter extended to the southern Maritime Province. It is more likely that the shell mound culture no longer existed in Han times and that its disappearance in the Maritime Province is linked directly to the Chinese campaigns in the north. It was at precisely this time that shell mounds also disappeared in southern Manchuria,[27] and significant changes occurred in the economy and culture of the population of Korea, Manchuria, and Japan.[28] In this respect, A. V. Grebenshchikov's report, based on Chinese chronicles, is very interesting. He wrote: "During the period 800–300 B.C., the eastern part of northern Manchuria and our [Russian] adjoining border country is indicated on historical Chinese and Japanese atlases as the home of Tungus people and, in particular, the location of the Su-shêng. The Su-shêng moved northwest to the banks of the Sungari and Amur rivers during the 3rd and 2nd centuries B.C., i.e., at the time of the Ch'in and Han dynasties."[29]

*[Artem is located some 30 km to the northeast of Vladivostok.—Editor.]

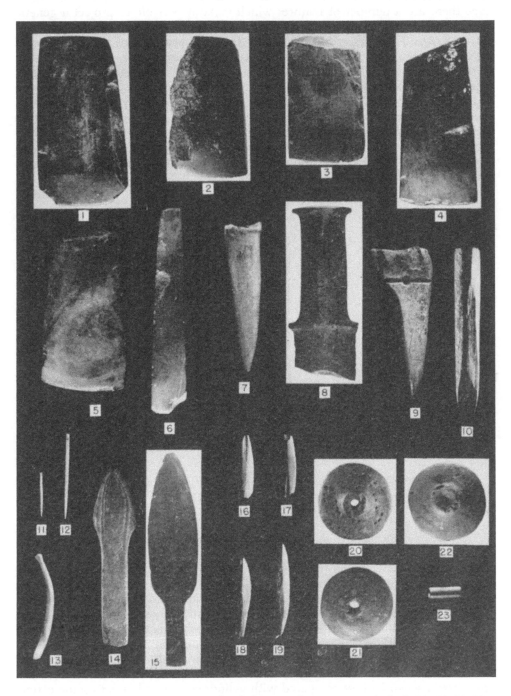

FIGURE 5. Artifacts from shell mounds of the second stage: items 1, 6, 8, 10, 14–19, 23—stone; items 7, 9, 11–13—bone; items 20–22—clay.

However, it would be incorrect to link the disappearance of shell mounds in the south of the Maritime Province with the different movements of tribes only, since there are a number of features which make it possible to project a genetic succession between shell mound pottery and that of sites transitional between the Bronze and Iron ages. Most probably the disappearance of the shell mound culture is associated with changes in the economy. The Chinese chronicles record, at the turn of our [Christian] era, the development of livestock-breeding and tillage for tribes which were supposedly located in the territory under discussion and somewhat to the north of it. The change in the economy led to movement from the coast to the interior, to places conductive to the pursuit of agriculture and animal husbandry and, in a number of cases, close to iron ore deposits. This economic change can be understood and explained only if we accept the premise that in the Maritime Province the use of iron begins just at that time, i.e., at the very end of the shell mound era.

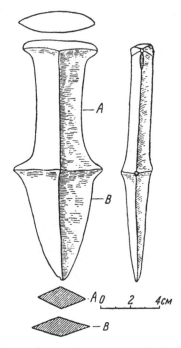

FIGURE 6. Stone dagger from a site on Sedlovidniy Cape.

The point of view, which prevailed until the 1930's, that a shell mound culture existed throughout the first half of the 1st millennium A.D. is thus untenable and was able to arise only because the Maritime Province had been studied so little.

The presence, in the late shell mound culture only, of such objects as polished stone daggers which imitate bronze originals furnishes us the date of the second stage. Daggers with crosspieces were known to exist during the so-called "Scythian period" over the very extensive zone of steppe lands stretching from central Europe to the shores of the Pacific.[30] The stone daggers found in the Maritime Province can be best compared with cylindrical, ribbed and, to some extent, with winged-hilt daggers of the Scythian-Tagar-Suiyüan period.[31] Finds of similar daggers in the Maritime Province can serve as a valid argument that these belong to the late stage and cannot be dated earlier than the 7th to 6th century B.C. As was mentioned earlier, the lack of objects which would link the shell mounds to China of the Han dynasty is also significant for dating purposes. Finds of perforated cylindrical jasper beads and jade ornaments in the mounds of the late stage do not contradict this. Thus, there is a basis for dating the shell mounds of the late period to a chronological span extending from the 7th or 6th century B.C. to the 3rd or 2nd century B.C.

The dating of the first stage, or better, its beginnings, is much more difficult. As we have already noted, sites datable to a period prior to the appearance of shell mounds in the Maritime Province have become known in recent years. Stratigraphic examination shows that the upper mixed layer of the site Zaysanovka I includes numerous potsherds which without doubt belong to the shell mound complex.[32] This is underlined by the close proximity of the site to the shell mound sites. Sites of the Zaysanovka I type have been dated to the 16th to 12th century B.C. As yet, we do not know of sites on the shores of the Gulf of Peter the Great which could be dated to the period intermediate between the sites of the Zaysanovka I type and the shell mounds. Moreover, their material contents are so different that they can be explained either by a sizable chronological break, or a hiatus, or a change of population in this territory. Now that considerable material has been gathered from the coastal area it seems to me that the absence of intermediate sites here does not provide the explanation, but that most probably the replacement of the Zaysanovka I type of culture by a shell mound culture may be explained by a change of population, especially so since the economies of the two cultures differ so much. The people of Zaysanovka were, for the most part, hunters and fishermen, while the tribes which left the shell mounds were primarily gatherers of sea mollusks, fishermen, and livestock breeders, and only secondarily hunters. The inhabitants of Zaysanovka I used obsidian tools, while the implements found in the shell mounds were made of soft stones.

The 12th to 11th century B.C. is also the period of large-scale movements of tribes after the fall of the Shang-Yin dynasty, indications of which we find in Chinese chronicles.[33] The question then arises: Is it possible to find confirmation that it was precisely at this time, i.e., at the end of the 2nd millennium B.C., that new tribes with a sea-oriented economy, heretofore unknown in these regions, arrived at the coast of the Maritime Province?

There is little to go on as yet; however, some support does exist. Among the materials published by A. P. Okladnikov from Popov Island and other places there are daggers of the Yin-Karasuk type.[34] V. P. Margaritov found a miniature polished knife reminiscent of the Karasuk angular knives.[35] In addition, the collections of the Vladivostok Museum contain quite a few haftless polished stone daggers with crosspieces which are comparable with bronze originals of the end of the 2nd and beginning of the 1st millennium B.C. In the Kansu sites[36] painted pottery is found throughout the entire 2nd millennium B.C. Painting is observed on a number of vessels of the first stage of the shell mounds.

On the basis of these data, it seems to me that the dating of the first stage of the shell mound culture has been substantiated, the period involved extending from the end of the 2nd millennium to the first half of the 1st millennium B.C., say, approximately to the 7th or 6th century.

Having examined the problems of dating and having established the two main stages, let us now turn to some questions concerning the economic activities of the people who left these shell mounds.

All investigators of the southern Maritime Province agree that gathering, hunting, and fishing characterized the economy of these peoples.

Gathering is well represented by preserved shells, 23 species of which have been determined preliminarily:[37] *Ostrea gigas* Thun; *Corbicula flumina* ( Müll. );

*Mytilus grayanus* Dunker; *Maktra sachalinensis* Schr.; *Chlamys laetus* (Gould); *Venerupis philippinarum* (Ad et Reeve); *Pecten yessoensis* Jay; *Rapana bezoar*; *Arca boucardi* Jouss; *Natica janthostoma* Desh; *Glycymeris albolineatus* (Lischke); *Mya arenaria* L.; *Saxicava arctica* (L.); *Litterina* sp.; *Venus stimsoni* (Gould); *Arca inflata* Reeve; *Balanus* (*rostratus* Pilsbry); *Aloidis amurinsis* (Schr.); *Neptunea* sp.; *Tellina lutea* Gray; *Protothaka staminea euglipta* (Sow.); *Chlamys swiftii* (Bern); and *Desinia japonica* (Reeve).

As one can see from this list of mollusks, which incidently coincides with that of O. A. Skarlato,[38] all of them are typical of littoral fauna; deep-sea mollusks are missing from the list. Crab remains, found in almost all shell mounds, are also quite numerous. In all probability, other marine fauna and flora were gathered, but either no trace of them has remained or none have been found as yet. In addition to sea products, nuts, grapes, berries, and roots were probably gathered.

We can judge about fishing both from finds of quantities of sinkers and fragments of shanks of composite fishhooks as well as by the considerable amount of fish bones which we collected in 1955. Although the ichthyological materials are still being processed, a number of other species, including freshwater fish, were detected together with those which A. Taranets[39] had already determined.[40]

Preliminary classificatory studies reveal an absence of demersal fish, which do not come close to the shore. In the opinion of specialists who are classifying the remains of the ichthyofauna, the fish represented by the remains are not of the kind caught by deep-sea fishing.

Although fish were caught and mollusks gathered near the shore (without recourse to deep-sea fishing), we assume that boats were used widely. At present, it is impossible to say what kind of boats were then used, but their structure was hardly as complicated as Okladnikov describes it.

Hunting may be judged by the numerous fragments of arrowheads, darts, and spear heads, and by osseous remains, among which have been found bones of the maral [*Cervus asiaticus*], roe deer, fox, bear, and wild boar.[41]

Three shell mounds contained one seal bone each. At present, it is difficult to judge whether this is evidence of shelf-ice hunting or whether the seals were cast onto the shore.

A small but very interesting amount of bird bones, mostly aquatic, was also collected.[42] The following species were determined: (1) bean goose (*Anser fabalis* Lath) (?); (2) duck (*Anas* sp.); (3) mallard (*Anas platyrhynchos* L., possibly *Anas poecilorhychna* Forster); (4) gadwall (*Anas strepera* L., possibly [falcated teal] *Anas falcata* Georgi); (5) spoonbill (*Anas clypeata* L.); (6) widgeon (*Anas penelope* L.); (7) goldeneye (*Clangula clangula* L.); (8) loon (*Gavia* sp.?); (9) old squaw (*Clangula hyemalis* L.); (10) Scaup duck [diver] (*Nyroca marila* L.); (11) black-necked grebe (*Colymbus nigricollis* Chr. L. Brehm); (12) great-crested grebe (*Colymbus cristatus* L.); (13) red-face cormorant (*Phalacrocorax urile* Gm.); (14) great [or common] cormorant (*Phalacrocorax carbo* L.); (15) gull (*Larus* sp.?; (16) grouse (*Falcipennis falcipennis* Hartlaub); (17) pheasant (*Phasianus colchichus* L.); (18) domestic chicken (?); (19) large-billed crow (*Corvus levaillantii* Leison). This list will probably be enlarged by future investigations.

Prior to Okladnikov's studies, no one had raised the question whether the people of the shell mound cultures were livestock breeders. Okladnikov expressed the opinion that pig breeding existed but thought that the pigs were only semi-domesticated. On the basis of V. I. Tsalkin's classification we arrive at the conclusion that domestication of pigs did take place during this time. Pig bones not only occur in the shell mounds of the Maritime Province but surpass quantitatively other animal bones.[43]

Dog bones are the second most frequently found bone remains.[44] In connection with this, Tsalkin thought it possible that dogs were bred for meat. A similar situation obtained in the Neolithic of northeastern China.[45] It is also known that in Korea dogs are used as food to this day.

Let us now consider the nature of agriculture during the shell mound era. Okladnikov points out that grinders, sickles, and hoes were found in the shell mounds. Reliably only grinders are known to me as having come from the shell mounds; the other artifacts are chance finds or [obtained] from extraneous fill. Also, the grinders could have served for the pulverization of roots or [mineral] paints.

The case of slate sickles or knives with perforations must also be regarded with caution since these implements were used differently in different regions at different times and under different conditions. The Eskimos used them for cutting meat and processing hides, while in China they were used as agricultural implements. In the Maritime Province these finds are rare; therefore we suggest that until the matter is clarified by additional finds, the interpretation be left open.

Those implements which Oklandikov considers as hoes could be excavating implements, and not only agricultural ones, i.e., they could have been used for building sod huts, digging pits, and so on.

The absence of grains and the small numbers of implements connected with soil cultivation (these, moreover, being chance finds only) do not give us grounds for denying the existence of agriculture, but the conditions are quite insufficient to resolve the problem. The question of the nature of agriculture is all the more debatable since it is hardly possible to compare agriculture in the Maritime Province during the shell mound era with the agriculture of Yangshao because of the chronological difference of approximately 1000 years between the two.[46] Moreover, the natural and climatic conditions of the two areas are quite different.

The problem of the economic activities of the shell mound people presents a series of interrelated questions which includes the development of a harpoon complex, sea-hunting as the basis of the economy, and the development of [deep-]sea fishing.

Okladnikov defines the people of the shell mounds as daring sea-mammal hunters who went far out to the open sea. This statement is based on data contained in publications by A. I. Razin and A. Taranets. The former mentions the presence of deep-sea shellfish among the various species and considers them to be the "wavy horns"* (*Buccinum undatum*). However, we know that this type of shellfish does not inhabit the Sea of Japan at present and, according to the specialists,[47] did not inhabit it in the past. This species is characteristic of northern seas and is typically littoral. We have noted earlier that all the types of shells identified are typical of the littoral zone, i.e., their gathering did not require trips to the open sea.

Taranets' publication, which deals with the study of fish remains from the shell mounds, mentions vertebrae of the tuna; in the author's opinion the catching of the tuna required open-sea fishing.[48] This information lay at the base of Okladnikov's suggestion that deep-sea fishing was fundamental to the shell mound economy.

In this connection we shall, first of all, take up the question of tuna. It is true that a small number of vertebrae belonging to two species of tuna were found in

*[The quotes are those of the author.—Editor.]

a number of shell mounds on the shores of Peter the Great Bay. However, was it really necessary to go out to the open sea to obtain tuna and what were the implements used to catch them? This is what, for example, A. I. Rumyantsev and I. V. Kizevetter say about the eastern tuna: "The eastern tuna prefers to swim against the current [of the outgoing tide] and during low tide go [find themselves] in bays and mouths of rivers, whence they swim against the incoming tide out to the open sea."[49] We find P. Orel's information even more interesting: "Observations made in August 1924, by the personnel of the Far Eastern State Fishing Industry, and from 1917 to 1925, by the personnel of the Industrial Inspectorate of the Far Eastern Administration of Fish Cultivation and Conservation, confirm the stories of fishermen that at a particular time of the year (from the end of May to September) tuna collect in large schools and come near Vladivostok."[50] And further, "In 1921 in Poset Bay on the 'Pallada' roadstead in the fishing grounds at Cape Astafyev of Krabbe peninsula [42° 45′ N., 131° 20′ E.] several scores of tuna fish approached a seine and some of them went in. In trying to save the seine (which in the end was torn), the fishermen, armed with stakes and axes, killed 22 tuna whose total weight was approximately 200 poods."[*51]

The number of tuna fish that sometimes approach the shore may be estimated from the following report: "On August 20, 1924, we saw, from the lighthouse on Askold Island, many large tuna weighing probably 10 poods or more. When they reached Nayezdnik Bay (of Askold Island) they filled the entire bay and additionally reached about a mile into the open sea. We observed this school of fish for about half an hour, and then it disappeared in a southerly direction."[52]

It was not by chance that in 1925 the Japanese firm Fujihiko Kudō, through its representative in Vladivostok, tried to acquire rights for the region of Peter the Great Bay for 5–10 years for the purpose of seine fishing of tuna. In Okladnikov's opinion the shell mound people used toggle harpoons for catching tuna. However, this is what P. Orel writes about tuna fishing in the 20th century: "The use of harpoons by the local population in hunting tuna had no positive results since the harpoons tore out the flesh, and could not hold the tuna. Tuna flesh is much too tender for such implements."[53]

The fishing rod, in wide use at the present time for catching tuna, could hardly have existed at the time of the shell mound people. It seems that the tuna was not a basic economic animal for them, a fact supported by the rarity of the finds of its bony remains.

Thus, we cannot agree with Okladnikov when he assumes

that the acquisition of deep-sea mollusks and fish was not possible with going to the open sea. It required appropriate technical gear—primarily, large seaworthy boats, possibly with a sail or stabilizing outrigger. Fishnets were necessary as well as special fishing rods with sinkers which could be lowered to a great depth, and other equipment —this being the basis of fishing by the various tribes of the Pacific at the time of their initial contact with Europeans.[54]

We have already noted that open-sea fishing was not required to obtain the type of fish whose remains were found in the shell mounds. Certain types of fish, particularly cod and gobies, could have been caught with a rod; flounder and a number of other fish could have been caught with nets in the bays; freshwater and certain sea fish could have been caught in nets deployed in the mouths of rivers. Bow and sleeve nets could also have been placed in the rivers and "weirs"

*[One pood (*pud*) is equal to 16.3 kg or 36.1 lb.—Editor.]

and other intercepting devices constructed. Boats with a sail or outrigger were not necessary for catching any of the fish mentioned above.

The last problem connected with the fishing economy of the shell mound era concerns the presence of a developed harpoon complex in the Maritime Province. We have made it clear that a harpoon was not necessary for catching tuna. The small number of seal bones convincingly indicates that seal hunting was in no way a widespread occupation of the population. And for the occasional killing of seals on shelf-ice, no complicated [toggle] harpoon was necessary. To this day no harpoon heads with open or closed sockets, or harpoons of simple types, have been found in the shell mounds. In those instances where sea fishing played an important role or constituted the basis of an economy, large numbers of projectile points were usually found in the sites together with bones of the animals.[55] Nothing of this kind occurs in shell mounds.

Okladnikov considers occasional finds of slate points with one or two perforations to be blades of toggle harpoon heads.[56] He compares them with some of the blades found in sites of the Old Bering Sea culture. S. I. Rudenko considers them [end] blades of harpoon heads, knives, or spear heads.[57] Specialists interested in fishing and, in particular, harpoons do not regard these finds to be [associated with] toggle harpoons.[58]

Thus, it sems to us that there are as yet too few data to confirm the presence of a developed harpoon complex in the shell mound era and that sea-mammal hunting constituted the basic economy of the tribes who inhabited the southern Maritime Province at that time.

To summarize: Shell mounds appear in the south of Maritime Province evidently not earlier than the end of the 2nd millennium B.C., and continue throughout almost the whole length of the 1st millennium B.C. (up to the Han dynasty). The economy of the tribes who left the shell mounds was varied. Gathering and fishing (both riverine and littoral), hunting, and pig breeding—all of these formed the basis of their economy. The possibility of incipient agriculture, though different from that of Yangshao, is not ruled out. The use of metal probably begins at the end of the 3rd millennium B.C.

If such skills as weaving, the manufacture of beautifully burnished earthenware, and the building of boats, and a number of other skills such as the processing of hides and the sewing of clothes, are taken into account, it becomes obvious that the people of the shell mounds were on a rather high level of cultural development.

If we did not take into consideration all the facts that have been mentioned above, and, additionally, the links that must have existed with the cultural center of China, it would be impossible to explain the rise of the first organized state in southern Maritime Province and northeastern China, the P'o-hai, as early as the turn of the 7th–8th century A.D.

## Notes and References

1. A. P. Okladnikov, Arkheologicheskiye issledovaniya v Primorye v 1953 g. (Archaeological investigations in the Maritime Province in 1953), *Soobshcheniya dalnevostochnogo filiala Akademii nauk SSSR im. V. L. Komarova*, no. 8, Vladivostok, 1955; *idem*, Primorye v I tysyacheletii do n.e.; po materialam poseleniy s rakovinnymi kuchami (The Maritime Province during the 1st millenium B.C.; based on materials from sites

with shell mounds), *Sovetskaya arkheologiya*, no. 26, 1956, pp. 54–96; G. I. Andreyev, Poseleniye Zaysanovka I v Primorye (The site Zaysanovka I in the Maritime Province), *Sovetskaya arkheologiya*, no. 2, 1957, pp. 122–146.

2. M. I. Yankovskiy, Kukhonnyye ostatki i kamennyye orudiya, naydennyye na beregu Amurskogo zaliva na p-ve, lezhashchem mezhdu Slavyanskoy bukhtoy i ustyem r. Sidemi (Kitchen middens and stone implements found on the shore of the Amur Bay on the peninsula which lies between Slavyanskaya Bay and the mouth of the Sidemi river), *Izvestiya Vostochno-Sibirskogo otdeleniya Russkogo Geograficheskogo obshchestva* (hence *IVSORGO*), vol. 12, nos. 2–3, Irkutsk, 1881, pp. 92ff.

3. I. S. Polyakov, Otvet ob issledovaniyakh na o-ve Sakhaline, v Yuzhno-Ussuriyskom kraye i v Yaponii (Report on the investigations on Sakhalin Island, in the southern Ussuri region, and in Japan), Supplement to vol. 48 of *Zapiski Akademii nauk*, no. 6, St. Petersburg, 1884, p. 16.

4. V. P. Margaritov, *Kukhonnyye ostatki, naydennyye na beregu Amurskogo zaliva, bliz reki Sidemi* (Kitchen middens found on the shore of Amur Bay near the Sidemi river), Vladivostok, 1887, p. 5.

5. F. F. Busse, Ostatki drevnostey v dolinakh rek Lefu, Daubikhe i Ulakhe (Remains of antiquities in the valleys of the Lefu, Daubikhe, and Ulakhe rivers), *Zapiski Obshchestva Izucheniya Amurskogo kraye* (hence *ZOIAK*), vol. 1, Vladivostok, 1888.

6. V. K. Arsenyev, Materialy po izucheniyu drevneyshey istorii Ussuriyskogo kraya (Materials for the study of the ancient history of the Ussuri region), *Zapiski Priamur-skogo otdela Obshchestva Vostokovedeniya*, no. 1, 1913, p. 31; idem, Obsledovaniye Ussuriyskogo kraya v arkheologicheskom i arkheograficheskom otnosheniyakh (Exploration of the Ussuri region in its archaeological and archaeographical aspects), *Izvestiya Yuzhno-Ussuriyskogo otdeleniya Priamurskogo otdela RGO*, Nikolsk-Ussuriyski, no. 4, April, 1922, pp. 55, 56; idem, Otchet o raskopkakh v Primorskom kraye (A report on excavations in Maritime *kray*), *Arkhiv Leningradskoye otdeleniye Instituta istorii materialnoy kultury* (hence *LOIIMK*), 1924, file no. 161.

7. A. I. Razin, Arkheologicheskaya razvedka na beregu Ussuriyskogo zaliva (Archaeo-logical survey on the shores of Ussuri Bay), *Sovetskoye Primorye*, no. 8, Vladivostok, 1925; idem, Stoyanki kamennogo veka na beregu Ussuriyskogo zaliva (Stone Age sites on the shores of Ussuri Bay), *Sovetskoye Primorye*, nos. 3–4, 1926.

8. N. G. Kharlamov, Arkheologicheskaya poyezdka v Sibir i na Dalniy Vostok v 1928 godu (An archaeological sojourn in Siberia and the Far East in 1928), Diary, *Arkhiv LOIIMK*, 1929, no. 135.

9. A. P. Okladnikov, K arkheologicheskim issledovaniyam v 1935 g. na Amure (Archaeological investigations in 1935 in the Amur region), *Sovetskaya arkheologiya*, no. 1, 1936; idem, Neoliticheskiye pamyatniki kak istochniki po etnologii Sibiri i Dalnego Vostoka (Neolithic remains as sources for the ethnology of Siberia and the Far East), *Kratkiye soobshcheniya Instituta istorii materialnoy kultury* (hence *KSIIMK*), no. 9, 1941, p. 12.

10. L. N. Ivanyev, Izucheniye drevnostey Primorya (A study of the antiquities of Maritime Province), *Krasnoye Znamya*, no. 120, Vladivostok, May 24, 1951.

11. Okladnikov, Neoliticheskiye pamyatniki kak istochniki . . . , p. 12.

12. L. N. Ivanyev, Rakovinnyye kuchi Primorskogo kraya (Shell mounds of the Maritime Province), *KSIIMK*, no. 47, 1952, p. 136.

13. G. I. Andreyev, Keramika Primorya i Priamurya do XIII veka n.e. (Pottery of the Martime Province and the Amur region in the 13th century A.D.), p. 11. Dissertation defended in May 1953 in the Section of Archaeology, Department of History, Moscow State University. Deposited in the Section of Archaeology, Moscow State University.

14. A. P. Okladnikov, Arkheologicheskiye issledovaniya v Primorye v 1953 g.; idem, Primorye v I tysyacheletii do n.e.; idem, U istokov kultury narodov Dalnego Vostoka (Origins of the cultures of Far Eastern peoples). Collection: *Po sledam drevnikh kultur*, Moscow, 1954.

15. M. V. Vorobyev, Kamenny vek stran Yaponskogo morya (The Stone Age of countries about the Sea of Japan). Author's summary of dissertation, Leningrad, 1953.

16. Based on materials from my excavations in 1955 as well as the collections of the Vladivostok Regional Museum.

17. The same materials.

18. Materials from surveys made by the author in 1954 and 1955; deposited in the Institute of the History of Material Culture of the Academy of Sciences, U.S.S.R.

19. Collection 412, nos. 14, 69, and 71, in the Vladivostok Regional Museum.

20. Collection 412, nos. 78 and 108, in the Vladivostok Regional Museum. Margaritov, *Kukhonnye ostatki* . . . , Plates 2–3, Fig. 42.

21. Okladnikov, U istokov kultury . . . , p. 252; *idem*, Primorye v I tysyacheletii do n.e., Fig. 18.

22. Excavations of Zh. V. Andreyeva in 1955. She also dates similar sites to the late stage of development. (Cf. Andreyeva's report at the session of the archaeological department of the State Historical Museum, November 26, 1956.)

23. Collected by the author in 1955. Now at the Institut istorii materialnoy kultury Akademii nauk SSSR.

24. Collected by the author in 1955.

25. Collection 514 in the Vladivostok Regional Museum.

26. Okladnikov, Primorye v I tysyacheletii do n.e., Fig. 5, items 1 and 2.

27. Mu-Yang Ch'ĕng, Han and pre-Han sites at the foot of Mount Lao' Tien in south Manchuria, *Archaeologia Orientalis*, vols. 1, 2, Tokyo–Kyoto, 1931, pp. 12, 13.

28. A. P. Okladnikov, O drevneyshem naselenii Yaponskikh ostrovov (The ancient population of the Japanese islands), *Sovetskaya etnografiya*, no. 4, 1946; Vorobyev, Kamenny vek stran Yaponskogo. . . .

29. A. V. Grebenshchikov, K izucheniyu istorii Amurskogo kraya po dannym arkheologii (A study of the Amur kray on the basis of archaeological data). Commemorative collection, no. 25, Museum Society for Study of the Amur Region. The first 25 years. Vladivostok, 1916, p. 51.

30. O. Janse, L'empire des steppes et les relations entre l'Europe et l'Extrême Orient dans l'antiquité, *Revue des Arts Asiatiques*, vol. 9, no. 1, Paris, 1935.

31. S. V. Kiselev, *Drevnyaya istoriya Yuzhnoy Sibiri* (The early history of southern Siberia), MIA, no. 9, 1951, Plate 25; T. R. Martin, *L'âge de bronze au Musée du Minoussinsk*, Stockholm, 1893; [?], Inner Mongolia and the region of the Great Wall, *Archaeologia Orientalis*, ser. B, vol. 1, Tokyo–Kyoto, 1935, Corpus no. 2, Plate 39, Fig. 17.

32. Andreyev, Poseleniye Zaysanovka I v Primorye.

33. N. Ya. Bichurin (Iakinf), *Sobraniye svedeni o narodakh, obitavshikh v Sredney Azii v drevniye vremena* (Collection of data on the peoples who inhabited Central Asia in ancient times), Moscow–Leningrad, 1950 ed., vol. 2, p. 8.

34. Okladnikov, U istokov kultury . . . , p. 252; *idem*, Primorye v I tysyacheletii do n.e.

35. Margaritov, *Kukhonnyye ostatki* . . . , Plate 3, no. 42.

36. J. G. Andersson, Preliminary report on archaeological research in Kansu, *The Geological Survey of China, Memoirs*, ser. A, no. 5, Peking, 1925; N. Palmgren, Kansu mortuary urns of the Pan Shan and Ma Chang groups, *The Geological Survey of China, Palaeonthologia Sinica*, ser. D, vol. 3, Fig. 1.

37. Ya. I. Starobogatov has analyzed the shell mollusks and other types of fauna which I collected in 1955.

38. G. S. Ganeshin and A. P. Okladnikov, O nekotorykh arkheologicheskikh pamyatnikakh Primorya i ikh geologicheskom znachenii (Selected archaeological sites in the Maritime Province and their geological significance), *Geologiya i poleznyye iskopayemyye*, no. 1, Moscow, 1956.

39. A. Taranets, O kostyakh ryb, naydennykh v kukhonnykh ostatkakh plemeni Ilou (Fish bones found in the kitchen middens of the I-lu tribe), *Vestnik Dalnevostochnogo filiala Akademii nauk SSSR*, no. 18, Vladivostok, 1936.

40. The classification of fish remains is being done at the Institute of Oceanology of the Academy of Sciences, U.S.S.R., by L. N. Besednov under the supervision of Professor T. S. Rass.

41. Bones of mammalia were analyzed by V. I. Tsalkin, to whom I herewith express my gratitude.

42. Bird bones were analyzed by M. A. Voinstvenski, Assistant Professor in the Section of Vertebrates, Biology Department, Kiev State University. I hereby express my deep gratitude to him.

43. G. I. Andreyev, Otchet o raskopkakh v Primorye v 1955 (An account of excavations in the Maritime Province in 1955). Deposited in the archives of the Institute of the History of Material Culture.

44. *Ibid.*

45. *Bolshaya Sovetskaya Entsiklopediya*, vol. 21, 2nd ed., p. 197, Kitay, arkheologicheski ocherk (China, archeological survey).

46. Both A. P. Okladnikov and I think that the shell mound era began at the end of the 2nd millennium B.C. since the Yangshao culture terminated at the end of the 3rd millennium B.C. Cf. R. F. Its, Kultury Yanshao i Lunshan i ikh sootnosheniye (The Yangshao and Lungshan cultures and their relationship), *Sovetskaya arkheologiya*, no. 24, 1955.

47. For this, see the classifications of O. A. Skarlato and Ya. I. Starobogatov.

48. Taranets, O kostyakh ryb, naydennykh. . . .

49. A. I. Rumyantsev and I. V. Kizevetter, *Tuntsy* (The tuna), Vladivostok, 1949.

50. P. Orel, Tuntsovy promysel (The tuna industry), *Sovetskoye Primorye*, no. 5, 1926, p. 59.

51. *Ibid.*, p. 60.

52. *Ibid.*, p. 61.

53. *Ibid.*

54. Okladnikov, U istokov kultury . . . , p. 250.

55. S. I. Rudenko, *Drevnyaya kultura Beringova morya i eskimosskaya problema* (The ancient culture of the Bering Sea and the Eskimo problem), Moscow–Leningrad, 1947, Plates 5, 7, 8, 9, 10 [English translation published by the University of Toronto Press for the Arctic Institute of North America as *Anthropology of the North: Translations from Russian Sources*, no. 1]; *idem*, Drevniye nakonechniki garpunov aziatskikh eskimosov (Ancient harpoon heads of Asiatic Eskimos), *Trudy Instituta etnografii im. N. N. Miklukho-Makhlaya*, n.s., vol. 11, Moscow–Leningrad, 1947.

56. Okladnikov, U istokov kultury . . . , p. 249.

57. Rudenko, *Drevnyaya kultura Beringova morya* . . . , Plate 17, item 5; Plate 20, item 5.

58. Opinion voiced by V. V. Fedorov at the session of the Paleolithic section in Leningrad dealing with A. P. Okladnikov's report on the shell mounds in the Martime Province. In private conversation, V. N. Chernetsov also expressed doubts on their use as toggle harpoons.

S. I. RUDENKO

# THE CULTURE OF THE PREHISTORIC
# POPULATION OF KAMCHATKA*

THE USE OF METAL, more precisely iron, in Kamchatka, began very late. Krasheninnikov wrote: "They say of Kamchatka that iron tools were known to the Kamchadals even before their subjugation by the Russian state, and that they obtained them from the Japanese, who used to come to the Kurile Islands."[1] It was supposed by Torii[2] that individual metal objects could, even in antiquity, have reached Kamchatka from Japan through the Kuriles, who used to obtain Japanese commodities from their brothers-in-trade, the Ainus. And true enough, while excavating at Lake Kurilskoye in southern Kamchatka, Jochelson[3] discovered Japanese copper coins from the 11th century A.D. However, individual metal objects penetrating into Kamchatka were so few in number that up to the 17th century, until the arrival of the Russians, the population of Kamchatka was actually living in a Stone Age environment. This is evidenced by numerous reports from persons in government service when they first visited the region.

In former years the northern third of the peninsula was occupied by the Koryaks. They lived along the Pacific coastline north of the Uka river and on Karagin Island, and along the coast of the Sea of Okhotsk north of the Tigil river. The southern two-thirds of Kamchatka, as well as Shumshu Island (the nearest of the Kurile group to Kamchatka), was occupied by the Kamchadals or Itelmens. The Kuriles or Ainu lived on Paramushir, the second nearest island. At the time of Bering's expedition in the first half of the 18th century, the inhabitants of the southernmost part of Kamchatka (the Kurilskaya Lopatka†) and of Shumshu were called the Near Kuriles to distinguish them from the Remote Kuriles, who lived on Paramushir and [also] on the more southerly islands. However, Krasheninnikov, a participant in Bering's expedition, was insistent in his notes that "although the inhabitants discovered on the first island and on the Kurilskaya Lopatka do differ somewhat in their language, customs, and physical appearance from the Kamchadals, it is known that they are descended from the Kamchadals . . . , and the difference remarked upon is due to propinquity, association, and intermarriage with full-blooded Kuriles."[4]

The first investigations of old settlements on Kamchatka were carried out in 1852 by Ditmar. He wrote:

Not far from the bank of the Vakhil river, at some distance from one another, were located square pits of regular shape, about 20 feet on a side. . . . These were remains of old Kamchadal yurts, mostly with one trench-like entrance, less frequently with two. In many of them, refuse lay as much as three feet deep, and it always contained charcoal, bones, shells, and a few worked stones.

And further,

. . . from Cape Nalacheva and starting even further west, from the mouth of the Nalacheva river to Bichevinskaya Bay and to Cape Shipunskiy, the shore was covered

*Translated from Sovetskaya etnografiya, no. 1, 1948, pp. 153–179.
†[The area adjacent to Cape Lopatka.]

by a large number of yurts, the dwellings of hundreds of people. With two assistants, I undertook the excavation of the old Kamchadal yurts. . . . The five pits examined by me yielded absolutely identical finds. Everywhere the yurts were half-filled with all kinds of refuse which contained charcoal, fragments of bone (including the lower jawbone of a bear), parts of horns of the mountain goat and antlers of reindeer, shells, and half-decayed wood. Implements of worked stone, or chips struck off in their manufacture, were encountered very rarely. Nevertheless, little by little we dug up a sufficient quantity of such artifacts. On very rare occasions, we found bone spear heads and small-sized pottery of very primitive workmanship.[5]

One of the inhabitants of Stary Ostrog [Old Stockade] told Ditmar that old sod yurts "are frequently encountered on the eastern coast of Kamchatka, and that if one excavates them various objects are found, such as stone implements, walrus tusks, bones, fragments of very crude clay vessels, stakes, and pieces of wood." The objects brought to Ditmar consisted of obsidian and jasper arrow points, and flat, oblong implements made of the same material, with a curved, sharpened edge on one side. "Tools quite similar to these," wrote Ditmar, "I subsequently found to be in full use by the Koryaks. Koryak women use such stones to scrape hides when dressing leather. Obsidian, dark gray-green jasper and other kinds of quartzite, quartzite and dioritic slates—such are the types of stone particularly liked by the ancient inhabitants of this land for the manufacture of tools of this sort."[6]

In 1910, the ancient settlements in Kamchatka were investigated by K. D. Loginovskiy, who was commissioned by the Ethnographic Section of the Russian Museum, where the collections from his excavations were subsequently sent. According to Loginovskiy's report, ancient Kamchadal yurt settlements are scattered all over Kamchatka; they are encountered both on high ground and near rivers and lakes. Such settlements were observed by him near the city of Petropavlovsk on the slope of Mishennaya Gora [Mountain], near the large village of Kharlaktyrka, near the village of Savoyko, on one of the islands in the mouth of the Avacha river, and on the shore of Tarinskaya Guba [Bay]. Remnants of yurt settlements are found along the headwaters of the Bystraya river near the communities of Malka and Ganaly, and along the lower reaches of the Kamchatka river near the village of Kluychevskoye [modern Klyuchi], Kamaki, and Ust-Kamchatsk. According to Loganovskiy, stone arrowheads and spear heads, as well as stone lamps (oil lamps), were frequently found by local residents at the site of the old yurt settlements near the village of Kamaki. Farther to the north, on the knolls near the shores of Karaginskoye ozero [Lake Karagin], many stone implements were found, most numerous among them being arrowheads, adzes, and wedges. Obsidian arrowheads were found in blowouts in Korfa Bay, between Sibir Bay and the Koryak camp of Tilichek. Seven dwellings were excavated by Loginovskiy near the village of Malka, and three near the village of Ganaly. He surveyed the surface at the old yurt settlements on the shore of Kichiga Bay, collecting several obsidian arrowheads, bone spear heads, and fragments of pottery. Excavations of more substantial nature were carried out by him on Karaginskiy ostrov [Karagin Island], which is separated from the mainland by Litke Strait.

In the west-central part of the island Bukhta Lozhnykh Vestey [False News Bay] is located. Old settlements, marked by groups of dwellings, were located mainly along the shores of the bay. They are situated partly at the very shore and partly on [adjacent] hillslopes, the dwellings on the hillslopes being more ancient than those at the shore. Only stone and bone objects were found in the hillside

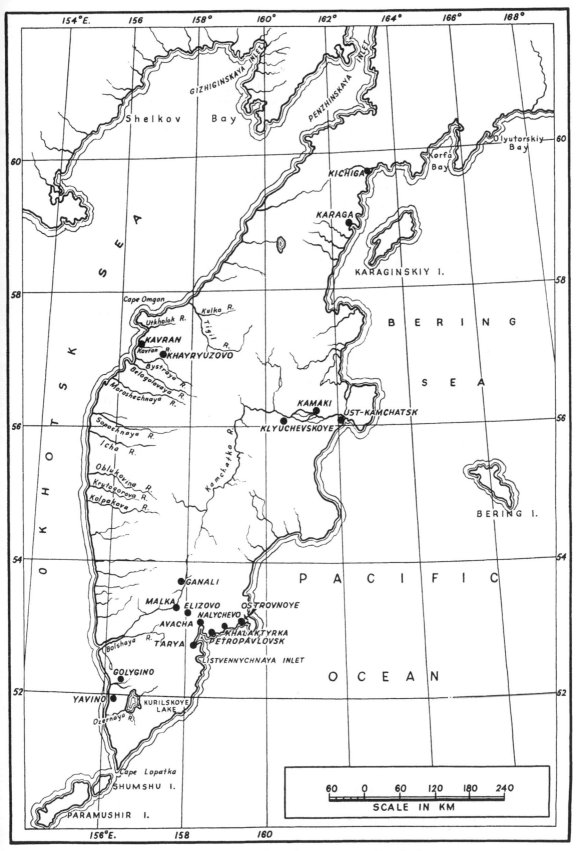

Localities of archaeological excavations in Kamchatka.

dwellings, while in the shore dwellings, in addition to stone and bone objects, iron arrowheads were found together with a fragment of an iron ax. There, in Kenlu, Loginovskiy excavated three hillside dwellings and one shore dwelling. In the hillside site spear heads, arrowheads, adzes, awls, and scrapers, of stone or bone, were found. Among the arrowheads were some of obsidian, which was in all probability brought from the mainland, since no obsidian is found on the island. In the shore dwelling, in addition to the iron implements already mentioned were found sledge runners of bone such as are used by Koryaks to this day. In the kitchen middens of Karagin Island an abundance of seal bones were found, as well as a particularly large quantity of shells of edible mollusks. Near each dwelling, traces of pits lined with [matted] twigs, used for pickling seal meat, were found.

In 1911 and 1912, excavations were made in Kamchatka by V. I. Jochelson, whose report was published in 1928. Jochelson excavated near the city of Petropavlovsk at four points on Tarya Bay: at the village of Savoyko on the left bank of the Avacha river, on the Nalycheva river, and on the shores of Lake Nalycheva. In the central part of the western coast of the peninsula, excavations were carried out by him on the Kulki river, half a kilometer from its confluence with the Tigil, and also on the bank of the Kavran river, not far from its mouth. In southern Kamchatka, excavations were conducted on the banks of the Ozernaya river 3 km from its source in Lake Kurilskoye, as well as on the shores of this lake. In 1926, a small-scale survey was carried out by Ye. P. Orlova on the Khayryuzovaya river, which enters the Sea of Okhtosk south of the Kavran river [57° 05′ N., 156° 45′ E.]. At Avachinskaya Bay on the Tarya peninsula, where Jochelson had excavated in 1911, excavation were made in 1920 by a Swedish expedition. In 1922, the same expedition carried out excavations on the eastern coast of Kamchatka, at Listvennichnaya [Listvennaya?], a bay 200 km north of Cape Lopatka. The finds of the Swedish expedition have been published in part by Bergman and Schnell.[7] Archaeological investigations at Ust-Kamchatsk, which is also on the eastern coast of Kamchatka, were carried out by Nakayama.[8] A small collection from the excavations at Tarya Bay in 1934 was delivered by Assistant Captain N. A. Guryev to the Museum of Anthropology and Ethnography of the Academy of Sciences, and has been published in part by Lev.[9] Of great interest is the collection from the dune campsite on Cape Lopatka, now at Kamchatka Regional Museum in Petropavlovsk. It consists of bone and stone arrowheads, stone knives, scrapers, spoke-shaves, and adzes. No less interesting are the objects given to the Vladivostok Museum by A. K. Vernarder, which came from a cave on Omayan Bay, south of Anadyr Bay, explored by him in 1930. These objects are, however, more characteristic of the littoral culture of the Bering Sea than of the Neolithic culture of Kamchatka. Also worthy of mention is the considerable number of fortuitous finds of stone implements, particularly those from the Yavina and Golygino settlements in southern Kamchatka, now in the Kamchatka and Vladivostok Regional Museums.

In addition to the publications mentioned above I have had the opportunity to familiarize myself with the collections from the Kamchatka excavations of Loginovskiy, Jochelson, and Orlova, and, during my 1945 expedition to the Chukchi peninsula, with the collections from Kamchatka in the Petropavlovsk-Kamchatskiy and Vladivostok museums. A summary of the published materials on the archaeology of Kamchatka was published by Davis.[10]

The archaeological materials of Kamchatka now number thousands of items, and permit a clear concept of the basic types of stone and, in part, bone implements characteristic of the culture of the population of this peninsula before the arrival of the Russians. Nevertheless, this material is insufficient to establish a relative chronology of the sites investigated and the successive stages of development in this culture. However, two and one-half centuries ago, when the Russians first came to Kamchatka, its people were using exactly the same implements and weapons as those found in the ancient settlements. In addition, extremely valuable written information on the life of the people of Kamchatka, at a time when they were not yet using metal, is available to us. It is in the light of this information that we shall examine briefly the material culture of the ancient population of Kamchatka in connection with its ethnic origins.

## Dwellings

According to Krasheninnikov, the Kamchadals built their dwellings in the following manner:

They dig out the earth to a depth of some two *arshins* [56 inches]; the length and width depend on the number of occupants. Four stout poles are placed in the pit, almost at the very center of it, about one *sazhen* [7 feet] and more from one another. Thick crossbeams are placed on the poles, and the roof is set over these, leaving in the center a rectangular opening which serves as window, door, and chimney. Logs are then leaned against the above-mentioned crossbeams, their lower ends being secured in the ground, and after poles are laid crosswise on them to form a grid, it is covered with grass, and dirt is heaped on it so that from the outside the yurt resembles a small round hillock. Inside it is rectangular, yet almost always two of the walls are long and two short. Near one of the long walls between the upright poles there is usually a hearth, with a conduit [above it], the exterior opening of which is much lower than the rectangular opening mentioned above. This outlet is so made that by means of air entering through it the smoke is driven out from the yurt through the higher opening. Inside the yurt, along the walls, they construct shelves, on which one family sleeps next to another. Only in front of the hearth are there no shelves, for there usually stand the family utensils, cups and troughs of wood, in which they cook food for themselves and for the dogs. And in yurts where there are no shelves, logs are put about the places where they sleep and the places themselves are covered with mats. The yurts are entered by means of [notched] ladders, under which the hearth is usually located. Charred logs are tossed out through the upper opening. . . . In these yurts the Kamchadals live from autumn until spring, and then move out into their wooden summer dwellings.[11]

Excavations in Kamchatka have provided supplementary information on the type of ancient dwellings. Almost all of the sod huts excavated were rectangular in shape, elongated or almost square. Only in the northern part of the peninsula are they of irregular rounded shape. In the center, or closer to the wall opposite the entrance, the remnants of the hearth are usually located. It should be noted that every sod hut excavated in the southern half of Kamchatka had a corridor-like depression which served as an entryway into the hut. One old Koryak man told Loganovskiy of a fact also mentioned by Slyunin.[12] In times past the entrance to the hut was on the side, through a door, and only subsequently was the hole closed and the dwelling entered through the "roof" opening. The purpose of this was to protect the dwelling against winter snowstorms and, primarily, to do away with the necessity of shoveling the door clear of snowdrifts. Let us recall that, according to Krasheninnikov, Kamchadal dwellings had, besides the overhead opening and in place of a vestibule or corridor, and underground passageway in the form of a narrow tunnel which started near the hearth and reached the exterior at one of the walls. The tunnel served to provide a draft and was kept open while the fire was burning. This tunnel, which according to Krasheninnikov was called by the Russians *zhupan*, was used for entry and egress by women, children, and the so-called *koyekchuchi* [male transvestites]. Groups of sod huts, including the lightly constructed summer huts, prior to the arrival of the Russians, had been surrounded by ramparts and palisades, for protection in internecine wars and wars with the Koryaks. Logs were hewn and dressed with stone axes with transverse blades or adzes; these will be described later. A bone mattock, probably used for digging, and a whalebone shovel were found by Jochelson on the Kavran river in central Kamchatka.

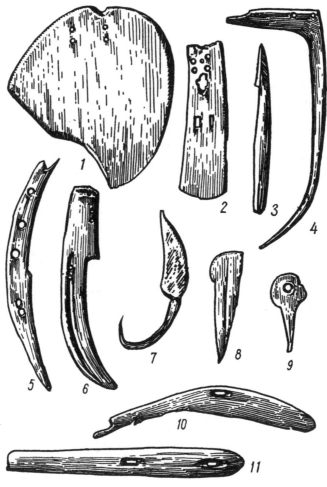

FIGURE 1.

The shovel (Fig. 1, item 1), which is oval in shape, is of a type well known among the inhabitants of the remote northeast of Asia. It was made with a stone adze, the marks of which are clearly visible on its surface. Its lower edge and its sides are sharpened. Its upper edge has two pairs of drilled holes to fasten it to a wooden handle. Identical shovels were found by me in excavations on the

Chukchi peninsula,[13] also by Geist[14] and Collins[15] on St. Lawrence Island. The same type of shovel is still used by the Chukchis[16] and Eskimos.[17] Similar shovels were used by Chukchis and Eskimos for cutting sod for the construction of dwellings, and also for cutting through snowdrifts and for clearing snow away

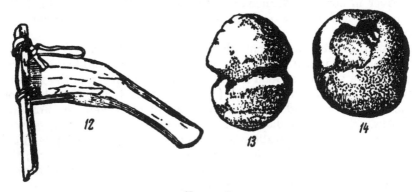

FIGURE 2.

from dwellings. Bone mattocks, similar to the one depicted in Figure 1, item 2, are also known from the Chukchi peninsula and St. Lawrence Island,[18] where they were used for digging and leveling the site selected for a dwelling. The manner of attachment to the handle can be seen from the example of a similar Koryak mattock illustrated in Figure 2, item 12.

## Gathering Economy

The gathering of edible roots was one of the principal occupations of Kamchadal women, with supplies dug from mouse burrows constituting a significant part of their harvest. Krasheninnikov wrote that mice (*Myodes oeconomus* Pall.) store in their nests "roots and other matter which includes martagon, the roots of fireweed (*Anacamferos*), knotweed (*Polygonum bistorta*), and wartwort, and also burnet, crowfoot, and pine cones. With special rites and rejoicing, the Kamchadals remove these edibles from the mouse holes in the fall."[19] Martagon (*sarana*) is the name given in Kamchatka to the rootstock of several species of the lily plant (*Lilium martogon, Fritillaria kamtschatica,* and others). Since ancient times, a special pick was used for digging a highly variegated assortment of edible roots in addition to mouse burrows. Three such picks are known from excavations, one from the north, from the mouth of the Kichiga river, and two from central Kamchatka, from the Kavran river. The pick from the Kavran river, very simple in design, is shown in Figure 1, item 4. It is a tine of a reindeer antler with round holes drilled in the base for fastening it to a handle. The pick from the Kichiga river (Fig. 1, item 5) is of an improved type and is more efficient. It is also made from an antler tine with a curvature characteristic of such implements and several holes for secure fastening to the handle by means of thongs. Typologically similar to it, but of more highly developed shape, in the second pick from the Kavran river (Plate 1, item 6). This rather sturdy whalebone implement is of interest because of the presence of a bilateral, longitudinal groove, which thins the tool considerably and facilitates the drilling of holes required for tying the implement to a handle. The types of picks described above have been used by the inhabitants of northeast Asia up to very recent times. The handles, which no doubt were of wood, have not been preserved. Only one preserved handle of whalebone, possibly intended for a pick, comes from excavations at the Kichiga river in northern Kamchatka (Fig. 1, item 11). Longitudinal apertures in the middle and at the end of this bone shaft served for fastening the implement.

Besides edible roots, the ancient inhabitants of Kamchatka also collected edible mollusks, shells of which have been found in kitchen middens. According to Steller, the main sources of food in his day were scallops from the group *Pecten pacobaeus* and the common mussel (*Mytilus edulis* L.).[20] The shells of these same mollusks were found also in excavations of old dwellings. For extracting the mollusks, principally scallops, from the shells special bone gouges were used. Two of these, both from excavations in northern Kamchatka, are depicted in Figure 1, items 8 and 9. One of them is a small flat wedge of bone (item 8); the other is shaped like a short ear pick, with its end blunted, and there is a hole drilled in its upper part for suspension (item 9). Sharpened bone splinters were also used for extracting mollusks from their shells. Such primitive implements for the extraction of mollusks from their shells are known wherever mollusks have been used as food.

## Fishing and Hunting Implements

Gathering, although significant in supplying the ancient population of Kamchatka with food, had probably always been an auxiliary occupation. Hunting, and in particular fishing, were of primary importance. "No one goes out to hunt or on a journey," wrote Kracheninnikov of the Kamchadals, "without reliable sustenance in the form of dried roe."[21] And it stands to reason, in a land like Kamchatka, where the rivers abound in fish, and where in the spawning season untold numbers of fish enter the rivers in order to spawn, that roe must always have constituted a principal source of food. Nevertheless, excavations have yielded few fishing implements. This is probably to be explained by the fact that previously, as well as at present, the indigenous population of the peninsula fished almost exclusively by use of weirs. From excavations, we have only one bone fishhook barb, one iron fishhook set in a bone shank, and bone platelets with holes for suspension of spoon bait. All these objects come from the Lozhnykh Vestey [False News] Bay of Karagin Island and additionally from the excavations of a late Koryak camping site. The bone point (Fig. 1, item 3) is apparently part of a double-hooked shank with one inside barb at the point and a flat, roughly trimmed haft. The large iron fishhook (Fig. 1, item 7), unquestionably obtained from the Russians, is a typical casting hook. Bone spoon baits similar to the one shown in Figure 1, item 10, are used by the Koryaks in sea-fishing through the ice. Stone net sinkers of two types, grooved (Fig. 2, item 13) and perforated (Fig. 2, item 14), have been found in large numbers in excavations in central and southern Kamchatka.

First among the implements of armament and hunting is the bow and arrow. Among the objects excavated, arrows are particularly numerous, and these exhibit very great variety in material and form. No bows or remnants of bows were found, but very likely they were simple wooden ones; otherwise bone reinforcing plates, which are usual in complex bows, would have been present. In shooting their bows, the natives of Kamchatka made use of bone wrist guards similar to those in widespread use throughout Siberia, from the Urals to the Pacific coast. One such wrist guard (Fig. 3, item 15) was found in central Kamchatka on the Kavran river. Although it is in a poor state of preservation, transverse rows of ornamental identations remain on its outer surface. Such wrist guards are known on the Chukchi peninsula from dwellings belonging to the Punuk stage of Eskimo culture. The second wrist guard, of very crude workmanship (item 16), is from Karagin Island; it was found in a complex of objects of recent make that included iron implements; it is undoubtedly Koryak. Of the arrows used by the Kamchadals, Krasheninnikov wrote that they "are usually an arshin and three-quarters long, with bone or stone points, and they are called by different names according to the type of point. An arrow with a narrow bone point is a *pensh*, with a wide bone point an *aglkpynsh*; one with a stone point is a *kauglach*; an arrow with a blunt bone point or [in Siberian Russian] a *tomar* is a *kom*; a wooden *tomar* is a *tylahkur*."[22] In spite of the fact that, as we shall see below, the points of Kamchadal arrows were usually of small size, they were used for attacking both foe and dangerous beast. "Their arrows," we read in Krasheninnikov, "although of rather poor quality, are nevertheless dangerous in combat, for they are smeared with a poison from which the wounded person immediately swells up."[23] Koryaks, Yukagirs, and Chukchis, recounts the same author,

smear their arrows with the crushed root of the cursed crowfoot, so that wounds inflicted with them are incurable to the foe; and it is the very truth that the wound from such an arrow turns blue right away and the flesh swells up, and upon the passage of two days death will most assuredly follow if the requisite precaution be not observed, which consists only in sucking the poison out of the wound. The largest whales and eared seals, even if only lightly wounded [with such a poisoned arrow], cannot remain in the sea, but with a terrible roar hurl themselves onto the shore and perish miserably.

The poison, according to Krasheninnikov, is obtained from the root of the (cursed) crowfoot; interesting in this connection is the report of the same author that the Kamchadals previously used to obtain this poison from inhabitants of the first Kurile island, i.e., from the Ainu.[24] Concerning the Kuriles or Ainu, Krasheninnikov writes that

in the vicinity of [Cape] Lopatka and their islands they travel about in boats and seek out such places where whales usually sleep, and upon finding them shoot them with poisoned arrows. And although the wound from an arrow is at first imperceptible to such a large animal, soon afterwards it is the cause of intolerable pain, which they manifest by threshing about in all directions and by a most horrible roar; thereafter in a short time they swell up and expire.[25]

It is quite probable that this use of poisoned arrows was borrowed from the Ainus since, with the exception of the tribes enumerated, the other peoples of Siberia did not poison their arrows.

The bone arrowhead from excavations in Kamchatka may be subdivided into three groups: those with simple points, barbed points, and blunted points. The simplest of the types is shown in Figure 3, item 17. This laurel-leaf-shaped bone arrowhead (from the region of the Kavran river) had a thinned base for insertion into a split shaft. The same site also yielded thick, elongated, stemmed arrowheads, ellipsoid and flat rhomboid in cross-section (items 18 and 19). Arrowheads of the same type have also been found in the north, in a cave on Omayan Bay (item 20). A third type is represented by very narrow and long

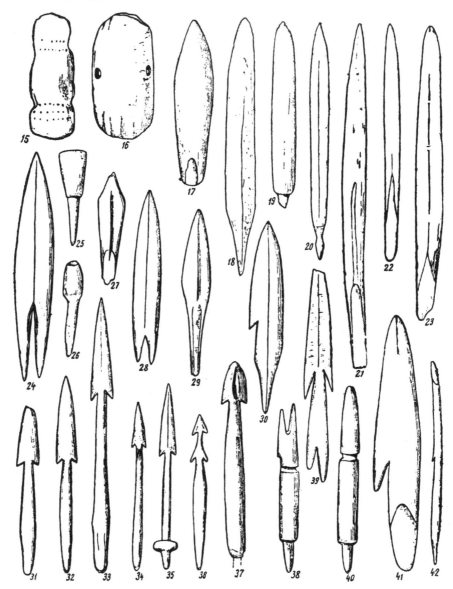

FIGURE 3.

bone arrowheads, with a wedge-shaped base inserted into the shaft. Arrowheads of this type have been found also on the Kavran river (item 21) and on the northeast coast, in the Omayan cave (item 22) and at Olyutorsk Bay (item 23). Arrowheads of a fourth type, flat and wide with a forked base, also came from Olyutorsk Bay (item 24) and Omayan Bay (item 28) on the northeast coast. The only example of a laurel-leaf-shaped arrowhead with a flat stem (also from northeastern Kamchatka) comes from the vicinity of the Kichiga river (item 29). Thus, the area of distribution of simple bone arrowheads is the northern half of Kamchatka, north of the Kavran river, in other words, a region inhabited predominantly by Koryaks.

No less diverse are the bone arrowheads of the second group, those with barbed points. The first type of arrowheads belonging to this group, broad and flat, differ from simple arrowheads only by the presence of one or two small barbs. They occur in two variants, with a flat, wedge-shaped base, or stemmed. Such arrowheads are known only from the northeast, at Olyutorsk Bay and at Omayan Bay (items 30, 39, and 41). Arrowheads of the second type, with narrow and long barbed points, vary as to detail but are all stemmed (items 31 to 36 and 42). Such arrowheads have been found only at Cape Lopatka and belong to a group of arrowheads characteristic of the Ainu culture. Similar arrowheads have also been found by Torii in the course of excavations in the northern Kurile Islands.[26] Of the same group also are the thick arrowheads from excavations in southern Kamchatka at Lake Kurilskoye (item 37), with the [long] shaft of the head of almost circular cross-section, two barbs at the very tip, and a thinned base annular in cross-section. One of the flattened sides of the leaf-shaped tip has a deep groove. This groove could have served as a repository for poison. Arrowheads of this type are genetically linked to ones provided with bone [inset] blades, found by Torii in the northern Kurile Islands.[27] The bone arrowheads or points of arrowheads found at Lake Kurilskoye (items 38 and 40) are of the same category. These arrowheads are circular in cross-section, and have a [pointed] stem for insertion into the shaft; they have an annular groove about the middle [of the shank] and a slanted plane on one of the sides. The forward and, into which the blade was inserted, was pointed or forked. These arrowheads of special form may be linked typologically only to similar Eskimo arrowheads.

Blunt bone arrowheads are very few in number, and are apparently not typical of the cultures under consideration. We know of only three such arrowheads: one from Omayan Bay, in the extreme northeast (item 26), one from the Kavran river region in central Kamchatka (item 27), and one from Cape Lopatka (item 25). The last is similar to one found by Torii on one of the northern Kurile Islands.[28]

Stone arrowheads from Kamchatka are very numerous, and are highly diversified in form. Some are truly miniature ones, 1.5–2 cm in length; others are so large that they could be taken for darts or spear points. To distinguish precisely between arrowheads and spear heads is not easy; for Kamchatka the most reliable criterion is apparently not so much dimension but massiveness. With rare exceptions, arrowheads are light, narrow, and long, and in those cases where they are short, they are nevertheless light and flat. Dart points and spear heads are thick, massive, and heavy. The classification of Kamchatka arrowheads is made difficult because of the lack of consistency in their shapes, as well as their variability and the presence of transitional forms among the individual types. Most of the Kamchatka arrowheads at our disposal do not come from excavations, but are chance finds. For this reason chronological changes in form and local variations cannot be established and it becomes difficult to study them in detail. In 1897, Wilson, dealing with American stone arrowheads, and in 1932, Schnell, dealing with those from the Kurile Islands, adopted a typological division of three basic groups: leaf-shaped (most frequently in the shape of a willow leaf), triangular, and stemmed. According to our calculations, the leaf-shaped arrowheads are the most numerous in Kamchatka. Approximately two-thirds of all the arrowheads are of this category. Stemmed and triangular ones are encountered less frequently. The great majority of the leaf-shaped stone arrowheads are of laurel-leaf (Fig. 4, items 43 and 44) or willow-leaf shape (items 45, 46, and 49). The stemmed arrowheads are usually rhomboid (items 47, 48, 53, and 54). The

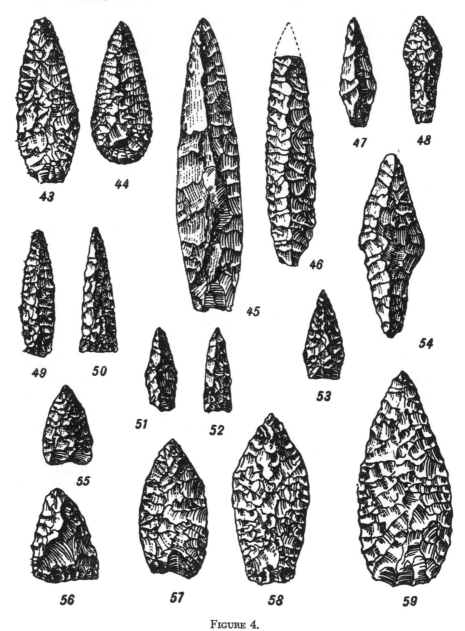

FIGURE 4.

triangular ones are more frequently elongated, with a concave base (items 50, 52), and in rarer instances are wide, with a straight or concave base (items 55 and 56). We repeat that among these types there are also transitional forms. Judging by the thick bases and stems of the stone arrowheads, they were attached directly to the wooden shaft. Only the flat ones with a straight or concave base (items 50, 52, 53, and 55) were probably inserted as blades in the slits of bone arrowheads. Arrowheads of this type are particularly characteristic of southern Kamchatka and the neighboring Kurile Islands.

The aggregate of Kamchatka arrowheads is typically Neolithic and has much in common with arrowheads found to the west, in [administrative] eastern Siberia starting with the Cis-Baykal region,[29] and to the south, on the Kurile and Japanese islands.[30] Kamchatka arrowheads are not typical of the extreme northeast of Asia, in the area of Eskimo culture.

Stone spear heads and dart points differ from each other only in that the former are more massive, and the latter considerably lighter in weight. In the absence of shafts, a sharp distinction cannot be drawn between them and, therefore, they will be considered together.

Of the numerous series of dart points from Kamchatka, only three are illustrated here: one each from the Kávran river (item 58) and the Kula river in central Kamchatka (item 59), and one from Lake Kurilskoye in southern Kamchatka (item 57). All are of obsidian, leaf-shaped, and excellently finished on both sides with pressure retouch. Such points are encountered everywhere in Kamchatka.

Spear heads are considerably larger than dart points and present a greater variety. Some of them are leaf-shaped, and have a thinned base that is inserted into the shaft; such is, for example, the one shown in Figure 5, item 62, from Lake Kurilskoye, or item 64, from Tarya Bay. Frequently they are rhomboid in form with a distinctly separated stem (item 60, from Tarya Bay), sometimes with a constriction above the stem (item 61, from the Kulka river). There are also intermediate forms. The spear heads are flatter than the dart points.

Arrowheads, dart points, and spear heads were made of obsidian, flint, chalcedony, and, less frequently, of jasper, silicified tuff, and flinty slate.

In ancient times, sea mammals used to frequent the coast of Kamchatka in great quantities. Krasheninnikov wrote, of seals, that there is "in those seas an indescribable multitude of them, particularly at the time when the fish come from the sea to go upriver, in pursuit of which they not only go into the estuaries, but also follow the river far upstream, and in such herds that there is not a river island close to the sea that is not covered by them as if by driftwood." According to Krasheninnikov the Kamchadals used to kill seals, both the common and eared varieties, with harpoons at breathing holes. He remarks, however, that "the Kamchadals never attack an eared seal at sea, knowing that it overturns and sinks manned boats. They shoot seals sleeping [on ice-floes] at sea with poisoned arrows, and then retire. Unable to endure the pain caused by the inflammation of the wound by the seawater the seals come out on shore, where they are either speared or, if the site is unsuited for slaughter, die by themselves."[31] Harpoons, according to Krasheninnikov, were used for hunting seals, sea otters, and sea cows, the last being hunted with iron-tipped harpoons. It should, however, be noted that the Kamchadals apparently never hunted sea mammals to any great extent, and went to sea in [wooden] boats rather than skin boats. The real open-sea hunters of sea mammals were the Kuriles, who "in the vicinity of both [Cape] Lopatka and their islands travel about in skin boats and seek out such places where whales usually sleep, and upon finding them they shoot them with poisoned arrows."[32] Regarding the Kamchadals, in comparison, Krasheninnikov wrote that they "do not possess sea[-worthy] vessels."[33] In spite of the abundance of sea mammals, hunting them was always of but tertiary importance to the economy of the indigenous population of Kamchatka, since, as Krasheninnikov wrote, "all of Kamchatka is nourished by fish alone."[34] This no doubt explains the extreme paucity of harpoon heads in Kamchatka. Among the many hundreds of stone implements, and the many dozen types of implements known to us, only few

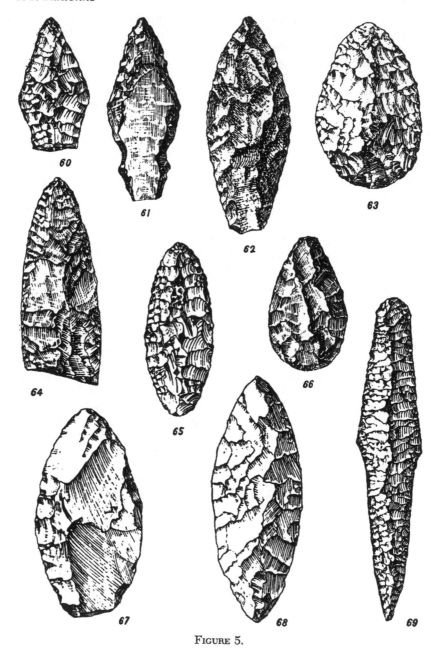

FIGURE 5.

specimens of bone harpoon heads were found. A barbed harpoon head from
Tarya Bay, illustrated by Schnell (Fig. 7, item 88), has two asymmetric barbs.[35]
However, this is not a toggling harpoon head, but is firmly fixed to a shaft, and
could not have been used for the hunting of sea mammals. This is the point of a
fish spear for spearing salmon. There is a distal half of a harpoon head from
Cape Lopatka (item 89) which has a line hole and could be a toggling one.

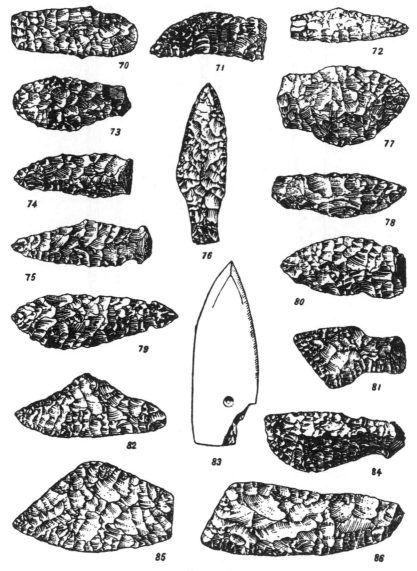

FIGURE 6.

However, this head probably belonged not to the Itelmens, but to the Kuriles. Barbed toggling harpoon heads have been found in excavations on the neighboring islands of the Kurile group.[36] A typical harpoon head shaft (Fig. 7, item 90) was found in the course of excavations at Lake Kurilskoye in southern Kamchatka, and two bone socket pieces (items 91 and 92) were found in central Kamchatka on the Kavran river. This constitutes all the objects of the [toggling] harpoon complex found in Kamchatka. In this connection, it should be stated that all these objects belong typologically to a comparatively late period, that is, by analogy with Eskimo harpoon heads, they may be dated to the 2nd millennium A.D.

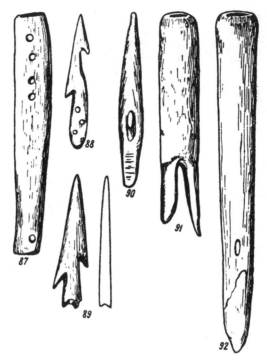

FIGURE 7.

## Stone Knives

Krasheninnikov wrote concerning the ancient Kamchadals: "They make their knives from greenish and smoke-colored rock crystal, sharp-pointed like lances, and set them in wooden hafts."[37] We know, however, that stone knives were used also without special handles, being gripped directly by the hand. Thus, "the women do all of their work wearing gloves made without fingers, and never take them off."[38] Stone knives were used for a great variety of tasks: for skinning animals and carving their carcasses, for filleting fish, for tailoring skins for clothing, for woodworking, and so on. This makes for a large variety of knife types and shapes. Of archaic appearance are two implements from Lake Kurilskoye, one of obsidian (Fig. 5, item 63) and one of silicified tuff (item 66), and an implement of smoky quartz from the Kavran river (item 67). Because of their massiveness and the thickness of their bases, they could not have been fastened to a handle, and they were probably used by being grasped directly in the hand. Crescent-shaped knives, resembling those shown in Figure 6, items 82 and 85, were also used without handles; both these items are of obsidian, from Tarya Bay. Of unusual form and magnificently crafted is a large knife from the Khayruzovaya river, made of smoky quartz (Fig. 5, item 69). According to the Itelmens, such knives were used for filleting fish and were held directly in the hand; both of the ends, the pointed and the blunted, were used although the major part of the work was done with the blunted end. In spite of its finely finished cutting edges, this knife is easily held and, with a gloved hand, absolutely safe. The leaf-shaped knives of silicified tuff (Fig. 5, item 62) from Lake Kurilskoye, and of flint (item 65) from the Kulka river, were probably also used without special handles.

Notable among wedge-shaped blades insertable into a wooden handle are the lamellar [roughly] rectangular ones. They are thoroughly retouched on both surfaces, and could have been used as side blades in bone or wooden handles. However, such side blades are encountered very rarely in Kamchatka, and altogether we know of only two or three. The rectangular obsidian knife blade from Tarya Bay (Fig. 6, item 70) is of such form.

The series of asymmetrical knives is numerous and varied; several types may be distinguished:

1. Knives with a more or less straight dorsum and a convex cutting edge, both narrow-bladed and wide-bladed, with and without a hafting stem. An example of a stemless, narrow-bladed flint knife from Lake Nalychevo is shown in Figure 6, item 72, and a wide-bladed one from Tarya Bay is illustrated by item 86. Items 76 and 78 show, respectively, a narrow-bladed stemmed knife of silicified tuff from the Kavran river, and an obsidian one from Tarya Bay. Wide-bladed stemmed knives are particularly numerous; item 73 in Figure 6, a knife of silicified tuff from the Kavran river, and item 77, from Lake Kurilskoye, may serve as examples.

2. The second type of knife, in essence differing very little from the first, consists of those with a convex dorsum and a straight cutting edge; these also are found as narrow-bladed and wide-bladed, and may be represented by two obsidian knives from southern Kamchatka (item 74 from Tarya Bay and item 80 from Savoyko village). The knives just described may be called butcher knives. Those of cruder workmanship served for dismembering carcasses and for cutting meat; those with a better finish were used for skinning animals and scraping hides.

3. The "humped" type of knife is comparatively rare. The criterion which makes such knives classifiable as a separate type is the presence of a finely retouched concave cutting edge as, for example, in the obsidian knife from Lake Kurilskoye (item 71).

4. Knives with a constriction at the stem, with a slightly convex dorsum and a strongly convex cutting edge, are also encountered in two variants, narrow- and wide-bladed. The narrow-bladed knives of this type become narrower towards the point and, as for instance, the knives of silicified tuff from the Kulka river (item 75), while the wide-bladed ones usually become wider towards the point, as, for instance, the obsidian blade from Tarya Bay (item 81) and one from southern Kamchatka (item 84). This type of knife is numerous in Kamchatka, being encountered both in the central part as well as in the southern part of the island. It is also known outside of Kamchatka, particularly in Japan[39] and in the Aleutian Islands.[40] The use of such knives is the same as that of the butcher knives, but it is not certain whether they were hafted. In some of them the constriction facilitated firmer fastening by means of special lashing. Some of these knives, on the other hand, were not haftable, either because they had a very thick dorsum (item 75), or because of the small size of the base and, consequently, its fragility. The latter was true especially of knives with a thin collar as, for instance, of the blade of smoky quartz from the Kavran river (item 79). The constriction in such knives more likely served for the fastening of a leather thong for suspension. Similar knives are known also in Japan.[41] The almost total absence of polished slate knives in Kamchatka is noteworthy. Women's tailoring knives (*ulu*), so typical of the late stages of Eskimo culture, have not been found

there, and only two examples of men's slate knives are known: one complete stem-like knife with a biconically drilled hole close to the base (item 83) from Avachan Bay and a fragment of another from Kavran river. Of interest too is a bone knife handle (Fig. 7, item 87) from the Kavran river, the only one of its kind known to us. A hole for suspension has been drilled at one end and, at the other, near the blade socket, four ornamental holes. These holes were drilled with a metal bit. Typologically the handle closely resembles the well-known late Eskimo handles from the Chukchi peninsula. The absence of polished slate knives in Kamchatka is all the more striking, because their distribution is extensive in regions peripheral to Kamchatka, the Japanese and the Philippine islands, the mainland coast of the Sea of Okhotsk, and particularly in the north where they are typical of the late stages of Eskimo culture.

## Stone Scrapers

These are very numerous in excavations in Kamchatka. Krasheninnikov, describing in detail the techniques of dressing furs and skins, recounts that the Kamchadals removed the sheathing and tendons with a "stone fastened to the center of a stick,"[42] after which they scraped them with a knife. Round, disklike scrapers are rarely encountered in Kamchatka. One such obsidian scraper is known to us from the upper reaches of the Sopochnaya river in central Kamchatka. It is of comparatively small size, roughly chipped on both surfaces. Besides discoid scrapers, one may also distinguish end-scrapers worked out [on one end] of long flakes, trapezoid or pear-shaped ones, ellipsoidal ones, sometimes almost round ones, and scrapers or special shape. End-scrapers on long flakes constitute one of the most extensively distributed types in Kamchatka. The flakes are split off a core, usually with a well-defined striking platform and bulb of percussion on the proximate, flat side. The convex distal surface is roughly chipped and retouched. With their characteristic unilateral trim, some of these scrapers show careful chipping and fine retouching of the edges, while others, and these constitute the majority, are rather crudely shaped. As examples of such scrapers we show one of silicified tuff from Karagin Island (Fig. 8, item 94), another from Tarya Bay (item 93), and one of quartzite (item 95) and another of silicified tuff (item 96) from Lake Kurilskoye. Scrapers of this type are encountered all over Kamchatka. Trapezoidal scrapers are usually of elongated form, with a steep, almost rectangular, working edge. Good examples of this type are the obsidian scrapers from Avachinskaya Bay (items 97 and 98). By analogy with similar Eskimo scrapers, it may be projected that these scrapers were set in a short handle. The elongated obsidian scraper from Tarya Bay (item 107) is of the same type. Pear-shaped scrapers are particularly numerous, and are typical of the Kamchatka Neolithic. We estimate that there are over sixty such scrapers from Jochelson's excavations alone. While this group has a number of general features (one side is flat, the other is convex with a longitudinal ridge in the middle which gives the scraper a triangular cross-section, a broad and a steeply retouched working edge, a pointed base for insertion into a handle), many variations exist. Some are in the form of elongated leaves with almost parallel ribs; such is the scraper of silicified tuff from Lake Kurilskoye (item 99). Others are almost triangular, like the obsidian one from the same region (item 100). Still

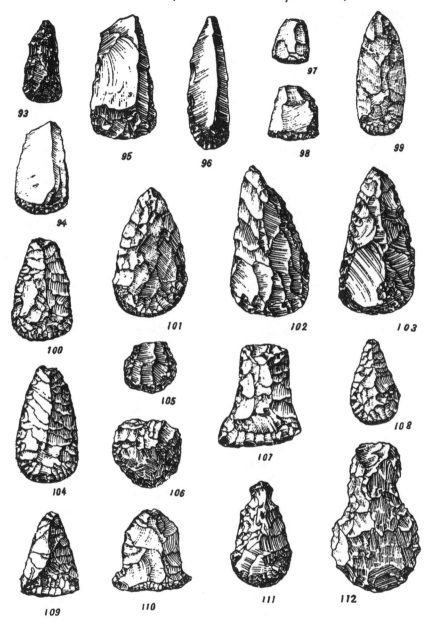

FIGURE 8.

others are almost regularly pear-shaped (item 101, from Lake Kurilskoye), or are others it is carefully refinished with secondary retouch (item 104, from the Kulka river). In a few scrapers of this type both surfaces are retouched (item 108 from Avachinskaya Bay). Ditmar wrote about the scrapers from the yurt complexes which he excavated: "The scrapers made by chipping are similar to those that are even today used by the Koryaks for scraping raw hides. . . . They are of elongated pear-shaped form . . . the broader end serves for scraping. . . . The other, narrower end is, by means of narrow thongs, clamped into a handle consisting of two pieces of wood. . . ."[43] The obsidian scraper from Lake Kurilskoye asymmetrically pear-shaped (item 102, from the same locality). The convex surface of some is almost without retouch (item 103, from Sovoyka village); in (item 105) and the one of silicified tuff from the Kavran river (item 106) are round, of smaller dimensions, with a flat ventral and a convex dorsal side, with the working edge retouched along almost the entire circumference. The narrow, triangular-shaped obsidian scraper from Cape Nalycheva (item 109) is rare. Equally rare is the bell-shaped scraper of silicified tuff from Lake Kurilskoye (item 110). The ventral side is flat; the dorsal side is convex and its entire surface is finished by pressure flaking. In spite of their unusual shape, made so as a result of a deliberate widening of the working edge, such scrapers are widely distributed. Similar scrapers occur to the south, in Japan, among the Eskimos of North America, and in large numbers in Greenland.[44] The appearance of such a scraper in southern Kamchatka may be the result of Ainu influence. Of singular shape are the flint scrapers from Avachinskaya Bay and from the Kulka river (items 111 and 112), with a constriction for suspension on a thong. In shape, these scrapers are reminiscent of the knives with a constriction near the stem, but their profile, and the unilateral chipping of the working edge, leave no doubt that they are scrapers. With the exception of the bell-shaped scraper, which is presumably connected with the southern influence of the Ainu, all the basic types of scrapers are distributed throughout Kamchatka; the most typical of these are the "two-handled" pear-shaped scrapers inserted into a wooden handle.

## Adze-like Scraping Tools

A large and diversified group of stone implements, very typical of Neolithic Kamchatka, consists of adze-like tools of elongated shape, sometimes approaching the triangular, with slightly cambered ribs, with the entire surface carefully polished. These were made not only from varieties of hard stone, such as jasper and felsitic porphyry, but also from soft varieties, such as sandstone and clayey shale. As noted earlier, implements of this type are characterized by a triangular shape, sometimes very elongated (Fig. 9, items 113 and 114), more rarely squat (item 117). In some instances they are almost regularly triangular (item 118) with slightly convex sides. One of the largest such instruments is comparatively flat (item 113), as is also item 118, while in the case of the implements represented by items 116 and 117, one of the flat sides forms such a steep angle with the other two that the cross-section has almost the shape of an isosceles triangle. The purpose of these implements is unknown, since their use for wood shaving, in view of the availability of true adzes, would not have been very expedient. Besides, these implements rarely have a jagged or blunted cutting edge from long use. Nevertheless, they have an extensive distribution in Kamchatka. They occurred everywhere in excavated dwellings, and are the most common object in surface collections. The Itelmens called such implements *aut*, which is a scraping tool used in the processing of hides. The only example is an implement of flinty slate (item 115), which is somewhat similar in shape to those described above, but narrower and flatter, carefully ground and polished at both ends.

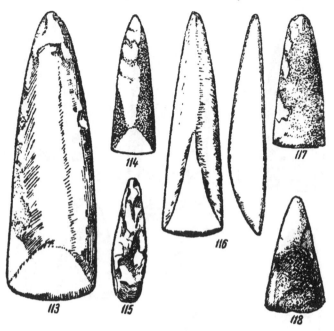

FIGURE 9.

*Adzes*

These are also typical of Neolithic Kamchatka. Krasheninnikov wrote, about Kamchadal "axes": "They were made of reindeer and whale bones [?], also of jasper, in the form of a wedge, and were tied by straps to curved hafts, such as adzes are with us [i.e., among the Russians]. With them they hollowed out their boats, bowls, troughs, and other things. . . ."[45] The extensive collection of such stone implements can, in spite of the great diversity of shapes and sizes, be classified as a single group of axes with a transverse blade, that is, adzes. All of them are made of hard rock. Some of them are rectangular in shape; some are of a trapezoidal shape approaching the triangular. Some are thick, others thin. They also vary greatly in size, ranging from 4.3 cm to 16.4 cm in length. The cutting edge of all the adzes is sharpened from both sides, but asymetrically. With the exception of the unfinished tools, the surface is almost completely polished. On the basis of shape, and to a certain extent purpose, three subgroups of adzes may be distinguished. To the first belong adzes of large size, which were hafted. These are either very thick, rectangular in cross-section, with a steeply chipped cutting edge (Fig. 10, items 119 and 120), or flat thin ones with a less steeply chipped cutting edge (items 121–123). To the second subgroup belong adzes of medium dimensions (items 124, 125, and 127). The third subgroup comprises carefully finished miniature adzes which were hafted by inserting the butt into an antler collar or directly into the socket of the handle (item 126). Such small adzes were more likely used for working bone than wood. Adzes occur throughout Kamchatka, and are particularly frequent as chance finds.

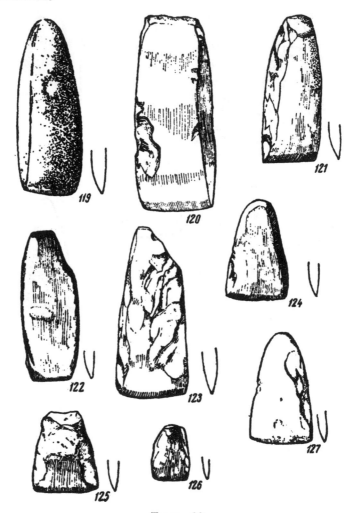

FIGURE 10.

*Stone Reamers and Chips Used as Tools*

Both hand reamers (Fig. 11, item 128) and drill bits (items 129 and 130), mostly of obsidian, were used for drilling holes. Although not as numerous as knives, scrapers, or adzes, drill bits have nevertheless been found regularly in excavations all over Kamchatka. In addition to the easily recognizable types of tools, the Kamchadals undoubtedly made use of stone chips in their manufacture of various objects of wood and bone. Unfortunately, little attention was paid to such chips during excavations, and they are completely lacking in such large collections as those of Jochelson and Loginovskiy. However, a quantity of such chips (items 131–134) was found in the small collection from the Khayruzovaya river, gathered by Orlova. From a careful examination of these it can be seen

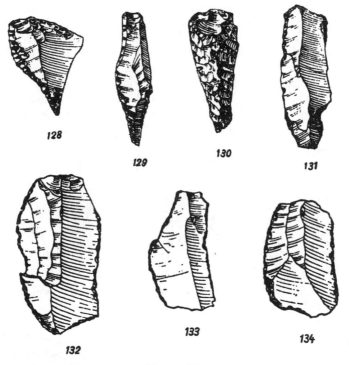

128

129

130

131

133

134

132

FIGURE 11.

clearly that some of them with sharp cutting edges could have been put to immediate use as tools; indeed, some of them bear signs of use, while on others shallow retouching is clearly discernible.

## Food Preparation and Utensils

The number of household objects found in the course of excavations is very small. These include hooks of reindeer tines used in suspending vessels over the fire or in the dwelling (Fig. 12, item 140). A common find is stone pestles, which are exemplified by items 141 and 142. Stone lamps, frequently of very primitive design (items 145 and 146), are often found. Thirty-six stone lamps were found by Jochelson alone in the course of his excavations of dwellings. They vary in size and shape: round, elliptical, egg-shaped; none, however, are rectangular. Some of them are simply pebbles with a depression for the oil (items 145 and 146); others have a special groove or ear for suspension (items 143 and 144). The majority are flat-bottomed, and stand without a supporting base. Unlike Eskimo oil lamps, the Kamchatkan ones have neither crossbars nor grooves for the wick and oil. None are of clay or soapstone, hard rock being the material used.

Two types of bone spoons are known. There are deep spoons with a very long straight handle, as for instance, from the excavations at the Kavran river (item 135). This type of spoon is typical among the tribes of northeastern Siberia, including the Eskimos. Spoons of the second type are flat, shallow, and with comparatively short handles (items 136 and 137); they are also from the Kavran river. Judging by the presence of drill bits, fire was started by [friction] drilling. Such a method of obtaining fire has been described in detail by Krasheninnikov.[46]

The inconsiderable amount of pot remains, particularly clay pots, found in the course of excavations of dwellings in Kamchatka, is apparently due to the common use of a special way of preparing food. According to Krasheninnikov, ornaments were roasted by the Kamchadals "in fire pits."[47] In the same pits, heated like an oven, seal meat was also "steamed."[48] The flesh of sea and land animals

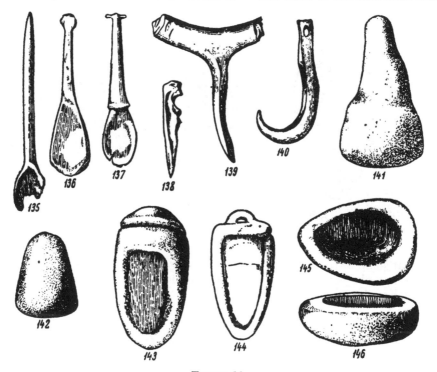

FIGURE 12.

was cooked in troughs, "and in such utensils they cooked fish and flesh by means of heated stones."[49] Kamchadal vessels, according to Krasheninnikov, consisted "of bowls, troughs, and birch-bark vessels." Krasheninnikov does not mention Kamchadal clay vessels and Schrenck thought that clay vessels were not known at all to Paleo-Asiatics. On the other hand Atlasov, the first of the Russians to visit Kamchatka, notes that clay vessels were made by Kamchadals.[50] Later Vitsen, basing himself on Cossack data, mentioned the presence of clay vessels among the Kamchadals. In 1852, Ditmar was the first to excavate clay vessels at Bichevinskaya Bay. These vessels were "of most primitive workmanship"[51] and were only lightly fired. The vessel described by him had a height of 10 cm and a maximum diameter of 14 cm. Fragments of clay utensils have been found in small number at excavations everywhere.

Kamchatkan pottery was divided by Jochelson and Schnell into three types; Quimby[52] distinguished eight types, mainly on the basis of ornamental motifs, which, in the final reckoning, may be reduced to the same three groups. To the first group belong comparatively thin-walled, shallow vessels, having a height of about 25 cm with a maximum transverse diameter of about 20 cm. Such vessels are fashioned by hand, and their entire outer surface is covered with pseudo-textile ornamentation. A good example of such a vessel from the coast of the Sea of Okhotsk (between the Nayakhan and the Gizhiga rivers) is reproduced in Jochelson's work on the Koryaks.[53] Sherds of vessels of this type have been found only in northern Kamchatka, along the Kulka and Kavran rivers. This type of pottery was widely distributed in the extreme northeast of Asia all the way to Bering Strait. Small, thin-walled vessels, with ornamentation consisting of a num-

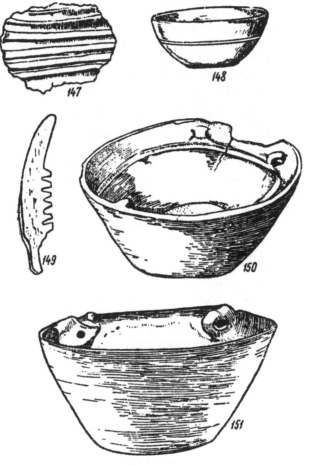

FIGURE 13.

ber of parallel lines with dots or dentates near the rim, are of the same group. The region where pottery of this type was found is limited to central and northern Kamchatka. Numerous examples of this pottery (from the Kulka and Kavran rivers) are reproduced in the work of Jochelson.[54]

The second type of pottery consists of vessels of coarser workmanship and thicker walls, frequently of large dimensions (up to 35 cm in diameter); at the lip they have four or five appliqué ornamental relief strips, separated by grooves (Fig. 13, item 147). Such pots are encountered principally in southern and central Kamchatka, but sherds of this type have been found also in the north. A bone implement (item 149) found by Jochelson at Lake Kurilskoye may have been used for the application of such grooved ornamentation. De Laguna is of the opinion that such ornamentation was produced by affixing strips of clay along the rim of the vessel.[55]

The third type consists of cauldron-like, flat-bottomed, thick-walled vessels of crude workmanship, unornamented, with lugs for suspension inside the vessel close to the lip. Their usual diameter along the wide rim is 30–40 cm and the height is 15–20 cm. These vessels with interior lugs (items 150 and 151), especially typical of southern Kamchatka, are of great interest. Aside from southern Kamchatka, they have been excavated on Sakhalin Island as well as in the Kurile Islands and Japan. The Japanese traveler Mamiya Rindzo, who visited Sakhalin early in the 19th century, saw such clay vessels with interior lugs still in use among the Ainus. They were hung over the hearth by thongs sheathed in birch bark for protection against the fire. On the basis of information given by the Ainu inhabitants of Shikotan Island (the Kurile group), Torii has described in detail the method of making such cauldron-like clay vessels.[56]

It should be recalled that the Kamchadals, as well as the Ainus, made extensive use of heated stones for cooking food in wooden vessel.[57] It is because of this that a comparatively small number of clay vessels have been excavated in comparison with other objects. Regarding the clay vessels with interior lugs for suspension, so typical of Ainu culture, the presence of similar metal vessels would seem to indicate their recent origin, possibly as copies of metal cauldrons. Especially notable is a bowl from the Kamchatka river valley, found near the village of Klyuchevski. This is a small, shallow vessel (item 148), with walls about 1 cm in thickness. The paste is well prepared, and is of fine texture and black color; the exterior surface is black and glossy. This vessel was made on a potter's wheel, and was well fired. It has much in common, in both form and material, with similar bowls of Ainu or Japanese origin. In his work on the Ainu of the Kurile Islands, Torii illustrates a similar bowl of wood.[58] Munro describes identical Japanese bowls.[59] It is quite possible that this bowl is of foreign origin, and was brought to central Kamchatka from Japan via the northern Kuriles, or on a Japanese ship driven by chance to this region.

## Miscellaneous Objects

For sewing clothes, the Kamchadals used "bone needles, and in place of thread they used reindeer sinews, which they split and twisted like thread, as thin or as thick as required."[60] The needles were kept in needlecases made "of the wing bones of albatrosses."

Awls of various sizes and shapes were used for stitching hides. Examples are awls from Lake Kurilskoye (Fig. 12, item 138) and from the Kavran river (item 139), the latter of a time. Pointed bone splinters, most frequently those of wing bones of geese and ducks, were also used as awls; numerous examples of these are shown in Jochelson's work.[61] The only accessories to clothing known are some bone disks which probably served as buckles. One of these, from excavations at Lake Kurilskoye, is ornamented (Fig. 14, item 158). This buckle is similar to one found by Torii while excavating at a Neolithic campsite on Shumashir Island.[62] The narrow bone combs with long teeth excavated by Jochelson at Lake Kurilskoye and by Torii on Shumashir Island are identical in form. Such combs (item 185) were more probably used for combing hair or grass than for applying ornamentation to clay vessels. Loginovskiy considered the button obtained from excavations on Karaginsky Island (item 160) as an accessory to children's clothing, since similar buttons are sewn by Koryaks to the trousers for the purpose of attaching to them a small moss-filled bag. Very few objects excavated on Kam-

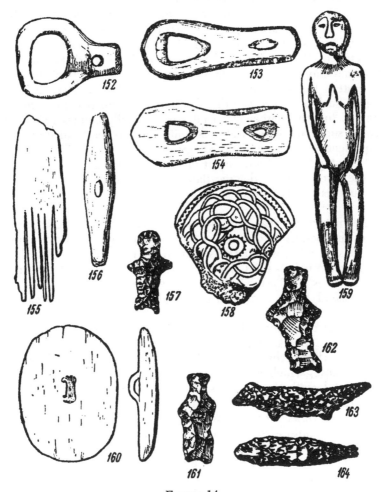

FIGURE 14.

chatka have to do with transportation. There are only either pegs (item 156) or blocks (items 152–154) of dog harnesses, or, finally, the bone runners found on the Kavran river and on Karagin Island.

Among cult objects there is a small tusk from Lake Kurilskoye with a hole for suspension; it is rather an amulet than an ornament.[63] Of interest is the figurine of a woman carved from a walrus tusk (item 159). It is a representation of a nude woman, with realistically depicted sexual characteristics and a crudely outlined face. The legs are bowed, the breasts udder-like; neither fingers nor toes are indicated; there is a longitudinal spinal groove characteristic of the human figure, but no indication of external ears. According to Loginovskiy, the statuette was acquired from Koryaks of the Kichiga river. Their tradition states that it arrived in a boat and was found there on the riverbank by a woman. Judging by the color of the bone and the absence of patina, it is unlikely that this figurine is particularly old; it is probably of Koryak origin and may be linked genetically to similar cult figurines of the Coastal Chukchis and Eskimos. Of special note are the miniature stone representations of man and animals. Items 157 and 161 to 164 comprise five small obsidian figures: a fish (item 164), a dog or fox (item 163), a seal (?) (item 161), and two human figures with outstretched arms (items 157 and 162). All are from Avachinskaya Bay, more precisely, from Tarya Inlet. The first two are in the Regional Museum in Vladivostok, the other three in the Museum of the Academy of Sciences in Leningrad. Similar human and animal figurines have long been known from the "developed" forest Neolithic culture of eastern Europe and the Neolithic cultures of western Europe and the United States. Such a widespread distribution in regions distant from one another and not linked together either ethnically or culturally suggests the independent emergence of such cult objects at a specific stage of cultural development and with the presence of the same ideological foundations. Closely analogous to figurines of this type, but of a different material and produced by a different technique, are the "protectors" and "guardians" of various sorts, so well known among the same Chukchis and Eskimos.

All of the materials excavated in Kamchatka indicate a comparatively late settlement of this peninsula by peoples with an already fully developed Neolithic culture. Jochelson believes that the southern complex of Lake Kurilskoye is the most ancient in Kamchatka.[64] I think that the most ancient of the known settlements are the sites of Tarya Inlet. However, even in these sites, we already encounter polished stone implements of highly perfected workmanship, adzes, and adze-like scraping tools.

While the level of cultural development was the same for all of the ancient populations of Kamchatka, three complexes, differing somewhat from one another according to the three ethnic groups that represent them, can be distinguished. The oldest of these, which is particularly well expressed in central Kamchatka, is the one we link with the Itelmens [Kamchadals], and the two later ones are the northern Koryak and the southern Ainu complexes.

The complex of objects typical of the ancient Itelmens has little in common with the Neolithic culture to the southwest, along the lower Amur, or with the Ainu culture to the south, or with the Eskimo culture of the Bering Sea coast. It is the culture of the fishermen and hunters of the taiga zone of Siberia, which is possibly linked genetically with the Neolithic cultures of the upper Lena river and the Cis-Baykal. The ancestors of the Itelmens moved in from the north to settle a peninsula which, prior to their arrival, was apparently uninhabited. The second, more recent wave of people consisted of the Kuriles or Ainus, who arrived from the south and, if we adhere to Krasheninnikov's view, in the course of their advance northwards encountered the Itelmens on the last of the Kurile Islands and at Cape Lopatka. However, their intermixture with the Itelmens took place a long time ago, not later than the end of the 1st millennium A.D. As a result of this intermixture, there emerged in the south of Kamchatka a culture which, under evident Ainu influence, differed somewhat from the culture of central Kamchatka. Thus I am inclined to regard this culture, which Quimby considers to be the initial Itelmen culture,[65] as a more recent one formed in the process of mixing of ancient Kamchadals with the Kuriles or Ainus.

The third and most recent wave came from the north and was that of peoples ancestral to the Koryaks; they gradually pushed the Itelmens out of the northern half of the peninsula. In their material culture they differed little from the Itelmens. In contrast with the square dwellings of the Itelmens, their dwellings were round; their finer earthenware bore some resemblance to the Late Neolithic pottery of the lower Amur region. As a result of military encounters between these two peoples, their ancient settlements were fortified by ramparts in the area of contact; this has not been observed in the south.

The economic basis of the ancient population of Kamchatka was fishing and, in part, hunting. Hunting of sea mammals was of tertiary significance. Sea mammals were hunted in the open sea only by the Ainus. The primitive social structure was retained in Kamchatka until the arrival of the Russians. According to Krasheninnikov, the population of the peninsula lived "in complete freedom, having no chiefs over themselves. . . . In each community old people and bold ones possessed an advantage, which, however, consisted only in the fact that their advice was [preferentially] considered; in everything else there was equality among them; no one could command another, and no one could punish another of his own accord."[66] Nor did cult leaders stand out against the general social background. "The Kamchadals," wrote Krasheninnikov, "have no special shamans, as do the other peoples there, but any woman, and particularly an old woman, as well as any fool is esteemed as a soothsayer or as a diviner of dreams.[67]

# Notes and References

1. P. Krasheninnikov, *Opisaniye zemli Kamchatki* (A description of Kamchatka), St. Petersburg, 1755, vol. 2, p. 34.

2. R. Torii, Etudes archéologiques et ethnologiques: Les Ainus des Isles Kouriles, *Journal Coll. Sci. Imp. Univ. Tokyo*, 1919, vol. 42, art. 1.

3. V. Jochelson, Archaeological investigations in Kamchatka, *Carnegie Institution of Washington, Publ.* no. 388, 1928.

4. Krasheninnikov, *Opisaniye zemli Kamchatki*, vol. 2, p. 3.

5. K. Ditmar, *Poyezdka i prebyvaniye v Kamchatke v 1851–1855 gg.* (Travel and residence in Kamchatka, 1851–1855), St. Petersburg, 1901, pp. 210–213.

6. *Ibid.*, p. 189.

7. S. Bergman, *Vulkane, Bären und Nomaden; Reisen und Erlebnisse im wilden Kamtschatka*, Stuttgart, 1926; I. Schnell, Prehistoric finds from the Island World of the Far East, *Bulletin of the Museum of Far Eastern Antiquities*, Stockholm, no. 4, 1932.

8. Nakayama, Excavations of semisubterranean dwellings at Ust-Kamchatka on the eastern coast of Kamchatka, *Journal of the Anthropological Society of Tokyo*, 1933, vol. 48 (in Japanese).

9. D. N. Lev, Novyye arkheologicheskiye pamyatniki Kamchatki (New archaeological remains from Kamchatka), *Sovetskaya etnografiya*, 1935, no. 4–5.

10. E. Mott Davis, Archaeology of northeastern Asia, *Papers of the Excavators' Club*, vol. 1, no. 1, Cambridge, 1940.

11. Krasheninnikov, *Opisaniye zemli Kamchatki*, vol. 2, pp. 25–28.

12. N. V. Slyunin, *Okhotsko-Kamchatskiy kray* (The Okhotsk-Kamchatka region), St. Petersburg, 1900, p. 384.

13. S. I. Rudenko, *Drevnaya kultura Beringova morya i eskimosskaya problema* (The ancient culture of the Bering Sea and the Eskimo problem), Moscow, Glavsermorput, 1948, Plate 15, item 18.

14. O. W. Geist and F. G. Rainey, *Archaeological excavation at Kukulik, St. Lawrence Island*, Washington, D.C., U.S. Govt. Print. Off., 1936, p. 105, Plate 24, items 6 and 7.

15. H. B. Collins, Archaeology of St. Lawrence Island, Alaska, *Smithsonian Miscellaneous Collections*, 1937, vol. 96, no. 1, Plate 60, item 12.

16. W. Bogoras, The Chukchee, *Memoires of the American Museum of Natural History*, 1904–1909, vol. 2, p. 79, Fig. 99a.

17. E. W. Nelson, The Eskimo about Bering Strait, *18th Annual Report of the Bureau of American Ethnology*, 1899, p. 78, Fig. 22.

18. Geist and Rainey, *Archaeological excavation . . .* , p. 104, Plate 24, items 3 and 4.

19. Krasheninnikov, *Opisaniye zemli Kamchatki*, vol. 2, p. 227.

20. G. W. Steller, *Beschreibung von dem Lande Kamtschatka*, 1793, p. 3.

21. Krasheninnikov, *opisaniye zemli Kamchatki*, vol. 2, p. 50.

22. *Ibid.*, pp. 65–67.

23. *Ibid.*, vol. 1, p. 209.

24. *Ibid.*, pp. 118–119.

25. *Ibid.*, p. 299.

26. Torii, Etudes archéologiques et ethnologiques . . . , Plate 30, items 7, 12, and 14.

27. *Ibid.*, Plate 31, items A and B.

28. *Ibid.*, Plate 30, item 18.

29. B. E. Petri, Neoliticheskiye nakhodki na beregu Baykala (Neolithic finds on the shores of Lake Baykal), *Sbornik Muzeya antropologii i etnografii*, 1916, vol. 3.

30. Torii, Etudes archéologiques et ethnologiques . . . , Plate 28; E. Akabori, Local differences among chipped stone arrowpoints in Japan, *Journal of the Anthropological Society of Tokyo*, 1931, vol. 46 (in Japanese); Schnell, Prehistoric finds from the Island World of the Far East, Plates 12 and 13.

31. Krasheninnikov, *Opisaniye zemli Kamchatki*, vol. 2, p. 271.

32. *Ibid.*, p. 299.

33. *Ibid.*, p. 69.

34. *Ibid.*, p. 310.

35. Schnell, Prehistoric finds . . . , Plate 17, item 7.

36. Torii, Etudes archéologiques et ethnologiques . . . , Plate 30, item 10.

37. Krasheninnikov, *Opisaniye zemli Kamchatki*, vol. 2, p. 32.

38. *Ibid.*, p. 48.

39. N. G. Munro, *Prehistoric Japan*, 1911, p. 92, Plate 18, item 16.

40. V. Jochelson, *Archaeological investigations on the Aleutian Islands*, Carnegie Institution, Washington, 1925, pp. 23–26, 36, and 39, Plates 15 and 61, Fig. 17.

41. S. Umehara, Official account of the excavation of a prehistoric campsite Kitashirako-Kokuroko in Kyoto, 1934 (in Japanese).

42. Krasheninnikov, *Opisaniye zemli Kamchatki*, vol. 2, pp. 40–41.

43. Ditmar, *Poyezdka i prebyvaniye v Kamchatke . . .* , p. 214.

44. Munro, *Prehistoric Japan*; O. Solborg, Beiträge zur Vorgeschichte der Osteskimo, *Vid.-Selsk.-Skrift, Hist.-Filos. Klasse*, vol. 2, no. 2, 1907, Fig. 5.

45. Krasheninnikov, *Opisaniye zemli Kamchatki*, vol. 2, p. 32.

46. *Ibid.*, pp. 32–33.

47. *Ibid.*, vol. 1, p. 336.

48. *Ibid.*, vol. 1, p. 266.

49. *Ibid.*, vol. 2, pp. 32 and 87.

50. N. Ogloblin, Dve "skazki" Vl. Atlasova ob otkrytii Kamchatki (Two "Tales" of Vl[adimir] Atlasov about the discovery of Kamchatka), *Chteniya Obshchestva istorii i drevnosti Rosii*, 1891, book 3, section 1, p. 14.

51. Ditmar, *Poyezdka i prebyvaniye v Kamchatke. . . .*

52. G. Quimby, The prehistory of Kamchatka, *American Antiquity*, 1947, vol. 12, no. 3, pt. 1.

53. V. Jochelson, The Koryak, *Memoirs of the American Museum of Natural History*, 1908, vol. 10, p. 640, Fig. 165.

54. Jochelson, Archaeological investigations in Kamchatka, Plate 19.

55. F. De Laguna, Eskimo lamps and pots, *Journal of the Anthropological Institute*, 1940, vol. 10, p. 62.

56. Torii, Études archéologiques et ethnologiques . . . , p. 188.

57. *Ibid.*, p. 193.

58. *Ibid.*, p. 187, Fig. 56.

59. Munro, *Prehistoric Japan*, p. 536, Fig. 311, and ff.

60. Krasheninnikov, *Opisaniye zemli Kamchatki*, vol. 2, p. 42.

61. Jochelson, The Koryak, Plate 12, items 1–10.

62. *Ibid.*, p. 71, Fig. 70 and Plate 16, item 9.

63. *Ibid.*, Plate 16, item 6.

64. *Ibid.*, p. 61.

65. Quimby, The prehistory of Kamchatka, p. 179.

66. Krasheninnikov, *Opisaniye zemli Kamchatki*, vol. 2, pt. 3, p. 14.

67. *Ibid.*, p. 81.

M. G. LEVIN

# FIELDWORK ON THE CHUKCHI
# PENINSULA IN 1957*

DURING THE SUMMER of 1955, in the settlement of Uelen, remains of ancient burials were found during incidental excavation work. We learned of these finds from I. P. Lavrov, who was working in Uelen as an art instructor in the well-known bone- and ivory-carving workshop.

During the same year D. A. Sergeyev, who was traveling with a group of school children from Urelik along the Chukchi coast, visited Uelen. At the site of the excavations, Sergeyev gathered several objects from the burials, which were later sent to the Museum of Anthropology and Ethnography in Leningrad. Some of these objects are highly characteristic of the Old Bering Sea culture.

In 1956, according to the information at hand, excavations of limited extent were carried out at Uelen by N. N. Dikov, Director of the Anadyr Regional Museum. From I. P. Lavrov we received a skull found by him in these excavations. The skull has pronounced Eskimo traits.

The significance of these finds is apparent. Burials of the Old Bering Sea culture had not been found previously on the Asiatic or American coast, and the discovery of craniological material from this period is very significant for the clarification of the problem of Eskimo origins.

As we shall see later, the burial site proved to be more complex: besides the Old Bering Sea graves, there were ones of different aspect, belonging to a much later period.

The excavation of the Uelen burials was the basic task of the Chukchi Detachment of the Northern Expedition organized by the Institute of Ethnography of the U.S.S.R. Academy of Sciences. The work was conducted by the author and by R. V. Chubarova, a member of the Institute for the History of Material Culture.

The second task was the collection of blood samples among the Chukchis and Eskimos. Data concerning the distribution of blood groups among the peoples of northeastern Asia were completely lacking. [Indeed, in this respect] the entire northeast of the U.S.S.R. was a blank area on world maps.

The work of the detachment proceeded as follows. On July 21, we left Provideniya [Providence] Bay for Uelen. Because of a storm and the ice conditions, the vessel could not get through Bering Strait and we were put ashore in [St.] Lawrence Bay. The stay was utilized for gathering blood samples.

On the morning of the 25th we left Lawrence Bay for Uelen in a whaleboat. In good weather it is possible to make this trip in 10 to 12 hours, but because of the storm we were forced to put ashore near the village of Nunyamo, some 35 km from Lawrence Bay. Nunyamo is the center of a Chukchi collective. Here we had the opportunity to acquaint ourselves with the ways of life of the population, to type the blood of about 50 people, and to gather archaeological materials. The small village is located at the site of an ancient settlement and even cursory

---

*Translated from *Sovetskaya etnografiya*, no. 6, 1958, pp. 128–134.

sampling resulted in the uncovering of a quantity of stone and bone tools and pottery (all of comparatively late date). Of interest among the objects gathered in the old cemetery (Chukchi and Eskimo cemeteries are located, as a rule, on a knoll near a village) were stone and clay oil lamps and vessels of various dimensions, made from coarse paste, jar-shaped, with a flat bottom and external lugs for suspension.

We left Nunyamo on July 26, traveling along the shore, as one generally does in a whaleboat so as to avoid the dangers of the open sea. The entire shore is steep and rocky, and only rarely is there a place where it is possible to put in. In such places one could always spot the remains of ancient houses from the boat, these sites being easily recognized because of the thicker and more luxuriant vegetation covering them.

Again, we were not able to make it to Uelen and had to land at the village of Naukan, which is situated on a high, steep bank. Even with the whaleboat landing in the windy weather was difficult. The site is hardly fit for habitation, and if, in spite of this, a large Eskimo settlement existed there over the centuries, it must have been because of the very favorable hunting conditions. Not so long ago there was a walrus rookery not far from Naukan.

The village of Naukan is rather unusual. It is commonly called the "Eskimo capital" of the Chukchi peninsula. There is even a popular saying: "Who has not visited Naukan has not seen the Chukchi peninsula." One of the remarkable sights of the village is the memorial lighthouse to Semen Dezhnev, constructed in 1948 to commemorate the 300th anniversary of his famous voyage.

Even today the village has preserved its ancient aspect. Because of the local topography, new construction is difficult and the Eskimos continue to live in [semisubterranean] sod houses (*yaranga*) of the old type.[1] These are covered with walrus hides, weighted with stones (strong winds are constant along the Bering Sea coast), and have corridor entranceways constructed with stone slabs. Near the houses are pits for storing "kopalkhen"—walrus meat fermented in a special way.

Some time ago S. I. Rudenko conducted an archaeological survey in Naukan. The local inhabitants gave us several ancient tools they had found. In the cemetery on the hill near the village, which is even more extensive than the one at Nunyamo, we also collected some oil lamps and clay vessels.

Taking advantage of our forced stay in Naukan, we did some blood typing there also, the first-aid station of the village making our work easier.

On July 27 we left Naukan and in two hours arrived at Uelen. During our entire stay there we enjoyed the hospitality of I. P. and V. M. Lavrov, both of whom participated in the excavations.

The village of Uelen is located on an elongated sandy spit, which separates a large lagoon from the sea. The well-known Uelen site is located on the slope of an elevation which inclines towards the lagoon, to the southeast of the present village. It was investigated in the summer of 1945 by S. I. Rudenko and I. P. Lavrov. To the north of this site, upslope, on the first knoll from the seashore— the one from which the spit emerges—is the locality containing the graves mentioned above. Here, there is a comparatively large natural platform. On it, and on the gentle slope which descends to the spit, there is a cemetery. Throughout it jutting structures made of whale bones and large stones are visible. Its surface is stony, here and there covered by thin turf.

Somewhat farther upslope from a trench dug in 1955 (the one which furnished the first artifacts from the ancient graves), we excavated an area of about 60 m².

Here, 20 graves were uncovered, at different levels. The majority of them were almost surface burials; several were topped by a structure of whale bones and stones, while others were not marked in any way.

The deep burials that were found in connection with complexly constructed graves were especially interesting. The orientation of the skeletons varied; in most the head pointed to the east, but in some it pointed to the north or northeast. In addition to single burials, double burials were also encountered.

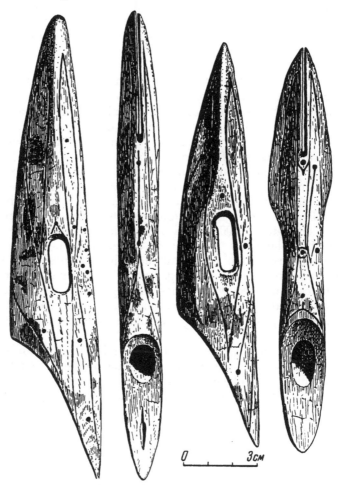

FIGURE 1. Harpoon heads from grave no. 1.

The inventory of the excavated graves dates them to different periods. Of all the artifacts extracted from ancient Eskimo sites harpoon heads have been best studied. Detailed studies of Danish and American archaeologists (Mathiassen, Collins, Jenness, Ford) provide us with a series of types of toggle harpoon heads characteristic of the different stages of ancient Eskimo cultures.

In the graves excavated by us harpoon heads were quite numerous. Among them were forms analogous to those from St. Lawrence Island and Alaska described as Old Bering Sea, Okvik, Birnirk, and Punuk. Among the finds characteristic of the Old Bering Sea culture, we should mention the "winged objects," and also harpoon heads and socket pieces decorated with typical Old Bering Sea ornamentation.

Apart from the harpoon heads and numerous harpoon foreshafts, various artifacts of walrus ivory and bone (dart points, fish spears, needlecases, stamp paddles, possibly used for the ornamentation of pottery, and others) were also found. Stone tools, polished and chipped, are represented by numerous knives, dart and arrow heads, adzes, and scrapers. Very characteristic are the figurines of birds and animals and also an anthropomorphic statuette found in one of the graves. There were no traces of metal in any of the graves.

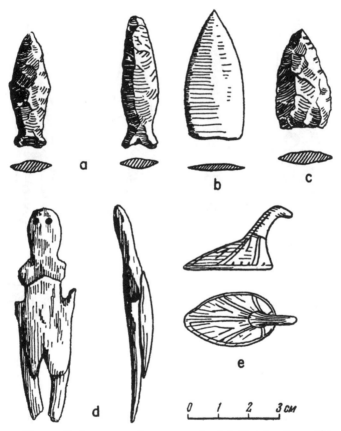

FIGURE 2. Inventory of grave no. 1: (*a*) flint figurines; (*b*) polished slate arrowhead; (*c*) flint arrowhead; (*d*) bone anthropomorphic figurine; (*e*) button of walrus ivory in the shape of a bird.

Since it is impossible in this short report to describe the characteristics and inventories of all the graves uncovered, we shall mention only a few.

Figures 1 and 2 show objects found in grave no. 1. The grave was marked on the surface by a rectangular structure of whale bones, 2 m by 1 m in area; on the west (short) side of the rectangle the bones were reinforced by stones. The skeleton lay directly under a thin layer of turf, with its head oriented to the east. The body was apparently placed on its right side and was flexed. The objects accompanying the skeleton were placed both at the same level with it and also under it. Notable among them are the very large, ornamented whale harpoon heads of Punuk type, an ornamented representation of a bird (in the form of a button), stone arrowheads, flint figures (representations of fish?), and an anthropomorphic figurine. Fourteen bola weights and ground slate knives were also found in this grave. By the general aspect of the inventory the grave may be dated to Punuk times.

Next to grave no. 1, somewhat farther north, directly under the same [rectangular] structure, was another grave (no. 4). Under the sod lay the scapula of a whale covering the skeleton, which was oriented to the east. The bones of the

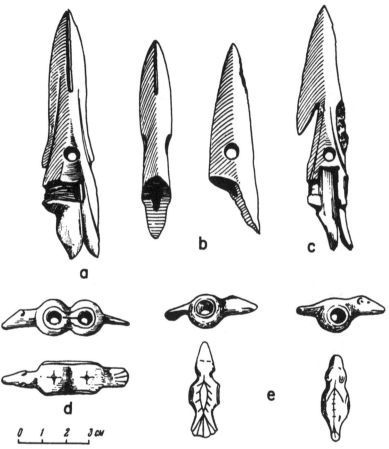

FIGURE 3. Inventory of grave no. 4: (a)–(c) harpoon heads; and (d, e) zoomorphic figurines (walrus ivory).

skeleton were located at a depth of 45 to 55 cm, lower than the skull, which lay base down. Some of the objects were placed under the skeleton. The inventory of this grave is illustrated in Figure 3. In it were found harpoon heads of different types, including those characteristic of the Birnirk type. Of interest are the ornamented figurines with animal heads and bird tails. Here were found also bola weights, a harpoon foreshaft, a ground slate blade of large size, and other objects.

Below graves nos. 1 and 4, at a depth of 70–75 cm from the surface, was located still another grave (no. 5). The skeleton lay on its back, oriented to the northeast. Slightly to the side of the head there was a so-called winged object (at a depth of 75 cm); near the chest was a harpoon socket-piece and near the hip an ornamented harpoon head. The inventory of this grave (Fig. 4) is quite characteristic of the Old Bering Sea culture.

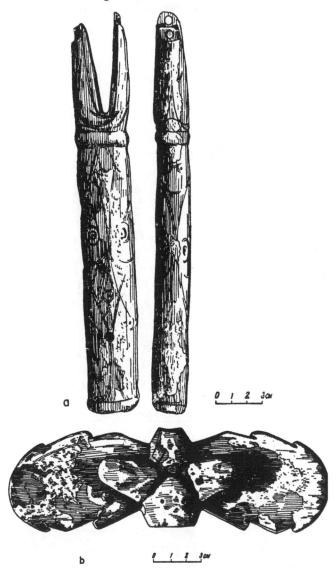

a

b

FIGURE 4. Inventory of grave no. 5: (a) socket piece;
(b) "winged object" (walrus ivory).

We shall also mention grave no. 10–11, which is especially interesting because of its construction. This is a double burial: one skeleton is that of a man of mature age, the other of a young person. Both skeletons were oriented to the north; they were supine and extended. The right arm of the male skeleton crossed the left hand of the other. The skeletons lay on a wooden flooring made of longitudinally placed planks. Under this flooring, at the heads and feet, was placed an additional layer of wooden planking at right angles [to that above it].

Under the floor, at a depth of about 1 m from the surface, was a level, neatly fitted layer of stones. The walls of the grave were lined by vertically positioned large, flat stones. The length of the grave was about 175 cm, the width 60–65 cm. The wooden flooring was apparently covered with fur; its remnants were clearly traceable over the entire length of the grave.

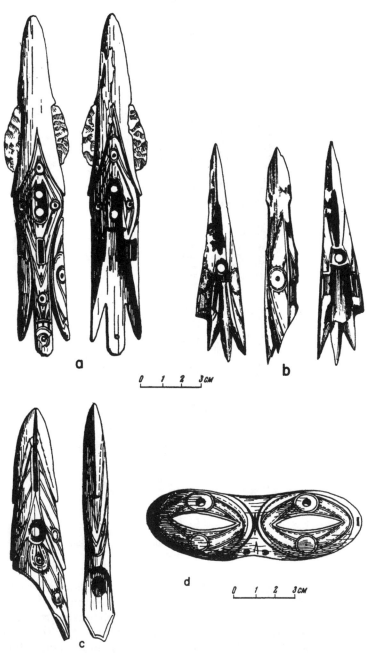

FIGURE 5. Inventory of grave no. 10–11: (*a*)–(*c*) harpoon heads; (*d*) snow goggles (walrus ivory).

The skeletons were accompanied by various objects (Fig. 5). Of interest are the harpoon heads, which are characteristic of the Old Bering Sea type in construction and ornamentation. With the two harpoon heads were found their foreshafts. Notable also are the snow goggles of walrus ivory, ornamented in Old Bering Sea style. Here also were found slate knives, fragments of pottery, and other artifacts.

The general appearance of this grave, as already indicated, dates it to the Old Bering Sea period.

In addition to grave no. 10–11, another double grave was also discovered (no. 13–14). In it a woman and a child were buried. The date of this grave appears to correlate with the Punuk period.

Graves nos. 7 and 18, in which harpoon heads of the Okvik type were found (Figs. 6 and 7), should also be mentioned.

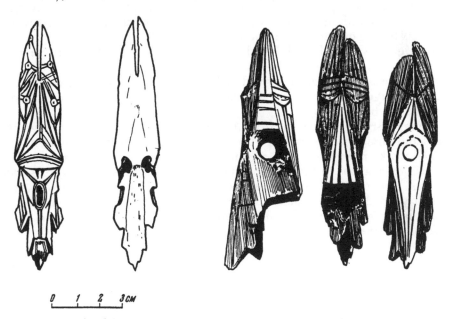

0   1   2   3 CM

FIGURE 6. Harpoon head from grave no. 7.    FIGURE 7. Harpoon head from grave no. 18.

The skulls excavated from the graves were largely broken (apparently when the stone roofing collapsed), but about 20 skulls could be restored successfully. All of the skulls possess markedly Eskimo traits and are distinctly dolichocephalic. This fact is all-important for the solution of the much discussed question of the aboriginal physical type of the ancient Eskimos.

Besides the skulls, the long bones were collected, and in some cases even entire skeletons.

As indicated, we excavated only a small area in one section of the burial ground (above the trench of 1955). There is reason to believe that with more extensive excavations new burials will be uncovered.

Contemporaneous with our work in the cemetery, we made small-scale excavations at the Uelen site which had been surveyed in 1945 by S. I. Rudenko; on the basis of the limited material collected here he noted its similarity with that of the Okvik culture and distinguished it as a special stage of the old Eskimo culture, the Uelen-Okvik stage. Over the course of two years I. P. Lavrov continued the survey of this site. Although we did not encounter permafrost in the cemetery area, with its stony soil, we did so at the [Uelen] site. The excavations progressed with the thawing, but we did not reach the underlying bedrock. The cultural layer is thick and spreads over a very large area. It must be taken into account that a considerable part of this area was dug over by the local inhabitants, the Chukchis, who for a long time procured artifacts of walrus ivory here. The old walrus tusks, long buried in the ground, were much valued in Alaska and were sold there by the Uelen people.

The excavations at the site yielded a large quantity of diverse objects. Besides, Lavrov turned his finds over to us. During clearing operations of one of the areas of the site, the remains of an ancient dwelling were revealed. The excavations of this must be included in plans for future work.

In Uelen, as well as at other points along the coast, there are many remains of old dwellings. The present village is built on the site of an ancient settlement. We brought back with us a sizable collection of objects gathered by Lavrov and also artifacts turned over to us by local inhabitants.

In Uelen we conducted blood typing, carrying out the work in the local hospital.

On August 15, a steamer arrived at Uelen, which we used for our return journey. On August 20 we arrived at Anadyr, where we stayed long enough to collect samples for blood typing.[2]

On August 25 we travelled from Anadyr to Magadan, where we stayed for two days and familiarized ourselves with the collections in the local museum.

The work on the Chukchi peninsula in 1957 was of a reconnaissance character. Future excavations in the Uelen cemetery are of great importance, as well as a serious archaeological reconnaissance along the coast. The cemetery at Uelen, which yielded highly valuable materials, despite the limited size of the excavations, occupies undoubtedly an important place among the remains of ancient Eskimo cultures.[3]

## Notes and References

1. In 1958 the greater part of the population was moved from Naukan to Nunyamo, where a new, well-organized village is being built.

2. Information about the blood groups of the Chukchis and Eskimos was published in *Sovetskaya etnografiya*, no. 5, 1958.

3. In 1958 the excavation of the cemetery was continued and yielded very interesting results. Especially noteworthy were the finds of ancient Eskimo art objects.

M. G. LEVIN

# AN EARLY ESKIMO CEMETERY AT UELEN*

## A PRELIMINARY REPORT ON THE EXCAVATIONS OF 1958

IN A PRELIMINARY REPORT on the work of the Chukchi [peninsula] Detachment of the Northern Expedition in 1957 we included descriptions of some of the materials found in an ancient cemetery at Uelen.[1] As a result of excavations carried out by R. V. Chubarova and the author of this report, 20 graves (in an area of about 60 m²) were uncovered in 1958. The materials excavated as well as the contents of small excavations conducted in the cemetery by D. A. Sergeyey in 1955 and by N. N. Dikov in 1956 made it possible to establish the presence of several types of graves.[2] As we have reported elsewhere, in the Uelen cemetery there were found in addition to graves with an inventory characteristic of the Old Bering Sea culture, ones characteristic of the Okvik and also the Birnirk and Punuk cultures.

In 1958 the Chukchi Detachment continued the excavations in the cemetery.[3] Let us review the topography. The cemetery is located at the settlement of Uelen, on a hill from which the Uelen spit continues, separating an extensive lagoon from the sea. On the slope of the hill, descending towards the lagoon, to the southeast of the present settlement, is located the well-known Uelen site which was investigated in the summer of 1945 by S. I. Rudenko and I. P. Lavrov. The cemetery is located to the north of this, separated from it by a ravine cut by a stream from which even now the people of Uelen obtain water.

The cemetery is located on the slope of the hill, at an elevation of 30 to 35 m above sea level, at a place where the steep slope of the hill becomes a gentle plateau. The eastern edge of the cemetery is almost coincident with the surface of the plateau, but the southern, western, and northern edges are above the surface of the slope, forming a mound about 1.5 m high.

As we have already reported, the cemetery was discovered in 1955 during construction work when a long trench (about 70 cm deep and up to 1 m wide was cut through the area, progressing in a slightly irregular line from southwest to northeast. Above the trench, from which the first objects from the ancient graves were obtained in 1955, excavations were continued in 1957. In 1958, work was carried out in the area between the trench and the 1957 excavations.

In the section of the cemetery excavated in 1957, there were large whale bones and stones projecting above the ground in different places, forming a sort of foundation. In the area excavated in 1958, such features were rarely encountered.

In 1958, an area of 100 m² was excavated, exposing 23 graves in different stages of preservation. In some burials well-preserved skulls and bones of the trunk and extremities were found; in others, only individual bones or fragments of bones were found. In addition to these graves there were found in several places (the excavations were carried out in levels with the surface area divided into quadrants) fragments of skulls, mainly those of children, and separate skeletal bones, an accumulation of artifacts, and several objects which could not have been connected with a specific grave.

*Translated from *Sovetskaya etnografiya*, no. 1, 1960, pp. 139–150.

Especially noteworthy was the discovery of a peculiar construction in one of the quadrants at a depth of 50 cm. In the circle of stones, a vertebra and two vertically placed whale shoulder blades, and also a walrus shoulder blade and rib, were found. On the walrus shoulder blade lay the skull of a dog; between the whale shoulder blades were two bone arrow points.

Besides the principal excavation work, [the contents of] three graves on the slope below the trench which had been partially disrupted by the construction of 1955 were removed.

As was the experience in the 1917 excavations, the graves were located at different depths—from almost the surface to a depth of over a meter. The burials differed in character. Some of them had no "construction"; others were formed with stones and large bones, partially or fully enclosing the grave (as indicated above, such structures were not always visible on the surface but were discovered at various depths). Some of the skeletons rested directly on the soil; others lay on wooden planking.

The orientation of the skeletons differed. In the 23 burials excavated in 1958, orientation of the head was to the northeast in 10 graves (13 skeletons, among them skeletons from 3 paired burials, had this orientation), to the north in 4 graves, to the east in 1, to the southeast in 3, and to the southwest in 4. The most common position of the skeleton (in the graves where the preservation of the bones permitted this to be determined) was dorsal with legs extended and arms slightly bent. In several graves the position of the bones suggests that the deceased was buried between stones and the posture was determined by the position of the stones. In four burials, together with the skeleton of an adult, fragments of children's skulls and bones were found, i.e., these were joint burials.

The find of a "gravestone" in one of the quadrants (between graves no. 16 and 17) was of special interest. It was a deliberately placed, slightly inclined flat bone (part of a whale rib), almost 80 cm in length and 50 cm in its widest part. The bone was perforated in three places, the round perforations forming a triangle. The upper edge of the "gravestone" was uncovered at a depth of about 30 cm after the layers of turf and stone debris had been removed.

In 1957 we had already noted graves overlapping each other at different levels. We encountered these also during the 1958 excavations. Side by side with individual burials were discovered, as in 1957, paired burials. Figure 1 shows the paired burials nos. 12a–b and 14–15. Burial 12a–b had a side wall of stone on the western side and of whale bone on the eastern side. The skeletons were positioned one over the other, the orientation of both being with head to the northeast. The upper skeleton was well preserved, with the bones in anatomical order; its position was dorsal, arms slightly bent at the elbows, hands in the region of the pelvis, legs extended. The depth of the burial was 0.45 m at the skull, 0.50 m at the shin bones. The lower skeleton was incomplete. The skull was missing and the lower jaw was lying to the right of the skull of the upper skeleton. The arms of the lower skeleton were slightly bent at the elbows and the legs were extended and somewhat displaced to the right from the legs of the upper skeleton. Under the skeletons, in the region of the chest, pelvis, and hip, were preserved remains of wood (the wooden platform of the grave). About the skeletons was found a rich and diverse inventory: harpoon heads and foreshafts, plugs for harpoon floats, pick-mattocks, a trough-shaped scraper of walrus ivory, a needlecase (with bone needles in it), a small clay vessel with round bottom (crushed), a woman's large slate knife, a stone spear point, a stone adze, fragments of rock crystal, and other items. The distribution of the objects is shown in Figure 1.

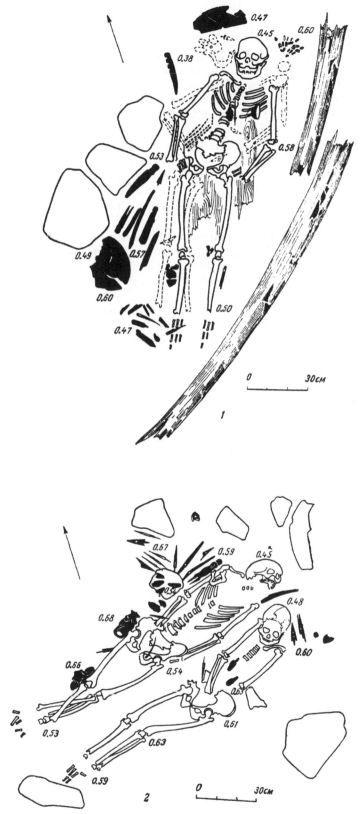

FIGURE 1. (1) Burial no. 12a–b. (2) Burial no. 14–15. The figures designate the depth of interment.

Burial no. 14–15 was bordered on the northeast by four small stones, and on the south by two larger stones. The two skeletons lay parallel to each other and were oriented to the northeast. Skeleton no. 14 was in satisfactory condition, but no. 15 was greatly damaged. The bones of both skeletons were in anatomical order. Their position was dorsal with the arms bent, and the wrist under the pelvic bones. The depth of the burial was 45–60 cm. The inventory of this paired burial included harpoon heads and foreshafts, a socket piece of harpoon, bone arrow points, a pick-mattock, a small shovel, and other items. The distribution of these objects is indicated in Figure 1. Of note is the so-called winged object (at the right knee of skeleton no. 15) and the quite original anthropomorphic figurine found near the same skeleton, to the right of the pelvis. It will be described more fully below.

The richest and most varied material was found in the third paired burial (no. 22a–b). The skeletons in this grave were placed partially one over the other. The depth of the skull of the upper skeleton was 50 cm, of the knees 70 cm; the depths of the lower were respectively 55 cm and 78 cm. The orientation of both skeletons was to the northeast.

Above the skulls, partly pressing on them, lay a large stone. (Skeletons in other quadrants also were weighted with such stones.) The skeletons were poorly preserved. Judging from the bones that were preserved the position of the skeletons was dorsal (the leg bones of the upper skeleton were displaced, the left thighbone and left tibia being twisted 180 degrees). Under the entire length of the skeletons were remains of a wooden flooring and on it were strewn pieces of fur (probably from fur clothing). Around the skeletons were found harpoon heads, foreshafts, [harpoon] ice picks, plugs for harpoon floats, stone adzes, points, scrapers, slate knives, and so on (altogether over 100 objects). Especially notable is the sculptured representation of a human face (found near the skull of the lower skeleton).

The excavations of 1958 resulted in the uncovering of a large quantity of varied objects. Since the materials are still in a stage of primary processing, we shall limit ourselves to a very short description of them in this preliminary report.

Artifacts manufactured from walrus tusk, bone, and antler (the last is significantly rare) include: harpoon heads, foreshafts, and socket pieces, plugs for harpoon floats, dart and fish-spear points, ice picks, pick-mattocks, and objects used by women (needlecases, fat scrapers, handles of women's knives, and so on). A large quantity of stone implements, both polished and chipped, were found. They include: knives, lance points, darts, arrows, side blades of harpoon heads, adzes, scrapers, and other implements.

Among the materials obtained, the harpoon heads are of the greatest interest since on their typology is constructed the basic classification of ancient Eskimo cultures. In 1958 we found over 50 harpoon heads. Examples from the various graves are shown in Figures 2–6.

The most characteristic heads are those of the Old Bering Sea type with wide, open sockets, two round line holes, a triple spur (with a long central and short side spurs), [two side blades], and two slots for lashing. In our 1958 collection different forms of harpoons of the Old Bering Sea type are present. Besides the "classical" ones (heads with stone side-blades), other forms were found. There were heads with one and more often a pair of symmetrical side-barbs. These Old Bering Sea heads differ also in style of ornamentation.

In the graves excavated in 1958, there were a relatively large number of harpoon heads which, on the basis of their form and ornamentation, may be classi-

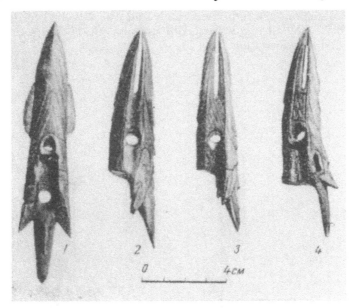

FIGURE 2. Harpoon heads from burial no. 5.

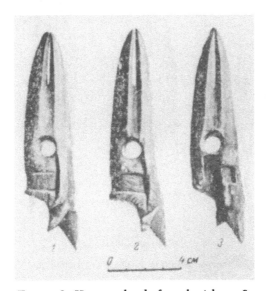

FIGURE 3. Harpoon heads from burial no. 6.

fied as Okvik. There were also representatives of the Birnirk type (in 1957 we found heads with stone side-blades and a unilateral barb; in 1958 heads of such form were not encountered). Of special note are very large harpoon heads (for whale hunting), which by their form and ornamentation are characteristic of the early Punuk stage. The number of forms and the finding in one and the same grave of heads of several [chronological] types make the materials collected in 1958 especially interesting.

Harpoon foreshafts were found separately and also attached to the heads. The 1958 excavations yielded two examples of harpoon socket-pieces decorated with rich carving and ornamentation (Fig. 7).

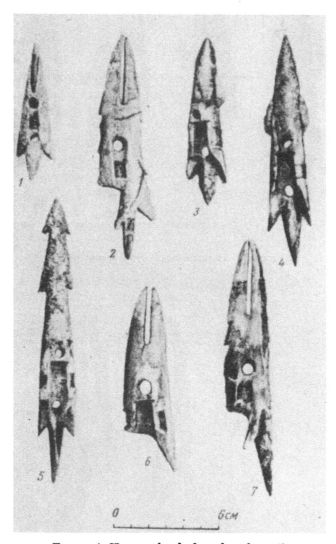

FIGURE 4. Harpoon heads from burial no. 7.

Arrow points of walrus ivory and bone comprise, together with harpoon heads, the most numerous items of the recovered materials. Among them are several types: with a stone end blade, with a sharp [unslotted] point, with varied cross-sections, with a different number of barbs, and so on. The points uncovered from the graves, [which in turn are] dated by the harpoon heads of various types, offer the possibility of associating these various forms of points with definite culture complexes. Points from one burial (no. 19) are shown in Figure 8.

The variety of the stone inventory is noted above. Especially numerous are slate knives of various dimensions (women's knives for cutting meat and scraping fat); several of them have round perforations (see Fig. 9).

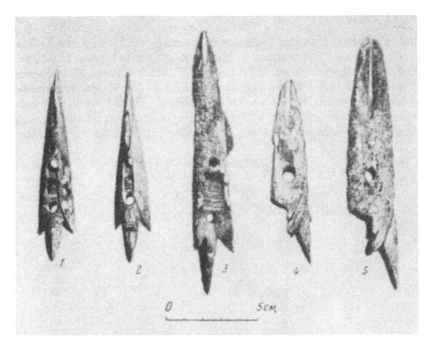

FIGURE 5. Harpoon heads from burial no. 14–15.

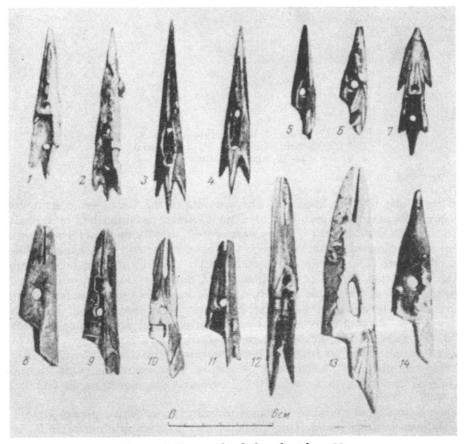

FIGURE 6. Harpoon heads from burial no. 22.

Earlier we referred to the Uelen site, investigated by S. I. Rudenko. In his opinion its inventory (in particular, the harpoon heads) reveals a close resemblance to that of the Okvik site described by Froelich Rainey. This does not sufficiently clarify the issue. Regarding the relation of the Uelen site to the Uelen graves, the case could be somewhat as follows. Thanks to the collections of Lavrov, we now have available a large number of harpoon heads from the Uelen site. No analogies between the types of harpoon heads obtained from the burials

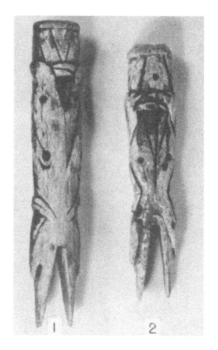

Figure 7. Harpoon socket-pieces:
(1) from burial no. 4; (2) from burial
no. 14–15. About ⅔ natural size.

and the Uelen site are apparent. The contents of the Uelen cemetery reveal a much greater resemblance to those of Cape Dezhnev (as published by Rudenko).

In the 1958 materials art is richly represented. Besides the sculptured figurines, there are also many utilitarian objects: harpoon heads and socket pieces, flakers, and other items. The different harpoon heads are so richly and carefully decorated with complicated ornamentation that doubt enters as to their utilitarian use. Possibly some of the heads had ritual significance. The carved socket pieces of walrus ivory mentioned earlier differ from everything else by their high artistic quality. The "winged objects" are exceptional works of art. Among the finds of 1958, the most perfect forms and ornamentations are represented by "winged objects" found in one of the disturbed graves near the trench (see Fig. 10). On their lower [obverse?] surface are zoomorphic representations, while the upper surface is covered with a very artfully executed ornamentation in Old Bering Sea style.

Several small and round plates made of walrus ivory with a groove around the circumference were found. Our Chukchi and Eskimo informants from Uelen defined them as objects used in tying up the openings of harpoon floats. On the upper, somewhat protuberant surface of these "plugs" there is different ornamentation. Especially interesting is one of the plugs with an incised line drawing of a human face executed in a realistic manner; lines on the cheeks and chin apparently represent tattooing (Fig. 11).

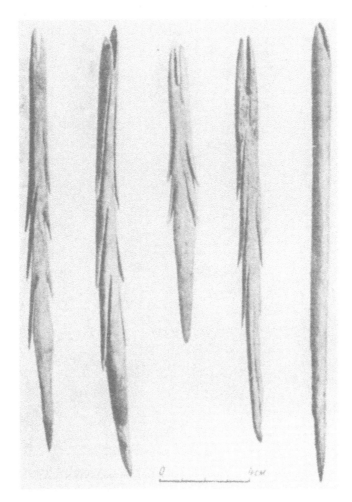

FIGURE 8. Arrowheads from burial no. 19.

Another representation of the human face was done in low relief (Fig. 12). The two upper round holes represent the eyes, the middle one the mouth; the two lower holes apparently are holes for labrets, which, as is known, were worn by the Eskimos until recent times.

Several sculptured zoomorphic representations of walrus ivory were found. Two of these were used as drill handles (Fig. 13, items 3 and 4).

Together with the realistic portrayals there are other figures of fanciful form. The most interesting one portrays a man in an odd pose, with legs pulled up to the stomach and chin propped up with the hands (Fig. 14). Such a pose is reminiscent of the position in which the Aleuts and Eskimos of southern Alaska buried the dead. The upper (dorsal) surface of the sculptured piece is covered with carefully executed ornamentation in Old Bering Sea style. In the back there is an opening which seems to have served for seizing the figure. The purpose of this figure remains unknown to us. We know of no near analogies to this figure in the published material with which we are familiar.

FIGURE 9. Stone inventory of burial no. 23.

The 1958 excavations resulted in a rich paleo-anthropological collection. The skulls are now being restored. As a result of the excavations carried out by the Chukchi Detachment, and also by N. N. Dikov, we have gained a sufficiently large craniological series and considerable osteological material.

The Uelen cemetery represents an exceptional antiquity of its kind. At present it is the only known ancient Eskimo cemetery on the Asiatic coast of the Bering Sea. So far, cemeteries of this type have not been uncovered on the American coast. On St. Lawrence Island, on the Punuk Islands, at different points along the Alaskan coast, many sites have been uncovered and studied by foreign archaeologists and dated to various of the ancient Eskimo cultures. Burials have also been excavated, among them the Ipiutak cemetery on Point Hope, which is of wide repute. But, in no cemetery have graves with so many different cultural complexes

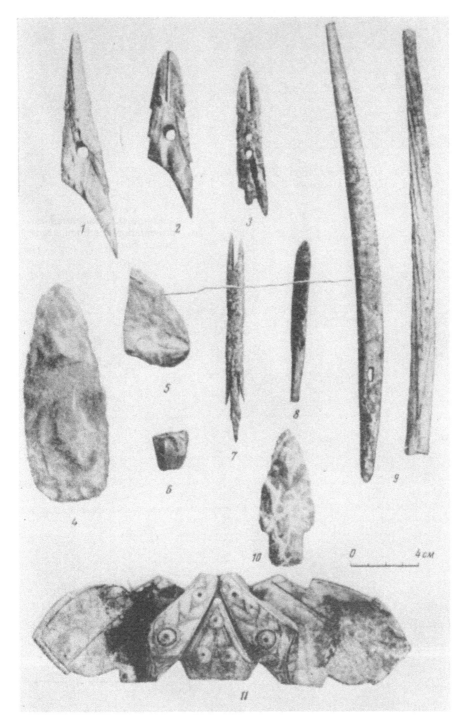

FIGURE 10. Inventory from burial A, at the trench: 1–3, harpoon heads; 4, stone adze; 5, fragment of a stone implement; 6, piece of ochre; 7, arrowhead; 8, bone artifact; 9, harpoon foreshafts; 10, spear point; 11, "winged object."

FIGURE 11. Representation of a human face on a plug for a harpoon float.

FIGURE 12. Sculptured representation of a human face from a burial.

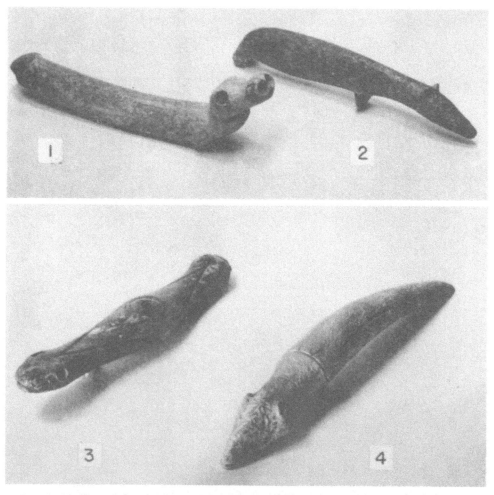

FIGURE 13. Zoomorphic figurines: 1, handle of a flaker (not from burial); 2, figurine of an animal (burial no. 13); 3, handle of a drill (burial no. 7); 4, handle of a drill (burial no. 13). Items 1 and 2 are about ½ natural size, 3 and 4 about ⅓ natural size.

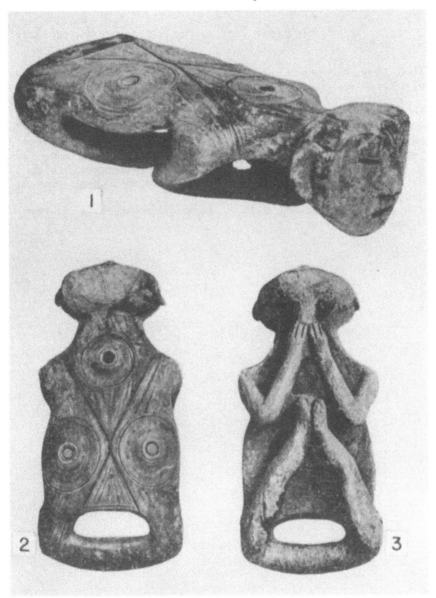

FIGURE 14. Anthropomorphic figurine from burial no. 14–15: 1, side view (about ⅔ natural size); 2, view from above (about ½ natural size); 3, view from below (about ½ natural size).

been encountered as at Uelen. (Burials that can be connected with both the Old Bering Sea and Okvik cultures were first found in the Uelen cemetery.)

Careful study of all recovered materials and further excavations of the Uelen cemetery will undoubtedly permit a better definition and possibly a modification of the present classification and chornonology of the Eskimo cultures of the Bering Sea. It will take time. This short report is intended to inform specialists about some of the results of the excavations, even though the definitive processing has not yet been completed.

Notes and References

1. M. G. Levin, Raboty na Chukotke v 1957 g. (Fieldwork on the Chukchi peninsula in 1957), *Sovetskaya etnografiya*, 1958, no. 6, pp. 128–134. [For English translation see pp. 296–304 of this book.]

2. D. Sergeyev, Pervye drevneberingomorskiye pogrebeniye na Chukotke (The first Old Bering Sea burials on the Chukchi peninsula), *Kratkiye soobshcheniya Instituta etnografii Akademii nauk SSSR*, 1959, no. 31, pp. 68–75; N. N. Dikov, Predvaritelnyy otchet o rabote arkheologicheskoy ekspeditsii Chukotskogo krayevedcheskogo muzeya v 1956 g. (Preliminary report on the work of the archaeological expedition of the Chukchi Regional Museum in 1956), *Zapiski Chukotskogo krayevedchskogo muzeya*, no. 1, Magadan, 1958.

3. The detachment consisted of M. G. Levin (Director), S. A. Arutyunov, I. P. Lavrov, and D. A. Sergeyev.

M. G. LEVIN AND D. A. SERGEYEV

# THE PENETRATION OF IRON
# INTO THE ARCTIC

## THE FIRST FIND OF AN IRON IMPLEMENT IN A SITE
## OF THE OLD BERING SEA CULTURE*

ARCHAEOLOGICAL INVESTIGATIONS in the Far North of America and in Greenland, which have been conducted intensively by foreign scholars during the past three decades, and the newer work of Soviet investigators in the northeast of our country, have covered numerous ancient remains and led to the elucidation of many questions connected with the history of the early settling of the littorals of the arctic seas and the inland regions of the tundra. Most of the publications are concerned with the problem of the origin of the Eskimos and the historical and chronological correlation of early Eskimo cultures. Connected inseparably to this complicated problem are questions of the ethnogenesis of other peoples of the Arctic and their cultural ties with populations of more southerly regions. One of the essential questions of this problem seems to be that of the time and path of penetration of iron to the ancient peoples of the Arctic.

Even a quarter of a century ago the belief still prevailed that iron penetrated into the Arctic, the region most remote from the centers of civilization, only very recently. During excavations on St. Lawrence Island in 1930, Collins found, among artifacts from sites dated to the Punuk period, engraving instruments equipped with iron gravers and knife handles with remnants of iron blades.† Collins wrote: ". . . iron in small quantities might possibly have reached the Bering Strait and St. Lawrence Island more than a thousand years ago."[1] Later, facts accumulated to testify to the earlier appearance of iron in this region. Already in 1941, Rainey, when describing the materials from the Okvik site, expressed the opinion that the Eskimos of the Okvik period knew the use of iron. No iron objects were actually found in the Okvik site, but he formulated his opinion by considering the construction of the knife handles, the slot of which indicated the use of a very thin blade, which could only have been made of iron.[2]

Rudenko defended the idea that metal was used by the early inhabitants of the Bering Sea coast during the Uelen-Okvik and Old Bering Sea period: "No evidence of metal has been found yet at either of these two oldest stages of Bering Sea culture, though we find it difficult to conceive how the slots for the socket lashings on the harpoon heads of that time could have been made without a metal tool."[3]

A remarkable discovery was made by Larsen and Rainey in their investigations of the Ipiutak cemetery at Point Hope. Here they found an engraving tool with

*Translated from *Sovetskaya etnografiya*, no. 3, 1960, pp. 116–122.

†[Collins wrote: ". . . ivory knife handles . . . , the sockets of which seemed designed for metal blades."—Editor.]

the remains of an iron blade in it.[4] Microchemical and spectrum analyses showed that the iron was not of meteorological origin.[5] The find in the Ipiutak cemetery is a persuasive confirmation of the view regarding the early penetration of iron to the Far North. Larsen and Rainey justifiably wrote that it is indirect evidence of the possibility that iron was also present in the Old Bering Sea period. Today, we have direct evidence of this.

In 1959 during excavations in the Uelen cemetery one of the authors of this report (D. A. Sergeyev) found an engraving tool equipped with an iron blade (Fig. 1). Burial no. 6, which contained the tool, was conspicuous by the richness and variety of its inventory. It contained many arrowheads and harpoon heads. pick-mattocks, stone heads for arrows and darts, women's slate knives (two of these with round perforations), and other objects. Most interesting are the toggle harpoon heads, socket pieces, and a "winged object."

FIGURE 1. An engraving tool with an iron blade, ⅝ natural size.

As is known, the classification of ancient Eskimo cultures accepted at present is based mainly on the typology of harpoon heads. As we have stated elsewhere, the burials at Uelen contained harpoon heads characteristic of the various periods —Old Bering Sea, Okvik, Birnirk, and Punuk.[6] In burial no. 6, examined by us, there were eight harpoon heads, both complete and broken (Fig. 2). Five of them (items 2, 3, 6, 7, and 8) belong to the type characteristic of the Old Bering Sea culture. The distinguishing traits of this "classical" type are: a wide, open socket, two round line-holes, a triple spur with a long central and short side projections, two lashing slots, and stone side-blades. Harpoon heads 2 and 3 have grooves for side blades. It may be assumed that harpoon heads 6, 7, and 8 (of which only the lower parts are preserved) also were similarly constructed.

The shape of the large harpoon head (no. 1) is unusual, but even it may be related to the Old Bering Sea type.[7] Harpoon heads with closed sockets (nos. 4 and 5) found in burial no. 6 also belong to the group which is attributable to the Old Bering Sea culture. Although the harpoon socket-pieces (Fig. 3) and "winged object" (Fig. 4) found in the burial cannot be considered (aside from the ornamentation) as bases for dating, they are nonetheless most widely distributed among the remains of the Old Bering Sea period.

In favor of an Old Bering Sea age is also the ornamentation on the harpoon heads and "winged object" from burial no. 6. Without going into an analysis of the complicated question of the different styles characteristic of the various periods of the early Eskimo cultures of the Bering Sea region,[8] let me simply note that the ornamentation on these objects conforms in general to the developed Old Bering Sea style. The ornamentation on the harpoon socket-pieces, consisting of bold interrupted lines, differs from this style and indicates a somewhat later stage. On the whole, however, the inventory of burial no. 6, in which the engraving tool with an iron point was found, is datable to the Old Bering Sea period.

The solitary find of an iron blade in the Uelen cemetery, where already more

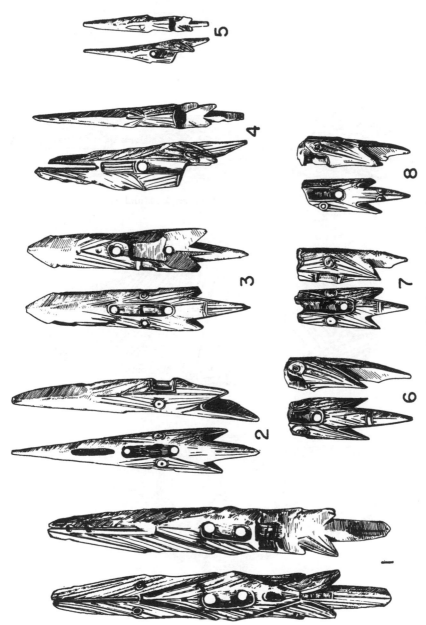

FIGURE 2. Harpoon heads (about ⅔ natural size).

FIGURE 3. Harpoon socket-pieces. ⅔ natural size.

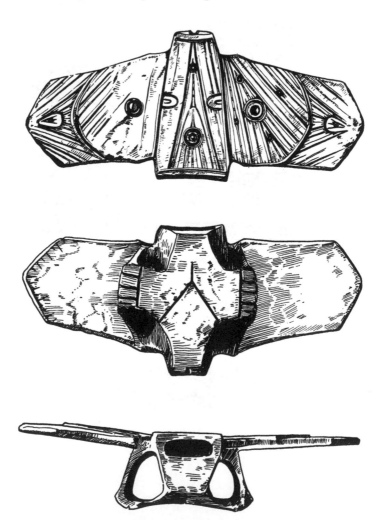

FIGURE 4. "Winged object," ⅔ natural size.

than one hundred graves have been opened, testifies to the fact that metal penetrated to this region in very limited quantity, and that not only during the Old Bering Sea period, but also later, iron was indeed the "diamond of the north," in both use and value. It was used mainly in drills and engraving tools (gravers), and sometimes also for making knife blades. The relative chronology of the different early Eskimo cultures and, even to higher degree, the absolute chronology is, at present, in a very unsatisfactory state. Significant differences of opinion about these questions exist among students of the early cultures of the Arctic. This pertains also to the early Eskimo sites that interest us, primarily those of northern Alaska and the Bering Strait region.

In the monograph of Larsen and Rainey mentioned earlier, the scheme shown in Table 1 is presented.[9]

TABLE 1*

THE POSITION OF THE IPIUTAK CULTURE IN THE RELATIVE TIME SEQUENCE OF ESKIMO CULTURE

|  | Southeast Alaska | Bering Strait | North Alaska | Arctic Canada | Greenland |
|---|---|---|---|---|---|
| 1900 A.D. | Modern | *Modern†* | *Modern* | Central Eskimo | *Modern* |
|  |  |  |  |  | *Intermediate* |
| 1500 A.D. | Kachemak Bay III | *Recent Prehistoric* | *Tigara* | *Eastern Thule* | *Inugsuk* |
|  |  |  |  |  | *Eastern Thule* |
| 1000 A.D. | Kachemak Bay II | *Punuk* | *Western Thule* | Dorset and *Western Thule* | Dorset? |
|  |  | *Early Punuk* | *Birnirk* |  |  |
| 500 A.D. |  |  | ? | Dorset |  |
|  |  | *Old Bering Sea* |  |  |  |
|  | Kachemak Bay I |  | Near Ipiutak |  |  |
|  |  | *Okvik* |  |  |  |
| 1 A.D. |  |  | Ipiutak |  |  |

*[The table as presented by Levin and Sergeyev was incomplete. The original table of Larsen and Rainey is printed here.—Editor.]

†Italics indicate phases of the Neo-Eskimo or Arctic Whale Hunting culture.

In the more recent article of Giddings,[10] concerning the problems of archaeology of the Bering Sea, the relative and absolute chronologies of the early cultures are presented as given in Table 2.

In recent years works have appeared based on physical-chemical methods of determining the age of ancient sites, i.e., radiocarbon analysis. In an article

## TABLE 2*
### ESTIMATED CHRONOLOGY—BERING STRAIT REGION

| Date | Asia | America | |
|------|------|---------|---|
| | | Bering Sea | Chukchi Sea |
| 1900 | Recent and Late Prehistoric | Recent | Recent |
| 1700 | | | |
| | | Nukleet III | |
| 1500 | | | Tigara—Arctic Woodland |
| | | Nukleet II | |
| 1300 | "Thule—Punuk" | | |
| | | | Western Thule— Arctic Woodland |
| 1100 | | Nukleet I | |
| | Punuk | | |
| 900 | | | |
| | Early Punuk | | |
| 700 | | ? | Birnirk |
| | "Birnirk" | | |
| 500 | | | |
| | Old Bering Sea III | | |
| 300 | | | |
| | Old Bering Sea II | | |
| A.D. 100 | | Norton | Ipiutak and "Near Ipiutak" |
| B.C. 100 | Okvik II— Old Bering Sea I | | |
| 300 | | | |
| | Okvik I | | |
| 500 | | | |
| | ? | | |
| 700 | | | Choris |
| 900 | | | |
| 1100 | | | |
| | | | ? |
| 1300 | | | |
| | | ? | |
| 1500 | | | |
| 1700 | | | Krusenstern notched point assemblage |
| 1900 | | | |
| 2100 | | | ? |
| 2300 | | | |
| 2500 | | | |
| 2700 | | | |
| | | Denbigh Flint complex | Denbigh Flint complex |
| 2900 | | | |
| 3100 | | | |
| | | ? | ? |
| Much Older | | | Palisades assemblage |

*[The table as presented by Levin and Sergeyev was incomplete. The original table of Giddings is printed here.—Editor.]

published in 1959 by Rainey and Ralph[11] the following data were given for the different cultural complexes (periods):

[NEO-ESKIMO HORIZON]
Punuk Period     A.D. 1000
Birnirk Period     A.D. 800
Kachemak Bay III Period     A.D. 600
Okvik Period     A.D. 500
Old Bering Sea Period     A.D. 300

[PALEO-ESKIMO HORIZON]
Ipiutak Period     A.D. 300
Norton Period     300 B.C.
Dorset Period     200–700 B.C.
Kachemak Bay I Period     700 B.C.
Choris Period     700 B.C.

[ANCIENT ARCTIC HORIZON]
Firth River
(Early Mountain Phase)     1300 B.C.
Sarqaq     400–2000 B.C.
Denbigh Flint
Complex     Earlier than 2000 B.C.

However tempting the results obtained from radiocarbon dating are for the solution of the problems of Arctic archaeology, one must keep in mind the discrepancies which may arise through contamination of the samples. The dates presented above are quite generalized [approximate?].

For the different samples from the sites with an Old Bering Sea inventory, the following dates were obtained: 1630 ± 230 [B.P.] (A.D. 327); 1700 ± 150 (A.D. 257); 1380 ± 118 (A.D. 577); 1398 ± 116 (A.D. 559); 1296 ± 108 (A.D. 661); 1002 ± 107 (A.D. 955).[12] Rainey and Ralph regard the first two figures as being the trustworthy.

If the facts presented here are taken into account and the necessary caution shown in view of our [imperfect] present-day knowledge of the chronology of ancient Eskimo cultures, then it is possible to date burial no. 6, which we have described above, to the middle of the 1st millennium A.D. But from where and how did iron penetrate to the Bering Sea region during this period? As pointed out by Okladnikov earlier, there are two possibilities: the early inhabitants of the Bering Sea coast could have obtained iron from the southwest, from the tribes of the lower Lena, or from the southeast, from the Amur.[13] In both regions (in the case of Yakutia as far north as Bulun and Chokurovka in the lower reaches of the Lena iron was already present by the middle of the first millennium A.D.[14] With regard to the second possibility, the Amur route, the find made in 1931 during excavations on Olski [Zavyalov] Island in the Okhotsk Sea opposite Nogayev Bay[15] is of considerable interest. Here, in the "upper site," in the shell layer, an engraving tool equipped with an iron point was found, together with stone implements.[16] The Olski Island site is undoubtedly of great age; no pottery was found, and the stone implements are of Neolithic types.

The existence of historical ties between the early population in far northeastern Siberia and the southerly regions of eastern Asia, even in the early stages of Eskimo culture, has been noted by many authors. It is quite possible that the route by which these ties were effected was along the Okhotsk coast. Naturally, the single find on Olski Island of an engraving tool with an iron point, which so far has not been dated precisely, cannot serve as a convincing argument that iron penetrated to the Bering Sea specifically along this route. Above we noted still another possibility, penetration from the northern reaches of Yakutia. As yet, the evidence is insufficient for a final decision on the question in point.

## *Notes and References*

1. H. B. Collins, Archaeology of St. Lawrence Island, Alaska, *Smithsonian Miscellaneous Collections*, vol. 96, no. 1, 1937, p. 305.

2. F. G. Rainey, Eskimo prehistory: The Okvik site on the Punuk Islands, *Anthropological Papers, American Museum of Natural History*, vol. 37, 1946, pt. 4.

3. S. I. Rudenko, *Drevnyaya kultura Beringova morya i eskimosskaya problema* (The ancient culture of the Bering Sea and the Eskimo problem), Moscow–Leningrad, 1947, p. 112. [Rudenko's book has been translated into English and published under the same title in the series *Anthropology of the North: Translations from Russian Sources*, no. 1, University of Toronto Press, 1961. See p. 177.]

4. H. Larsen and F. Rainey, Ipiutak and the Arctic Whale Hunting culture, *Anthropological Papers of the American Museum of Natural History*, vol. 42, New York, 1948, p. 83.

5. *Ibid.*, p. 254.

6. See M. G. Levin, Uelenskiy mogilnik (The Uelen cemetery), *Sovetskaya etnografiya*, no. 1, 1960.

7. Harpoon heads of this type were found also in other burials with typical Old Bering Sea inventories in the Uelen cemetery.

8. This question is considered at length in Collins, Archaeology of St. Lawrence Island, Alaska.

9. Larsen and Rainey, Ipiutak and the Arctic Whale Hunting culture, p. 155.

10. J. L. Giddings, The archaeology of Bering Strait, *Current Anthropology*, vol. 1, no. 2, 1960, p. 123.

11. F. Rainey and E. Ralph, Radiocarbon dating in the Arctic, *American Antiquity*, vol. 24, no. 4, 1959, p. 373.

12. *Ibid.*, p. 369.

13. See: Narody Sibiri (Peoples of Siberia), in the series *Narody mira*, Moscow–Leningrad, 1956, p. 101. [Published in English by the University of Chicago Press, 1964.]

15. A. P. Okladnikov, Yakutiya do prisoyedineniya k russkoma gosudarstvu (Yakutia before its annexation to the Russian state), in *Istoriya Yakutii*, vol. 1, Moscow, 1955, p. 199. [To be published in English as no. 6 of the series *Anthropology of the North: Translations from Russian Sources*.]

15. Concerning the archaeological materials excavated on Olski Island by V. I. Levin and M. G. Levin, see: M. G. Levin, Etnicheskaya antropologiya i problema etnogeneza narodov Dalnego Vostoka (Ethnic anthropology and the problem of the ethnogenesis of the peoples of the Far East), *Trudy Instituta etnografii, Akademii nauk SSSR*, n. s., vol. 36, Moscow, 1958, p. 225, note 26. [Translated into English and published as: Ethnic origins of the peoples of northeastern Asia, Arctic Institute of North America, *Anthropology of the North: Translations from Russian Sources*, no. 3, University of Toronto Press, 1963. See note 26, pp. 230–231.]

16. These materials have not been published; they are in the Museum of Physical Anthropology of Moscow State University, Collection no. 311, Specimen no. 105.

S. A. ARUTYUNOV AND D. A. SERGEYEV

# NEW FINDS IN THE OLD BERING SEA CEMETERY AT UELEN*

RESULTS OF THE EXCAVATIONS conducted by the Chukchi Detachment of the Northern Expedition of the Institute of Ethnography† in the Uelen cemetery have been published on several occasions.[1] This ancient Eskimo cemetery, unique among those thus far discovered on the Asiatic continent, has yielded already, in the years 1957–1959, new materials for the study of early Eskimo cultures. It will suffice to recall the finds of an iron[-bladed] graver in this cemetery of Neolithic aspect, the unique anthropomorphic figurines, the finds in one grave of various types of harpoon heads which formerly were regarded as separated by several centuries.

No less interesting were the finds of the 1960 season, which were concentrated in a single burial complex. This complex consisted of three identically oriented superimposed burials—A, B, and C. Multilayered coverings of burials have been encountered in the Uelen cemetery before, but in this case we are dealing with a complex where the graves are united by a common stone walling, yet neatly separated from each other. Thus, the lowest burial, C, is separated from the one above it (burial B) not only by a 0.50-m thickness of earth but also by a wooden covering; burial B rests on a wooden platform, which, in turn, is laid on specially positioned slate flagstones. The distance between burials A and B is not as great as between B and C, being only 0.10 m; yet this cannot be regarded as a paired burial, since the burials are also separated by a layer of earth and a wooden interlayering (Fig. 1). Judging by the inventory which accompanied the deceased, both were males. The upper burial A contained a group of implements for both sea and land hunting, while B contained such objects as arrowheads, a fragment of an arrow shaft, and a man's knife.

The arrangement of the lowest burial, C, was of particular interest. It was delimited by a rough walling of oval shape, constructed of large stones. The depth of the walling was from 0.10 m to 1.05 m below the surface. The skeleton was oriented to the east-by-southeast. The head was resting at 0.85 m, the feet at 0.75 m below the surface. As was mentioned earlier, traces of a wooden covering were found above the burial and above it a layer of earth separating it from burial B. Under the skeleton in burial C there was a decomposed polar-bear skin[2] and a bedding of brush cuttings; the later was placed below the skin almost at the level of the base of the stone walling,[3] which had been specially constructed for burial C. With the passage of time alluvial deposits of earth, sand, and small stones filled in the spaces in the stone walling and the wooden planking above burial C. Turf developed within the walling; after some time, perhaps a few centuries, slate flagstones were laid above it, on top of them a wooden platform, and another deceased was buried (burial B). Alluvial deposits again allowed the development of turf, which covered the stones of the walling in

*Translated from Sovetskaya etnografiya, no. 6, 1961, pp. 120–124.
†Affiliated with the Academy of Sciences of the U.S.S.R.

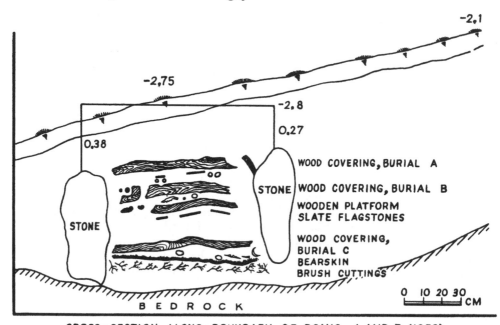

CROSS−SECTION ALONG BOUNDARY OF ROWS 4 AND 5 (1958)

FIGURE 1. Cross-sectional view of a triple interment.

several places. Where this had happened, later on several small stones were positioned, thus "renewing" the walling within which yet another deceased was laid (burial A). Thus the stratigraphy indicates that in this case we are dealing with three chronologically different graves which are united by common walling.

The observation that the burials were made over a considerable span of time is supported by their inventories. Thus, in the lowest burial C a "winged object" of a type widely encountered in the Old Bering Sea culture, with classical Old Bering Sea ornamentation (Fig. 2, a), was found. In burial A a "winged object" was also found (Fig. 2, b). Its shape clearly reflects deterioration, and the ornamentation has acquired traits which are close to Punuk with the typical pitting in the center of the circle. Characteristic of the Old Bering Sea style is a socket piece from burial C, while a socket piece of considerably later time from burial A is not only simple in construction but lacks ornamentation. True, in burial A there was found, together with a toggle harpoon head, also clearly of late type (Fig. 3, a), a typical Old Bering Sea head with ornamentation identical in style with that of the "winged object" from burial C (Fig. 3, b). However, this seeming contradiction may be clarified by introducing ethnographic materials. In the same year, 1960, D. A. Sergeyev was able to show that the Eskimos of Sirenik [Sirhenik] and Naukan villages utilized Punuk whale harpoon heads that they had found, after they had removed the stone end-blade and replaced it with an iron one. It would seem natural that chance finds of harpoon heads in earlier times would also be reutilized. This is supported by the fact that the coloring of the walrus ivory, from which the Old Bering Sea head found in burial A was made, is considerably darker than that of the other harpoon head and other artifacts of the same material found in the inventory of this burial. Thus, the finding in this burial of an Old Bering Sea harpoon head does not indicate a great age for the burial but merely genetic ties with earlier cultures.

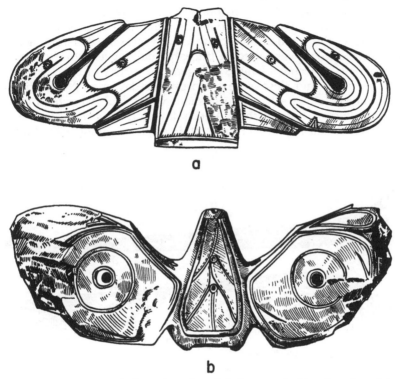

FIGURE 2. *a*, winged object from burial C; *b*, winged object from burial A (⅝ natural size).

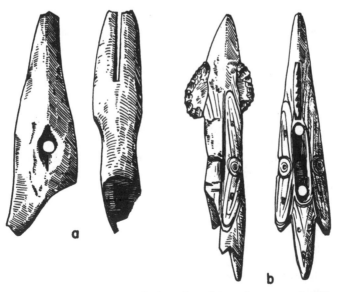

FIGURE 3. Harpoon heads from burial A: *a*, late type; *b*, Old Bering Sea type (⅝ natural size).

Several objects from burial C were mentioned earlier. Of no less interest is the remainder of its inventory. Because of the formation of a lens of permafrosted ground in the grave, the wooden objects were fully preserved here, whereas in the other burials the wood was rotted of traceable only as a texture surrounded by humus. In burial C a wooden handle of a *gatka*, a variety of adze, was found. It was carefully formed and close in shape to present-day ones. Next to it was a second wooden object, either a handle or a shaft, which had two lateral grooves; its use remains unknown. Both of these objects lay atop a third, large one (Fig. 4), a piece of bent wood superficially resembling a boomerang with the differ-

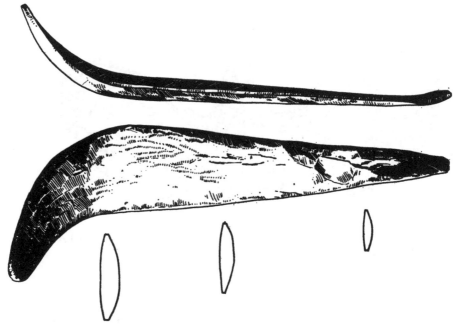

FIGURE 4. Boomerang-like artifact from burial C (⅙ natural size).

ence that its shoulders [sides] are not equal and its thickness is a bit greater than in a usual boomerang. In dimensions it agrees with the most widespread types of these weapons. The artifact was made from a specially selected piece of bent wood. The bending of the wood fibres was natural, that is, it is not to be explained by warping of the piece either before or after it was placed in the grave. This must be so, first, because the cutting and whittling marks still visible on its surface and made during its manufacture, remain straight, that is, have not lost their configuration, and second, the adze handle and the shaft lying next to it have not been warped in the least. In section the artifact has a streamlined shape close to that of an airplane wing. On its longer shoulder [side] there is a cut which makes it convenient to grasp the weapon with the right hand in the way that a usual boomerang is held just before it is thrown. Perhaps it would not be too bold to suggest that we are dealing here with a throwing club. Such a suggestion is well supported by the fact that even today a similar weapon exists among the Chukchis and Yukagirs of the Kolyma lowlands, the staff-boomerang (in the Chukchi language *tynvychgyn*), which is used nowadays mostly in herding reindeer but sometimes also in hunting partridge, ducks, and geese.[4]

In the stone inventory, particularly notable is a large slate knife, which is unfortunately broken and incomplete (Fig. 5, *a*). In contrast to all the other knives found in the Uelen cemetery it has a cross-section of complex configuration. Near the edges the blade is thickened; towards the center it thins out only to become thick again at the very center. The configuration is further complicated by a groove running almost the entire length of the [central part of the] blade. A cross-section of this shape is illogical for a stone weapon and also dele-

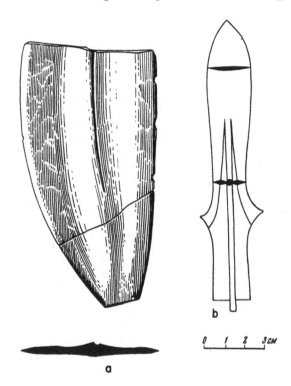

FIGURE 5. *a*, slate knife from burial C; *b*, bronze halberd point from Japan, beginning of our era.

terious to its strength. However, this can be easily explained if we suppose that this is a stone imitation of a metal weapon, most likely a sword of cast bronze. We have mentioned previously the fact of the penetration of metal (iron) to the Bering Sea area in antiquity, proposing the possible path of this penetration, namely from the Amur [river basin] area. We might look for parallels of this stone knife in nearly the same area, among the copper and bronze swords and daggers of Korea, Japan, and China of the end of the 1st millennium B.C. and the initial centuries of our era. As seen in Fig. 5, *b*, the cross-section of such a metal weapon is fairly close to that of the knife found in burial C.

In connection with the above it is interesting to note that the word for "copper" in the Eskimo language is *kanuya*, and that in Japanese "metal" in general and "copper" in particular are both expressed by the word *kane*.

The bone inventory of burial C is typical of Old Bering Sea interments. The "winged object," ornamented plaque of walrus ivory, and other artifacts as well, all have many parallels in other Old Bering Sea burials. Only one bone tool, encountered for the first time in the Uelen cemetery, is an exception to this—a beater made of walrus ivory.[5]

Besides the burial complex described in this report, two other burials, of no particular interest, were uncovered in 1960. Altogether an area of 182 m² was excavated during this season. The absence of interments in the areas surrounding the level space that was excavated earlier and the local relief of the place indicate that there is little hope that new interments will be found in any numbers in the territory adjacent to the cemetery excavated in 1957–1960.

## Notes and References

1. M. G. Levin, Raboty na Chukotke v 1957 g. (Fieldwork on the Chukchi peninsula in 1957), *Sovetskaya etnografiya*, no. 6, 1958, pp. 128–134 [for English translation, see pp. 296–304]; D. A. Sergeyev, Pervyye drevnoberingomorskiye pogrebeniya na Chukotke (The first Old Bering Sea burials on the Chukchi peninsula), *Kratkiye soobshcheniya Instituta etnografii Akademii nauk SSSR*, no. 31, 1959, pp. 68–75; M. G. Levin, Drevneeskimosskiy mogilnik v Uelene; predvaritelnoye soobshcheniye o raskopkakh v 1958 g. (An early Eskimo cemetery at Uelen: A preliminary report on the excavations of 1958), *Sovetskaya etnografiya*, no. 1, 1960, pp. 139–150 [for English translation, see pp. 305–318]; M. G. Levin and D. A. Sergeyev, Drevneberingomorskiy mogilnik v Uelene (The Old Bering Sea cemetery at Uelen), *Nauchnaya konferentsiya po istorii Sibiri i Dalnego Vostoka (Tezisy dokladov)*, Irkutsk, 1960; M. G. Levin and D. A. Sergeyev, K voprosu o vremeni proniknoveniya zheleza v Arktiku; pervaya nakhodka zheleznogo orudiya v pamyatnike drevneberingomorskoy kultury (The penetration of iron into the Arctic: The first find of an iron implement in a site of the Old Bering Sea culture), *Sovetskaya etnografiya*, no. 3, 1960, pp. 116–122 [for English translation, see pp. 319–326].

2. Remnants of fur covering the entire length of the grave were also found in 1957 in burial no. 10–11 (see Levin, Raboty na Chukotke v 1957 g., p. 131).

3. Until recent times brush cuttings were used by the local population to improve the bedding; it was put under a walrus hide. In regard to the bear skin, it clearly had served as the bedding of the deceased and had been buried with him. (Nowadays they place the deceased on new reindeer-skin bedding, but it is not carried to the grave and, after interment, it is given to one of the relatives or to a close friend.) Ethnographic data indicate that relatively recently, in the 17th and 18th centuries, the Eskimos about Bering Strait used the skins of polar bears for bedding and the trimming of bed-curtains; reindeer skins were used only later.

4. Cf. I. S. Gurvich, Metatelnoye orudiye na Kolyme (Throwing weapons of the Kolyme region), *Kratkiye soobshcheniya Instituta etnografii Akademii nauk SSSR*, no. 18, 1953, pp. 47–49.

5. Usually similar tools, intended for beating off crystallized snow from fur clothing, were made of antler.

S. A. ARUTYUNOV, M. G. LEVIN, AND D. A. SERGEYEV

# ANCIENT BURIALS ON THE CHUKCHI PENINSULA*

THE COMPLICATED PROBLEM concerning the origin of the Eskimo are very closely linked to a wide range of questions dealing with the time and area of formation of the economic and cultural type of arctic sea mammal hunters. A great deal still remains obscure. The correct approach to these problems is still hampered by inadequate archaeological knowledge of the arctic regions of northeastern Asia, northwestern North America, and the Chukchi peninsula coast. Among the studies of the Chukchi peninsula, we mention S. I. Rudenko's book[1] reporting the results of his archaeological survey of the coastal area of the Chukchi peninsula in 1945, N. N. Dikov's fieldwork in the interior of the peninsula,[2] the publications by A. P. Okladnikov,[3] and several other works.

Among the archaeological remains of the Chukchi peninsula, a special place is occupied by the ancient Eskimo burials, discovered and studied during the past few years.

The Uelen cemetery, which was discovered in 1955, and is already known from several preliminary descriptions, was excavated during four field seasons (1957–1960) by the Chukchi Detachment of the Northern Expedition sponsored by the Institute of Ethnography of the Academy of Sciences of the U.S.S.R.[4] The excavations yielded extensive materials, which are now under study. They have been described partially in the publications mentioned above as well as in a number of papers by other authors.[5]

The aim of this paper is to present a summary report of certain, though also preliminary, results of the excavations at the Uelen cemetery since fieldwork there has been completed.

In all, 76 graves have been uncovered in the Uelen burial grounds by the Chukchi Detachment.

The Uelen cemetery represents archaeological remains which took a long period of time to accumulate. This is apparent from the presence of superimposed graves, as well as from the typological differences in burial inventories. One of the best examples of this is the sector excavated in 1960.[6] However, an overwhelming majority of the burials can be placed within the confines of one lengthy era, that of the Old Bering Sea culture.

As is already known, the basic criterion for the classification and periodization of ancient Eskimo cultures is the typology of structural details and ornamentation of particular parts of the [so-called] harpoon complex and primarily of toggle harpoon heads. The latter were found in 37 graves within the burial complex excavated by the Chukchi Detachment. Altogether there were 171 toggle harpoon heads.

The most detailed classification of ancient Eskimo harpoon heads has been made by H. B. Collins.[7] This classification, like any other, is to a considerable

*Translated from *Kratkiye soobshcheniya Instituta etnografii Akademii nauk SSSR*, no. 38, 1963, pp. 57–69.

extent conditional, and as shown by the present material, it is also not quite complete. Nevertheless, an attempt to determine the [quantitative] distribution of a particular type among the harpoon heads of the Uelen cemetery is of certain interest.

We shall first describe the harpoon heads with an open shaft-socket. Twenty-seven of these heads belong to type $Ix$, which is characterized by two line-holes (one above, the other below the shaft socket), a symmetrical, trifurcated spur, two lashing slots, and side blades parallel with the line-holes. Thirteen heads belong to type $Iy$, which is distinguished from the preceding only by the position of the side blades, which are [inserted] at right angles to the line-holes.

Three heads belong to category $IIx$, which is differentiated by having a single line-hole and an asymmetrical, trifurcated, lateral spur, and also two side blades parallel with the line-hole and two lashing slots. No specimens of type $IIy$, differing from $IIx$ by the alignment of the side blades (i.e., at right angles to the line-hole), have been found in the Uelen burials.

Twenty-five heads belong to type $IIIx$, which is characterized by an irregular, asymmetrical spur, a single line-hole, and an end blade in line with the line-hole. Type $IIIy$, with the end blade at right angles to the line-hole, does not occur at Uelen.

Among the harpoon heads with an open socket, two or three heads are representative of types $II(a)x$, $III(a)x$, $II(d)x$, and $II(c)y$ (slight modifications of the basic types described above).* Six heads of the open-socket variety do not fit any of Collins' types.

In Figs. 1$a$, 1$b$, and 1$c$, type $Ix$ is represented by item 2, type $Iy$ by item 3, type $IIx$ by item 8, type $IIIx$ by items 17 and 18,† type $II(a)x$ by item 16, type $III(a)x$ by item 19. Heads which do not fit into any of the classified types are shown in items 1, 4, 5, 6, and 7.

Type $IIIx$ is the type of harpoon head with closed socket seen most frequently; (represented in Fig. 1 by items 10, 11) belong to this type, which is distinguished by the symmetrical, trifurcated spur with shortened lateral prongs, the single line hole, and the end blade parallel to it. A few specimens of this series represent types $IVy$, $Vx$, $V(a)x$, $Vy$. These are related types, characterized by a single, plain and straight spur, one line hole, an end point parallel to the line hole (in group $x$) or at right angles to it (in group $y$). Among these, the type $Vy$ is comparatively numerous (Fig. 1, items 12, 20), six of this type having been found.

Standing apart are five heads with open sockets, not included in Collins' classification and belonging to types characteristic of the Birnirk culture[8] (Fig. 1, items 14, 15).

The remainder (some 50 heads) cannot be classified with any degree of accuracy either because of their poor state of preservation or because they had been placed in the graves in an unfinished state (as blanks for harpoon heads).

The generally accepted periodization of ancient Eskimo cultures includes, as a rule, the following stages: Old Bering Sea, Okvik, Birnirk, and Punuk. All these cultures are represented in the inventories of Uelen graves. Over one-third of all toggle harpoon heads may be considered characteristic of the Old Bering Sea culture. To determine which belong to the Okvik culture presents more difficulties. In many specimens (approximately 25), it is rather difficult to decide

---

*[Type $II(d)x$ does not occur in Collins' classification; type $II(g)x$ is probably meant.—Editor.]

†[While items 8, 17, and 18 conform to Collins' types $IIx$ and $IIIx$ in basic features, their spurs are of thick asymmetrical type characteristic of Okvik harpoon heads, which were not included in Collins' classification.—Editor.]

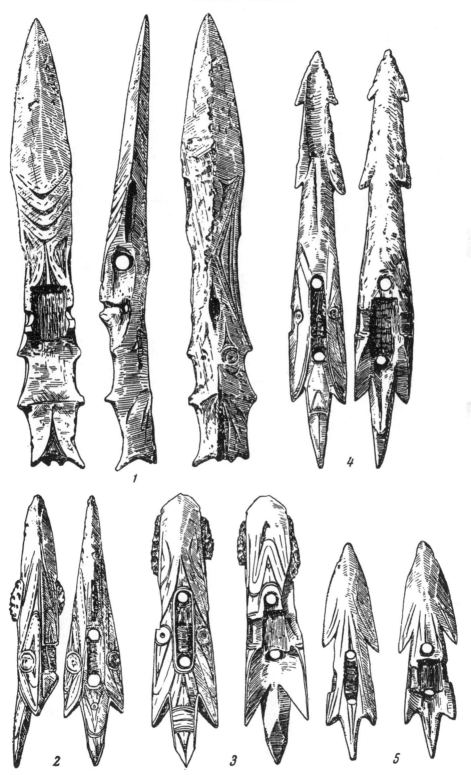

FIGURE 1. *a*. Toggle harpoon heads from the Uelen cemetery (size reduced), items 1–5.

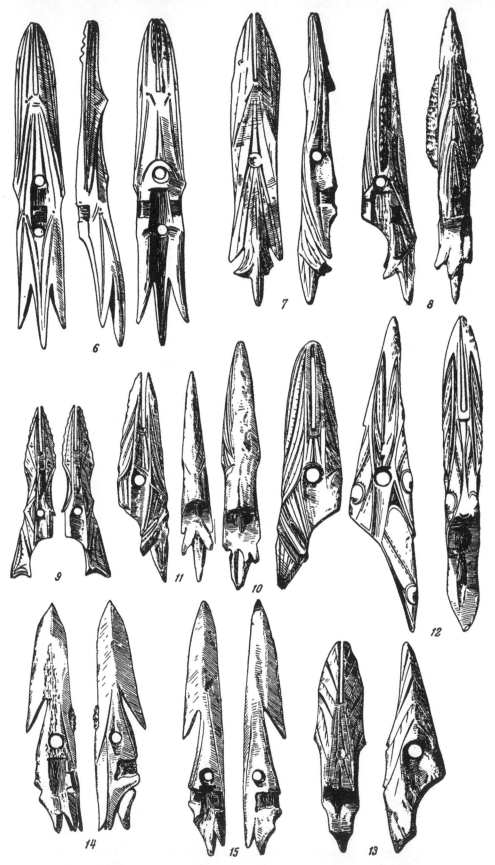

FIGURE 1. *b*. Toggle harpoon heads from the Uelen cemetery (size reduced), items 6–15.

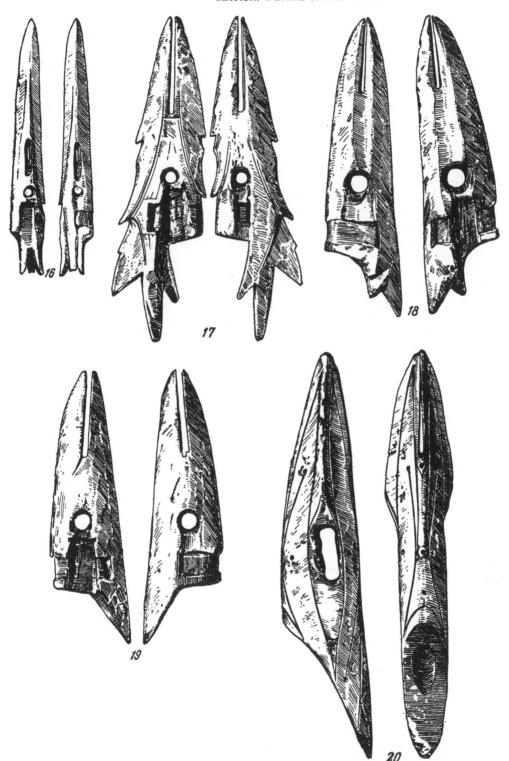

FIGURE 1. *c.* Toggle harpoon heads from the Uelen cemetery (size reduced), items 16–20.

whether a given head should be classified as Old Bering Sea or Okvik. About the same number of heads can be classified, with more or less certainty, as of the Okvik culture. Seven heads have been typed as Birnirk and about as many as Punuk. (More than 40 heads cannot be typed with any degree of certainty because of the poor preservation, unfinished state, or lack of ornamentation.)

In most graves, the inventory belongs to the Old Bering Sea type. However, in addition to inventories characteristic of this culture, objects considered typical of the Okvik culture have also been found. This combination is not surprising, as these two stages are contiguous (according to all systems of periodization) and a transition from one to the other would be quite natural. However, the crux of the problem is concerned with precedence: which culture preceded the other? Old Bering Sea or Okvik? A widely accepted opinion is that Okvik is the older of the two. This view is held by H. E. Larsen and F. G. Rainey.[9] According to J. L. Giddings' tabulation, Okvik I is to be dated to the 5th–3rd century B.C., and Okvik II–III to the 1st century B.C.–1st century A.D.[10]

In recent years analyses of ancient Eskimo cultures based on the radiocarbon dating method have appeared. According to the data, Okvik is dated later than Old Bering Sea, the Old Bering Sea being dated to the 3rd century A.D. and the Okvik to the 5th century A.D.[11] The exactness of these findings should not be overrated; the considerable divergences among the radiocarbon analyses, due, primarily, to the contamination of specimens, must be considered. Yet, on the whole, it seems quite likely that the Old Bering Sea culture is older than the Okvik. The analysis of the Uelen materials tends to support this view. It should be emphasized, however, that in our opinion there existed a definite and quite lengthy stage when the Old Bering Sea and Okvik cultures existed side by side. This may be deduced with comparative certainty from the presence in a particular grave of inventories characteristic of both cultures or even objects which combine ornamental traits peculiar to both.

Unfortunately, we still have an insufficient knowledge of the Okvik culture. Materials from the Okvik site, excavated over 25 years ago, have been published only in part. Nevertheless, it seems clear that the Old Bering Sea culture could hardly have developed directly from Okvik, or vice versa. Both are closely related and we may presume that in this case we are dealing with local variants which prevailed among different groups of the ancient Eskimo population of the Bering Sea region. We cannot as yet indicate the foci from which these local peculiarities originated. However, it appears very probable that, within the territory under discussion, the cultural and primarily the artistic and stylistic influences of the two foci permeated each other. The existence even in our times of several Eskimo dialects on the Chukchi peninsula indicates that even among geographically close groups strong local differences are possible. On the other hand, we cannot exclude the probability of intercourse between territorially distant groups, when we consider the way of life of the ancient Eskimos with their frequent and long sea voyages. In this way certain cultural elements could have been diffused.

Yet another possibility should be kept in mind in analyzing the inventories of individual graves. It is a well-known fact that, till recent times, the Eskimos made use of Punuk whaling harpoon heads after removing the stone end-blades and supplanting them with iron blades.[12] This could have occurred in the more distant past, too.

A few graves datable to the Birnirk and Punuk cultures have also been discovered in the Uelen cemetery. There are no differences of opinion about their dating. The Birnirk culture follows the Okvik–Old Bering Sea stage and is followed, in turn, by the Punuk. Birnirk is dated by the radiocarbon method to the 8th century A.D. and Punuk to the 10th century A.D.

Burials belonging to these late cultures are the exception at Uelen; they present a completely alien aspect in relation to the interrelated graves of the cemetery, which for the great majority and apparently represent a large tribal burial plot.

In some parts of the cemetery there were superimposed burials. To a certain extent this helped to clarify the relative chronology of individual aspects and complexes of the inventories. A particularly interesting sector was excavated in 1960.[13] It enabled us to trace differences between the early and late variants of such artifacts as harpoon heads and "winged objects."

The 1959 find of a graving tool with an iron blade was of great significance; it was dated to the middle of the 1st millennium A.D.[14] This is the oldest iron object found in the Arctic.

In this publication we shall not concern ourselves with the original and rich art of the people of ancient Uelen, which was in part reported in previous articles. Let us just note the richness of the inventory of the Uelen cemetery—18 "winged objects" alone were found.

In 1960, when the last sectors of the Uelen burial ground had been excavated by D. A. Sergeyev, information was received that human bones had been discovered along the right bank of the Eylyukeu river in the area of Cape Verblyuzhiy. An archaeological survey in these areas conducted by Sergeyev in 1961 showed that the cemetery in the area of the Tunytlen settlement dated to the 19th century. A similar cemetery was found 2–3 km from the coast in the area of Cape Verblyuzhiy.

An ancient Eskimo cemetery extending over two knolls was discovered near the ruins of the ancient settlement of Ekven, 800 m from the seacoast. Burials were indicated by the whale bone and stones protruding through the surface of the turf. Fifteen graves were excavated on the western knoll and three on the eastern.

Work on the Ekven cemetery has just begun. A final judgment about this burial compound must await further excavations. The 18 graves, mentioned above as marked by the presence of protruding stones and whale bones, yielded an inventory characteristic of the Old Bering Sea–Okvik stage. There were comparatively few harpoon heads with two line holes. Points of types *IIIx* and *IIIy* prevail.

The objects from the Ekven cemetery are related to the major part of the Uelen inventory. At the same time, however, there can be no doubt of the presence of strongly expressed, local, original elements. This originality appears in the details of construction and ornamentation of toggle harpoon heads, and in the rich ornamentation, unusual refinement, and fanciful shapes of artifacts. Harpoon foreshafts—objects usually left unadorned—are covered here by complex ornamentation (Fig. 3). Six "winged objects," some of them fanciful to the point of being grotesque, were found in the graves. Of particular interest is an amulet of walrus ivory (Fig. 4), on which the ancient talented carver succeeded in reproducing almost the entire economically usable fauna: the sculptured heads of a walrus, mountain sheep, seal, polar bear, and some sort of cetacean are combined in such a way that a particular detail, when looked at from one side, represents the horns of a sheep, and, when turned around, the tusks of a walrus, and so on.

Anthropomorphic representations portray faces almost exclusively. (Only one

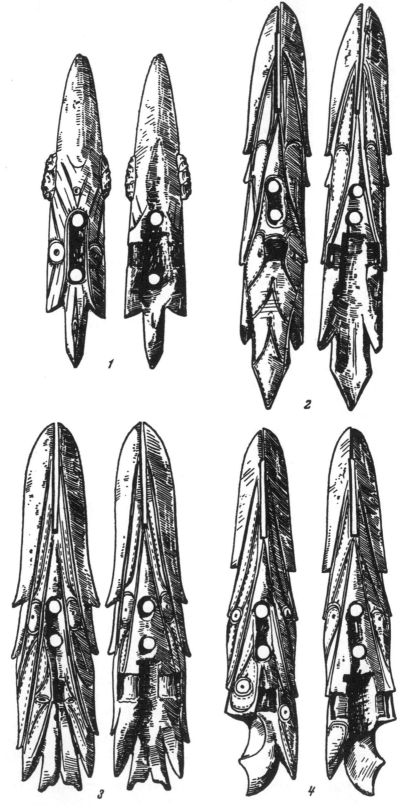

FIGURE 2. *a.* Toggle harpoon heads from the Ekven cemetery (size reduced), items 1–4.

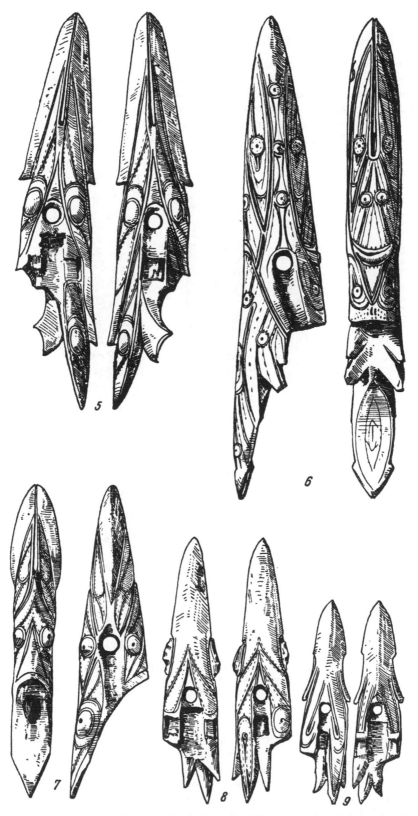

FIGURE 2. *b*. Toggle harpoon heads from the Ekven cemetery (size reduced), items 5–9.

statuette shows the full human figure.) They differ greatly in style; some are quite realistic, others are of schematic design. One specimen has tattoo marks on the cheekbones in the shape of bird tracks and also shows cheek labrets (Fig. 5).

Particularly notable is a representation of a human face on an ornamented walrus ivory object of unknown use. It consists of a combination of curvilinear ornamental elements of the Bering Sea style (Fig. 6).

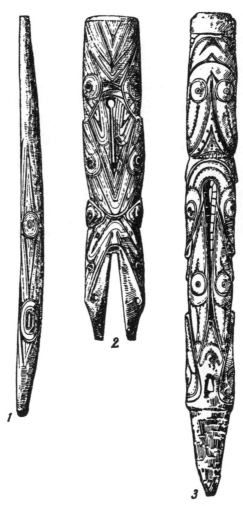

FIGURE 3. Harpoon foreshaft (item 1)
and harpoon socket-pieces (items 2 and 3)
[from the Ekven cemetery] (size reduced).

Interesting, too, are models of kayaks (Fig. 7), adze handles, and of a clay vessel found in the different graves; all of the models are made of walrus ivory.

Several potsherds with textile impressions were also found.

Worth noting is the comparatively large number of objects made of bone and horn [antler?]: punches, small troughs, paddles. Many well-preserved wooden objects were discovered in the Ekven cemetery. A large number of them were in grave no. 12. Here were found the wooden part of a bow identical with those of bows still used by Eskimos in the 19th century, a fragment of a wooden spear-

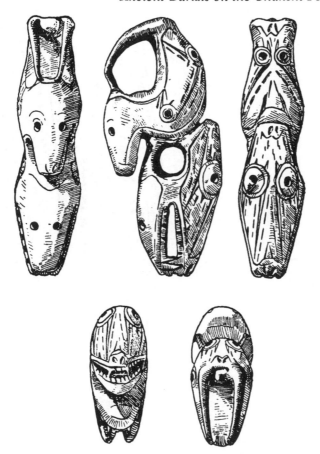

FIGURE 4. Amulet of walrus ivory [from the Ekven cemetery].

FIGURE 5. Anthropomorphic sculpture of walrus ivory [from the Ekven cemetery].

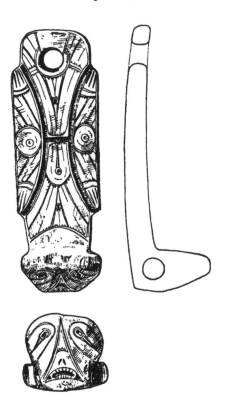

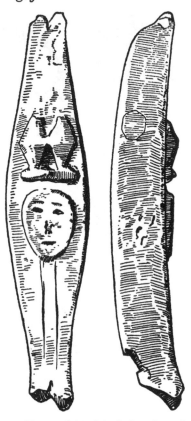

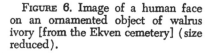

FIGURE 6. Image of a human face on an ornamented object of walrus ivory [from the Ekven cemetery] (size reduced).

FIGURE 7. Model of a kayak made of walrus ivory [from the Ekven cemetery] (size reduced).

thrower, and a wooden shaft, inserted in the socket of a "winged object"—the first find of its kind and very important for determining the purpose [or use] of this mysterious object (Fig. 8).

Noteworthy too is the complex structure of some of the graves. Wood was used extensively as a building material: thick logs formed the flooring and side walls, thinner ones were used for roofing. Of even greater importance in grave construction were whale bones, particularly the ribs, jaws, and scapulae. Thus, grave no. 9 had its side walls constructed with whale jaws, the end walls with shoulder blades, and the roofing with ribs. Two large whale jaws resting on transversely placed ribs formed the roofing of grave no. 15, and so on.

Excavations of the Ekven cemetery have just started. We have every reason to believe that future work at this site will reveal new, important data reflecting the ancient culture of Asiatic Eskimos.

The excavations of the Uelen and Ekven burial grounds yielded considerable craniological material. The craniological series of the Uelen cemetery consists of about 70 skulls in different states of preservation; the Ekven cemetery, where skeletal remains have been very well preserved, has already yielded 16 skulls. These are the first craniological series datable to the Old Bering Sea and Okvik cultures; only very few skulls are known from Collins' excavations on St. Lawrence Island.

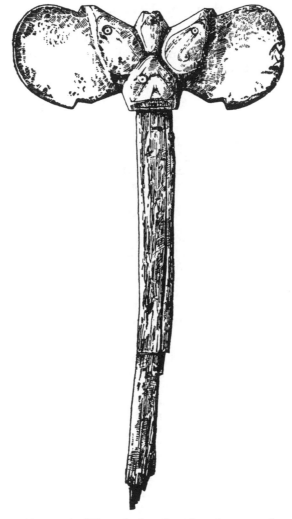

FIGURE 8. "Winged object" with a fragment of a wooden shaft [from the Ekven cemetery] (size reduced).

Let us examine the basic traits of the two Chukchi peninsula series. In basic diagnostic traits the craniological type of the ancient population who buried their dead in the Uelen and Ekven cemeteries is very close to that of present-day Eskimos. The series from the [two] cemeteries differ from the present-day Eskimo series of the Chukchi peninsula and Alaska mainly by the structure of the cranium: the ancient skulls are considerably longer, narrower, and higher. The cephalic index of the Uelen series is 71.6 (male) and 70.8 (female) and of the Ekven series, 69.2 and 69.0, respectively. Generally speaking, the ancient series of the Chukchi peninsula are relatively close to those of present-day Labrador and Greenland Eskimos.

The series from the Uelen and Ekven cemeteries are not quite identical. The crania from Ekven are considerably longer and have a larger maxillary width; there are other differences too.

The considerable antiquity of a strongly dolichocephalic type on the Asiatic coast of the Bering Strait, shown by the skulls from these cemeteries, is a new and rather important argument in favor of the theory that this type preceded the mesocephalic "western" type even in the western reaches of the Eskimo area.

The discussion of these problems, as well as the description of the ancient [craniological] series, lies outside the scope of this preliminary report.

## Notes and References

1. S. I. Rudenko, *Drevnyaya kultura Beringova morya i eskimosskaya problema* (The ancient culture of the Bering Sea and the Eskimo problem), Moscow, 1947. [Translated into English and published under the same title in the series *Anthropology of the North: Translations from Russian Sources*, no. 1, University of Toronto Press, 1961.]

2. Cf. *Zapiski Chukotskogo krayevedcheskogo muzeya*, Magadan, nos. 1 and 2, 1958 and 1961.

3. A. P. Okladnikov, O pervonachalnom zaselenii chelovekom vnutrenney chasti Chukotskogo poluostrova (On the initial peopling of the interior of the Chukchi peninsula), *Izvestiya Vsesoyuznogo geograficheskogo obshchestva*, vol. 85, no. 4, 1953; A. P. Okladnikov and I. A. Nekrasov, Novyye sledy kontinentalnoy neoliticheskoy kultury na Chukotke; nakhodki u oz. Elgytkhyn (New traces of an island Neolithic culture on the Chukchi peninsula; the Lake Elgytkhyn finds), *Sovetskaya arkheologiya*, 1957, no. 2.

4. Some sectors of the cemetery were excavated by N. N. Dikov in 1956 and 1958, independently of the fieldwork of the Institute of Ethnography of the Academy of Sciences of the U.S.S.R.

5. M. G. Levin, Raboty na Chukotke v 1957 g. (Fieldwork on the Chukchi peninsula in 1957), *Sovetskaya etnografiya*, no. 6, 1958, pp. 128–134 [see pp. 296–304 for English translation]; D. A. Sergeyev, Pervyye drevneberingomorskiye pogrebeniya na Chukotke (The first Old Bering Sea burials on the Chukchi peninsula), *Kratkiye soobshcheniya Instituta etnografii Akademii nauk SSSR*, no. 31, 1959, pp. 68–75; M. G. Levin, Drevneeskimosskiy mogilnik v Uelene; predvaritelnoye soobshcheniye o raskopkakh v 1958 g. (An early Eskimo cemetery at Uelen: A preliminary report on the excavations of 1958), *Sovetskaya etnografiya*, no. 1, 1960, pp. 139–150 [see pp. 305–318 for English translation]; M. G. Levin and D. A. Sergeyev, K voprosu o vremeni proniknoveniya zheleza v Arktiku; pervaya nakhodka zheleznogo orudiya v pamyatnike drevneberingomorskoy kultury (The penetration of iron into the Arctic: The first find of an iron implement in a site of the Old Bering Sea culture), *Sovetskaya etnografiya*, no. 3, 1960, pp. 116–122 [see pp. 319–326 for English translation].

6. Cf. S. A. Arutyunov and D. A. Sergeyev, Novyye nakhodki v drevneberingomorskom mogilnike v Uelene (New finds in the Old Bering Sea cemetery at Uelen), *Sovetskaya etnografiya*, no. 6, 1961, pp. 120–124 [see pp. 327–332 for English translation].

7. H. B. Collins, Archaeology of St. Lawrence Island, Alaska, *Smithsonian Miscellaneous Collections*, vol. 96, no. 1, 1937.

8. Cf. J. A. Ford, Eskimo prehistory in the vicinity of Point Barrow, Alaska, *Anthropological Papers of the American Museum of Natural History*, vol. 47, pt. 1, 1959, pp. 75–79.

9. H. E. Larsen and F. G. Rainey, Ipiutak and the Arctic Whale Hunting culture, *Anthropological Papers of the American Museum of Natural History*, vol. 42, 1948, p. 155.

10. J. L. Giddings, The archaeology of Bering Strait, *Current Anthropology*, vol. 1, no. 2, 1960, p. 123.

11. F. G. Rainey and E. Ralph, Radiocarbon dating in the Arctic, *American Antiquity*, vol. 24, no. 4, 1959, p. 373.

12. Cf. Arutyunov and Sergeyev, Novyye nakhodki. . . .

13. *Ibid.*

14. Levin and Sergeyev, K voprosu o vremeni. . . .

# THE GEOMORPHOLOGY OF NORTHERN ASIA: SELECTED WORKS

*Translated by* DAVID KRAUS

# LIST OF SYMBOLS

## SYMBOLS USED IN THE SPORE-POLLEN DIAGRAMS:

- Total pollen of tree species
- Total pollen of grassy plants
- Total spores of the higher spore plants

Larch (*Larix*)

Fir (*Abies*)

Spruce (*Picea*)

Pine (*Pinus*)°

Siberian pine (*Pinus sibirica*)

Birch (*Betula*)

Alder (*Alnus*)

Willow (*Salix*)

Hornbeam (*Carpinus*)

Total pollen of the oak (*Quercus*), linden (*Tilia*), and elm (*Ulmus*)

Grasses (Gramineae)

Sedges (Cyperaceae)

Orachs (Chenopodiaceae)

Heathers (Ericales)

Wormwood (*Artemisia*)

Motley grasses

Green mosses (Bryales)

Sphagnum mosses (Sphagnales)

Pteridophytes (Filicales)

Club mosses (Lycopodiaceae)

°In the spore-pollen diagrams in which the species composition of the pines is shown, this symbol is used to indicate the common pine (*Pinus silvestris*).

A. I. POPOV

# THE QUATERNARY PERIOD IN
# WESTERN SIBERIA*

THE WEST SIBERIAN LOWLAND is the world's largest alluvial marine plain. Its primary subsidence during the Mesocenozoic was responsible for its major geologic and topographic features, viz., its very thick unconsolidated Mesocenozoic deposits, its flatness, and its strictly regular alternation of topographic zones, which is expressed more distinctly here than anywhere else on earth.

The thick layer of Quaternary deposits that covers the other Mesocenozoic deposits affords a quite complete record of the stages and conditions of development of the West Siberian Lowland during the Quaternary period.

An acquaintance with the Quaternary deposits of western Siberia will show that neither the lithologic character, nor the paleontologic, geomorphologic, or any other characteristics of the deposits, can serve as the sole criterion for their stratigraphic zonation. The criterion can be found only by joint examination of all the data, the paleogeographic data in the broad sense. This approach to the question of the stratigraphic zonation of the Quaternary deposits of the northern part of the lowland will permit an objective examination of all the data that have been accumulated to date and will suggest conclusions that may not always substantiate the hypotheses and systems generally accepted at present.

The nature of the western Siberian Quaternary deposits is such that the stratigraphic and paleogeographic schemes of the Quaternary period assumed for the [European] Russian Plain cannot be applied to them. The relative lack of knowledge of western Siberia makes a comparison between the stages of its development during the Quaternary period with the stages of development of the Russian Plain difficult at present.

In our opinion, the attempt to find confirmation of the stratigraphic scheme of quaternary deposits of the European U.S.S.R. everywhere in Siberia (Saks, 1948, 1955) detracts from objective analysis of the facies of the Quaternary deposits and interferes with proper recognition of the physicogeographic conditions of the Quaternary period. Therefore, the present schemes of development of the West Siberian Lowland during the Quaternary contain many contradictory propositions, doubtful interpretations of the essence of the physicogeographic phenomena, and so on.

In seeking a rational means of determining the Quaternary history of western Siberia, we have tried a new method of analyzing the data. In this article, we have proposed a variant of the solution, which, we believe, will provide a satisfactory answer to a number of difficult questions concerning the ancient glaciation of western Siberia and the Quaternary history of this region in general. The method proposed here should be regarded merely as a working hypothesis which may be modified substantially as facts accumulate and lend themselves to new

*Translated from K. K. Markov and A. I. Popov (Editors), *Lednikovyy period na territorii yevropeyskoy chasti SSSR i Sibiri*, Moskovskiy gosudarstvennyy Universitet, Geograficheskiy Fakultet i Muzey Zemlevedeniya, 1959, pp. 360–384.

interpretations, but we feel that any attempt to shed new light on an unsolved scientific problem is of value in its own time. In this article, we have re-examined critically several propositions we had formerly accepted (A. T. Popov, 1949).

In our opinion, an analysis of the available data justifies the division of the entire, comparatively uniform layer of Quaternary deposits of the lowland into three principal parts:

1. Preglacial formations, represented by alluvial-lacustrine sediments in the central part of the plain and by marine and deltaic-alluvial sediments in the north.

2. Glacial and late glacial formations, represented by glacial, fluvioglacial, and in particular by glaciomarine and marine sediments north of 58° N. lat., and deltaic-alluvial and lacustrine sediments south of 58° N. lat.

3. Postglacial formations, represented by lacustrine, lacustrine-paludal (swamp), and alluvial sediments, and to a much lesser extent by marine sediments in the far northern section of the plain.

This division is advisable because, as we shall see, the physicogeographic conditions of the preglacial, glacial, and postglacial periods differ substantially.

## Preglacial Deposits

Although relatively little information is available on the preglacial deposits, it has been observed that wherever an entire layer of Quaternary deposits has been uncovered, the lower part consists of alluvial formations, and of marine formations near the mouths of the Ob and Yenisey rivers. There is no convincing evidence that the clays, aleurites (silts), or sands with scattered boulders and pebbles comprising the lower part of the layer are of glacial origin. The clastic matter scattered in deposits and even the pebble gravel accumulations that are many meters thick (e.g., in cores from the Yenisey delta in the ancient deposits of the valleys or the marine estuaries of such powerful streams as the Yenisey and the Ob cannot be regarded as moraine material and, what is more, they are not necessarily products of moraine outwash.

Over a broad area of the lowland south of the Polar Circle, boulder beds usually attributed to the glacial maximum are underlain everywhere by a clearly alluvial-lacustrine layer of sediments. Undoubtedly this stratum, which must be regarded as preglacial, began to accumulate when the sea level was much lower than at present. It should also be noted that the sediments accumulated during a period when the base level of erosion was steadily rising, during a gradual rise of the sea level.

Unfortunately, very little study has been devoted to the preglacial deposits and therefore our knowledge of them is still quite inadequate.

## Glacial Deposits

It had long been thought that the West Siberian Lowland was once covered with a major ice sheet and that this ice sheet consisted of two parts, the Siberian and the Ural. All investigators left open the question of the coalescence of these two ice sheets, but did not doubt that the two were in very close proximity to each other. The concept of such extensive development of the ice sheets was based on the known occurrence of boulder beds over an enormous area of western Siberia north of 58° N. lat.

N. I. Mikhaylov (1947) was the first to reject the ice sheet explanation of the broad distribution of boulders in the West Siberian Lowland; he rejected the sheet glaciation hypothesis for the Central Siberian Plateau. He held that there had been no ice sheet on this plateau and thus there could not have been an ice sheet in western Siberia. Later Yu. M. Sheynman (1948) also rejected the sheet glaciation hypothesis with regard to the North Siberian Plateau. Now the concept that the glacial maximum had limited distribution in the northern part of the Siberian Plateau and was chiefly of the mountain-valley type is gaining more and more adherents.

Having examined the available data on the question of the ancient glaciation of western Siberia, we have concluded that the hypothesis of sheet glaciation on the grand scale assumed thus far is untenable. The following facts, concerning principally the Siberian ice sheet, refute this hypothesis:

1. The Central Siberian Plateau bears no traces of an active sheet glaciation which could have ensured the existence of an extensive ice sheet extending over the West Siberian Lowland to the meridian of Surgut on the Ob. Only active sheet glaciation of the plateau could have transported boulders far to the west.

2. The petrographic composition of the boulders indicates that only the Putorana mountain network could have been the center of nourishment of the Siberian ice sheet. Geologic and geomorphologic evidence on the plateau itself bears witness to this and limits the region of ancient glaciation to the northwestern part (north of the Nizhnyaya [Lower] Tunguska river) of the lowland. A glance at the map will show that the area of the alimentation region clearly does not correspond to the broad area of the lowland ice sheet, which stretched so far to the west. There can be no question that the Taymyr could have been a source of alimentation, because there is almost no Taymyr boulder material in western Siberia.

3. The facts indicate that the Siberian and Ural boulder material formed two tongues (Fig. 1). In general, the very center of the lowland has no deposits with boulders. The boulder beds represented by these "tongues" can scarcely have corresponded to the contours of former ice sheets.

4. There are no moraine formations on the Gydan peninsula. Further, the ice sheet, having descended from the Central Siberian Plateau, should have extended, in conformity with the topography, uniformly both to the southwest and west and to the northwest, i.e., into the region of the Gydan peninsula, and even more to the northwest than to the southwest. If the absence of a moraine in the north is to be attributed to its subsequent complete erosion, why did the boulder beds (that is to say, the moraine of the glacial maximum) remain so intact farther south, in the Yenisey-Taz interfluve and the middle Ob?

5. Hilly-moraine relief, corresponding to the regions of former glaciation, is observed only in the peripheral areas of the lowland, not in the interior. This difference cannot be explained by differences in the time of glaciation in the two regions, because the relief is fresh and there is no talus mantle on the slopes in either area, i.e., erosion of the watersheds is equally slight in the interior and in the peripheral areas of the lowland (Popov, 1949).

6. It is difficult to imagine the movement of the thin (as all investigators agree) Siberian ice sheet along a plane horizontal surface and, more than likely, uphill.

7. The development of extensive ice sheets, even if they did not completely

coalesce, would indubitably have dammed up the large rivers on a grand scale. The geologic evidence, however, indicates that the northward drainage did not cease but continued satisfactorily the entire time. It is highly improbable that the water of the powerful rivers flowed beneath the ice sheet, as proposed by V. N. Saks (1948). The drainage problem is one of the most vulnerable points of the hypothesis that western Siberia was extensively glaciated.

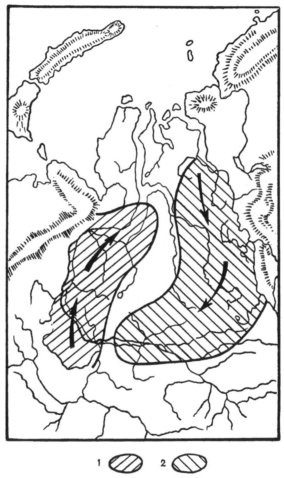

FIGURE 1. Areal limits of the Ural (1) and Siberian (2) boulder material in the West Siberian Lowland.

Thus, all the conditions stated above, taken together, show decisively that the glaciation of the West Siberian Lowland was not as extensive as has been assumed.

What can be offered in place of the unconvincing hypothesis of sheet glaciation? Let us turn to the facts for an answer and focus our attention on several of the less-studied regions regarded as belonging to the area of the maximum Siberian ice sheet.

In cross-section of the watershed plateau of the Yenisey-Taz interfluve, bluish-gray, dark clays, loams and sandy loams, in all more than 70–80 m thick, lie above the preglacial sands (Fig. 2). In many places they grade upward into sands. The degree of stratification in the clays and loams varies; sometimes they are varved. Boulders and pebbles of crystalline rock, chiefly traps, are present in the lower part of the stratum and often in the upper part as well. Sometimes the boulders form accumulations of the moraine type. In this layer in the Turukhan and Yenisey rivers and the Turukhan-Taz interfluve, boulder horizons (sometimes there are several) combine, often alternating with non-bouldery marine clays.

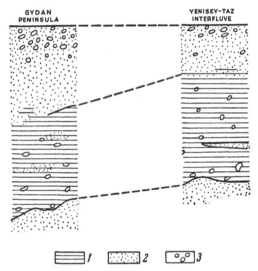

FIGURE 2. Schematic sections of Quaternary deposits of the watershed plateau of the Gydan peninsula and the Yenisey-Taz interfluve: 1, clays and loams; 2, sands; 3, boulders and pebbles. 1 cm = approximately 20 m.

On the Yenisey, in the region between Igarka and Dudinka, almost all horizons and facies of the deposits contain various amounts of clastic material. Only the varved clays are almost free of clastic materials. Marine faunas have been found in nearly all horizons (except the alleged moraine horizons), even in the varved clays.

We do not have sufficient grounds for establishing the stratification of this entire friable layer, because most of the lithologically differentiated deposits lie on marker beds, which are sometimes identical and sometimes different, but often similar, and there are no clear erosion contacts between them; further, there are usually gradual transitions between these deposits. Evidently, there is more reason to regard this as a quite complex system of facies transitions of some deposits into others.

The almost constant presence of clastic material, a special combination of marine, fluvioglacial, and glacial facies, and the scarcity of organic remains inclines us to regard the entire stratum of the Yenisey-Taz interfluve and the Yenisey as a single, but complexly constructed fluvioglacial assemblage.

A cross-section from the watershed plateau in the Yeloguy-Kellog basin (left bank of the Yenisey) shows a similar, though simpler structure. Here gray loams, boulder and non-boulder, lie above the white preglacial sands. Farther west, in

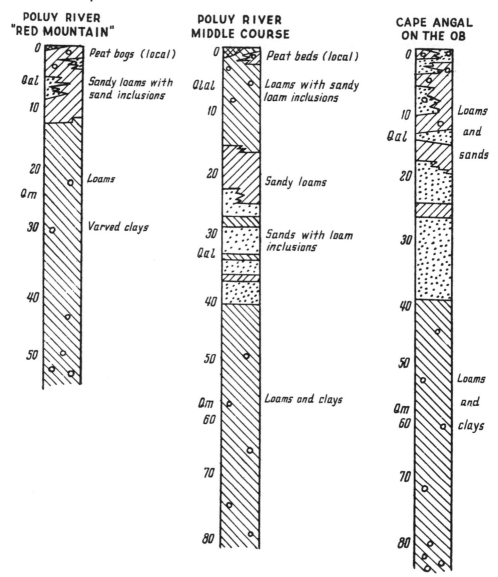

FIGURE 3. Columnar sections of Quaternary deposits of the lower Ob.

the basin of the Vakh, Bolshoy Yugan, and the latitudinal* section of the Ob river, the boulder deposits become thinner. However, even here the boulder facies are closely associated with the non-boulder and form transitions.

On the lower Ob (Fig. 3), on the Mesozoic clays, there is a layer 80–100 m thick consisting of thinly stratified dark clays and loams with scattered gravel, pebbles, and boulders, and also fine plant remains. Thin interlayers of whitish aleurite, which give the rock a ribbon-like character, are commonplace. Sometimes the aleurite is isolated in bands 0.5–2.0 m thick, sometimes 10–20 m thick. Pebbles and interlayers of gravel also occur here. Often the layers of loam and

*[I.e., that section of the Ob flowing from east to west along latitude 61° N.—Editor.]

aleurite form fine, very intricate folds, microfaults, microdisplacements, and the like.

Higher, there are light-gray fine-grained sands about 20 m thick. The strata within the sands are intricately folded, forming numerous small faults. Still higher, there is a layer of dark loam as much as 50–60 m thick, sometimes forming a tight alternation of thin layers of loam and sandy loam or fine sand and sometimes forming massive strata. The stratification often becomes banded. Microfolds are frequent. Individual, uncoordinated boulders, pebbles, and rubble of Ural crystalline and metamorphic rock, not forming any perceptible accumulations, are found in the loam layers. Comminuted plant remains of the river "chaff" type are also observed; they are scattered through the rock and do not form accumulations. Gradually, then more abruptly, the loam stratum becomes fine-grained sand and sandy loam interbedded with loam. Uncoordinated boulders, pebbles, and gravel or, more rarely, accumulations of such are encountered sporadically. Sometimes, the upper horizons are enriched with cobble, but this is far from universal. The sandy layer is 40–50 m thick.

It should be noted that the contact between the upper dark loams and the sand underlying them is complicated by the intrusion of contorted sands into the loams and, vice versa, of loams into the sands, by fault benches, etc. This type of contact was produced by the peculiar processes of underwater unheaval and interpenetration of masses of sand and soft plastic or liquescent freshly deposited silt. Numerous fine faults and dislocations formed under these conditions. The fine structure of the main stratum, the sporadic finds of comminuted plant material ("chaff"), the finds of fine ferruginous pipes (evidently, the product of the activity of marine organisms), the presence of only small boulders and pebbles scattered in the rock in most cases, the frequently well-defined thin and coarser stratification often of the band type, all refute the hypothesis that these deposits originated from moraines. There is more reason to assume that they are of marine, perhaps glaciomarine origin, and some of them may be of deltaic-alluvial origin.

The sections from the Ob and the Yenisey rivers do not actually differ basically; it should merely be noted that the entire section in the eastern part of the lowland is enriched to a great extent by boulder material.

According to V. N. Saks, there are no typical glacial deposits in the Taz-Nadym region, in the central part of the lowland, and what is more, the are no boulder deposits whatever in the cross-sections from the watershed pleateaus.

If this region had been covered with an ice sheet in the past (as implied by the theory of sheet glaciation of the plain), the moraine, by analogy with the Vakh basin and other southern regions, surely would not have been completely destroyed by erosion. In the latter case, the products of moraine erosion should have remained in the form of boulder accumulations in the alluvium, but this is not the case.

The loams and clays of this region are regarded by Saks as lacustrine deposits. In our opinion, the very broad development of such deposits in the Pur basin and beyond, their comparatively great thickness, and also the absence of freshwater fossils in them bear witness to their non-lacustrine origin. Marine mollusks appear in the lower reaches of the Pur in loams of analogous stratigraphic position. Here, on the lower Pur, according to Saks, marine Sanchugovka sediments* are represented by relatively benthic facies, of the order of 100 m. To the south, farther up the Pur, they are replaced suddenly by continental, according to Saks, lacustrine sediments.

*[Named for the Sanchugovka river.—Translator.]

However, under the conditions of the West Siberian Shelf basin, where the Sanchugovka Sea must have been, there could hardly have been a steep bottom ledge, suddenly dropping off to a depth of about 100 m. By analogy with the present Kara Sea shelf, and in agreement with the contemporary relief of the West Siberian Lowland, it should be held that the Sanchugovka Sea, being as much as 100 m deep in the region of the lower Pur, should have stretched even farther south, where it became gradually shallower and where the shallower-water sediments were deposited. Therefore, we feel that the loams in the middle reaches of the Pur, in the central part of the lowland, regarded as lacustrine by Saks, are more likely shallow-water marine sediments.

Further, boulder formations, which can be classified quite reliably as truly continental moraines, have been observed in the marginal regions of the lowland, in the Ural and Yenisey regions, where there is a clearly defined hilly relief. Actually, moraine formations are observed in the Pre-Ural* region, in the Severnaya Sosva basin, in the region of Samarov on the Irtysh, on the right bank of the Ob between Belogorye and Berezovo, and somewhat to the north of Berezovo. True moraine material is also found on the right bank of the Yenisey between Verkhne-Imbatskoye and Dudinka, at Turukhan, Mangutikha, and elsewhere.

However, in the middle section of the lowland, between the Ob and the Yenisey, where the watersheds are not hilly, there are somewhat similar boulder formations, which differ from the preceding as follows:

1. They contain coarse clastic material, but relatively little of it; they consist of loam similar in structure to true moraines, but with sparsely scattered fine fragments and pebbles of crystalline rock; large boulders are infrequent.

2. They have more clearly defined stratification than true moraines.

3. They contain a fairly large amount of fine plant remains, evenly distributed through these deposits.

4. They contain frustules, transported from the Tertiary deposits often in an undisturbed state.

These features of the boulder formations of the middle lowland justify their classification as fluvioglacial, perhaps glaciomarine accumulations.

As was mentioned earlier, the boulder deposits form two tongues on the lowland. One stretches southwest from the northwestern part of the Central Siberian Plateau; here one finds boulders exclusively from the northwestern part of the plateau. The other tongue extends northeastward from the northern Urals and contains coarse clastic material exclusively of Ural origin. There are no boulder beds at all in the very center of the territory examined. Here the Quaternary deposits consist of boulderless loams, sandy loams, and fine-grained sands.

Thus, in the regions studied, upward along the cross-section, there are clays and loams with varying amounts of boulders, which sometimes gradually and sometimes abruptly become sandy loams and fine-grained sands with scattered rubble and boulders. Our observations indicate that this lithologic change upward along the section is explained by the change in facies due to the gradual shoaling of the basin. There are no lithologic and biostratigraphic indications of interglacial deposits that are either analogous to or even slightly reminiscent of the interglacial deposits of the Russian Plain. Thus, the substitution of boulder for non-boulder deposits, and vice versa, is more important from the standpoint of facies than of stratigraphy.

The presence of but one glacial complex on the margins of the lowland, where there are true moraines, and but one fluvioglacial complex within the country,

*[I.e., along the eastern slopes of the Urals.]

where boulder pseudo-morainic facies and boulderless facies form mutual transitions both horizontally and vertically, compels us to synchronize the two complexes and to refer their formation to the same glacial period.

The morphologic youth and freshness of the glacial topography of the marginal areas of the lowland (the Pre-Ural region and the zone adjacent to the Yenisey) harmonize with the morphologic youth of the flat watershed plateaus of the internal regions of this country. Here, too, the symmetrical watersheds, the long gentle slopes, and the corresponding talus trails are witness to the effect of erosion on the relief. Thus, the geomorphologic characteristics also indicate the uniform topographic youth of the marginal and internal regions of the lowland.

Now let us examine the relationship between the strata in the regions examined and in the region farther north, the Gydan peninsula.

Within the Gydan peninsula, above the so-called Messo sands,* there are marine Sanchugovka clays, sometimes with boulders, which give it a moraine-like character. Saks attributes the boulder enrichment of these clays to the erosion of the moraines of the glacial maximum preceding the Messo period. However, it is difficult to explain the abundance of clastic materials in the Sanchugovka marine sediments by the erosion of ancient moraines which never actually existed beneath either the marine Sanchugovka clays or beneath the continental, so-called Messo sands. These moraines could hardly have been completely washed away, without a trace, especially since they are well preserved farther south.

All this indicates rather that the boulders in the Sanchugovka clays are not the product of the erosion of ancient moraines, but of the breaking off of [boulder-carrying] "icebergs" from the glaciers of the Siberian Plateau which ultimately descended to the sea.

The boulder content of the Sanchugovka clays does not increase towards the escarpment of the Central Siberian Plateau. Saks considers this proof that there is no connection between the marine deposits and the moraines of the plateau. In our opinion, it proves just the opposite. Glaciers descending into a sea basin leave little clastic material about their edges, because the glacier substance is expended chiefly on the calving of icebergs and their removal. The icebergs, moving quickly with the current, melt only slightly at first and, of course, do not drop their boulders immediately; thus the bottom deposits in their path of initial movement are only slightly enriched with coarse clastic material. This material begins to appear in the bottom sediments only at some distance from the point of initial iceberg movement. Therefore, the marine sediments near the edge of the glacier should have little, if any, coarse clastic material.

In our opinion, the arctic nature of the fauna in the Sanchugovka clays and, probably, the distribution of boulders by icebergs indicate that the Sanchugovka Sea and the glaciation of the peripheral areas of the lowland existed simultaneously, i.e., they support the earlier views of Saks (1936), from which he regrettably departed in later years.

As we know from the works of Saks, above the Sanchugovka clays are the so-called Kazantseva marine strata,† which are arenaceous and of a shallower-water type than the Sanchugovka deposits, with thermophilic, evidently coastal fauna. Higher in the sections of the watershed plateaus there is a layer of Zyryanka sands and cobbles.‡

---

*[Named for the Messo river.—Translator.]
†[Named for the Kazantseva river.—Translator.]
‡[Named for the Zyryanka river.—Translator.]

The surface of the highest portions of the Gydan peninsula (e.g., the Olkhov range, the *Olenii Rog* [Deer Antler] upland on the Tanan river) often consists of sands with boulders, whose formation Saks attributes to the glacial maximum in some places and to the Zyryanka glaciation in others. The principles upon which he makes this division are obscure. It would be more natural to refer all such formations on the uplands of the Gydan peninsula to the Zyryanka horizon.

The descriptions in the numerous works of Saks indicate that the Zyryanka sands of the Yenisey basin and the Gydan peninsula often are associated with the Kazantseva horizon by gradual transitions; sometimes, however, they are separated by a denudation line. It should be mentioned that the Kazantseva and Zyryanka strata substitute for each other in the sections in many cases and do not differ from each other substantially. Earlier, in pointing out that marine fauna were also found in the Zyryanka sands, Saks held that the Zyryanka glaciation was synchronous with the marine transgression, whose level (according to his earlier concept) was 100 m higher than the present sea level. However, now he considers these formations to be the product of fluvioglacial streams which originated in the Zyryanka glacier, and he reduces the sea level of that time by 200 m, i.e., he places it 100 m below the present sea level. One might say that this departure of Saks from his earlier views is also disappointing.

The arenaceous formations classified by Saks as fluvioglacial deposits of the Zyryanka era, are observed as a thick layer on the high watersheds. However, if the country had actually been 100 m higher during the Zyryanka era than at present, it should have experienced considerable erosional dissection, in which case the fluvioglacial sands would have been deposited primarily in the valleys, not on the watersheds. But this is not the case. The adaptation of the Zyryanka fluvioglacial sands to the valleys should have been particularly well defined in view of the remoteness of the small Zyryanka glacier from the region examined.

In all probability the Sanchugovka, Kazantseva, and Zyryanka horizons should be considered elements of a single complex of marine, coastal-marine, and deltaic-alluvial formations which developed during the glaciation of the Siberian Plateau. The succession of clays and sands upward, complicated sometimes by the appearance and sometimes by the disappearance of boulders and also by the change in composition of the marine fauna, indicates a gradual shoaling of the basin during the succession of various facies both horizontally and vertically, depending on the change in position of the ice margin, the distribution of currents, and resultant conditions. This shoaling continued to the delta and river-valley stage. According to Saks, the considerable freshening of the Sanchugovka Sea waters (fresher than the present Kara Sea waters) is further evidence of the simultaneous existence of glaciers, whose melting always freshens sea water considerably.

The appearance of Sanchugovka clays in the preglacial alluvial formations in sections of the watershed plateaus of the Gydan peninsula and the evidence of a gradual shoaling of the basin, recorded upward in the stratum, agree well with the sand-enrichment of the upper horizons, above the clayey stratum found in the sections of the watershed plateaus of the Yenisey-Taz interfluve. This leads us to regard the two series of sediments in the sections from the Gydan peninsula and the Yenisey-Taz interfluve as synchronous and as part of a single glacio-marine complex (Fig. 2).

In tracing this complex of sediments, which occupies a specific stratigraphic position between the preglacial deposits and the postglacial deposits covering it, southward and southwestward, we find a basis for correlating it with the boulder (although, probably not the morainal) formations in the basin of the Vakh and Yugan rivers and the latitudinal [east-west] section of the Ob. All investigators regard these boulder formations as belonging to the glacial maximum. However, if the so-called Messo era had followed the glacial maximum, during which these boulder formations were deposited in the Vakh, Yugan, and Ob basins, followed by the Sanchugovka and Kazantseva transgressions, followed by the Zyryanka glaciation, the moraine of the glacial maximum should have been strongly eroded in this region. What is even more significant, there are no continental facies at these southern points corresponding to the Messo, Sanchugovka, Kazantseva, and Zyryanka periods. The adherents of Saks are completely silent on this difficult question.

In an earlier work (A. I. Popov, 1949), we held that the main stratum of boulder and boulderless loams and clays of the Yenisey-Taz interfluve and the Vakh basin was synchronous with the so-called Sanchugovka horizon on the Gydan peninsula and the lower Yenisey. However, then it appeared that the deposits of the so-called glacial maximum in the central part of the lowland were actually glacial deposits and, since the Sanchugovka formations are undoubtedly marine, we concluded that the glaciation and the marine transgression were synchronous, i.e., we arrived at a conclusion formerly held by Saks, but which he later rejected.

Later, in 1950, a comparatively detailed examination of the question of the glaciation of western Siberia and the uplands on its eastern margin (the Norilsk and Putorana mountains) led us to another conclusion. We found is impossible to regard the boulder horizons of the middle portion of the lowland as glacial formations. We were inclined to regard these boulder horizons, which are synchronous with the so-called Sanchugovka horizon, as marine sediments and as a southward continuation of the Sanchugovka sediments, comprising with them a unified stratigraphic horizon. The existence of glacial deposits proper is also possible, but evidently they are limited to the periphery of the lowland and are synchronous with the marine deposits.

All that has been said pertains to the lower Ob as well. One very important point has become clear: in the sections of the ancient watershed plateau at the point where the Ob passes only 60 km from the eastern slope of the Ural range there are no deposits which can be classified as glacial moraine or which may be regarded as fluvioglacial. Further, there is no topography which can be regarded as glacial.

First there was a gradual subsidence of the land (aleurites, loams, and clays comprising the lower parts of the Quaternary layer), which may be related to the preglacial period; then the land rose somewhat (middle horizons, sands), which probably led to desiccation in places (sometimes traces of erosion at the contact with the higher loams); next there was a new subsidence (upper, dark loams), followed by a gradual uplifting (sands and sandy loams completing the watershed stratum). Consequently, marine sediments are replaced by deltaic-alluvial and alluvial sediments.

The upper, fine-grained sands and sandy loams, stratified horizontally, cannot be classified as fluvioglacial formations. The boulders scattered in them, and not always present at that, do not alter the situation essentially. The coarse clastic material in these deposits could have been transported by shore and river ice, while the accumulation of cobbles and pebble gravel at some points can be associated with the mechanical action of the ice, as can be seen at present on the banks of the Ob and the Yenisey.

Thus, the entire layer of sediments in the lower Ob basin should be regarded more or less as a single layer, reflecting two stages of relative land subsidence and two stages of relative uplifting, probably with partial, but not complete, local desiccation. The reported lithologic variations are essentially facies variations, which are definitely related through the regular alternation of facies.

This picture of the structure of the Quarternary deposits in the western part of the lowland does not differ in principle from that of the Gydan peninsula and the Yenisey-Taz interfluve. It is important to note that here, too, we are dealing with a single stage of glaciation and not with traces of repeated glaciation.

Evidently the following conclusions can be drawn from what has been said above:

1. There is evidence of only one glaciation along the periphery of the West Siberian Lowland.

2. There is evidence of one major Kara Sea transgression within the confines of the West Siberian Lowland.

3. There is a firm basis for assuming that the glaciation and the West Siberian Gulf existed simultaneously or nearly simultaneously.

South of 58° N. lat., where the entire layer of Quaternary deposits consists of alluvial sediments, part of the alluvium belongs to the glacial era. In the southern part of the lowland it is difficult to distinguish between the alluvial deposits of the preglacial and the glacial periods. Consequently, south of the region of maximum glaciation and the marine transgression an alluvial plain persisted throughout the Pleistocene, during both the preglacial and the glacial periods.

The facts indicate very thick alluvial deposits in the sections of the watershed plateaus along the Ob, above the junction with the Irtysh, and on the Irtysh itself. All the lithologic differences in these sections (Vysotskiy, 1896; Sukachev, 1936; Nikitin, 1938; Vasilyev, 1946; et al.) fully support current concepts that the facies of river alluvium, from streambed to floodplain to oxbow, combine regularly in a single complex.

The alluvium in the plateau sections is undoubtedly thicker than normal (Shantser, 1951), i.e., it is thicker than the possible thickness of all alluvium facies for a constant base level of erosion and tectonic calm. The normal alluvium thickness for the Ob river under these conditions could hardly have exceeded 15–20 m, but we find alluvium layers here 50–60 m thick and more.

The above conclusion is based not only on the thickest alluvium layers, but also on the alternation of facies, indicating a former subsidence of the region during the accumulation of alluvium: channel sands and pebble gravels often cover the floodplain sediments, which are alternating layers of loam and peat.

Ye. V. Shantser showed that it was primarily the subsidence, not other factors, that determined the great thickness of the alluvium, which exceeds the normal thickness for the valleys of lowland rivers. He asserts that even a relatively small-amplitude subsidence or uplifting of the base level of erosion would have had a conspicuous influence on the behavior of the alluvium accumulation, while "the alluvium thickness of lowland rivers is only slightly dependent on climatic fluctuations. Therefore, in most cases abnormal alluvium thickness can be interpreted reliably as evidence of tectonic subsidence or uplifting of the base level of erosion" (Shantser, 1951, p. 235).

At the village of Demyanskoye on the Irtysh, in a section of the original plateau, in a layer of so-called diagonal sands about 30 m thick, V. N. Sukachev (1910) discovered remains of arctic flora in the peaty interlayers amid the diagonal sands, representing riverbed and inshore-shoal deposits. Four species of plants (*Salix polaris, S. herbacea, Dryas octopetala, Pachypleurum alpinum*) were found there, "being typical plants of the arctic and alpine regions; they are not found outside these regions" (Sukachev, 1910*a*).

Keeping in mind that all the plants found are characteristic of both the taiga and the tundra, it would be quite proper to conclude that there was tundra, or at least forested tundra, in the Demyanskoye region or nearby during the deposition of the diagonal sands.

The good state of preservation of the plant remains would seem to rule out the possibility that they were transported there and, what is more, there is no source from which they could have been transported.

Thus, the Quaternary deposits in the afore-mentioned portions of the Ob and Irtysh basins show that the ancient alluvium in the lower horizons contains plant remains, which gives evidence of an abundant forest vegetation very nearly like the present vegetation in these latitudes; higher up the section, the alluvium horizons are impoverished of forest vegetation and indicate the existence of areas of forested tundra or tundra here, i.e., they indicate a distinct cooling of the climate or at least of the summer climate. The terrace alluvium here again indicates forestation similar to the present forestation and, consequently, a modification of the climate.

There is every reason to associate the layer of alluvium containing arctic flora at Demyanskoye on the Irtysh and at Viskov-Yar on the Ob with the epoch of maximum glaciation of the West Siberian Lowland. The deposits which underlie this layer do not correspond to the conditions of the glacial period; the composition of their plant remains is quite similar to that of the terrace deposits and recent deposits. Thus, we conclude that only alluvial and alluvial-lacustrine Quaternary deposits appear in the southern part of the West Siberian Lowland and these deposits, like those of the northern part of the lowland, may be subdivided into preglacial, glacial, and postglacial on the basis of their time of formation.

One may draw a very important conclusion from the above. If the accumulation of alluvium in the southern part of the lowland during the glacial epoch took place during the subsidence of the plain and was parallel to and closely associated with the first accumulation of boulder formations to the north, the maximum glaciation of the West Siberian Lowland must have coincided with the subsidence of the land, not with the uplift, as most investigators have asserted. This is essential for explaining the true cause of the glacial maximum of the Putorana upland and the northern Urals, which also included the West Siberian Lowland.

## Postglacial Deposits

The deposits covering the glaciomarine complex of sediments indicate a different regimen for the plain.

Upward in the stratigraphic column the glaciomarine and marine deposits give place to lacustrine clays, and these latter become peaty swamp deposits. Relict peat mounds in the taiga and forested tundra zones of the West Siberian Lowland have been treated extensively in the literature; they are of postglacial origin and some of them belong to the postglacial thermal maximum. Further, all deposits of the river terraces in the valleys can also be classified as postglacial.

(As a rule, two terraces above the floodplain are observed, the first 10–12 m high, the second 20–25 m above the present river level.)

Postglacial deposits, sometimes thick alluvial accumulations composed of sands, sandy loams, loams, and very often peaty-silt and peat formations, are observed in the extensve river valleys of the Yamal and Gydan peninsulas. These peat and peaty-silt formations are deposits of the high floodplain facies and comprise thick layers of sediments covering enormous expanses of the river valleys. The presence of fissured fossil ice, which is sometimes found in several layers separated by stratified peat-silt floodplain formations, is a characteristic feature of these deposits. This fissured ice may be of varied thickness, depending on the thickness of the accumulated alluvial layers (A. I. Popov, 1953 a, b, 1955). Thus, the main difficulty in explaining the origin of this thick and multilayered fossil ice has been removed; it had been the chief stumbling block to a solution of the problem until very recently. There is no reason to assume this fossil ice to be buried firn, as had been done previously.

It must be noted that the indicated method of accumulation of fossil ice is admissible under only one condition, viz., that the upper limit of permafrost rose as the sediments accumulated. In a severe continental climate, the accumulation of sediments is actually accompanied by the accumulation of frozen ground both on the floodplains and in the small lakes and swamps outside the floodplains. This process creates the generally observed regular distribution of thin horizontal interlayers of ice with depth in the alluvial and lacustrine-paludal deposits. This produces the characteristic tetragonal relief of such surfaces (polygonal mounds and their derivatives, the *baydzharakhi* [baidzherakhi], flat-mound peat beds, and so on).

## History of the Development of the West Siberian Lowland in the Quaternary Period

How can the events of Quaternary history be portrayed in the light of what has been discussed above?

Apparently the physical geography of western Siberia during the early Quaternary period differed little from that during the Pliocene. The characteristics of an alluvial plain were preserved here during both the Pliocene and the Pleistocene. During this long time-period, different sections of the lowland experienced epeirogenetic movements of various kinds, dominated by a gradual, very slow land subsidence. It should be noted that the amplitudes of the vertical movements in the southern part of the plain were smaller than in the north, where they reached several tens of meters. Since the strata of preglacial alluvium lie 60 m or more below the present river level in the northern part of the lowland, it may be assumed that the shoreline was 60–100 m lower than at present at some period during this long time-interval.

The West Siberian land mass extended far to the north at that time, including, apparently, the entire present-day shelf zone of the Kara Sea.

The Pliocene-Pleistocene land area of western Siberia was characterized by the broad development of an erosional-accumulation activity of rivers which flowed northward, as at present. There are no indications of a glacial advance in western Siberia during the entire time-interval. The climate of the plain was somewhat more continental than at present.

Later in the Pleistocene, however, because of tectonic activity, the northern part (almost half) of the lowland became inundated with seawater and formed a large gulf of the Kara Sea. The waters of this gulf reached the Urals in the west and the scarp of the Siberian Plateau in the east. The southern shore of the gulf was low-lying and followed, approximately, the latitude of the Vakh-Yugan-Berezovo [rivers].

The West Siberian basin was relatively shallow, but in the north (approximately north of 63°–64° N. lat.) it was several tens of meters deep and perhaps even more than 100 m deep, as indicated by the sediments which formed there.

The southward encroachment of the Arctic basin modified the climate and caused a relative descent of the snow line along the western edge of the Siberian Plateau (chiefly in the Putorana mountains), owing to the decrease of ablation in the transition from land to sea; more solid precipitation appeared in winter and the summer was colder.

In our opinion, the Arctic Sea transgression in the West Siberian Lowland, resulting from the subsidence of the land, is the most probable and real physico-geographic cause of the glaciation. This is in direct contradiction to the prevailing point of view that the glaciation occurred because of the uplifting of the land.

Actually, a 100–200 m upheaval of the Norilsk and Putorana mountains and of the polar Urals, with simultaneous uplifting of western Siberia, could hardly have induced glaciation. Vertical zonality would have completely cancelled the effect of some cooling of the climate in this case through an attendant increase of continentality: the colder winter would have been less snowy and the summer undoubtedly would have been hotter than at present. There is no evidence of a considerable uplifting (several hundred meters) of the mountains surrounding the West Siberian Lowland which could possibly have caused glaciation; indeed all the facts indicate just the opposite. Thus, it seems less and less probable that a 100–200 m uplifting of the mountains could have caused glaciation.

Subsidence of the land, however, even a subsidence of 20–50 m, would have had a collosal modifying influence on the climate. More of solid precipitation would have fallen than at present (primarily in the mountains) and the snow-melt would have been less intense than at present, because the summer would have been colder. One need but look at the chart of July isotherms in western Siberia to see that even such a comparatively small body of water as the Gulf of Ob exerts a great influence on the summer air temperatures, reducing them. The July isotherms, which converge along the southern edge of the gulf, diverge east-ward and northeastward, owing to the increasing continentality of the climate. What must the cooling effect have been if the cold Kara waters had inundated the broad northern part of the lowland! However, a 20–50 m subsidence of the uplands adjacent to the Central Siberian Plateau and the polar Urals would not, in itself, have had any substantial influence on the climate of these mountains.

Thus, the neighboring cold sea, as an additional source of moisture, but chiefly as an impetus for lowering the snow line, induced a more intensive accumulation of ice on the uplands of the western margin of the Siberian Plateau (Putorana). The literature indicates that there are no signs of ancient glaciation farther to the east and southeast of the Kotuy valley on the Siberian Plateau. The same applies to the eastern slope of the Urals and, apparently, to a greater extent because of the greater moistening and the greater accumulation of ice in the Urals.

The Urals bear definite traces of a quite considerable sheet glaciation. The Central Siberian Plateau, however, has no clear traces of sheet glaciation, as has been noted by a number of authors (Mikhaylov, 1947; Sheynman, 1948).

Having increased in size, the glaciers themselves began to exert an influence on the climate and to cause a further drop of the snow line. The glacier tongues descended onto the plain, appeared on the shores of the West Siberian Gulf, formed a quite considerable piedmont glaciation, and occupied the shallowest part of the basin (Fig. 4).

The Ural ice sheet occupied a larger area and the Siberian ice sheet a smaller area, as indicated by the paleogeographic map (Fig. 4). The relationship between the western Siberian ice sheets answers more to the laws of decreasing glaciation from west to east, with decreasing Atlantic influence in that direction, than to the concept of continuous sheet glaciation.

Obviously, the Ural ice sheet and particularly the Siberian ice sheet were thin. Probably the Ural ice sheet did not reach a maximum thickness of much more than 300 m, while the Siberian ice sheet did not become this thick. The edge of the Siberian sheet was scarcely more than 50 m thick; probably the ice wall at the edge of this sheet was less than 50 m high, most likely of the order of 20–30 m. Consequently, the icebergs produced at the edge of this ice sheet must have been small.

In the present arctic seas, icebergs rarely reach a height of 40–60 m and usually they are much smaller. Since the broader and denser part of an iceberg rests in the water, usually one-fifth or even one-third of it appears above water, not one-tenth, as commonly assumed. Thus, the Siberian icebergs in the West Siberian Gulf, being 20–50 m thick, on breaking off from the ice sheet, should not have sunk more than 18–45 m into the sea, and most often only 10–20 m. Thus, the small Siberian icebergs would not have required great depths to move in the sea water. Probably the icebergs of the Ural sheet were somewhat larger, but they were carried off northward into increasingly deeper water.

Icebergs which broke off the edges of the ice sheets were carried clockwise by currents which evidently existed in the gulf at that time as at present in the Kara Sea. In the West Siberian Gulf this clockwise direction of the currents (Fig. 4) should have been even more pronounced, because of the pear shape of the gulf.

The icebergs, coming from the Siberian ice sheet, were caught up in the influx of Kara waters and gradually melted while moving southward and southwestward, i.e., in a direction where they encountered the relatively warm waters entering the gulf from the large rivers of the ancient Yenisey and Ob systems (Fig. 4).

The warm water of the ancient Ob, which entered the gulf approximately in the middle of the lowland, defined the limit of westward penetration of the Siberian icebergs. This is why Siberian boulders are never, or practically never, observed west of the meridian of Surgut on the Ob. The icebergs calving from the Ural ice sheet immediately entered a region with a relatively warm current (the waters of the great rivers) and thus melted quickly. Hence, two tongues of boulder formations appeared: a longer, southwestward tongue from the Siberian plateau and a shorter, northeastward tongue from the Urals.

As already stated, the shallowness of the southern part of the gulf, probably 10–25 m, did not prevent the movement of the small Siberian icebergs. Where the depths were inadequate, the icebergs became grounded and melted, unloading their boulder material. This applies especially to the largest icebergs. We must also keep in mind that the Siberian icebergs gradually melted as they moved

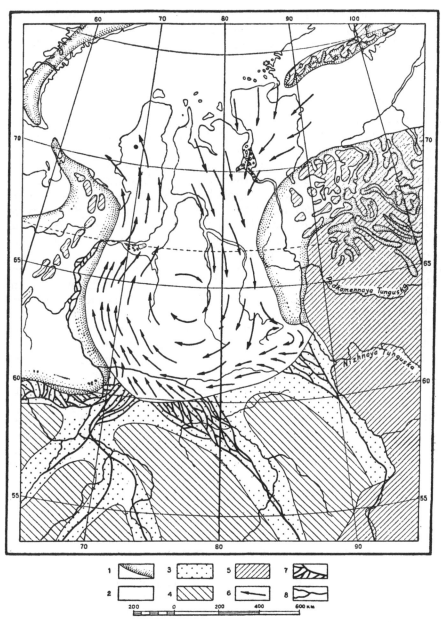

FIGURE 4. Paleogeographic map of western Siberia during the epoch of the marine transgression and the continental glaciation: 1, glaciation; 2, sea; 3, large river valleys; 4, watershed plateaus; 5, portions of the Central Siberian Plateau, the Byrranga and Ural mountains not covered with ice; 6, presumed direction of currents in the gulf; 7, rivers and their deltas; 8, present hydrographic network.

southward and southwestward, becoming smaller and therefore requiring ever shallower ocean depths for movement. Thus, as the depths of the gulf decreased southward, so did the size of the icebergs and the depths required for their free movement. This effect should have been particularly evident at points where the warming effect of the large streams of the ancient Yenisey and especially the Ob systems began to be felt.

The problem of iceberg distribution is by no means as important for the icebergs of the Ural ice sheet, which moved northward into ever deeper waters, as it is for the Siberian icebergs.

The melting of the ice in the West Siberian Gulf in spring and, probably in part, in summer and also the melting of icebergs in summer cooled the air considerably during the summer.

Judging by the finds of arctic flora at Demyanskoye on the Irtysh and at Viskov-Yar on the Ob, the forested tundra, and in places perhaps tundra as well, on the southern shore of the gulf probably resulted from the incursion of the cold sea. Thus, the appearance of this basin entailed an abrupt southward displacement of the zones, compared with the preceding stage. V. N. Sukachev's (1933) discovery of traces of frozen ground processes in corresponding deposits on the Ob merely strengthen this proposal.

The considerable cooling and, at the same time, moistening of the climate at a certain stage of the Quaternary era in Kazakhstan, the Urals, and the southern part of western Siberia (Gerasimov and Markov, 1939) are also probably associated with the indicated transgression and glaciation.

The movement of water from north to south in the eastern part of the gulf, in definite analogy with the present Kara Sea, explains the relative scarcity of organic remains in the sediments of the eastern part of western Siberia adjacent to the Yenisey river and, on the other hand, the relatively abundant organic material in these sediments in the western region adjacent to the Urals. The cold water penetrating from the north did not provide the necessary conditions for rich development of life on the edges of the Central Siberian Plateau, especially in the presence of the glaciers which descended from there. In the western part of the basin, where the ancient Ob and Yenisey rivers entered, the waters were warmer, but what is more important, the organic remains were brought in by the fresh water of the large rivers along the western margin of the basin (Fig. 5).

Examination of a pollen diagram of the Ob-Poluy watershed plateau (Fig. 5) indicates, first of all, a relatively uniform vertical distribution of tree pollen, non-tree pollen, and spores, and also a clear predominance of Siberian pine pollen over other forest species, and this prevalence is maintained with depth. Fir, birch, and pine are found throughout the section in more or less equal amounts.

The quantity of non-tree pollen is not great, but the composition of these species is so diverse that it is difficult to draw any conclusions from them. However, the absence of pollen of true arctic plants is striking, as is the considerable role played by sphagnum moss and fern pollens, which evidently indicates the arrival of pollen chiefly from the taiga. However, the pollen composition of the non-tree vegetation does not exclude the possibility that forested tundra or even tundra conditions existed here.

The most important conclusion which may be drawn from an analysis of the pollen diagram is that the thick clayey and sandy layer contains a nearly uniform pollen distribution throughout the section, which indicates that northern taiga or forested tundra conditions existed along the shores of the basin, bu does not fully reflect the character of the plant associations comprising it. The comparatively

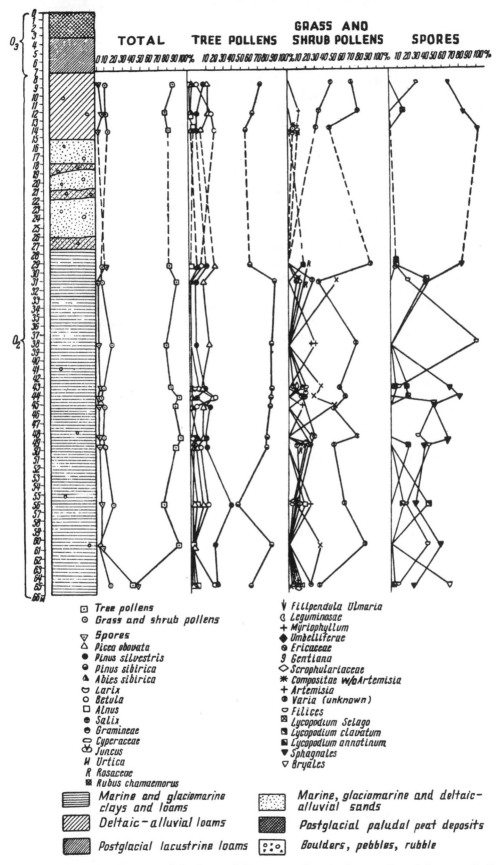

FIGURE 5. Pollen diagram of the layer of Quaternary deposits in a section of the watershed plateau (lower Ob basin). (After S. V. Kats.)

benthic nature of the sediments and also the mixed, although vertically uniform composition of the pollen evidently indicates that this pollen was transported into the basin by the wind and chiefly by the rivers and then was transported in a suspended state by the waters of the basin, gradually settling to the bottom. This is also indicated by the constant presence of very fine plant remains of the "chaff" type, which could be borne by even weak currents in a comparatively deep and broad basin. The composition of the plant remains (green and sphagnum mosses, bushes, etc.) adds nothing to a reconstruction of the topography of the shores of the basin and does not exclude the possibility of northern taiga, forested tundra, or even tundra conditions.

The absence of marine mollusks in the southern part of the basin (they are found in the northern part) evidently can be explained by the considerable freshening of this region by the large masses of fresh river waters from the ancient Yenisey and Ob rivers, and also by the melting of glacier ice that descended into the gulf.

Evidently the maximum glaciation of the West Siberian Lowland was due not so much to the general cooling of the climate as to the establishment of an optimum ratio of solid precipitation to ablation during the marine transgression, which reduced the continentality of the climate somewhat.

Although the general cooling of the climate under optimum conditions of the North Atlantic caused intense (maximum) glaciation of Europe, it could not have caused glaciation in Siberia, because of the strong climatic effect exerted by the continent, which extended far to the north. Only the transgression of the Arctic Sea caused the maximum glaciation of central and western Siberia. Glaciation became possible during the general climatic cooling, but the glaciation did not reach great proportions and did not extend to western Siberia until the influence of the Asiatic pressure high had been suppressed by the boreal transgression.

A cold marine transgression had to occur before any considerable glaciation could develop in central and western Siberia and the retreat of the glaciation almost certainly was due to the disappearance of this factor, i.e., to the regression of the sea and the increased continentality.

In connection with the southward displacement of geographic zones during the transgression and glaciation, permafrost probably developed along the southern edge of the gulf under forested tundra or tundra conditions. The manifestations of frozen ground processes in the very region of the ancient Ob delta, noted by V. N. Sukachev, are proof of the activity of permafrost-forming factors at that time on the southern margin of the West Siberian Gulf. Permafrost surely did not exist beneath the broad water expanses of this greatly freshened gulf, which was more than 20 m deep everywhere and was 50–100 m deep and deeper to the north.

After the marine regression, the West Siberian Lowland became a flat lacustrine plain, more or less hilly in the east and west. As the ice melted and the sea retreated, the rivers flowed more freely northward and gradually formed independent valleys. The intensive persistent marine regression and the rapid drop in the base level of erosion led to rapid downcutting by the rivers; terraces formed, and the valleys widened considerably. Thenceforward the watershed plateaus began to exist as an independent relief form. The large late-glacial bodies of water shoaled and became shallow lakes and later became swamps. Only the deepest of these water bodies have been preserved to date as relics, containing stunted marine forms (e.g., *Mysidae* and others in lakes on the watershed plateau of the Yenisey-Taz interfluve, according to P. L. Pirozhnikov, 1931). This indicates that after the sea had shoaled, the relictal local bodies of water were not covered by the ice sheet; gradually degrading, these have been preserved to our time.

As the region became drier and the continentality increased, the desiccated sectors became covered with vegetation. The former tundra expanses along the southern periphery of the gulf were quickly crowded out by taiga. Because of the rapidly increasing continentality, the tundra plant associations could not establish themselves on the newly dried sectors to the north and were rapidly crowded out by taiga. Only the stabilization of the base level of erosion, which approached the present level, permitted a sufficiently stable zonal distribution of vegetation to become established. A tundra zone began to form and the northern limit of the taiga became more or less stabilized.

The present topography of the northern part of the West Siberian Lowland derives from those primary level expanses into which the embryonic rivers cut as the surface became dry and ice-free. At first flatland plant complexes, which we find comparatively unchanged today, developed on the flat and weakly dissected watershed plateaus. Then, as the fluvial downcutting progressed, the valleys became broader and terraces formed, creating valley topography, natural second-order complexes.

The youth of the topography of western Siberia and the mixed, unstable nature of the flora and fauna of this region have been stressed by many investigators and are explained by the relatively late desiccation and deglaciation of the region.

Let us consider the nature of the so-called thermal maximum which occurred during the postglacial period.

According to the data assembled thus far, we know that the thermal maximum, i.e., the time of greatest northward advance of the forest, coincided with the marine regression and the deposition of the lower layers of sediments in the cross-section of the first terrace above the floodplain in the river valleys of the lowland. The data indicate that the peat deposits of the first terrace above the floodplain, both autochthonous and allochthonous, correspond to this stage of the postglacial era and in some places are now found at sea level (e.g., in northern Yamal, on Belyy Island). Undoubtedly they formed at approximately 15–20 m above the present sea level, judging by the similarity of facies here and in analogous formations that develop now. The data for Belyy Island, now separate from the Yamal peninsula, will serve as an example.

Thus, we have reason to assume that the greatest postglacial northward displacement of the geographic zones took place at the same time as and in connection with the regression of a sea whose level was 15–20 m (and perhaps more) lower than the present sea level. A considerable portion of the relief became desiccated and the climate became more continental; the winters became colder and less snowy and the summers became warmer and drier, which probably resulted in a northward advance of the forests.

Our paleogeographic map of the northern part of the West Siberian Lowland during the thermal maximum (Fig. 6) reflects the geographic conditions of that time and allows a comparison with present conditions. The sea level was 20 m lower than it is now. A large portion of the shelf became land and the natural zones were displaced northward. The broad tundra areas of the thermal maximum are now covered by the sea.

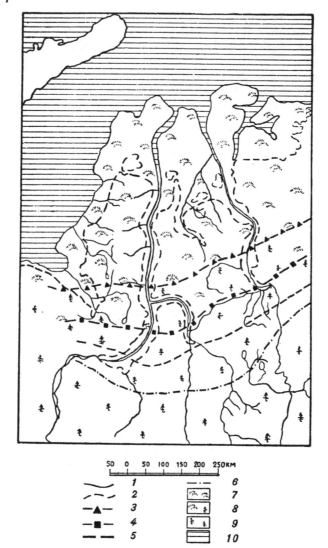

FIGURE 6. Paleogeographic map of the northern part of western Siberia during the epoch of the postglacial thermal maximum: 1, coastline during the thermal maximum; 2, modern coastline; 3, northern limit of the forested tundra during the thermal maximum; 4, northern limit of the taiga during the thermal maximum; 5, modern northern limit of the forested tundra; 6, modern northern limit of the taiga; 7, tundra zone; 8, forested tundra zone; 9, taiga zone; 10, sea.

In our opinion, the so-called thermal maximum does not represent a warming of the climate as a whole, but an increase of continentality due to the marine regression with warmer summers—a warmer growing period.

The extensive development of the afore-mentioned frozen ground processes in connection with the accumulation of sediments is not at variance with the concept of a northward displacement of the geographic zones: both were produced by the same factor, viz. increased continentality due to the marine regression.

In the period following the thermal maximum, as a number of authors have stated, there was renewed cooling of the climate, the forests again retreated southward, and the tundra encroached from the north.

There are many reasons for believing that this phenomenon also coincided with a new ingression of the sea, resulting in the inundation of part of the shelf and the formation of the Ob, Taz, and Yenisey gulfs, which, however, occupied an area larger than at present. This is indicated by frequent finds of whale remains in the frozen sand and of shells of marine mollusks at some distance from the shores of the present Gulf of Ob, and eliminates the possibility that they were transported here recently by onshore winds. Such whale and mollusk finds on the Yamal and Gydan peninsulas may be explained only by a higher sea level in the not-so-distant past.

The subsequent lowering of the sea level (the uplifting of the northern edge of the lowland) reduced the area of the Ob, Taz, and Yenisey gulfs to their present size. Hence, in this case the displacement of the natural zones evidently was connected with a change in the position of the coastline and the degree of continentality of the climate.

The transgressions and regressions of the sea in western Siberia produced opposite effects. The glacial, glaciomarine, and marine deposits in the northern half of the lowland were left by a major marine transgression, while the lacustrine, paludal, and alluvial deposits, which contain fossil ice in the north, were left by the marine regression.

One may speak with confidence of two stages of development of the northern part of the lowland as an alluvial plain, the preglacial and the postglacial. The preglacial stage probably had much in common with the postglacial, but remains largely unstudied. The postglacial stage, the time of broad development of alluvial processes on the desiccated plain, which had experienced oscillating epeirogenetic movements of comparatively small amplitude, was the time when permafrost and fossil ice accumulated under conditions of increased continentality of the climate.

We may also speak of one stage of a major transgression of the Kara Sea into the West Siberian Lowland, which caused glaciation of the neighboring northern uplands and the margins of the plain itself. The transgression suppressed the climatic influence of the continent and promoted an eastward advance of the western maritime influence.

It is difficult to say whether the physicogeographic changes which occurred during the Quaternary period in western Siberia required considerable general climatic displacements. Most likely the changes in the sea and land areas produced the afore-mentioned effect only on a background of the general cooling of the climate during the glacial epoch, although possibly this effect could have been produced by relatively slight fluctuations of the general climatic situation, similar to those which occur today.

S. S. VOSKRESENSKIY

# THE MAIN FEATURES OF THE QUATERNARY HISTORY OF THE SOUTHWESTERN BAYKAL REGION*

SUFFICIENT DATA have been assembled to date to outline the Quaternary history of the southwestern Baykal region, be it only in the form of a working hypothesis. Data are available on the area and the limit of glaciation in the mountains of this region and on their relationship to the terraces of the southeastern shore of Lake Baykal. Investigations in the upper Angara basin have given an idea of the order of stratification of the Quaternary deposits in this region and their relative age, according to spore-pollen analyses and finds of fauna and prehistoric human settlements.

The southwestern Baykal region consists of the mountain ranges that frame the southern basin of the lake, the Irkutsk amphitheater, the southeastern spur of the Central Siberian Plateau, and the adjacent portion of the Eastern Sayan range. Inasmuch as changes in the natural medium of contiguous regions are closely related, study of a territory with a highly diverse natural history makes for a better understanding of the entire pattern of changes in natural landscapes.

Numerous investigators have worked in the southwestern Baykal region since the early investigations of I. D. Cherskiy (1877–1889). Many summaries are available which reflect specific views of the Quaternary history of this part of eastern Siberia. The major works in this field are those of I. V. Arembovskiy (1951), N. V. Dumitrashko (1948), V. V. Lamakin (1952), V. P. Maslov (1939), Ye. V. Pavlovskiy (1948a), and N. I. Sokolov (1938).

Recently a interdisciplinary expedition of the Department of Geography of Moscow State University, led by Professor N. N. Kolosovskiy, worked in the southwestern Baykal region. The data upon which this article is based were obtained by the author during his activities with the geomorphology teams of the expedition. Observations of other members of the expedition have also been employed in writing this article.

Let us examine the geomorphology of the southwestern Baykal and the data on the friable deposits, and on this basis outline the main stages of the Quaternary history of the region.

## Geomorphology of the Khamar-Daban and Primorskiy Ranges and of the South Baykal Basin

The basin at the southern end of Lake Baykal is framed by the Khamar-Daban and the Primorskiy ranges. These morphostructural units represent a genetic

*Translated from K. K. Markov and A. I. Popov (Editors), *Lednikovyy period na territorii yevropeyskoy chasti SSSR i Sibiri*, Moskovskiy gosudarstvennyy Universitet, Geograficheskiy Fakultet i Muzey Zemlevedeniya, 1959, pp. 422–441.

entity, a link in the chain of high ridges and deep intermontaine depressions that stretches from the Eastern Sayan range to the Sea of Okhotsk. The highest peaks of the Khamar-Daban lie 20–30 km from Lake Baykal. West of the Snezhnaya river they reach heights of 2000–3100 m, they rise to 2200–2300 m between the Snezhnaya and Mishikha rivers, and then drop abruptly to 1750–1500 m east of the Mishikha. The morphology of the mountain range varies with the elevation. Dome-shaped peaks with gentle slopes prevail to the east of the Mishikha, while sharp peaks, jagged ridges, and steep-sloped valleys, i.e., troughs whose form is associated with ancient glaciation, prevail to the west of the Mishikha. Between the Slyudyanka and Snezhnaya rivers, the glacial forms are confined exclusively to the highest parts of the range. Well-preserved cirques (with lakes) are found here at elevations of 1650–1800 m.

The most recent glacial forms are found in the middle section of the Khamar-Daban, where almost any randomly chosen point is on the bottom or the side of a trough, and the wall or bottom of a cirque, or on a ridge separating the two. The floors of the lowest cirques are at an elevation of 1400 m. Glacial forms disappear very rapidly beyond the Levaya Mishikha river. There are no glacial forms whatsoever in the eastern Khamar-Daban.

Sharp differences between the eastern and western Khamar-Daban are evident in every detail of the relief. For example, the cirques of the west are replaced by sinkholes typical of eroded mountains in the east; the sharp, jagged ridges of the west are replaced in the east by slightly convex water divides with a wavy line profile in elevation. The mouth bars of the tributaries disappear in the east. The change in the morphology of the valleys is particularly abrupt. The transverse section of the Pereyemnaya, Vydrinaya, Anosovka, and other river valleys has a typical trough shape, while large, broad-bottomed valleys are typical of the lower reaches of these rivers, where later the bed of the ancient trough was overlain with alluvium. The bends of the valleys are smooth. The slight general concavity of the river valleys, which are steepest in the middle course of the river, is readily apparent in the longitudinal profile (Fig. 1). Breaks in the profile, characteristic of glacial valleys, are also apparent.

The river valleys of the eastern Khamar-Daban, e.g., the valleys of the Mishikha and the Ivanovka rivers, have different characteristics. Their transverse section forms a trapezoid with a narrow bottom, even in the lower course of the rivers right up to the point where they emerge from the mountains. The valley consists of a close succession of sharp bends. In the upper reaches of the rivers, in the ridge portion of the range, the slopes are gentle and not in the least reminiscent of the slopes of trough valleys. The longitudinal section (Fig. 1) is smooth, convex, and least steep in the middle course. The convexity of the longitudinal section is due to continuing upwarping of the range, similar to a fold of large radius.

The ancient surface of planation can be detected both in the east and in the west. In the west, its remaining elements arc bounded by steep cirque walls, while in the east, at some distance from the main erosional arteries and valley floors, the interfluves have not yet experienced a new erosional downcutting and have smooth, undisturbed contours.

There are definite reasons for these very sharp differences between the relief features of the eastern and western parts of the Khamar-Daban. The western Khamar-Daban* receives two to three times more precipitation than the eastern part. This is indicated by the data of meteorological stations and particularly by

*We refer to the portion of the Khamar-Daban range adjacent to Lake Baykal.

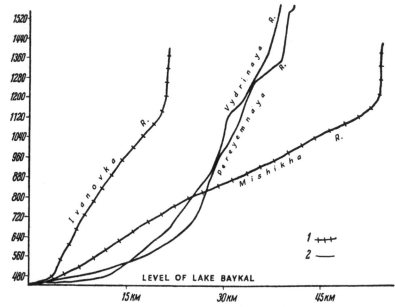

FIGURE 1. Longitudinal profiles of the rivers of the Khamar-Daban: 1, profiles of the water-eroded valleys (convex in the lower and middle part); 2, profiles of the glacial valleys (concave in the lower and middle part).

the flow modulus of the rivers. The precipitation pattern is associated with the great height of the range and, further, the part of the range between the Snezhnaya and the Mishikha rivers lies across the path of air masses which move eastward along the northeastern slope of the Eastern Sayan range. Undoubtedly, both these factors played a major role in the intensive glaciation of the western Khamar-Daban in the past, during the glacial epoch. Ancient moraines have been preserved in the river valleys of the western Khamar-Daban, e.g., the valleys of the Pereyemnaya, Vydrinaya, Anosovka [rivers], and many others. They denote specific stages of retardation or readvance of the glaciers. Terminal moraines were noted near the mouths of the Pereyemnaya and Vydrinaya rivers, at the point of emergence of the Anosovka and Osinovka rivers from the mountains (between the Vydrinaya and the Snezhnaya), and also of the Snezhnaya. In addition to the terminal moraine formations on the Pereyemnaya river, a second moraine line has been observed at its mouth, blocking the valleys at the junction with its tributary the Isakovka (12 km from the mouth of the Pereyemnaya). A third moraine line is clearly evident on the upper Pereyemnaya and its large tributaries. Lakes lie beyond some of these moraines (e.g., at Nemskiy Klyuch). A similar picture is observed on the Anosovka and Vydrinaya rivers, where the moraine lines also represent phases of the gradual retreat of the glacier.

The level of the floors of the lowest cirques is the lower limit of glacier formation. Calculation of the area lying above the level of these floors gives the following distribution: Pereyemnaya river basin, 270 km²; Vydrinaya, 175 km²; Anosovka, 45 km²; Osinovka, 25 km².

Thus, about 20–25 per cent of the entire Khamar-Daban territory adjacent to Lake Baykal which lies above the level of the cirque floors is in the basin of these four rivers. During the glacial epoch, 80–85 per cent of the territory in this basin lay above the cirque floors and served as an alimentation area, an ice reservoir. This percentage is much smaller for other rivers of the Kamar-Daban. The glaciers reached Lake Baykal only along the Pereyemnaya and Vydrinaya rivers, and emerged onto the piedmont plain along the Anosovka and the Osinovka. However, the glaciers ended deep in the mountain valleys of the other large streams, the Khara-Murin, Utulik, Solzan, and Slyudyanka.*

Summarizing our examination of the data on the Khamar-Daban region adjacent to Lake Baykal, we may say:

( *a* ) The glaciation embraced the highest portion of the mountain range. There are no traces of glaciation east of the Mishikha river; consequently, the moraines observed at the foot of the Khamar-Daban correspond to the maximum advance of the ice.

( *b* ) The valley moraines represent three stages of retardation or readvance of the ice.

( *c* ) The longitudinal sections of the river valleys and the entire morphology of the mountain range not included in the glaciation indicate that it was uplifted during the Quaternary period.

The Primorskiy range, which stretches along the northern shore of the South Baykal basin, is much lower than the Khamar-Daban. Its highest peaks reach 900–1200 m and only to the north, towards the headwaters of the Buguldeyka river, do they begin to become higher. Glacial forms of relief appear here. V. P. Maslov ( 1939 ) established the existence of several limits of ice distribution farther north, in the Baykal range. Except for its northern extremity, the Primorskiy range shows no traces of glaciation and its relief is similar to that of the eastern Khamar-Daban. The ancient surface of planation, lifted unevenly by subsequent movements, is well preserved in the Primorskiy range, as it is in the Khamar-Daban. It is uplifted least near the source of the Angara river, i.e., on the extension of the Sayan marginal faulting of the Siberian platform. This is very important for understanding why the discharge from Lake Baykal, which was much farther north in the past, became displaced to the region of least uplifting of the range in later times (the source region of the Angara).

As a rule nearly all the large rivers, such as the Goloustnaya, the Buguldeyka, and the Sarma, intersect the Primorskiy range, beginning at the Onot upland. The passes from the Baykal basin to the Angara and Lena basins lie in saddles, slightly depressed into the ridge of the Onot upland. The saddles of the passes become deeper only in the northward extension of the ridge. The following figures indicate the height of the passes (north to south, above the lake): Chanur-Onguren, 530 m; Strednyaya-Sarma, 660 m; Levaya Ilikta-Sarma, 540 m; Buguldeyka-Ungura, 261 m; passes south of the upper Buguldeyka, 350–450 m.

The pass from the Ungura river to the Buguldeyka stands out sharply because of its low hypsometric position. The ancient valley noted earlier by Ye. V. Pavlovskiy passed through here. Ancient alluvium, composing the flat floor of a valley approximately 1 km wide, is well preserved on the pass itself, and also along the Ungura and the upper Buguldeyka. A longitudinal section along the floor of this valley indicates that the floor is curved and becomes highest at the pass itself. The present valleys of the Ungura and Buguldeyka rivers are incised in ancient alluvial deposits, and the depth of incision increases with distance from the pass. Evidently, the waters of Lake Baykal once drained along this valley into the Lena valley.

*The glacial forms and deposits noted by P. P. Pilipenko ( 1930 ) at the water line of Lake Baykal in the Slyudyanka region are formations of different origin. Pertinent information can be found in the works of V. V. Lamakin ( 1935 ) and Ye. V. Pavlovskiy ( 1948*b* ).

The South Baykal basin, between the Khamar-Daban and Primorskiy ranges, reaches depths of nearly 2000 m and has a distinctly asymmetric structure. The steep scarps of the Primorskiy range, as much as 300–400 m high, continue below the lake level to depths of 1200–1300 m. The underwater slope of the scarp at Cape Tolstoy and Marituy is particularly steep, with the slope angle reaching 45°. For the most part the slope of the basin, although steep, has an average angle of 30°–20°. The steep side of the basin is articulated with the comparatively level floor by means of a quite sharp discontinuity. The lake depth are fairly uniform for a distance of 20–35 km from the northern shore. The southern slope of the depression is relatively gentle. Part of the floor of the basin now lies above the water line of Lake Baykal; these are its ancient terraces, which flank the foot of the Khamar-Daban. Tertiary deposits, consisting of friable sandstones, siltstones, clays with interbedded coal, and in places conglomerates crop out in the river valleys from beneath ancient Baykal pebble gravels; the Tertiary deposits are classified as Oligocence-Pliocene. Tertiary deposits at least 600 m thick rest on the uneven Precambrian surface. The gradual subsidence of the crystalline base from the foot of the Khamar-Daban range to the Baykal deeps is clearly evident. The same general direction and approximately the same angle are followed by the Tertiary strata, which form fairly steep folds, sometimes complicated by faults with an amplitude of several tens of meters, on the general background of the monoclinal plunge.

It is important to note that the broad ancient terraces of the southern depression of Lake Baykal appear in the main where they have a base of friable Tertiary rocks; elsewhere, however, they are merely very narrow bands. Opinions differ as to the number of terraces that formed on the southern shore of Lake Baykal. Some investigators have distinguished a large number of terraces, apparently on the basis of measurements of the height of the terrace brows, without considering that the height of the brow of a lake terrace is not uniform even at the time of its formation. The suture lines, i.e., the ancient wave-cut lines, and not the brow lines of the terrace maintained a uniform height during formation. The terrace surface always slopes. The slope and width of the terrace are irregular even during terrace formation; therefore, measurement of the heights of terrace brows cannot present any sort of reliable picture. Further, the terrace heights are modified irregularly by subsequent crustal movements and agencies of denudation.

By analysis of aerial photographs and maps and by comparing field data, we have arrived at the small number of terrace levels given below.

At the foot of the Khamar-Daban, there are four terraces:

1. The present-day littoral terrace, which lies below lake level, except for the beach.

2. The first ancient terrace (I). Height of the suture line, the ancient beach, 35–40 m (terrace brow at heights of 7.5–25 m).

3. The second ancient terrace (II). Height of the suture line, 120–150 m (brow at 40–70 m).

4. The third ancient terrace (III). Height of the suture line, 180–250 m (brow at 170–230 m).

The terrace widths vary. The strip of ancient terraces at the mouth of the Snezhnaya river reaches a width of 15 m. The ancient terraces I and II are widest, up to 5–8 km each. The highest terrace is preserved in only a few places as a narrow (0.5–0.7 km) band.

As a rule, the terrace surfaces are level, sloping slightly towards the lake.

Generally, the various terraces were built in much the same manner. A socle of Tertiary rocks forms the base of the terrace deposits; on these Tertiary rocks is a 5–10–20 m layer of coarse, poorly rounded pebble gravel, material carried by rivers from the mountains and redeposited by the lake waters. The stratification is poorly defined but is visible all the same because of the orientation of the pebbles. A thin layer of loam lies on the pebbles.

The lake terraces are dissected by broad, though shallow river valleys, whose floodplain terraces merge with the flat alluvial fans of the deltas and, through these, with the present lake terrace which lies below lake level.

The surface of ancient terrace I has hilly or hill and ridge relief in places. Small lakes are found here and there in the depressions between the hills. There are boulders on the surface of the hills. There is an acuate embankment with its convex side towards Lake Baykal on the Anosovka river. The sectors with hilly relief are arranged strictly uniformly, opposite river valley outlets, in whose basins one may see clear traces of ancient glaciation. Such sectors appear at the points of emergence of the Pereyemnaya, Anosovka, Vydrinaya, Osinovka, and Snezhnaya rivers from the mountains. The structure of these sections of Baykal terrace I can be deduced from sections near the mouth of the Pereyemnaya, which have been described by many investigators, beginning with I. D. Cherskiy.

An unstratified layer of sandy loam with sand, pebbles, rubble, and boulders appears in these sections. The largest boulders are 3 m in diameter and poorly rolled (only slightly smoothed, to be more exact). As already indicated by Cherskiy (1878), this layer is wedged out by a lacustrine pebble horizon on the Pereyemnaya river. At the base there is also stratified pebble gravel which contains typical taiga pollen, but with a small admixture of broadleaf species.

Thus, we may state:

1. There is a sharp difference in the granulometric composition of the Tertiary deposits and the terrace deposits, which indicates a period of sharp change in the geomorphologic situation, i.e., intensive uplifting of the Khamar-Daban began.

2. The dislocational character of the Tertiary deposits, including the Pliocene, indicates that the upliftings of the Khamar-Daban and the intensive downwarping of the Baykal basin occurred in the Late-Pliocene and the Post-Pliocene period.

3. Since the terraces at the foot of the Khamar-Daban emerged from below the lake level as a result of the uplifting of the peripheral part of the ridge and not as a result of a drop of the lake level, the terrace heights give an idea of the magnitude of the uplifting of the range (the height of the Baykal terraces decreases abruptly in the Selenga delta region, and Kultuk and other places where the uplifting was retarded or did not occur).

4. The time of maximum glaciation coincides with the time of formation of a large part of the layer of ancient terrace I deposits, when the glaciers moved towards Lake Baykal.

## *Geomorphology of the Irkutsk Amphitheater*

In the southeast the Central Siberian Plateau juts out as a deep "bay" between the mountain systems of the region adjacent to Lake Baykal and the Eastern Sayan. South of Balagansk, the Central Siberian Plateau is an undulating plain, dissected by deep river valleys. Except for the Kuda basin, which belongs to the

Baykal mountain district, the main features of the upland topography are quite uniform.

The following levels may be differentiated on the basis of orography: (1) the summit plains of the interfluvial plateaus with heights of 550–900 m, (2) the valley level, with heights of as much as 480–520 m, and finally (3) river valley floors. These levels are vastly different geomorphologically, i.e., they differ profoundly in their morphology and morphometry, in their material composition, origin, and age.

The interfluvial plateaus, more than 550 m high, are covered with bedrock eluvium 1–3 m thick. Diluvial (talus) deposits are found only on the slopes of the troughs, becoming ravine alluvium on the bottoms of the troughs. The summit plains of the plateaus are slightly convex. At the actual water divide they are almost flat, but the slopes gradually become steeper with distance from this line and merge with the slopes of the eroded troughs. Morphology and material composition show that the interfluvial relief owes its origin to slope processes associated with the processes of formation of erosional forms, that is, the river and creek valleys.

There are no traces of glaciation on the high interfluves, except for the "glaciodislocation" in the Cheremkhovo ravine and some other places. The dislocations are of uncertain origin and there is no conclusive evidence that they are glacial. Disturbances in the bedding of the Jurassic deposits in the Cheremkhovo ravine could not have been made by a glacier descending from the Eastern Sayan, because the Cheremkhovo ravine is a narrow valley of a small tributary of the Angara and there are no signs of glacial activity either here or in the surrounding area. Evidently, the dislocations are associated with the sagging or collapse of strata above the ancient karst caverns, or with ancient landslides, which occurred extensively along the steep slopes comprised of Jurassic deposits.

Level, almost flat surfaces are found at a height of about 500 m (at points of most recent anticlinal doming at 520 m, at points of subsidence at 480 m). They can be traced close to the river valleys and are very widely distributed in the Irkutsk amphitheater. Their level surfaces are associated with the previously described higher interfluves by means of concave slope discontinuities. The surfaces of the 500-m level have unique material composition. Layers of friable Quaternary deposits 30 m and even 40 m thick are observed here. They differ in character from place to place. They have an irregular base and fill in the deep depressions of the ancient erosion relief.

In the broad depression of the Kaya river valley, on the Angara-Ushakovka interfluve, along the Irkut and Olkha interfluve, along the left bank of the Irkut river near the village of Maksmovshchina, and in many other places, these Quaternary deposits are monotonous strata of loams and sandy loams.

Along the Kitoy-Belaya interfluve and near the valleys of these rivers, the structure of the deposits comprising the 500-m level is much more complex. Here, sandy deposits make up the surface layer and within the sandy layers are lenses and strata of loams and sandy loams as much as 16–20 m thick. The diversity and coarseness of the material in this region are associated with deposition of a mass of coarse clastic material by the large rivers, the Kitoy and the Belaya, on leaving the mountains. A. P. Bozhinskiy (1939) held that the sandy massifs in this region are outwash plains. However, there is no actual foundation for this assumption. The structure of the stratum definitely indicates the contrary. Farther west on the plateau at Kocherikovo (Angara-Belaya interfluve), borings have revealed multicolored, thin, fatty clays including large quantities of ferruginous oolitic globules at the 500-m level beneath layers of loam and sandy loam deposits. This composition of the deposits stems from the proximity of outcroppings of Cambrian kaolines and carbonates, whose washout and redeposition produced the multicolored clays with oolites.

The deposits of the 500-m level are of complex origin. Fluvial deposits, formed under conditions of increased accumulation of alluvium, both deltaic and lacustrine, prevail. Diluvial loams of later age are widely developed. It is interesting to note that basically these are the same facies of sediments that were noted in the Jurassic deposits. The contours of the Jurassic and the deposits of the 500-m level are very similar; both are confined to the marginal fault along the geosyncline. History, as it were, repeated itself. In both the Jurassic and the Quaternary, layers of continental sediments accumulated here; however, the phenomena occurred on a different scale in the two periods.

N. I. Sokolov (1938) and I. V. Arembovskiy (1951) noted the broad distribution of loesses within the Irkutsk amphitheater and both authors interpreted them as eolian. However, there is no basis for considering the loams of the Angara region as eolian deposits. The thin stratification sometimes observed in the deposits of the 500-m level, the finds of freshwater mollusk shells in these deposits, the remains of pelagic, plankton *Pediastrum* algae, their confinement to this level, and their bedding in the depressions of the relief indicate they are of aquatic origin. The prevalence of loams and sandy loams and their great thickness distinguish these deposits from the alluvial terraces of the Angara and its tributaries.

In the depressions of the ancient relief concealed beneath the layer of deposits at the 500-m level, there are sands and pebble gravels, the alluvium of ancient rivers which formed the relief before accumulation processes replaced the erosional processes due to the downwarping of the Pre-Sayan.

Thus, within the limits of the 500-m level of the upper Angara region there developed: (*a*) fluvial deposits of the ancient drainage system; (*b*) fluvial, lacustrine, and paludal sediments which filled the depressions of the old topography and covered the low secondary water divides; and (*c*) diluvial deposits.

All the irregularities were filled in to a height of 520–480 m and only later were the valleys of the present river system cut into the accumulation plain.

In beginning our investigation of the Angara valley, we depended wholly on the Sokolov-Kamanin concept of the multiplicity of terraces, which, according to N. I. Sokolov, numbered 15. However, the data of our large-scale survey, fortified by data from several thousand boreholes and pits, led us to reject a number of hypotheses proposed by these men. Table 1 summarizes briefly the basic information on the terraces of the upper Angara river valley in the sector between Balagansk and Irkutsk.

All the terraces have a socle. The high terraces (60–45 m) are usually narrow and are preserved in just a few places. They record brief interruptions of downcutting by the rivers. The middle terraces (30–35 m) and, particularly, the 18–22-m terraces are broad and widely distributed in the Angara valley. Clearly, the river continued to exist for a long time at the level of these terraces. The terrace at the 9–10-m level is the most distinct and best developed of the low terraces, while the 14-m terrace appears in only small sectors. Evidently, the river remained at the 14-m level only a short time, while it remained at the 9–10-m level for a fairly long stage in the development of the valley.

TABLE 1

TERRACES OF THE UPPER REACHES OF THE ANGARA RIVER

| Terrace height (in m) | Morphologic status, extent and width of the terrace | Material comprising the alluvium | Normal thickness of the alluvium |
|---|---|---|---|
| 60 | Preserved, in a few places much disturbed; width up to 4 km | Sands, sandy loams, pebble gravels | Up to 10–12 m |
| 45 | Poorly preserved and in few places only; width 200–300 m | Pebbles (floodplain facies destroyed) | 3–5 m (disintegrated) |
| 30–35 | Width up to 3 km; brows smoothed and sutures washed away | Sands, sandy loams, pebble gravels | Up to 10–12 m |
| 18–22 | Very well defined; width up to 6 km | Sands, sandy loams, pebble gravels | Mostly 12–14 m, but in places more (up to 20 m) |
| 12–14 | Well defined, found in a few places; width 1–2 km | Sands, sandy loams, pebble gravels | 10–12 m |
| 9–10 | Well defined, fairly widely distributed; up to 3 km wide | Sands, sandy loams, pebble gravels | 10–12 m |
| 2–6 (floodplain) | For the most part, islands; width of the floodplain, up to 4 km | Sands, sandy loams, pebble gravels | 10–12 m |

The floodplain of the Angara river is confined, to a considerable extent, to the islands so characteristic of that part of the valley. Several of the floodplain levels stand out, usually two or three of them, reflecting the diverse conditions of sediment accumulation connected with the different causes of rise of the level of river waters. On the Angara, the water level rose as a result of floods caused by increased discharge from Lake Baykal, floods caused by the overflowing of the Sayan rivers, and by the spring and winter ice jams which raised the water level.

The terraces of the lower reaches of the large tributaries are well coordinated with the Angara terraces. It is difficult to associate these terraces with the terraces of the upper reaches of the tributaries, because the terraces have been bent by the most recent crustal movements within the Pre-Sayan trough.

The portion of the Irkutsk amphitheater adjacent to the edge of the Baykal mountain district, the region of the Kuda river basin, has a different form of relief, because of its peculiar geologic structure. Here one cannot trace a single level of the summit plains of the interfluvial plateaus, nor can one trace the 500-m accretion level.

West of this region, there is a plateau deeply incised with river valleys, having a single level of summit plains on the interfluve; to the southeast is the Onot upland. Between them, in a sector 60–80 km wide and stretching parallel to the edge of the Baykal mountain district, are low and quite narrow ridges rising 120–180 m above narrow depressions of the same magnitude.

The ridges have anticlinal structure, while the depressions separating them are synclinal. The ridges contain Cambrian, Jurassic, and, in places, Pliocene deposits. Quaternary deposits lie in the synclines. It is especially significant that sandy clays, friable sandstones, and Pliocene coquinas are found both on the summits of the ridges and in the synclinal depressions. Consequently, fold formation continued at the end of the Pliocene and during the Quaternary period. The difference in the height of the base level of the Pliocene coquina deposits, i.e., 100–120 m, indicates the considerable scope of the movements.

Some investigators have assumed that the synclinal basins are a system of ancient river valleys. The absurdity of this position is indicated not only by the improbability of the development of a whole series of enormous valleys in one locale, but also by an examination of the morphometric data. The heights of the basin floors and the heights of the intervening isthmuses along the alleged ancient valleys differ greatly. Only one depression stretches continuously, connecting the Angara and Lena basins. This is an ancient valley along which Pre-Sayan waters once drained into the Lena basin. It was curved by the uplifting in the vicinity of the village of Bayanday, where not only its floor but also all the surrounding territory was lifted somewhat. Now the Kuda, Murin, and Manzurka rivers flow along the bottom of this ancient valley. The connection between the distribution of erratic boulders and Jurassic conglomerates is particularly clear in this part of the Irkutsk amphitheater. It may be assumed that the largest boulders of the Baykal mountain region and the Eastern Sayan are of glacial origin, but in this case it should also be assumed that the entire Irkutsk amphitheater was covered with a glacier. Inasmuch as boulders of the same type are found in the Jurassic conglomerates, while the contours of boulder distribution coincide with the contours of the Jurassic conglomerate outcroppings, it becomes apparent that they are not of glacial origin. Evidently the extremely rare finds of boulders on the Angara river terraces can be attributed to transport by river ice.

In addition to eluvial, fluvial, and lacustrine-paludal deposits, diluvial deposits are widely distributed in the Irkutsk amphitheater. Diluvial deposits smooth out the contact lines between the 500-m level and the higher level, and also the suture lines of the terraces cut both in them and in the higher interfluves. As a rule, the brows of the terraces, especially of the ancient terraces, are greatly eroded, while layers of diluvial loams and sandy loams lie at the suture lines. The diluvium may be quite thick and in this case the role of the diluvial process in the transformation of the topography can be judged from the diluvial layer. For example, even on the comparatively young 18–22-m terrace of the Angara river the diluvium is 20 m thick in places. The diluvium is usually porous, with columnar jointing.

The stratigraphic implications of loams are a matter of great interest. Often the variety of colors of diluvial loams is pointed out: they range from brown to red to yellow-brown. As yet we do not have sufficient evidence to establish age differences of loams on the basis of color, although it would be desirable to attribute stratigraphic significance to these differences in color. Quite often the change of color of diluvial loam (sandy loam) downward along a profile is connected with soil formation processes; the upper, yellow-brown horizon is rich in carbonates; the somewhat reddish horizon represents a zone of contemporary and ancient inwashing of sesquioxides. In the lowest strata, the diluvium is usually quite moist, with the brownish color of the deposit associated with the moisture. Finally, for the most part, there is no regular color sequence of diluvium in a vertical section. The layers which gradually succeed each other have different granulometric composition and, although they differ in color, the changes cannot be traced from profile to profile.

Horizons strongly enriched with humus are encountered, denoting, evidently, a delay in the accumulation of diluvium. These horizons are particularly distinct in the diluvial trains which descend to the 18–22-m terrace near Sivirsk.

## Archaeological, Paleofaunistic, and Paleobotanic Data Connected with the Age of the Relief and the Quaternary Deposits

Archaeological data, finds of fauna, and the results of spore and pollen analysis may be used to determine the age of the relief and the Quaternary deposits.

Archaeological and paleofaunistic data help in deciphering only the last pages of Quaternary history, because only the finds of relatively young fauna, the remains of the hairy rhinoceros and the mammoth, have a completely clear stratigraphic position. The most ancient camps of primitive man belong to the Upper Paleolithic. The spore-pollen analysis data pertain to the time span from the Upper Pliocene to the present.

The archaeological, paleofaunistic, and paleobotanic data relate well to each other and to the results of analysis of the land forms and the stratification sequence of the friable layers.

We have followed A. P. Okladnikov's interpretation (1950) of the archaeological material. He divides the history of man's conquest of this part of Siberia into two stages. He places the sites Malta, Buret, and Voyennyy Gospital [Military Hospital] in the first stage and Kayskaya Gora, Ust-Belaya (Boday-Cheremushnik), and others in the second stage.

In characterizing the site Malyy Kot, I. V. Arembovskiy (1951) remarks that the archaeological objects appear to be early Mousterian. This is the only indication in the literature of finds of implements of a culture earlier than Malta and Buret.

Implements of a Late Solutrean culture and remnants of dwellings and of vertebrate fauna were found at Malta and Buret. The construction features of the dwellings indicate the severity of the climate of that time. These structures formed enduring settlements. They were half buried in the ground, used a minimum of wood, and were quite similar to the dwellings of the former inhabitants of northeastern Siberia.[*]

The discovery of a primitive statuette, which depicts a man in parka-type clothing, also indicates that the climate was severe during that epoch; it indicates a treeless terrain and strong winds. Primitive man adapted himself to the situation by using air-tight clothing and half-buried dwellings.

Encampments of this type are found on the 18–23-m terrace of the Angara and on the corresponding terrace of the Belaya river. According to V. I. Gromov (1948), they are syngenetic with the upper alluvium of this terrace. There are no settlements with remnants of the Late Solutrean culture on the lower terraces.

Analyzing these settlements, A. P. Okladnikov (1948, p. 155) writes: "It is quite plausible that the original settlement of Siberia by man proceeded from west to east, from established regions of "Solutrean" culture of the arctic hunters of the great Russian plain. . . ."

The similarity of the culture and the relative similarity of the physicogeogra-

[*][See M. M. Gerasimov's article on pp. 3–32.]

phic situation make it highly probable that the deposits of the 18–22-m terrace are synchronous with the Late Solutrean or the beginning of the Magdalenian of the Russian Plain, since human culture apparently developed somewhat earlier on the Russian Plain.

The remains discovered in the numerous camps of the second stage of development of human culture are completely different from the first: these were not permanent settlements, but the temporary camps of nomadic hunters. They include the camps Kayskaya Gora near the present city of Irkutsk, containing remains of Magdalenian culture, and Ust-Belaya (Boday-Cheremushnik), whose archaeological material indicates that it belongs to the end of the Azilian (Arembovskiy, 1951).

Kayskaya Gora is associated in time with the alluvial deposits of the 14-m Angara terrace, with the period of considerable forestation of the country. Ust-Belaya belongs to the period during which the 9–10-m terrace dried and the modern floodplain alluvium began to form.

In our discussion of the fossil fauna of the southern Angara region, we shall employ the specifications and conclusions of M. M. Gerasimov, V. I. Gromov, and, particularly, I. V. Arembovskiy, who studied this region.

Finds of the most ancient Quaternary fauna of the Angara region are relatively rare. Remains of *Elephas trogontherii* and *Bison priscus longicornis* have been found, indicating that a fauna older than the mammoth existed here. In the Angara region, evidently the development of the animal world was closely connected with that of the other regions of the present Soviet Union. Unfortunately, however, we cannot associate the finds of trogontherium fauna with specific deposits and land forms with any degree of confidence, since they are limited to the diluvial mantles, which descended on the land forms, and to deposits not associated with the Angara terraces and deposits of the 500-m level. It may be assumed that the thermophilic fauna (the trogontherium and the long-horned bison) was replaced subsequently by a mammoth fauna, at first in the regions with a cold climate. With warming, it [the thermophilic fauna] then partially resettled the regions which it had previously abandoned, while the mammoth, gradually gaining dominance, moved northward during this [warmer] interglacial period.

Faunistic remains of *Rhinoceros antiquitatis*, *Elephas primigenius*, *Bison priscus deminutus*, *Rangifer tarandus*, *Equus* sp., *Vulpes lagopus*, *Ovis nivicola*, etc.* were found in the strata associated with man at the Malta and Buret sites (Arembovskiy, 1951; Tyumentsev, 1941). The finds of arctic fox and snow sheep indicate a cold climate, which agrees well with the archaeological data. Remains of *Alces machilis* and *Bos* sp. were found at Kayskaya Gora together with *Elephas primigenius*, *Equus* sp., and *Rangifer tarandus*. It is significant that the hairy rhinoceros had not yet appeared here, but that the elk, a typical forest animal, had appeared. The fauna of the Ust-Belaya settlement in the highest alluvial horizons is almost Holocene in appearance: *Cervus elaphus*, *Capreolus pigargus*, and *Castor fiber*. A great many finds have been made of other osteologic material of the same complex. However, individual finds of separate skeletal parts in alluvium may very well represent redeposited material. Most often the bone remains are discovered in diluvial loams. In such cases, one may speak only of the age of a given portion of the loam stratum at a given point. In other words, such finds may be associated with the main reference points, but they

---

*Bones of other animals were discovered in the upper stratum associated with man at Malta, along with later archaeological finds.

contribute little to a detailed paleofaunistic characterization of the Quaternary deposits of the Angara region. This requires a point of alignment with the terrace steps of the Angara or with the strata of deposits comprising the 500-m level.

The spore-pollen analysis data cover practically the entire Quaternary period. For a long time, spore-pollen data yielded very insignificant results in eastern Siberia because very many samples had no pollen and spores or had very little, but mainly because the stratification order of the Quaternary deposits and their genesis had not been clarified.

As material began to be accumulated, not only from the Irkutsk amphitheater, but from the entire Angara basin, it became possible to deduce the evolution of the Quaternary vegetation for the entire region. Such work, based on an analysis of numerous cross-sections (soil profiles), was carried out by M. P. Grichuk from samples taken by the East Siberian Expedition of Moscow State University and, in part, from the data presented by other organizations. In this work, the spore-pollen diagrams of aqueous origin (river and lake) were compared. The process by which the pollen was mixed while being buried was essentially the same in the case of rivers and lakes. However, it must be borne in mind that the results of analysis of the deposits of the diluvium mantles cannot be compared with corresponding alluvium deposits, because the diluvium deposits can give only a "local" pollen spectrum.

According to Grichuk's scheme (Fig. 2), the Quaternary history of the flora of the southern Angara region can be divided into a number of stages. Their general characteristics are evident from the Grichuk scheme, in which data of many individual overlapping or partially overlapping cross-sections [profiles] are generalized and systematized. Although this scheme is tentative, it gives the essence of the Quaternary history of the south Angara vegetation.

In this scheme it is evident how gradually the remains of Pliocene vegetation, the *Pterocarya, Tsuga,* pines of the section *Strobus,* etc., disappeared. A certain rhythm in the changes of the vegetation becomes evident. It is important to note the appearance of an admixture of broadleaf trees of the species *Quercus, Ilmus, Tilia,* and *Corylus* during the formation of the alluvial strata of the middle terraces of the Angara and its tributaries, noted in dozens of diagrams. It indicates a considerable warming and the return of broadleaf species to the Angara basin during a specific poch of the Quaternary period. On the other hand, the epochs of prevalence of birch and grassy plant pollens indicate steppe formation and, in certain combinations, cooling.

## Brief History of the Quaternary Period in the Southwestern Baykal Region

In this part of our paper we shall attempt to interrelate data on land forms, Quaternary deposits, the culture of primitive man, and the fauna and flora.

At present we still do not have sufficient data on the southwestern part of the Baykal region to distinguish between the Quaternary and the Pliocene periods with sufficient accuracy. The tectonic movements which created the mountains of southern Siberia in place of the low mountains and gently rolling plains began much earlier than the end of the Pliocene and, consequently, the beginning of the new tectonic stage of the territory cannot be taken as the beginning of the

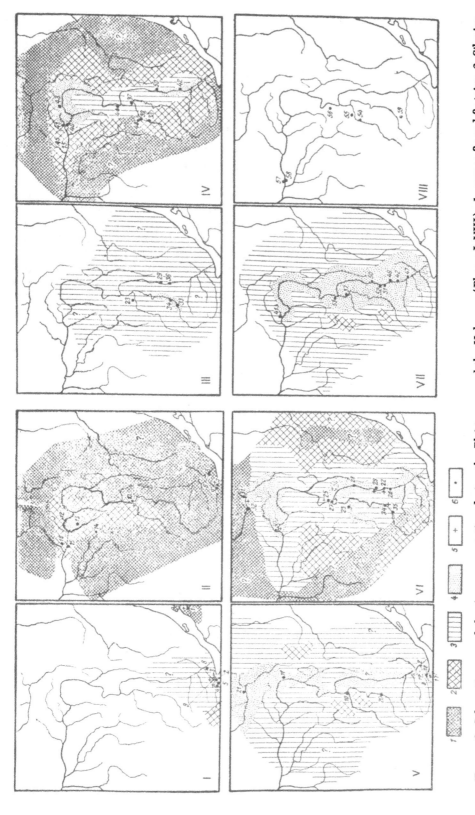

FIGURE 2. Schematic map of the Angara vegetation during the Pleistocene and the Holocene (Phases I–VIII): 1, spruce-fir and fir taiga; 2, Siberian pine–common pine and Siberian pine taiga; 3, light-coniferous and sparse birch forests; 4, broadleaf forests; 5, wooded steppe and steppe; 6, locations of sections investigated (cross section numbers are those in Grichuk's article "Results of paleobotanic investigation of Quaternary deposits of the Angara region," which is included in this collection).

Quaternary period. Data on the faunistic changes do not provide much along these lines either. A more distinct boundary between the two eras can be drawn on the basis of changes in the vegetation and climate, expressed in the substitution of taiga flora for the thermophilic coniferous and broadleaf flora. We have accepted this boundary, arbitrarily, as the lower limit of the Quaternary period.

A deeply incised valley network formed in the Irkutsk amphitheatre at the beginning of the Quaternary period. Its main channel ran from the present mouth of the Kitoy river across the Kuda, Murin, and Manzurka rivers into the Lena basin. The process of formation of this Pre-Quaternary river valley network was highly complex.

The presence of broad, smoothed surfaces in the mountains and the monotonous evenness of the heights of the interfluvial ridges on the upland indicate a long Pre-Pliocene stage of calm tectonic development of the territory, during which the agencies of denudation smoothed out the irregularities created by the tectonics of the past epochs of morphotectogenesis. At the end of the Pliocene, the mountains of the Khamar-Daban and Eastern Sayan had low absolute and relative heights. Intensive orogenic movements began in the Eastern Sayan and in the Baykal mountain region during the Pliocene, leading to a compensatory subsidence of the edge of the platform and to the formation in the Kuda basin of shallow depressions occupied by lakes, whose existence is indicated by the Pliocene deposits found there. However, the mountains were still low and could not have been a serious obstacle to the free movement of air masses. The taiga of the late Pliocene preserved Tertiary elements and was diversified in its species composition.

We propose that the Quaternary history of the southwestern Baykal region may be divided into the following stages:

1. The preglacial epoch, $Q_1$.
2. The first glaciation (perhaps a single glaciation, perhaps not, but consisting of several stages), $Q_2$.
3. The interglacial epoch, $Q_3^1$.
4. The second glaciation, consisting of three stages, $Q_3^{2-4}$.
5. The present epoch, $Q_4$.

This scheme is merely a working hypothesis, but all facts known to date fit into it. Let us examine each of the stages in turn (see Table 2).

## THE PREGLACIAL EPOCH, $Q_1$

The uplifting of the mountain ridges that frame Lake Baykal and the depression of the lake basin that began in the Pliocene continued in the preglacial epoch. The regions of mountain uplifting and the upland territory were included in the young erosional incision. By the end of the epoch, as a result of the irregularity of the upliftings that bent the longitudinal profile of the river artery that drained the Pre-Sayan part of the upland, the Pre-Sayan valleys began filling with lacustrine, fluvial, and deltaic sediments. This process was promoted by the downwarping of the territory at the foot of the Eastern Sayan (a compensatory downwarping at the edge of the uplifted mountain country). Nevertheless, as before, the Pre-Sayan drained into the Lena river through the valley now occupied by the Kuda, Mirin, and Manzurka rivers. Terrace III of Lake Baykal began to form at the foot of the Khamar-Daban.

During this epoch the forests which covered the upland and the mountain slopes consisted of fir, spruce, Siberian pine, and various pines of the section

*Strobus.* An admixture of broadleaf species, the elm, oak, linden, and hazelnut, was preserved in the forest tracts. Birch forests occurred here. and there and steppe areas appeared on slopes with southern exposure. Then, as the climate became colder, coniferous forests began to play an ever larger role, aided by the appearance of considerable lake and swamp massifs in the Pre-Sayan.

THE FIRST GLACIAL EPOCH, $Q_2$

This epoch encompasses a very long time-span, during which the Central Siberian Plateau continued to rise and ever more intensive differential movements occurred in the Baykal mountain region and in the Eastern Sayan. The lake basin became deeper and the Khamar-Daban mountains higher, but they were still far lower than at present.

This may be deduced, *inter alia*, by the fact that terrace III of Lake Baykal, which formed in crystalline rocks, is now 250 m above lake level. In the axial part of the range, where the ancient surface of planation is now considerably higher than in the peripheral region of the range (the zone of formation of terrace III), the subsequent uplifting was still greater. At present, near the zone of formation of terrace III, the ancient surface of planation lies 1000–1200 m above lake level, while towards the central part it gradually rises 1800–2100 m above lake level, i.e., to almost twice the height. Therefore, it may be assumed that the axial part of the Khamar-Daban has risen 400–500 m since that time [the pre-glacial epoch]. Thus, the Khamar-Daban mountains had still not become very high during the first glacial epoch, and the conditions for development of glaciation were less favorable then than later on, when the mountains became considerably higher. Therefore, the territory included in the first glaciation was smaller than that included in subsequent glaciation. In the future, it probably will be possible to subdivide the first glacial epoch.

The downwarping of the upland regions adjacent to the mountains caused a considerable broadening of the region of alluvial accumulation in the Pre-Sayan, and there were lakes in this region. Thick layers of different lithologic character accumulated. Sands and pebble gravels were deposited at the points of emergence of large rivers from the mountains; in other places, where the material was transported by small streams, there are sands, sandy loams, loams, and clays. Gradually a broad aggradation plain formed, remnants of which now appear as the 480–520-m level.

The plain of the Irkutsk amphitheater was covered with a sparse light coniferous taiga. Between the light coniferous forests, chiefly larch, lay broad treeless areas with arctophilic grassy vegetation.

By the end of the epoch, the continued uplifting reconstructed the river system. The upper reaches of the ancient Angara river, then located in the Ust-Uda region, received their drainage from the alluvial plain of the Pre-Sayan. The increased volume of flow led to downcutting of both the main river and its tributaries. The 500-m accretion level began to the dissected. The 60-m and 45-m terraces of the Angara appeared, indicating that the river was arrested briefly at these levels.

THE INTERGLACIAL EPOCH, $Q_3{}^1$

During this epoch, apparently the uplifting within the platform continued, as indicated by the formation of the lower, 30-35-m and 18–22-m terraces, which

TABLE 2. THE QUATERNARY HISTORY OF THE SOUTHWESTERN BAYKAL REGION

| Principal stages of Quaternary history | Geomorphology (relief, Quaternary deposits, and neotectonics) | | | Biosphere | | | Hydrography and permafrost |
|---|---|---|---|---|---|---|---|
| | Mountains | Baykal terraces | Main stages of geomorphologic development of platform | Vegetation | Vertebrate fauna | Sites of ancient man | Phases of development of river network; and permafrost |
| $Q_4^2$ Complete disappearance of glaciation | Destruction of glacial forms | — | Low floodplain | Present-day flora | Present-day fauna | — | Disappearance of permafrost |
| $Q_4^1$ Last cirque glaciers | Shrinking of glaciers. Moraines on upper Pereyemmaya and Vydrinaya | | High floodplain | Light coniferous birch, and also dark coniferous forests | Beaver, Siberian stag | "Ust-Belaya" (Azilian) | Spread of permafrost |
| $Q_3^4$ New glacial advance | New advance of the glaciers. Moraines at river mouths | Contemporary littoral. Formation of deltas and broad floodplains of rivers within limits of ancient terraces | 9–10-m terrace. Eolation of surface of highest terraces | Sparse light coniferous forests. Steppe areas, many arctophilic types | Mammoth, elk, short-horned bison | — | — |
| $Q_3^3$ Considerable glacier shrinkage | Considerable destruction of glacial forms in connection with considerable reduction of glaciation | | 14-m terrace | Light coniferous and dark coniferous forests | | "Kayskaya Gora" (Magdalenian) | — |
| $Q_3^2$ Epoch of maximum distribution of ice (last glacial epoch | Mountains almost as high as at present. Valley and cirque glaciers | Ancient terrace I of Lake Baykal. Moraines at outlets of glaciers from mountains along valleys of Pereyemnaya, Vydrinaya, Snezhnaya, etc. | Erosion and incision. Upper strata of the 18–22-m terrace | Sparse light coniferous forests, steppe formation (possible propagation of forested tundra vegetation, abundant birch pollen) | Rhinoceros, mammoth, snow sheep, Arctic fox, short-horned bison | Malta; Buret; Voyemnyy Gospital. (End of Solutrean) | Development of the drainage of Lake Baykal through the Angara. Cessation of drainage from Lake Baykal along valleys of present Ungura and Bugul- deyka rivers |

| Epoch | Tectonics | Baykal terrace | Alluvium | Vegetation | Fauna | Archaeology | Drainage |
|---|---|---|---|---|---|---|---|
| $Q_3^1$ Interglacial epoch | Continued uplifting. Deep erosional dissection | Ancient terrace II of Lake Baykal | Lower strata of the alluvium of the 18–22-m terrace; 30–35-m terrace | Dark coniferous forests with admixture of broadleaf trees | *Trogontherium*, long-horned bison | Malyy Kot (doubtful remains of Early Mousterian) | Formation of broad terraces with thick layers of alluvium |
| $Q_2$ Epoch of primary glaciation (perhaps more than one) | Uplifting of mountains considerable but mountains 300–500 m lower than today. Glaciation of mountains, ice caps, valley and cirque glaciers | Ancient Baykal terrace III | 45- and 60-m Angara terraces | Light coniferous and birch forests. Some steppe formation at the end of the phase | ? | ? | Change of direction of discharge from the Pre-Sayan. Drainage into the Yenisey rather than into the Lena river |
| $Q_1$ Preglacial epoch | Continued uplifting, bending, faults, and erosional action on the established relief | — | Formation of friable deposits of the 500-m accretion level. Ancient alluvium of buried valleys | Dark coniferous forests. Dark coniferous forests with small amount of pines of the section *Strobus*, *Pterocarya*, *Tsuga*, linden, elm, oak | ? | ? | Alluvial plains. Lake basins in the Pre-Sayan. Drainage into the Lena. Deep erosional downcutting of the river valleys |
| Pliocene | Beginning of upliftings in the mountains of the Baykal region | — | Lacustrine sediments of the Kuda basin | — | — | — | — |

belong to the middle complex, on the Angara and its tributaries. The deepening of the Baykal basin and the uplifting of the mountains surrounding the lake continued. Baykal terrace II formed at the foot of the Khamar-Daban. The irregular lifting of the ranges of the western framework of Lake Baykal was greater in the north, especially in the region of the ancient valley which accommodated the drainage of Lake Baykal (the Buguldeyka and Ungura rivers) and was smaller in the region of the present source of the Angara. This led to another reconstruction of the river system. The intensive uplifting of the threshold of drainage from the lake to the position of the present upper Ungura river caused the waters of Lake Baykal to pour through the narrow watershed between a small river which emptied into Lake Baykal and a river of the Ushakovka type, which flowed where the upper Angara river now flows. Subsequently, the river system assumed its present profile.

Paleofloristic data indicate the presence of an epoch of substantial warming of the climate during the formation of the Angara terraces of the middle complex (30–35 m and 18–22 m). Pollen of broadleaf species, such as the hazelnut, linden, and oak, which had disappeared earlier now appeared in the alluvium of these terraces, as indicated by a number of cross-sections. The occurrence of vegetation which was more thermophilic than the present indicates that the climate was warm during the interglacial epoch.

The broadleaf species formed only a small admixture in the forests of that time; Siberian pine, fir, and spruce predominated. The dark coniferous taiga covered the uplands and the mountain slopes. The climate was not only warmer than at present but was more humid.

Probably elements of thermophilic fauna, the trogontherium and the long-horned bison, survived until then in the Angara region. At present, it is difficult to say whether there were human settlements in the Angara region during the interglacial epoch, although the stone material found in the valley of the Malyy Kot river (Sokolov and Tyumentsev, 1949) is somewhat reminiscent of the primitive implements of the Early Mousterian.

### THE LAST GLACIAL EPOCH, $Q_3^{2-4}$

This epoch is divided into three parts: (1) the maximum advance of the ice (the "Pereyemnaya" stage), (2) the period of glacier shrinkage, and (3) the period of a second, less spectacular advance (the "Isakovka" stage) with subsequent glacier retreat, which is subdivided into further stages.

During the last glaciation, the Khamar-Daban mountains were only slightly lower than at present, which created favourable orographic conditions for significant development of glaciation. The glacial erosion deformed both the longitudinal and the transverse profiles of the valleys. The results of glacier activity can also be observed in the thick moraines, cirques, trough valleys, and the south bars of rivers.

During the glacial maximum, the largest glaciers descended from the gorges of the Khamar-Daban to the very shore of Lake Baykal. At the same time, the lake waters formed a narrow ancient terrace, indicated by the wedging out of the glacial deposits by lake deposits.

The formation of the 18–22-m Angara terrace ceased on the upland at that time and the upper horizons of the deposits comprising this terrace formed. The landscape of that time was like the present landscape of northeastern Siberia, where the forested tundra vegetation on the lowlands combines with the mountain tundra and with the mountain glaciers. The dark coniferous taiga was replaced by sparse light coniferous taiga and birch forests. Not only did the broadleaf trees disappear here, but the Siberian pine and the fir also disappeared. Animals capable of withstanding the severe climate—the hairy rhinoceros, the mammoth, the snow sheep, and the arctic fox-populated the forested tundra. All the circumstances of human life at that time, as A. P. Okladnikov (1950) writes, indicate man's adjustment to life in forested tundra. The characteristics of the dwellings, the implements of the arctic hunters of Malta and Buret, the character of the pollen spectrum, and the fauna all show the severity of the climate.

Later, during the shrinking of the mountain glaciers, which evidently occurred on the uplands as well, the climate became somewhat less severe. A considerable quantity of Siberian pine, fir, and spruce pollen has been found in the alluvium of the 14-m terrace of the Angara. However, there are no traces of broadleaf flora; the larch and the birch play a large role. The type of fauna also changed. Remains of mammoths, elks, and the short-horned bison have been found at the Magdalenian (Kayskaya Gora) campsites, evidently belonging to nomadic forest hunters (Late Magdalenian). Remains of the hairy rhinoceros are not found here.

The pollen spectrum of the alluvium of the youngest, the 10-m, terrace above the floodplain indicates a new cooling, the decline of the dark coniferous taiga, and the increasing importance of the birch and larch forests. The role of grassy plants, characteristic of the cold steppe, increased sharply.

During the formation of the 10-m terrace, the surface of the low and middle terraces experienced intensive eolation due to the sparseness of the vegetal cover. The new advance of mountain glaciers is associated with this period: This advance left moraines at the mouth of the Isakovka river on the Pereyemnaya, and at Lake Sobolinoye on the Selengushka river (right tributary of the Snezhnaya river).

## THE HOLOCENE, $Q_4$

During the Holocene, the movements of the earth's crust were very strong, as evidenced by the frequent, severe earthquakes of the time.

The glaciers of the Khamar-Daban and the Tunkinskiye, Kitoyskiye, and Belskiye Goltsy [Barrens] disappeared quite recently. This is indicated by the fresh cirques, on whose shaded slopes there are often large firm basins which survive the summer. In the upper Pereyemnaya and Vydrinaya rivers there are fresh moraines, beyond which are lakes. These moraines evidently denote one of the last stages of glaciation.

The Angara valley continued to deepen; a high and low floodplain formed. Permafrost gradually disappeared.

The vegetation changed during the Holocene. Even quite recently the forests of the Irkutsk amphitheatre occupied a large area, and they began to retreat only because of man's activities. At the same time, better conditions were created for dark coniferous taiga, as evidenced by the abundant new growth of fir and Siberian pine in the larch-pine forests of the Angara region, even though forest fires caused by man have been very destructive of the dark coniferous species. Usually the Siberian pine-fir forests are pushed far back into the interfluves near settled places. Apparently, one of the results of man's economic activities is the disappearance of animals such as the beaver and the Siberian stag, which were plentiful in the recent past. The climate is becoming milder in conjunction with the gradual warming process.

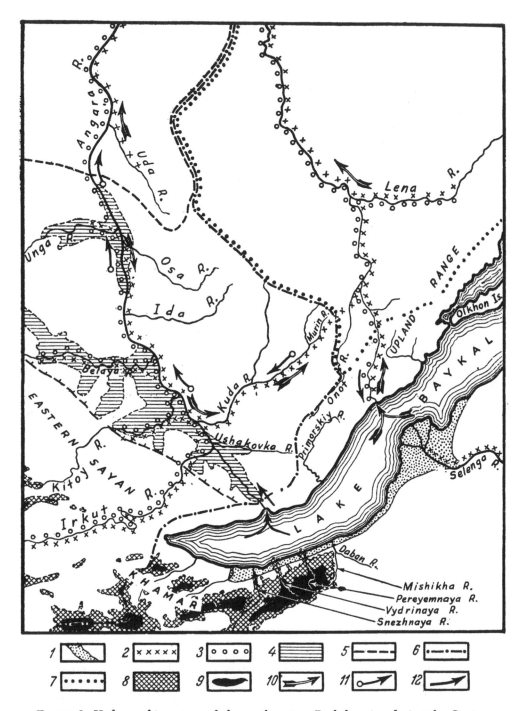

FIGURE 3. Hydrographic system of the southwestern Baykal region during the Quaternary period: 1, ancient Baykal terraces; 2, principal drainage arteries during the Lower Pleistocene; 3, the same, during the Middle Pleistocene; 4, territory of the aggradation plain during the Middle Pleistocene: 5, Angara-Lena watershed during the Lower Pleistocene; 6, the same, during the Middle Pleistocene; 7, the same, during the Upper Pleistocene and the Holocene; 8, boundary of the obliterated glacial forms; 9, boundary of fresh glacial forms; 10, direction of flow during the Lower Pleistocene; 11, the same, during the Middle Pleistocene; 12, the same, during the Upper Pleistocene.

*Conclusions*

Decisive changes took place in the entire nature of the southwestern Baykal region during the eventful Quaternary period.

In place of the low mountain country, mountains were lifted to heights of more than 2000 m, rivers deeply dissected the plain surface of the Central Siberian Plateau, and considerable areas of alluvial plains, i.e., the interior deltas of rivers originating in the Eastern Sayan mountains, developed at the boundary between the mountains and the plateau.

The drainage pattern changed. The "Baykal" orientation of the drainage (northeast, to the Lena) gave place to the "Sayan" orientation (northwest, to the Yenisey). (See Fig. 3.) Large lake-swamp massifs appeared and disappeared. The Baykal depression became very deep. The luxuriant and diversified Tertiary flora and fauna disappeared. The climate became continental and colder.

These changes were not of a gradual, evolutionary nature, as can be seen from the above. The climatic changes, the crustal movements, and the evolution of the plant and animal world are interrelated in a most complex manner. The life cycle of the components of the physiocogeographic medium was irregular, and the general trend of development of nature can hardly be presented in the form of repetitious cycles.

We feel we should refrain from synchronizing the individual stages of the Quaternary of the southwestern Baykal region with those of the territories to the west of it. This can be done only after local schemes have been worked out for each of them, schemes based on related data on the history of all the natural components, viz., climate, relief, conditions of sediment accumulation, hydrography, evolution of the flora and fauna, and the culture of ancient man.

Some parallels might be drawn even now, but at the present level of knowledge, errors might be made which would prove detrimental to the proper interpretation of the history of the Quaternary period.

M. P. GRICHUK

# RESULTS OF A PALEOBOTANIC INVESTIGATION OF THE QUATERNARY DEPOSITS OF THE ANGARA REGION*

*I*

THE SPORE-POLLEN METHOD of analysis was used to study the Quaternary deposits of the Angara region, in connection with the activities of the East Siberian Expedition of the Department of Geography, Moscow State University,† in that region. The work was hampered somewhat by the lack of a sufficiently well developed stratigraphic scheme of the Quaternary deposits. Further, relatively little was known about the history of the flora and vegetation. One reason why the history of the Angara flora remains somewhat obscure is that the Quaternary deposits contain few plant remains. The spore-pollen analysis method has shown that the concentration of pollen in the deposits is generally very slight; many of the samples analyzed had practically no pollen.

Nevertheless, after several years of work, data have been assembled from several dozen sections [profiles] of Quaternary deposits. A comparison of these data, with consideration of the conditions under which the samples were taken, has allowed us to draw a number of preliminary conclusions about the change in the general character of the Angara vegetation since the Tertiary period.

Most of the samples for analysis were taken from sections of terrace deposits of the Angara river and its tributaries the Oka, Iya, Kuda, and others; a few samples were taken from lake and swamp deposits.

Problems of the formation of the spore-pollen spectra have been treated in a number of systematic works (V. P. Grichuk, Ye. D. Zaklinskaya, Ye. V. Koreneva, Ye. A. Malgina, R. V. Fedorova, *et al.*). However, the doubts some investigators have expressed regarding the validity of the interpretation of the results of such analysis forced us to re-examine the problems of the formation of the spore-pollen spectra under various natural conditions. Pollen and spores are carried about by the wind and settle on the surface of bodies of water or on soil and plant surfaces. The pollen and spores are washed by rain and meltwater from the soil and plants into bodies of water, and they enter the sediment of the water bodies and eventually become buried. We know, too, that before the pollen and spores become buried, they are mixed in the water and air. Because of all this transportation, the pollen and spore composition of the deposits corresponds to the composition of the vegetation of a basin as a whole, or to a considerable part

*Translated from K. K. Markov and A. I. Popov (Editors), *Lednikovyy period na territorii yevropeyskoy chasti SSSR i Sibiri*, Moskovskiy gosudarstvennyy Universitet, Geograficheskiy Fakultet i Muzey Zemlevedeniya, 1959, pp. 442–497.

†[Vostochno-Sibirskaya Ekspeditsiya Geograficheskogo Fakulteta Moskovskogo Gosudarstvennogo Universiteta (MGU).]

of a large basin, and thus the spore-pollen spectrum is very much an average. Specially conducted systematic studies have shown that the degree of averaging of the spore-pollen spectra varies. The spectra of marine deposits and the alluvium of large rivers are more averaged than spectra of lake deposits and the alluvium of small rivers, and the spectra of swamp deposits are the least averaged.

The degree of averaging of the spore-pollen spectra in the present deposits of the basins of the Angara region may be shown by comparing the spore-pollen spectra of modern deposits with the present composition of the vegetation. Table 1 gives the results of analyses of nine samples from various localities indicated on the map in Fig. 20, where the contemporary vegetation is shown schematically, after the *Karta rastitelnosti SSSR* [Vegetation map of the USSR], 1939, edited by Ye. M. Lavrenko.

Sample no. 1 was taken from the floodplain surface of the Iya river near Burkhun, in a region of pine and pine-larch forests. Farther upriver there is a region of steppe and meadow-steppe vegetation with birch-pine* forests. The upper Iya flows through a region of mountain-taiga, dark-coniferous forests of Siberian pine,† spruce, and fir with an admixture of larch and Siberian pine-larch forests. Pollen and spores of all plants which predominate in the basin are found in the pollen and spore structure of this sample, with the exception of the larch, whose pollen is rarely preserved.

Sample no. 2 was taken from the surface of the floodplain of the Oka river near Podsochnaya in a region where meadow steppe with birch, pine, and larch is widely distributed at present. The upper reaches of the river intersect a belt of mountain-taiga, dark-coniferous, and Siberian pine-larch forests. Analysis of the sample showed a large quantity of grass pollens, which comprised 13 per cent of all the pollen and spores. The pollen composition of the tree species is similar to that of sample 1. The spore-pollen spectrum also corresponds to the vegetation of the entire basin.

Sample no. 3 was taken from the surface of the Kuda river floodplain near Granovshchina. Steppe vegetation predominates in the Kuda valley, and this is reflected in the pollen and spore composition of the sample. The grass pollen comprises 6 per cent of the total pollen in the sample. The grass pollen contains much sedge, cereal, and motley grasses. The upper Kuda and its tributaries are situated in a region of extensive pine and larch-pine forests; in places there are Siberian pine, fir, and spruce. Correspondingly, the pollen of the tree species contains more pine and less spruce and birch.

Sample no. 4 was taken from the surface of the Milka river floodplain near Savvate-yevka. The river basin is situated in a region of grassy-shrub, pine, and pine-larch forests. The Grass pollen comprises 3 per cent of the total spore and pollen structure, i.e., less than in samples 1 and 2. The composition of the tree pollen is similar to that of samples 1 and 2.

Sample no. 5 was taken from the floodplain of the Slyudyanka river. Mountain-taiga, Siberian pine-larch, and pine-Siberian pine forests prevail; in places they contain birch and fir. The Siberian pine pollen comprises 19 per cent of the tree pollen, and thus is slightly higher than in the preceding samples.

Sample no. 6 was taken from the surface of a swamp on the left slope of the Iya valley north of its tributary the Bolshoy Korabl. Pine and birch-aspen forests grow in the swampy region. The tree pollen composition of this sample differs from that of sample 1, which was taken from the floodplain of the Iya basin, in that it has a smaller spruce pollen count. Here the spectrum is formed, to a considerable extent, by local and sub-local components (V. P. Grichuk, 1948); therefore, the pollen structure of the grassy plants and spores of samples 1 and 6 are not much alike.

---

*[Pine, or common pine, identified as *Pinus* in the tables.—Translator.]

†[*Pinus sibirica*, called variously Siberian pine, Siberian cembra pine, stone pine, and Siberian cedar.—Translator.]

TABLE 1

RESULT OF SPORE-POLLEN ANALYSIS OF SURFACE SAMPLES OF
SWAMPS AND FLOODPLAINS IN THE ANGARA BASIN

| Sample no.: | 1 | 2 | 3 | 4 | 5 | 6 | 7 | 8 | 9 |
|---|---|---|---|---|---|---|---|---|---|
| Sampling depth in m: | 0.29 | 0.12 | 0.03 | 0.03 | 0.03 | 0.5 | 0.15 | 0.5 | 0.25 |
| Number of grains counted: | 190 | 266 | 569 | 303 | 324 | 216 | 483 | — | 268 |

SPORE AND POLLEN COMPOSITION (%)

| | 1 | 2 | 3 | 4 | 5 | 6 | 7 | 8 | 9 |
|---|---|---|---|---|---|---|---|---|---|
| **GENERAL COMPOSITION** | | | | | | | | | |
| Tree pollen | 68 | 64 | 84 | 95 | 57 | 63 | 62 | — | — |
| Grass and shrub pollen | 2 | 13 | 16 | 3 | 5 | 2 | 3 | — | — |
| Spores | 30 | 23 | 10 | 2 | 38 | 35 | 35 | — | — |
| **POLLEN OF TREE SPECIES** | | | | | | | | | |
| *Larix* | <1? | — | — | — | 1 | — | — | — | 4 |
| *Abies* | <1 | 2 | <1 | 1 | 2 | 1 | 3 | — | 10 |
| *Picea* | 9 | 8 | <1 | <1 | <1 | — | <1 | 3 | 11 |
| *Pinus sibirica* | few | few | 5 | 4 | 19 | 7 | 5 | — | — |
| *P. silvestris* | pre-dom. | pre-dom. | 92 | 93 | 62 | 78 | 73 | — | — |
| *Pinus* | 77 | 79 | — | — | 6 | — | — | 85 | 71 |
| *Betula* | 17 | 11 | 2 | 2 | 6 | 14 | 18 | 12 | 1 |
| *Alnus* | — | — | <1 | 1 | 3 | — | <1 | <1 | 1 |
| *Salix* | — | — | — | — | <1 | — | — | — | 2 |
| **POLLEN OF GRASSY PLANTS AND SHRUBS** | | | | | | | | | |
| Cyperaceae | — | 94 | 49 | 5* | 3* | — | — | — | — |
| Gramineae | — | — | 9 | 1* | — | — | 1* | — | — |
| Chenopodiaceae | — | 3 | 6 | — | — | — | — | — | — |
| Ericales | — | — | — | — | — | 2* | 6* | — | — |
| *Artemisia* | — | 3 | 2 | 1* | 3* | — | 6* | — | — |
| **MOTLEY GRASSES** | | | | | | | | | |
| Labiatae | — | — | — | — | 1* | — | — | — | — |
| *Polygonum* | — | — | 2 | — | — | — | — | — | — |
| Ranunculaceae | — | — | 1 | — | 1* | — | — | — | — |
| Caryophyllaceae | — | — | 2 | — | — | — | — | — | — |
| Leguminosae | — | — | 3 | — | — | — | — | — | — |
| Umbelliferae | — | — | 1 | — | 3* | — | — | — | — |
| Compositae | — | — | 10 | 4* | 4* | — | — | — | — |
| Unidentified dicotyledons | 5* | — | 15 | 2* | 2* | 2* | — | — | — |
| **SPORES** | | | | | | | | | |
| Bryales | 62 | 98 | 83 | 3* | 5 | 3 | 3 | — | — |
| *Sphagnum* | 18 | — | 4 | 1?* | <1 | 65 | 86 | — | — |
| *Lycopodium clavatum* | 2 | — | 3 | 2* | <1 | — | — | — | — |
| *L. cf. annotinum* | 9 | — | — | — | <1 | — | — | — | — |
| *Lycopodium* | — | — | — | — | — | 7 | — | — | — |
| Polypodiaceae | 9 | — | 9 | 1 | 93 | — | — | — | — |
| Filicales | — | 2 | 1 | — | — | 25 | 11 | — | — |
| *Equisetum* (?) | — | — | — | 1* | — | — | — | — | — |

*Numbers marked with an asterisk here and in the tables which follow indicate the number of pollen grains or spores found, not the percentage.

Sample no. 7 was taken from a swamp on the left slope of the Angara watershed near Padun village. The character of the vegetation is the same and the pollen and spore composition is nearly the same as in sample 6.

Sample no. 8 was taken from the Umykey swamp in the Oka river basin, where pine-larch and birch forests occur extensively, with spruce and Siberian pine in places. The pollen composition (according to M. N. Nikonov) is similar to the pollen-spore composition of samples 6 and 7.

Sample no. 9 was not taken from the Angara basin, but from the upper Angara valley (northeast of Lake Baykal), from the Uayan swamp on the second terrace above the floodplain (terrace II). The pollen composition of this sample (according to V. A. Povarnitsyn) differs from that of the surface samples of the Angara basin, in that it has a large quantity of spruce and fir pollen. Obviously the proximity of the large massifs of dark-coniferous taiga in the mountains to the northeast of Lake Baykal exerts an influence on the spectral composition of this basin.

A comparison of all spectra shows that the percentage relationship of the pollen of the individual tree species varies as the distribution of vegetation in the basin. However, the limits of the variation are negligible and the similarity is obvious. In all samples, pine pollen predominates; birch, spruce, or Siberian pine are second; and fir, alder, and willow are third. The pollen of tree species predominates in the general pollen composition of all samples. The similarity of the spore-pollen spectra is due to the averaging of the pollen and spore composition in forming the spectrum and particularly in forming the spectra in alluvium. Undoubtedly the ancient spore-pollen spectra were formed in the same manner. Therefore, it may be said that similar spectra formed under similar physicogeographic conditions and correspond to the same period of sediment accumulation.

Specialists in the field of spore-pollen analysis often point out that the ability of pollen to spread by air and water is a negative factor. In our opinion, this is an important positive factor for the forest and wooded-steppe zones, because it permits spore-pollen analysis to be used for stratigraphic purposes. With the considerable averaging of the spore-pollen composition in river alluvium, only the largest and most universal changes in the vegetation can be established, and the small, local changes smooth out.

## II

Few data are available on the Tertiary flora of the Angara region and even fewer on the Pleistocene flora. The works of I. V. Palibin (1936) and A. N. Krishtofovich (1929, 1945) give some idea of the Oligocene-Miocene flora of the southern and southwestern Baykal region. Broadleaf forests of the Turgay type, consisting of hornbeam, linden, elm, and birch with an admixture of coniferous trees, predominated in the Baykal region at the end of the Oligocene and the beginning of the Miocene. In the Miocene, according to I. M. Pokrovskaya's data (see S. V. Obruchev, 1946), coniferous-broadleaf forests with a preponderance of coniferous trees were widely distributed in the Baykal region. The conifers included *Keteleeria, Tsuga,* and *Taxodium*; the deciduous included *Magnolia, Fagus, Carya, Nyssa, Carpinus, Quercus, Tilia, et al.* Ye. D. Zaklinskaya (1950) studied the younger, Late Pliocene flora of the Barguzin valley near Alga (profile no. 1, Table 2, samples nos. 34 and 35). Coniferous forests with an admixture of deciduous trees, differing from the present forests in their floristic composition,

TABLE 2

Result of Spore-Pollen Analysis of Ancient Lake Deposits and the
Loams Covering Them at Alga in the Barguzin Valley, Section 1

| | Sample no.: 35 | 34 | 42 |
|---|---|---|---|
| Place samples collected: | Base of ancient lake stratum | | Diluvial loams |
| Number of grains counted: | 634 | 1044 | 42 |

### SPORE AND POLLEN COMPOSITION (%)

GENERAL COMPOSITION

| | | | |
|---|---|---|---|
| Tree pollen | 98 | 89 | 16* |
| Grass and shrub pollen | 0.3 | 0.5 | 25* |
| Spores | 1.7 | 10.5 | 2* |

POLLEN OF TREE SPECIES

| | | | |
|---|---|---|---|
| Abies | 6 | 3 | — |
| Picea | 9 | 11.5 | 2 |
| Pinus, section Cembra (a) | 11 | 10 | 10* |
| Pinus, section Cembra (b) | 2 | 5 | — |
| Pinus, section Strobus | 2 | 2 | — |
| Pinus subspecies Haploxylon | 12 | 8 | — |
| Pinus section Taeda (?) | — | 0.5 | — |
| Pinus section Eupitis | 4.5 | 3 | — |
| Tsuga | 8 | 0.5 | — |
| Cupressaceae | 0.5 | — | — |
| Lauraceae | — | <0.1 | — |
| Podocarpi typ. (Sacl.) | 3 | 0.1 | 2* |
| Alnus (two types) | 41 | 51 | — |
| Betula | 1 | 4 | 2* |
| Corylus | — | 0.1 | — |
| Carpinus | — | 0.1 | — |
| Pterocarya | — | 0.5 | — |
| Juglans | — | 0.1 | — |
| Salix | — | — | 4 |

POLLEN OF GRASSY PLANTS AND SHRUBS

| | | | |
|---|---|---|---|
| Myriophyllum | 2* | 4* | — |
| Potamogetonaceae | — | 2* | — |
| Gramineae | — | — | 2* |
| Typha | 1* | 1* | — |
| Artemisia | — | — | 11* |
| Ericaceae | — | 18* | — |
| Cyperaceae | — | 1* | — |
| Violaceae | — | — | 2* |
| Compositae | — | 2* | — |
| Cruciferae | — | 6* | 10* |
| Leguminosae | — | 2* | — |
| Unidentified | — | 14* | — |

SPORES

| | | | |
|---|---|---|---|
| Lycopodiaceae | — | — | 2* |
| Polypodiaceae | 12 | 87* | — |
| Bryales | — | 11* | — |
| Unidentified | — | 3* | — |

were widely distributed here. In addition to the tree species which grow here at present, these forests contained *Tsuga*, some species of *Pinus*, *Cupressaceae*, and the deciduous species *Pterocarya*, *Ulmus*, *Carpinus*, *Corylus*, and *Juglans*.

The flora identified in our laboratory on the basis of spore-pollen analysis of deposits taken from several sections in the Eastern Sayan mountains belong to a later period. Some of the results of an analysis performed by V. A. Stupishina and N. S. Sokolova have already been published in part in an article by Ye. M. Shcherbakova (1954).

Section [profile] 2 is situated on the left bank of the Bystraya river (Bystraya depression). According to Shcherbakova's data, cross-bedded, cemented sands, pebble gravels, and cobbles appeared in the outcropping of the high compound terrace near the river mouth. The spore-pollen analysis of the samples (Table 3, Figs. 1 and 2) shows that the deposits had formed under quite similar forest conditions. The flora may be classified as much depleted Tertiary. Evidently the depletion of the flora continued because of the gradual cooling of the climate, which may be judged by the disappearance of the species most requiring heat and moisture: *Carpinus*, *Fagus*, *Juglans*, and *Pterocarya*. The *Tilia*, *Ulmus*, *Carya*(?), *Ilex*, and *Tsuga* flora remained. The composition of the tree pollen in this sample is similar to that of the upper sample of section 1 at Alga (see Table 2) and is poorer than that of the lower samples. Therefore, the deposits of the Bystraya depression may undoubtedly be considered later than the Tertiary deposits at Alga. The diagram shows the quantitative relationships between the various pollen groups that are characteristic of the spore-pollen spectra of the forest type. Among the tree species, birch plays the largest role; in places there were pine forests of *Pinus silvestris* and *P. sibirica* with small admixtures of exotic pines, fir, and spruce. Pine forests occurred widely during the period of accumulation of the sands that now lie 132–133 m above the waterline. Let us reiterate that larch pollen, unfortunately, is seldom preserved in the deposits and therefore the extent of its participation in the Baykal forests remains obscure. Larch seeds were found in somewhat later deposits, which allows us to assume that the larch was present here. The spruce-fir taiga was no more extensive then than it is now in the Iya basin. Finds of *Selaginella selaginoides*, *Lycopodium*, and *Sphagnum* spores and a considerable amount of *Ericales* pollen indicate that taiga existed at that time. The constant presence of alder pollen is noteworthy, especially in samples from the upper part of the section, and this is also characteristic of section 1 at Alga. Such large quantities of alder pollen rarely appear in present-day spore-pollen spectra.

In contrast to the older deposits, where the pollen of the grassy plant group comprised 1–2 per cent of the total pollen, here it comprises 16–17 per cent. This may be connected with the development of open expanses in the mountains, especially since much *Artemisia* (up to 30%) appears in the grass pollen, in addition to motley grasses, Chenopodiaceae, and a considerable quantity of ephedra.

Remnants of similar flora were identified by V. A. Stupishina in the sandy-pebble gravel deposits comprising (according to Ye. M. Shcherbakova) the compound terrace of the Zamaraikha river (section no. 3, Table 3). The samples contained little pollen, but pollen of various types of pine (up to 98%), birch, fir, spruce, and also hemlock and linden were found in the sandy-pebble gravel deposits at depths of 12 and 69 m.

Ye. M. Shcherbakova (1954) analyzed two sections of the 100-m terrace of the Tory depression. This terrace is composed of cross-bedded sands. Figure 2 and

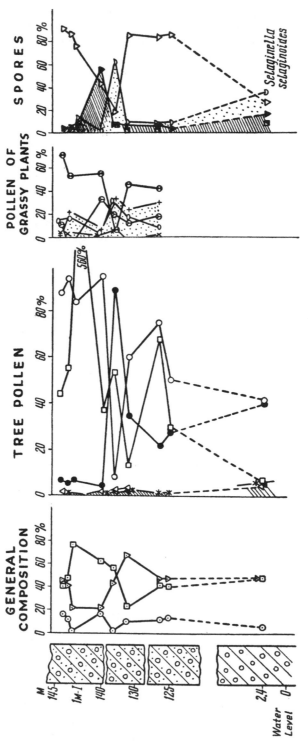

FIGURE 1. Spore-pollen diagram of section no. 2 from the mouth of the Bystraya river (Bystraya valley). For legend, see Figure 2.

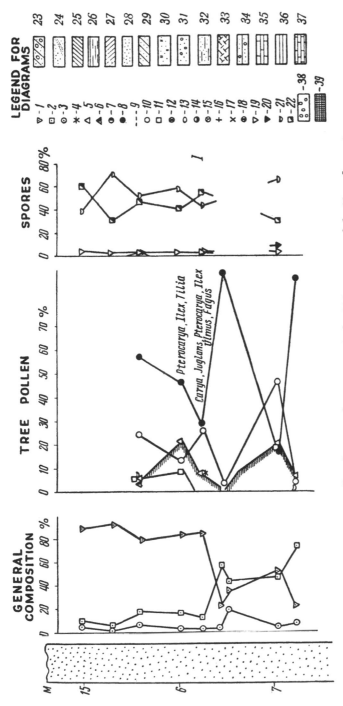

FIGURE 2. Spore-pollen diagram of section no. 4; the 100-m terrace of the Tory depression.

### LEGEND FOR ALL DIAGRAMS

1, total spores; 2, total tree pollen; 3, total grass and shrub pollen; 4, *Tsuga*; 5, *Picea*; 6, *Abies*; 7, *Pinus sibirica*; 8, common pine; 9, pine (exotic species); 10, *Betula*; 11, *Alnus*; 12, *Salix*; 13, Cyperaceae; 14, Gramineae; 15, Ericaceae; 16, *Artemisia*; 17, Chenopodiaceae; 18, motley grasses; 19, green mosses; 20, sphagnums; 21, Pteridophyta; 22, Lycopodiaceae; 23, cross-bedded sands with pebbles; 24, cross-bedded sands; 25, light loam; 26, medium loam; 27, heavy loam; 28, sand; 29, sandy loam; 30, peat; 31, sand with pebbles; 32, clayey sand; 33, tuff; 34, sandy loam with pebbles; 35, sandstone; 36, clay; 37, limestone; 38, pebble gravel; 39, sapropel.

This same 100-m terrace near Yelovka is dissected by ravines, forming isolated hills 70–80 m high. Shcherbakova took samples from the slope of one of these hills in connection with the discovery there of an antelope skull, which V. I. Gromov identified as *Spiroceros kiaktensis*. The following pollen and spores were found 15–20 m below the surface, within an interval of 1.5 m (section no. 5): *Picea*–7, *Pinus*–8, *Tsuga*–2, Gramineae–1, Chenopodiaceae–1, Bryales–1, and Polypodiaceae–2 grains.

The position of the layer in which the antelope skull was found led Shcherbakova to conclude that these deposits sloped towards the 100-m terrace, i.e., that they were later than the deposits of the 100-m terrace. From this we assume that the role of mountain steppes increased significantly as the thermophilic and hygrophilic species disappeared from the Baykal flora. Evidently the deposits adjacent to the 100-m terrace are similar in age to the deposits which comprise the high terrace of the Bystraya depression; this is indicated not only by the nature of the spectra of mixed forests with an admixture of exotic relictal species, but also by the preserved traces of mountain steppe. The deposits of the 100-m Tory terrace accumulated during the preceding period, which had a milder climate, as indicated by the existence of dark-coniferous taiga and the more frequent finds of the pollen of broadleaf species, including *Fagus*, *Pterocarya*, and *Juglans*.

Thus, during the period when the depressions of the Baykal region were being filled in, vegetation existed which differed from the Neogene vegetation in that the most thermophilic species of the Arcto-Tertiary flora gradually disappeared and Pleistocene flora began to develop.

The period of relatively abrupt change of flora is merely an expression of the general changes which took place in nature and particularly in the climate, which evidently coincided with the period of intensive uplifting of the Eastern Sayan (S. V. Obruchev, 1946; Dumitrashko, 1952; Shcherbakova, 1954, *et al.*). Arbitrarily we have called this phase of development of the Baykal vegetation Phase I, a phase of transition between the Neogene and Quaternary. We say "arbitrarily," because no general agreement has been reached on what constitutes the dividing line between the Tertiary and the Quaternary periods.

Two subphases of the initial phase of development of Quaternary vegetation can be noted: (1) coniferous forests, chiefly dark-coniferous with an admixture of broadleaf species, and (2) birch and pine-birch forests with spruce (possibly alder, as well) and with a small number of broadleaf species; there were open expanses, evidently mountain steppes. During this period the flora became sharply depleted, the role of small-leaf species and pines in the forests increased, and steppes appeared in connection with the increased continentality of the Baykal climate (xerophytation with some cooling). However, this division into subphases requires further verification.

We did not find deposits containing spore-pollen spectra of this type in the middle and northern portions of the Angara basin. However, few samples of the ancient terraces were taken for analysis.

Perhaps the deposits of this period have been preserved more or less in their entirety only in the depressions. It may be assumed that the vegetation north of the Sayan differed somewhat from that of the mountain regions, but the two should have been similar floristically because probably there were no barriers between the Angara region and the northern slopes and foothills of the Sayan mountains.

Younger deposits have been found in some regions; one such is section no. 6 near Maksimovshchina. S. S. Voskresenskiy took samples from a borehole in the

Irkut basin on the left slope of the interfluve plateau. The borehole passed through an 18-m stratum of interbedded light and heavy loams and the upper part of a stratum of fine-grained sands underlying the loams. Analyses showed that the loams apparently are of lacustrine origin, since they contained pollen of the hygrophytes and hydrophytes (*Sparganium, Typha, Potamogeton,* and Hydrocharitaceae) and algae (two species of *Pediastrum*). The spore-pollen diagram in Figure 3 shows that the basin existed under forest conditions. This section

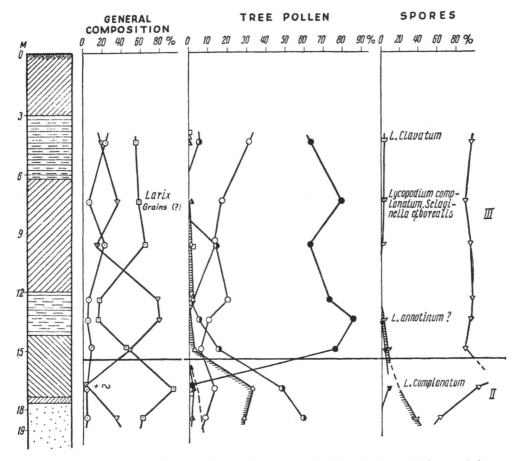

FIGURE 3. Spore-pollen diagram of the deposits in the Irkut basin at Maksimovshchina (section no. 6).

may be divided into two parts on the basis of the pollen composition. The lower part (sands and the lower part of the loams to a depth of 16 m) corresponds to the maximum of spruce and Siberian-pine pollen on the diagram. This provides a basis for assuming that continuous dark-coniferous taiga of Siberian pine, spruce, and fir predominated in the Sayan foothills during the formation of this horizon.

A comparison of the spore-pollen spectra of the lower part of the section with modern spore-pollen spectra of the forest zone shows that the spruce was more widely distributed in the upper Angara basin than it is at present in the Angara region or in the Ob basin. For example, the spruce pollen in the old lacustrine and alluvial deposits (sections 6 and 7) amounted to 33 per cent and 90 per cent, while in the modern floodplain and swamp deposits of the Angara basin, it comprises only 1–11 per cent. In the modern floodplain deposits of the Ob river, spruce comprises 1–5 per cent of the tree pollen (at Lokosovo, 2%; Surgut, 1%; Samarovo, 2%; Malyy Altym, 5%).

The presence of birch pollen in the samples indicates that birch groves existed in conjunction with the taiga, apparently occupying areas less favorable for taiga.

The negligible grass pollen content indicates that the grass cover was weakly developed beneath the canopy of dark-coniferous forests. The plant cover consisted essentially of green mosses and ferns. The small amounts of ephedra and wormwood pollen found here allows us to assume that these plants were also present, probably along the steep banks and talus of the basins.

The discovery of *Pinus* pollen, section *Strobus*, in the samples merits special attention. In sample no. 1, this pollen comprised as much as 6 per cent of the total tree pollen (see Table 5). The *Pinus*, section *Strobus*, which is not found at present in the Baykal flora, was preserved from Tertiary times and possibly was more widely distributed in eastern Siberia during the Early Quaternary. The discovery of *Pinus*, section *Strobus*, links the lacustrine deposits of Maksimovshchina with the deposits of the Tory and Bystraya depressions; however, no broadleaf pollen could be found in the Maksimovshchina register. By this time, the broadleaf forests had been completely crowded out by taiga.

Similar spore-pollen spectra were obtained from sections of strata comprising interfluves at several points in the Angara region. It is natural to assume that the period of broad distribution of dark-coniferous taiga was an independent stage of development of the Angara vegetation, viz., Phase II.

The conditions characteristic of dark-coniferous taiga have been described comprehensively and in detail by A. I. Tolmachev (1954). The wide distribution of taiga in the Angara region in the Early Quaternary together with its complete dislodgment of the mixed and broadleaf forests and its partial displacement of the steppes could have taken place only under optimum natural conditions for taiga. According to Tolmachev, these conditions are: (1) a moderately warm summer with average temperatures of the warmest month from $+13°$ to $+20°$ C and with moderately cold winters; (2) sufficient moisture, expressed by a generally positive moisture balance for the region, abundant summer precipitation, and high atmospheric humidity, especially in the summer months; (3) thermal differentiation of the seasons; (4) presence of and long retention of a thick snow cover during the cold season.

Thus, the formation of the dark-coniferous taiga and the extinction of the Tertiary flora helped reduce the summer and winter temperatures and increase the humidity. The Baykal climate of that time was milder than at present.

Examination of the borehole diagram (section no. 6) reveals that the pollen and spore composition changes greatly at the depth of the loam horizon. The curves of spruce and Siberian pine content decrease to a minimum, common pine is maximum, the amount of birch pollen increases, and larch pollen (?)[*] is present in sample no. 9. Spore diversity decreases noticeably owing to the disappearance of sphagnum mosses and the reduction of ferns; green mosses prevail. The amount of pollen of the grassy-plant group gradually increases, reaching 23 per cent in the upper part of the section. Pollen of plants associated primarily

---

[*][Here and elsewhere in the text, the question mark in parentheses is the author's notation indicating the supposed existence of larch, although the pollen has not been preserved.—Translator.]

with steppe localities, e.g., *Artemisia* (up to 52%), species of the family Cheno-podiaceae (9%), and *Ephedra* become noticeable in this group.

This change in the spore-pollen spectrum reflects the degradation of the dark-coniferous taiga and the thinning-out of the forests. The dark-coniferous taiga was gradually replaced by light-coniferous pine-larch (?) taiga, and birch was quite widely distributed. The change in the amount and composition of the grass pollens emphasizes the gradually increasing role of open landscapes of the meadow-steppe type and the lightening of the color of the forest, beneath whose canopy spring- and summer-blooming species of the families Umbelliferae, Primulaceae, Compositae, Caryophyllaceae, etc. could grow. However, the grassy cover here was incomparably less variegated than that which existed during Phase I and to the beginning of Phase II (see Table 5).

The spectra of the lower part of the section correspond to milder climatic conditions than those which exist at present. The spectra of the upper part of the section are similar to the present spore-pollen spectra (see Table 1), differing from the modern spectra in that they have somewhat more pollen of grassy plants. Apparently, the steppes then were more widely distributed than now. The climatic conditions were similar to those of the present, but were somewhat more continental or dry.

Thus, the loams formed under different natural conditions during the spread of the light-coniferous pine and pine-larch (?) taiga containing birch. This period may be designated Phase III in the development of the Angara vegetation.

Section no. 7 typifies deposits similar in age to the deposits of the Maksi-movshchina section; they were discovered by L. Polkanova in the Bystraya depression. Deep shafts cut into the watershed of the Bystraya river and Lake Baykal revealed the following deposits to a depth of 13.35 m: heavy, dark-gray loams; light, silty, gray loams; loams with interlayers of peat, greenish gray; sandy loam with interlayers of peat.

In small samples (about 100 g) of loams with interlayers of peat taken for spore-pollen analysis, the following were also found: *Picea*, 2 seeds; *Picea* (?), 1 seed; *Larix sibirica* Ledeb., 6 seeds; *Larix* (?), 2 seeds; Coniferae, 4 seeds; *Comarum palustre* L., 58 specimens of fruit; *Menyathes trifoliata* L., 17 seeds; *Carex*, many nuts. Large numbers of *Picea* needles were found (Fig. 4). At present, almost all the species enumerated are closely associated with taiga and with swamps in the taiga zone.

Figure 5 and Table 6 illustrate the results of the spore-pollen analyses. Complex spectra, including vegetation of dark-coniferous spruce-fir taiga with larch and of the richly variegated steppe grasses, appear in the lower part of the diagram. The curves of tree species are similar to those in the diagram of the Maksimovshchina section. Here, too, the spruce curve reaches a maximum higher up the section the Siberian pine becomes the maximum, and still higher the birch, which corresponds to Phase II of vegetation development exemplified by the Maksimovshchina section. In contrast to the Maksimovshchina section, however, a lower, earlier part of Phase II can be detected here, namely, the spruce maximum. This is extremely important, because the pollen of *Pinus*, section *Strobus*, *Quercus* cf. *dentata*, *Quercus* sp., *Corylus*, *Ulmus*, and also *Osmunda* spores, species which do not grow at present in the Baykal region, is found in the lower part of the section in the peat samples. Pollen of these species is not found higher up the section. The *Pinus*, section *Strobus*, was the last of these species to disappear. The proposition that the pollen of these species was transported here during the formation of the deposits can be dismissed, because we are dealing

## TABLE 5

### RESULT OF SPORE-POLLEN ANALYSIS OF SAMPLES FROM SECTION NO. 6 IN THE IRKUT BASIN AT MAKSIMOVSHCHINA

| Sample no.: | 1 | 2 | 3 | 4 | 5 | 6 | 8 | 9 | 11 |
|---|---|---|---|---|---|---|---|---|---|
| Sampling depth in m: | 19.20 | 16.7–17.4 | 15.0–16.7 | 14.5–15.0 | 14.5 | 12.1–14.5 | 8.5–11.0 | 6.3–8.5 | 3.1–6.0 |
| Number of grains counted: | 415 | 201+∞ | 8 | 180 | 383 | 255 | 258 | 72 | 408 |
| **SPORE AND POLLEN COMPOSITION (%)** | | | | | | | | | |
| **GENERAL COMPOSITION** | | | | | | | | | |
| Tree pollen | 62 | 93 | 4* | 45 | 16 | 17 | 64 | 58 | 57 |
| Grassy-plant and shrub pollen | 4 | 4 | 1* | 9 | 5 | 6 | 22 | 6 | 28 |
| Spores | 34 | 3 | 4* | 46 | 79 | 77 | 14 | 36 | 20 |
| Hydrophyte pollen | <1 | <1 | — | — | — | — | <1 | — | — |
| **POLLEN OF TREE SPECIES** | | | | | | | | | |
| *Larix* | — | — | — | — | — | — | — | 2 | — |
| *Abies* | 1 | 2 | — | — | — | 2 | — | 2 | <1 |
| *Picea* | 28 | 33 | 1* | 3 | — | 2 | 1 | — | <1 |
| *Pinus sibirica* | 59 | 48 | — | 15 | 5 | 2 | 14 | — | 5 |
| *P. silvestris* | <1 | 4 | 3* | 76 | 85 | 73 | 63 | 79 | 63 |
| *Pinus* (exotic species) | 6 | — | — | — | — | — | — | — | — |
| *Betula* | 8 | 13 | — | 6 | 10 | 20 | 13 | 17 | 31 |
| *Alnus* | — | <1 | — | — | — | 2 | 2 | — | 1 |
| **POLLEN OF GRASSY PLANTS AND SHRUBS** | | | | | | | | | |
| Gramineae | — | — | — | 3* | 2* | 3* | 9 | — | 24 |
| Cyperaceae | 3* | 3* | — | — | — | 2* | 4 | — | 2 |
| *Artemisia* | 12* | 4* | 1* | 9* | 6* | 2* | 52 (2 forms) | 4* | 30 |
| Chenopodiaceae | 1* | — | — | 2* | — | 3* | 9 (3 forms) | — | — |
| Ericales | — | 1* | — | — | — | — | — | — | — |
| *Ephedra* | — | 2* | — | — | — | 1* | — | — | — |

| | 1 | 2 | 3 | 4 | 5 | 6 | 7 | 8 |
|---|---|---|---|---|---|---|---|---|
| **MOTLEY GRASSES** | | | | | | | | |
| Borraginaceae | — | — | — | — | — | — | — | — |
| Compositae | 21 (3 sp.) | 10 (2 forms) | — | 2* (2 forms) | 1* | — | — | — |
| Caryophyllaceae | <1 | 5 | — | 2* | 1* | — | — | — |
| Primulaceae | — | 2 | — | 1* | — | — | — | — |
| Polygonum | 3 (2 sp.) | 4 (2 sp.) | — | — | — | — | — | — |
| Umbelliferae | 1 | — | — | 1* | — | — | — | — |
| Unidentified dicotyledons | 13 (10 forms) | 5 (3 forms) | 3* (3 sp.) | 7* (5 forms) | 1* | — | 9* (5 forms) | 10* (3 forms) |
| **SPORES** | | | | | | | | |
| Bryales | 95 | 94 | 96 | 95 | 89 | 4* | 1* | 63 |
| Sphagnales | — | — | — | <1 | 4 | — | 14* | 36 |
| Polypodiaceae | 4 | 3 | — | <1 | 4 | — | 1* | 1 |
| Other Filicales | 1 | 3 | 4 | 4 | 2 | — | — | — |
| Selaginella cf. borealis | — | — | — | — | — | — | — | — |
| Lycopodium | — | — | — | 1 | 1 | — | — | — |
| L. annotinum ? | — | — | — | <1 | — | — | — | — |
| L. clavatum | <1 | — | — | — | — | — | 1* | — |
| L. complanatum | — | — | — | — | — | — | 1* | — |
| Unidentified | — | — | — | — | — | — | — | — |
| **HYDROPHYTE POLLEN** | | | | | | | | |
| Sparganium | — | — | — | — | — | — | 1* | — |
| Potamogeton | — | 1 | — | — | — | — | — | — |
| Typha | — | — | — | — | — | — | 1* | 1* |
| Hydrocharitaceae | — | — | — | — | — | — | — | 1* |
| **ALGAE** | | | | | | | | |
| Pediastrum | 2 (2 sp.) | — | — | — | — | — | — | — |

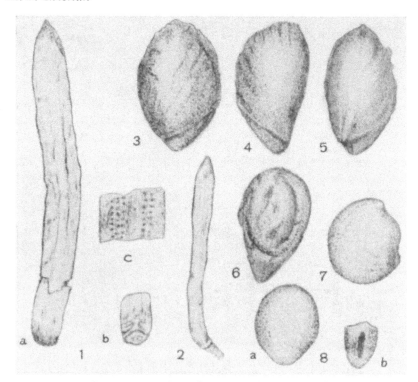

FIGURE 4. Plant remnants from deposits of the Bystraya depression (section no. 7): 1, *Picea* needle—*a*, general view; size 11.0 × 1 mm; *b*, leaf scar; *c*, arrangement of the stomata; 2, rudimentary spruce needle, size 6.5 × 0.7 mm, with leafstock; 3–5, seeds of the *Larix sibirica* Ledeb., 4.6 × 2.5 (3.0) mm; 6, *Picea* seed, 4.0 × 2.4 mm; 7, *Comarum palustre* L. seed, 1.8 × 1.2 mm; 8, *Menyanthes trifoliata* L. seed, 2.8 × 2.2 mm.

TABLE 6

RESULT OF SPORE-POLLEN ANALYSIS OF SAMPLES OF SECTION 7 OF THE
WATERSHED DEPOSITS OF THE BYSTRAYA RIVER

| Pit and sample no.: | p. 6–6 | p. 7–3 | p. 12–9 | p. 12–9a | p. 7–4 |
|---|---|---|---|---|---|
| Sampling depth in m: | 2.25–3.75 | 3.9–4.0 | 6.75–7.0 | 9.0–10.0 | 12.25–13.5 |
| Number of grains counted: | 264 | 108 | 333 | 320 | 736 |

SPORE AND POLLEN COMPOSITION (%)

| | | | | | |
|---|---|---|---|---|---|
| GENERAL COMPOSITION | | | | | |
| Tree pollen | 46 | 36 | 36 | 60 | 41 |
| Grass and shrub pollen | 27 | 3 | 14 | 31 | 29 |
| Spores | 27 | 61 | 50 | 9 | 30 |
| Hydrophyte pollen | <1 | — | <1 | <1 | — |
| POLLEN OF TREE SPECIES | | | | | |
| *Abies* | — | 5 | — | — | 14 |
| *Larix* | — | — | — | 2 | <1 |
| *Picea* | 21 | 18 | 90 | 89 | 71 |
| *Pinus silvestris* | 2 | — | — | — | — |
| *P. sibirica* | 2 | 77 | <1 | 2 | 5 |
| *Pinus* section *Strobus* | — | — | — | + | + |
| *Pinus* sp. | — | — | <1 | 4 | 1 |

TABLE 6 (*concluded*)

| Pit and sample no.:<br>Sampling depth in m:<br>Number of grains counted: | p. 6–6<br>2.25–<br>3.75<br>264 | p. 7–3<br>3.9–4.0<br>108 | p. 12–9<br>6.75–7.0<br>333 | p. 12–9a<br>9.0–10.0<br>320 | p. 7–4<br>12.25–<br>13.5<br>736 |
|---|---|---|---|---|---|
| *Betula* | 62 | — | 8 | 3 | 6 |
| *Alnus* | 6 | — | <1 | — | — |
| *Corylus* (?) | — | — | — | — | 2 |
| *Quercus* cf. *dentata* | — | — | — | — | 3 |
| *Quercus* sp. | — | — | — | — | <1 |
| *Ulmus* | — | — | — | — | <1 |
| *Salix* | 7 | — | 2<br>(2 sp.) | — | <1 |
| GRASS AND SHRUB POLLEN | | | | | |
| *Ephedra* (Leptocladae) | — | — | — | — | 1 |
| Gramineae | — | — | — | — | 1 |
| Cyperaceae | 61 | 1* | 29 | 84 | 45 |
| MOTLEY GRASSES | | | | | |
| Liliaceae | — | — | — | — | + |
| Ranunculaceae | — | — | — | — | + |
| *Thalictrum* | — | — | 1 | 1<br>(2 sp.) | 1 |
| *Polygonum* | — | — | — | — | 3 |
| Chenopodiaceae | 4 | — | 2<br>(2 sp.) | — | 3 |
| Caryophyllaceae | 1 | — | 1 | <1 | 2 |
| Rubiaceae | — | — | — | — | 1 |
| Dipsacaceae | — | — | — | — | 1 |
| Cruciferae | 1 | — | — | — | 8 |
| Umbelliferae | 1 | — | — | — | + |
| *Polemonium* | — | — | — | — | 1 |
| Ericales | — | — | — | — | 3 |
| Gramineae | — | — | 1 | — | — |
| Type Primulaceae | — | — | — | — | 1 |
| Compositae | 1 | 1* | 2 | — | 1 |
| *Artemisia* | 6 | — | 29<br>(4 sp.) | 9<br>(4 sp.) | 7 |
| Leguminosae | — | — | — | — | + |
| Unidentified dicotyledons | 23<br>(9 sp.) | 1* | 34<br>(12 sp.) | 5 | 20 |
| SPORES | | | | | |
| Bryales | 89 | 57 | 5 | 24 | 23 |
| *Sphagnum* | 3 | 2 | 93 | 76 | 60 |
| *Equisetum* | — | — | — | — | <1 |
| *Lycopodium clavatum* | — | — | — | — | +⎫ |
| *L. selago* | — | — | — | — | +⎬ 3 |
| *L. pungens* | — | — | — | — | +⎭ |
| *Lycopodium* | — | — | <1 | — | — |
| *Selaginella selaginoides* | — | — | — | — | +⎫ |
| *S.* cf. *sibirica* | — | — | — | — | +⎬ 3 |
| Polypodiaceae | 8 | 39 | <1 | — | +⎫ |
| *Athyrium rubripes* typ. | — | — | — | — | +⎬11 |
| *Osmunda* | — | — | — | — | +⎭ |
| Other Filicales | — | + | <1 | — | + |
| HYDROPHYTE POLLEN | | | | | |
| Sparganiaceae | 1* | — | 1* | — | — |
| Typhaceae | — | — | — | 1* | 1* |
| *Myriophyllum* | — | — | 1* | — | — |

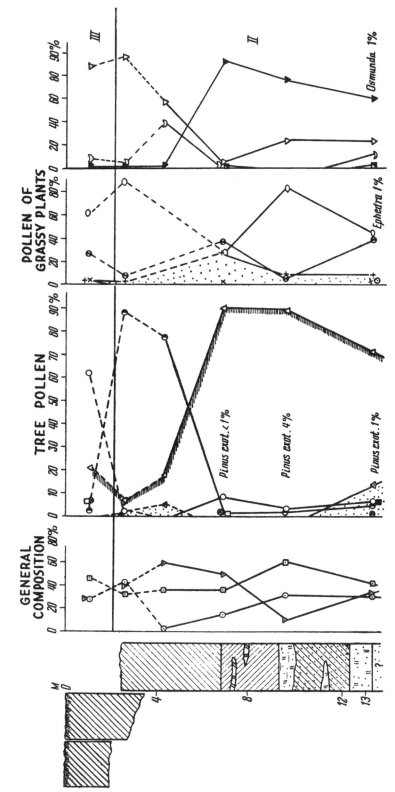

FIGURE 5. Spore-pollen diagram of ancient lake-swamp deposits (section no. 7). Watershed of the Bystraya river and Lake Baykal.

with ancient swamp deposits; furthermore, the pollen concentration of the indicated is so great that the idea of transportation is unacceptable. It is natural to
assume that these species are relics of the Tertiary flora. It may be assumed that
they played an even greater role in the Baykal vegetation during the period
immediately preceding the time of deposition. In the upper part of the section the
percentage of spruce in the spore-pollen spectra decreases conspicuously, the
quantity of fir pollen also decreases, and the amount of Siberian pine pollen
decreases, which probably should be associated with further cooling and drying
of the climate. The same changes were observed in the Maksimovshchina section.

The changes in the composition of the tree pollen agree with the changes in
the spore composition. The spruce maximum corresponds to the sphagnum spore
maximum. Up along the section, as the taiga becomes lighter-hued, the spore
composition also changes. The spore content of sphagnum and club mosses
decreases and the green moss spores comprise the maximum.

The amount of grass pollen decreased as the taiga expanded, but increased
again as the dark-coniferous taiga degraded and the forest color became lighter.
This group is very diversified in composition. The species saturation is especially
great in the lower part of the section. Among the 213 pollen grains of grassy
plants counted here, there were at least 40 species, not counting the diverse
Gramineae and Cyperaceae. We did not find such a species saturation in the
spectra of the younger Quaternary deposits or in the modern floodplain deposits
of the Angara. The species saturation decreases upward along the section, which
also emphasizes the depletion of flora in connection with the deterioration of the
climate.

Thus, this section associates the flora of the transitional Phase I more closely
with Phase II of dark-coniferous taiga forests and reflects the gradual evolution
of the vegetation, associated with the cooling of the climate and the increased
humidity.

Section no. 8, which, on the basis of spore-pollen spectra, may be regarded as
belonging to Phases I–III, was studied on the basis of a borehole (no. 305) sunk
into the slope of the interfluve plateau of the Kaya and Olkha rivers in the Irkut
basin. Heavy and light loams were found to a depth of 14.4 m. Only four samples
were taken for pollen analysis. The diagrams (Figs. 6 and 7) constructed on the
basis of this analysis show considerable changes in the vegetation during the
period in which this loam stratum formed. The stratum can be divided into three
sections, according to pollen composition: the 14.4–6(7)-m level, where birch
predominates; the 6(7)–2(3)-m level, where conifers with a spruce maximum
predominate; and the 2(3)–0-m level, where birch is again maximum. This pollen
composition reflects the wide range of birch forests containing pine, spruce, and
exotic pines; later the birch forests were replaced by dark-coniferous taiga (predominant) and then the dark-coniferous taiga became degraded, the exotic pines
disappeared, and birch forests became dominant. The fellow travelers of the
taiga, the *Lycopodium clavatum*, *L.* cf. *annotinum*, *L. inundatum*, and *Sphagnum* resettled as the taiga spread, increasing the role of this group of spore vegetation in the grass cover (Table 7). The *Pinus* pollen, section *Strobus*, comprised
no more than 5 per cent of the total tree pollen. The presence of pollen of this
species and the spruce pollen maximum allow us to consider the deposits of the
middle part of the section to be Older Quaternary and to regard them as belonging to Phase II. The upper section evidently corresponds to the beginning of
Phase III, and the lower part of the section to the end of Phase I. The participation of xerophyte pollen is also characteristic in the spectra of the period of taiga
propagation and the preceding period.

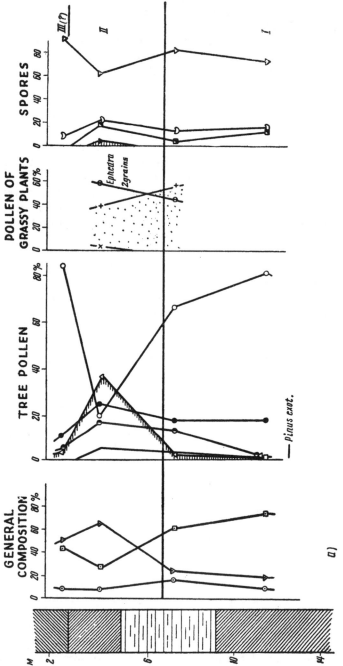

FIGURE 3. Spore-pollen diagram of section no. 8; Quaternary deposits of the Kaya river basin.

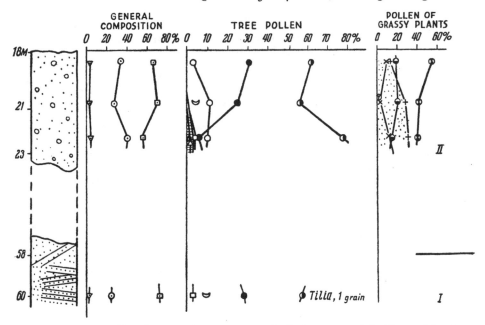

FIGURE 7. Spore-pollen diagram of section no. 9; the high terrace of the Zamaraikha river.

Section no. 9, representing the 120-m terrace of the right bank of the Zamaraikha river, has the following stratification, according to Ye. M. Shcherbakova's description:

| | | |
|---|---|---|
| 1. Loam | 0 | – 0.15 m |
| 2. Sandy-pebble gravel deposits | 0.15– | 5.0 |
| 3. Cross-bedded sands | 5.0 – | 9.8 |
| 4. Cemented sandstone with fine fragments of basalt | 9.8 | –12 |
| 5. Cross-bedded sands | 12 | –17 |
| 6. Sandy-pebble gravel deposits | 17 | –24 |
| 7. Cross-bedded sands | 24 | –60 |

The samples from layer 6 and the base of layer 7 (Fig. 7, Table 8) confirmed this analysis.

The time of formation of the sandy-pebble gravel horizon (layer 6) was similar to that of the deposits of the end of Phase II, i.e., the time of development of dark-coniferous forests. Here, evidently, at the end of the phase during which spruce and fir disappeared, the forests consisted primarily of Siberian pine and pine. Steppe areas also existed. Beginning at a depth of 60 m (sample no. 5), the lower, seventh horizon of this section is similar (according to the pollen analysis) to samples from the section of sandy-pebble gravel deposits of the high terrace of the Zamaraikha river (Table 3), where the pollen of the tree species consists basically of pine, while hemlock and linden comprise the exotic species. The tree pollen of sample no. 5 also consists basically of pine and Siberian pine, and linden pollen is found. This allows us to assume that the cross-bedded sands at a depth of 60 m in the section of the 120-m Zamaraikha river

## TABLE 7
### Result of Spore-Pollen Analysis of Samples from a Borehole in the Kama River Basin, Section No. 8

| | 1 | 2 | 3 | 4 |
|---|---|---|---|---|
| Sample no.: | 1 | 2 | 3 | 4 |
| Sampling depth in m: | 0.20–2.90 | 2.90–5.20 | 5.8–9.3 | 9.3–14.4 |
| Number of grains counted: | 100 | 412 | 194 | 172 |
| **Spore and Pollen Composition (%)** | | | | |
| GENERAL COMPOSITION | | | | |
| Tree pollen | 42 | 27 | 60 | 74 |
| Grass and shrub pollen | 8 | 8 | 16 | 8 |
| Spores | 50 | 65 | 24 | 18 |
| POLLEN OF TREE SPECIES | | | | |
| Picea | 2 | 36 | 2 | <1 |
| Pinus silvestris | 10 | 24 | — | — |
| Pinus subsp. Diploxylon | — | — | 17 | 17 |
| P. sibirica | 5 | 16 | 13 | 2 |
| Pinus sp. | — | 5 | 3 | + |
| Betula | 83 | 19 | 65 | 80 |
| Alnus | — | — | — | 1 |
| GRASS AND TREE POLLEN | | | | |
| Ephedra | — | 5 | — | — |
| Artemisia | 4* (2 sp.) | 39 | 56 (2 sp.) | 6* |
| Chenopodiaceae | 1* | 3 | — | — |
| MOTLEY GRASSES | | | | |
| Polygonaceae | — | 6 | 6 | — |
| Caryophyllaceae | — | 3 | — | — |
| Ranunculaceae | — | — | 3 | 1* |
| Cruciferae | — | — | 3 | — |
| Polygonum | — | 3 | — | — |
| Onagraceae | — | — | — | + |
| Compositae | — | 11 (2 sp.) | 3 | — |
| Unidentified dicotyledons | 3* (2 sp.) | 30 (5 sp.) | 29 (5 sp.) | 6* (2 sp.) |
| SPORES | | | | |
| Bryales | 92 | 63 | 83 | 73 |
| Sphagnum | — | 2 | — | — |
| Polypodiaceae | 4 | 7 | 9 | 12 |
| Other Filicales | 4 (2 sp.) | 11 (3 sp.) | 4 | 3 |
| Lycopodium | — | + ⎫ | + ⎫ 4 | + ⎫ 12 |
| L. clavatum | — | + ⎬ 16 | + ⎭ | + ⎭ |
| L. pungens (?) | — | + ⎮ | — | — |
| L. inundatum | — | + ⎭ | — | — |
| L. cf. annotinum | — | + | — | — |
| Selaginella sibirica | — | <1 | — | — |

terrace (sample no. 5) are older. They accumulated during the period of coniferous forests containing an admixture of exotic broadleaf species and coniferous species characteristic of Phase I. We cannot provide a more detailed analysis of the *Pinus* pollen in this case, because the original analysis was made in 1950 (V. A. Stupishina), before the morphologic differences of the *Pinus* pollen species were sufficiently well known.

TABLE 8

RESULT OF SPORE-POLLEN ANALYSIS OF SAMPLES FROM SECTION 9 OF THE
HIGH TERRACE OF THE ZAMARAIKHA RIVER

| | 2 | 3 | 4 | 5 |
|---|---|---|---|---|
| Sample no.: | 2 | 3 | 4 | 5 |
| Sampling depth in m: | 18.6 | 19.9 | 21.3 | 60.0 |
| Number of grains counted: | 124 | 101 | 241 | 48 |
| **SPORE AND POLLEN COMPOSITION (%)** | | | | |
| **GENERAL COMPOSITION** | | | | |
| Tree pollen | 66 | 70 | 56 | 73 |
| Grass and shrub pollen | 34 | 28 | 41 | 25 |
| Spores | <1 | 2 | 3 | 2 |
| Hydrophyte pollen | — | — | <1 | — |
| **POLLEN OF TREE SPECIES** | | | | |
| *Larix* | — | 4 | — | 9 |
| *Abies* | — | — | 1 | — |
| *Picea* | — | — | 1 | — |
| *Pinus* subsp. *Haploxylon* | 62 | 56 | 78 | 57 |
| *Pinus* subsp. *Diploxylon* | 31 | 25 | 6 | 28 |
| Deformed *Pinus* | 4 | 4 | — | — |
| *Betula* | 3 | 11 | 10 | — |
| *Alnus* | — | — | 4 | 3 |
| *Tilia* | — | — | — | 3 (singly) |
| **GRASS AND SHRUB POLLEN** | | | | |
| *Artemisia* | 14 | 29 | 31 | 5* |
| Chenopodiaceae | 12 | 3 | 13 | — |
| Gramineae | 19 | 22 | 15 | 2* |
| Ericaceae | — | 3 | — | — |
| **MOTLEY GRASSES** | | | | |
| Compositae | 10 | — | 1 | 1* |
| Primulaceae | — | 3 | — | — |
| Caryophyllaceae | — | — | 2 | — |
| Polygonaceae | — | — | 5 | — |
| Rubiaceae | — | — | — | 1* |
| Unidentified dicotyledons | 45 | 40 | 33 | 3* |
| **SPORES** | | | | |
| Bryales | — | — | 4* | — |
| Polypodiaceae | 1* | 1* | 3* | 1* |
| *Lycopodium* | — | 1* | — | — |
| **HYDROPHYTE POLLEN** | | | | |
| *Myriophyllum* | — | — | 1* | — |

Section no. 10, comprising the alluvium of the 40–45-m terrace of the Iya river near Kumeyka, represents the district north of the Baykal mountainous region. The samples were collected by V. A. Rastvorova in 1950. According to her description, the alluvium consists of a light, fine-grained, inequigranular, powdery sand with pebbles and gravel 0.5 m thick, above which is a 0.5-m layer of reddish-brown, fine-grained sand, and above this lies the present soil of the region. Table 9 shows the results of the spore-pollen analysis. The tree pollen consists basically of pine, Siberian pine, and birch. Spruce is not found, and fir pollen is found only occasionally. Pollen of the *Quercus* sp. (exotic species) was found in the lower sample, 0.79–0.81 m below the surface. These finds, and

TABLE 10

RESULTS OF SPORE-POLLEN ANALYSIS OF SAMPLES OF SECTION No. 11,
THE 35–40-M ANGARA TERRACE NEAR THE MOUTH OF THE KATA RIVER

| Sample no.: | 30 | 29 | 23 | 24 | 28 | 27 | 26 |
|---|---|---|---|---|---|---|---|
| Sampling depth in m: | 1.5 | 5.5 | 8.0 | 13.2 | 22.4 | 34.0 | 36.0 |
| Number of grains counted: | 62 | 186 | 20 | 6 | 132 | 192 | 39 |

SPORE AND POLLEN COMPOSITION (%)

| | | | | | | | |
|---|---|---|---|---|---|---|---|
| **GENERAL COMPOSITION** | | | | | | | |
| Tree pollen | 28 | 62 | 10* | 4* | 52 | 69 | 64 |
| Grass and shrub pollen | 25 | 24** | 1* | 1* | 29 | 5 | 18 |
| Spores | 47 | 16 | 9* | 1* | 19 | 26 | 18 |
| **POLLEN OF TREE SPECIES** | | | | | | | |
| Larix | — | — | — | — | — | 3? | 12 |
| Abies | — | 1 | — | — | — | — | — |
| Picea | 1* | 1 | — | 2* | 7 | 14 | 16 |
| Pinus sibirica | 1* | 6 | 5* | — | 20 | 82 | 32 |
| P. silvestris | — | 6 | — | — | — | — | — |
| Pinus subsp. Diploxylon | 72 | 86 | 4* | 2* | 64 | 1 | 36 |
| Betula (2 species) | 16 | 5 | 1* | — | 6 | — | 4 |
| Alnus | — | — | — | — | 3 | — | — |
| Quercus | — | 1 | — | — | — | — | — |
| **GRASS AND SHRUB POLLEN** | | | | | | | |
| Cyperaceae | — | 2 | — | — | — | — | — |
| Gramineae | 6* | 19 | 1* | 1* | 13 | — | — |
| Ephedra (?) | — | — | — | — | — | — | 1* |
| Chenopodiaceae | 1* | 10(?)† | — | — | 5 | — | 1* |
| Artemisia | 5* | 31 | — | — | 63(?)† (2 forms) | 1* | 2* |
| Ericales | — | 3(?)† | — | — | — | — | — |
| **MOTLEY GRASSES** | | | | | | | |
| Caryophyllaceae | — | 3(?)† | — | — | 3 (2 forms) | — | — |
| Polygonaceae | — | — | — | — | 5 | — | — |
| Rumex | — | 6 | — | — | — | — | — |
| Leguminosae | — | 1 | — | — | 3 | — | — |
| Umbelliferae | — | — | — | — | — | 1* | 1* |
| Compositae | 1* | 12(?)† | — | — | 5 | 1* | 1* |
| Unidentified dicotyledons | 3* | 14 | — | — | 3 | 2* | 1* |
| **SPORES** | | | | | | | |
| Bryales | 87 | 79 | 8* | 1* | 76 | 52 | 1* |
| Sphagnum | 3 | 7 | — | — | — | — | 1* |
| Polypodiaceae | 10 | 14 | — | — | — | 26 | 1* |
| Other Filicales | — | — | — | — | 20 | 15 | 2* |
| Lycopodium clavatum | — | — | 1* | — | 4 | 7 | 2* |

†Distorted spectrum, since the pollen of grassy plants was found in groups, in which 2–15 grains stuck together.

only here and there. A comparison of the percentage content of pollen from the horizon of fine-grained and clayey sands, beginning at a depth of 5.5 m, with the pollen composition of present-day surface samples also indicates a considerable preponderance of taiga. Further, the spectra of the upper part of the section are similar to the spectra of Phase III, which can be seen in the diagram of the Maksimovshchina deposits (see Fig. 3).

The diagram of the section from the mouth of the Kata river shows the same characteristic sequence in the alternation of species maxima upward along the section (spruce–Siberian pine–pine–birch) as the Baykal section of Early Quaternary deposits (Bystraya depression, Marksimovshchina). However, the mere general similarity of the spectra, without a thorough study of the section, does not provide a reliable comparison of the time of formation of this terrace with Phases II and III, although such a comparison certainly comes to mind.

Presumably, the deposits of the high terraces of the lower Angara basin may be referred to Phases I and II. R. V. Fedorova (1949) analyzed individual samples of terrace deposits at several points. Only the general nature of the spectra was revealed by the analysis. The samples were taken from the following sections:

No. 12, from the 50–60-m Angara terrace in the vicinity of the Koda and Kezhma rivers;

No. 13, from the 120-m terrace of the Mura river;

No. 14, from the 50-m terrace of the Chuna river;

No. 15, from the 110-m terrace of the Angara river at Boguchany;

No. 16, from the 80-m terrace of the ancient Yelchimo valley.

Table 11, which gives the results of spore-pollen analyses, shows that the samples contain fir, spruce, Siberian pine, pine, and birch pollen. Pollen of the coniferous species, Siberian pine and pine, predominates. The small quantities of spores and the prevalence of grassy-plant pollen are associated primarily with dry, open expanses (Chenopodiaceae and Artemisia). These deposits, which are relatively old as indicated by their bedding, evidently accumulated under conditions of gradual cooling and reduction of humidity, when the spruce-fir forests with Siberian pine were gradually displaced by Siberian pine and pine and when the forests were thinned out somewhat by the formation (primarily in the river valleys) of meadow-steppe patches. Presumably, the northeastern part of the Angara basin contained an especially large number of the meadow-steppe areas.

On examining several sections in which the dark-coniferous taiga (Phase II) can be detected in the Angara region, we see that the most complete sections are found in the mountain and piedmont depressions, where these depressions have filled in. In the northern Angara region, the deposits of this age are represented at a number of points by a very thin layer of alluvium on the high terraces. This alluvium lies on Jurassic or Tertiary deposits. Perhaps this is why we were unable to obtain complete sections containing pollen characteristics of Phases I and II.

The next stage of development of the vegetation was Phase III, the period of light-coniferous pine and pine-larch(?) forests and, later, pine-birch forests. In the upper part of the Maksimovshchina section, Phase III directly follows Phase II, which appears in the lower part of this same section. The beginning of Phase III has also been noted in the Kumeyka section of the 40–45-m Iya terrace. The deposits of the 60–80-m terrace of the Bolshaya Bystraya river at the point where it enters the Tunka depression (section no. 17) are similar in their spore-pollen spectra. Here the deposits are "weakly cemented boulder-pebble gravel deposits, which many investigators regard as fluvioglacial," according to Ye. M. Shcherbakova's (1954) description. Analyses conducted by N. S. Sokolova (Table 12) show that birch, pine, and Siberian pine (?) pollen predominated in the spore-pollen spectra of the upper part of the section, and that spruce and fir were absent. The pollen of exotic pines was noted, but not identified and described. Apparently, the spruce-fir taiga had already yielded to pine forests with Siberian pine and birch at the time the boulder-pebble gravel deposits formed. Possibly larch

TABLE 11

RESULT OF SPORE-POLLEN ANALYSIS OF SAMPLES FROM SECTIONS 12–16 OF
THE ANCIENT TERRACES OF THE ANGARA AND ITS TRIBUTARIES

| Section no.: | 12 | 13 | 14 | 15 | 16 |
|---|---|---|---|---|---|
| SPORE AND POLLEN COMPOSITION (%) | | | | | |
| GENERAL COMPOSITION | | | | | |
| Tree pollen | 55 | 59 | 97 | 75 | 85 |
| Grass and shrub pollen | 40 | 29 | 2 | 17 | 13 |
| Spores | 5 | 18 | 1 | 8 | 2 |
| POLLEN OF TREE SPECIES | | | | | |
| Abies | 3 | 1 | 2 | — | 1 |
| Picea | — | 1 | <1 | 6 | 4 |
| Pinus sibirica | 71 | 89(?) | 90 | 11 | — |
| P. silvestris | 6 | — | — | 74 | 79 |
| Pinus (unidentified) | 18 | 4 | — | — | 12 |
| Betula | 2 | 5 | 8 | 9 | 4 |
| GRASS AND SHRUB POLLEN | | | | | |
| Gramineae | 52 | 29 | 2* | 2* | 7* |
| Cyperaceae | — | 2 | — | 1* | 2* |
| Chenopodiaceae | 9 | 27 | 2* | — | — |
| Artemisia | 18 | 25 | 5* | 5* | 3* |
| MOTLEY GRASSES | | | | | |
| Labiatae | 3 | — | — | — | — |
| Caryophyllaceae | — | — | 1* | — | — |
| Umbelliferae | — | 2 | — | — | 2* |
| Compositae | — | 2 | — | — | — |
| Unidentified dicotyledons | 18 | 11 | 4* | — | — |
| SPORES | | | | | |
| Bryales | 6 | 5* | 3* | 4* | 2* |
| Sphagnum | — | 1* | 1* | — | — |
| Polypodiaceae | — | 4* | — | — | — |
| Lycopodium | — | 8* | — | — | — |
| NUMBER OF POLLEN GRAINS AND SPORES COUNTED | 85 | 150 | 194 | 47 | 94 |

forests also existed at that time. As the character of the forests changed, the grass cover also changed: the species saturation decreased owing to the disappearance of heather species and club-moss spores.

The spore-pollen spectra of the deposits of the 60–80-m terrace of the Bolshaya Bystraya river are similar to the loam deposits of the upper part of the Maksimov-shchina section (see Fig. 3) and may also be related to Phase III, the phase of light-coniferous and birch forests.

However, it is preferable to classify the sandy-pebble gravel deposits of the 60-m Ilcha Kultuchnaya terrace (section no. 18) as Phase III. Sokolova discovered a small quantity of pollen in six samples taken from the 4–8-m level here; pine and birch are the only tree species represented.

Section no. 19 is situated in the Irkut basin, not far from the Maksimovshchina section on the same slope of the interfluve plateau; it is represented by borehole 249, from which four samples were taken for analysis. Apparently the borehole passed through the same loam horizons as did the upper part of Maksimov-shchina borehole no. 246. Analysis showed that the pollen and spore composition

TABLE 12

| | 1 | 2 | 3 | 4 |
|---|---|---|---|---|
| Sample no.: | 1 | 2 | 3 | 4 |
| Sampling depth in m: | 6.6 | 7.5 | 9.0 | 10.0 |
| Number of grains counted: | 264 | 199 | 55 | 34 |

Spore and Pollen Composition (%)

| | 1 | 2 | 3 | 4 |
|---|---|---|---|---|
| **GENERAL COMPOSITION** | | | | |
| Tree pollen | 26 | 25 | 33 | 32 |
| Grass and shrub pollen | 6 | 7 | 12 | 6 |
| Spores | 68 | 68 | 55 | 62 |
| **POLLEN OF TREE SPECIES** | | | | |
| *Pinus*† | 75 | 70 | 10* | 11* |
| *Betula* | 16 | 20 | 7* | — |
| *Alnus* | 9 | 10 | 1* | — |
| **GRASS AND SHRUB POLLEN** | | | | |
| *Ephedra* | — | 1* | — | — |
| Cyperaceae | 1* | 3* | 1* | 1* |
| Chenopodiaceae | — | — | 1* | — |
| Ericaceae | — | — | 1* | 1* |
| *Artemisia* | 6* | 5* | 3* | — |
| **MOTLEY GRASSES** | | | | |
| Polygonaceae | 1* | — | — | — |
| Compositae | 8* | 3* | 1* | — |
| Unidentified dicotyledons | — | 2* | 1* | — |
| **SPORES** | | | | |
| Bryales | 15 | 9 | — | — |
| Sphagnales | <1 | — | — | — |
| Polypodiaceae | 85 | 88 | 97 | 16* |
| Lycopodiaceae | — | 1 | — | 5* |
| *Selaginella selaginoides* | — | 1 | 3 | — |
| Unidentified | — | 1 | — | — |

†The total includes *Pinus silvestris*, *P. sibirica*, and other species (exotic).

of the samples taken from these boreholes was very similar, i.e., it implies the same conditions of formation of the deposits at the two sites (Fig. 9a and Table 13). The tree pollen composition here is deficient in that fir and spruce are almost completely absent, while pine and birch predominate. Green mosses are preponderant among the spores. The diagram corresponds to Phase III, that of the light-coniferous and birch forests, and its upper part repeats and supplements the diagram of the upper part of the spectrum of the Maksimovshchina section.

Section no. 20, consisting of Quaternary deposits on the left slope of the Oka watershed near Zima, is analogous to section 19. The borehole passed through sands with pebbles to a depth of 4.4 m and then through sands to a depth of 11 m.

Figure 9b and Table 13 show clearly that pine and birch pollen predominated among the tree species, indicating the existence of pine-birch and larch (?) forests. Characteristically, as the forests became lighter, the diversity of the grassy plants increased and considerable wormwood appeared. Green mosses dominated the spores, the club mosses disappeared, and the number of ferns decreased.

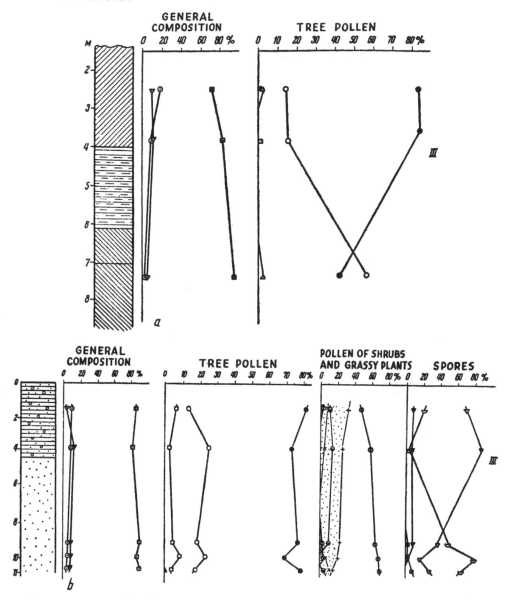

FIGURE 9. Spore-pollen diagrams: *a*, section no. 19, left bank of the Irkut river; *b*, section no. 20, watershed deposits at Zima on the left bank of the Oka river.

As already noted, the spectra of the end of Phase II in the northeastern Angara region indicated considerable distribution of steppes, which reached their greatest development in the coldest and driest period. Possibly, the 45-m terrace of the Podkamennaya Tunguska river at Panolik (section no. 21) formed during this cold and dry period following Phase II. The samples were taken by Yu. P. Parmuzin and were analyzed by A. N. Sladkov (Table 14). In these samples, grass pollen predominated (50–52%) in the total pollen and spore count, and spores did not appear at all or amounted to only about 2 per cent. This proportion of grass pollen can be found at present only in the wooded steppes. The

TABLE 13

RESULT OF SPORE-POLLEN ANALYSIS OF SAMPLES OF SECTION 20 ALONG BOREHOLE IN
OKA BASIN NEAR ZIMA AND SECTION 19 ALONG BOREHOLE IN IRKUT BASIN

| Sample no.: | 1 | 2 | 3 | 4 | 5 | 1 | 2 | 3 | 4 |
|---|---|---|---|---|---|---|---|---|---|
| Sampling depth in m: | 1.5 | 3.8 | 9.4 | 10.25 | 11.0 | 2.5 | 3.9 | 5.5 | 6.5 |
| No. of grains counted: | 392 | 302 | 506 | 210 | 514 | 122 | $108+\infty$ | $26+\infty$ | $60+\infty$ |

SPORE AND POLLEN COMPOSITION (%)

| GENERAL COMPOSITION | | | | | | | | | |
|---|---|---|---|---|---|---|---|---|---|
| Tree pollen | 85 | 81 | 89 | 86 | 89 | 72 | 83 | 8* | 95 |
| Grass and shrub pollen | 9 | 8 | 5 | 5 | 4 | 18 | 8 | 4* | 2 |
| Spores | 6 | 11 | 6 | 9 | 7 | 10 | $9+\infty$ | $14*+\infty$ | $3+\infty$ |
| Hydrophyte pollen | — | — | <1 | — | — | — | — | — | — |
| **POLLEN OF TREE SPECIES** | | | | | | | | | |
| Abies | — | — | — | — | <1 | 1 | — | — | — |
| Picea | — | — | — | — | — | 2 | — | — | — |
| Pinus | 81 | 73 | 77 | 69 | 79 | 83 | 84 | 5* | 42 |
| Betula | 13 | 25 | 18 | 23 | 17 | 14 | 15 | 2* | 56 |
| Alnus | 6 | 2 | 4 | 8 | 3 | — | 1 | 1* | — |
| **GRASS AND SHRUB POLLEN** | | | | | | | | | |
| Gramineae | 10 | 13 | 10 | — | 4 | 6* | 2* | — | — |
| Cyperaceae | — | — | — | — | — | — | 1* | 1* | — |
| Chenopodiaceae | 8 | 4 | — | 7 | 11 | 2* | 1* | — | — |
| Ericales | 3 | — | 2 | 4 | — | — | — | 1* | — |
| Artemisia | 32 | 25 | 23 | 21 | 15 | 7* | 1* | 1* | 1* |
| **MOTLEY GRASSES** | | | | | | | | | |
| Compositae | 3 | 8 | 13 | 14 | 33 | 4* | 3* | — | — |
| Caryophyllaceae | — | 8 | — | — | — | — | — | — | — |
| Labiatae | — | 8 | 4 | 4 | — | 1* | — | — | — |
| Leguminosae | — | 8 | 4 | — | — | — | — | — | — |
| Umbelliferae | 3 | — | 2 | — | — | — | — | — | — |
| Rosaceae | 3 | — | — | — | — | — | — | — | — |
| Gentianaceae (?) | 3 | — | — | — | — | — | — | — | — |
| Rubiaceae | — | — | 2 | — | — | — | — | — | — |
| Ranunculaceae | — | — | — | 4 | — | — | — | — | — |
| Unidentified | 35 | 25 | 40 | 46 | 37 | 2* | 1* | 1* | — |
| **SPORES** | | | | | | | | | |
| Bryales | 70 | 88 | 38 | 16 | 27 | 10* | $\infty$ | $\infty$ | $\infty$ |
| Sphagnum | 8 | 6 | 9 | 2 | — | 1* | — | — | — |
| Polypodiaceae | 20 | 6 | 45 | 65 | 49 | — | 8* | 2* | 2* |
| Other Filicales | — | — | 5 | 15 | 16 | 1* | 1* | 8* | — |
| Lycopodium | — | — | 3 | 2 | 8 | — | — | — | — |
| Unidentified | — | — | — | — | — | — | 1* | 3* | — |
| **HYDROPHYTE POLLEN** | | | | | | | | | |
| Myriophyllum | — | — | 1* | — | — | — | — | — | — |

composition of the grassy-plant pollen is also characteristic [of wooded steppes]. *Artemisia* pollen amounts to 64 per cent; Chenopodiaceae, 13 per cent. Of the tree pollens, 60 per cent are pine and 35 per cent birch. Fir and spruce pollen is also found, but these species played only a negligible role in the vegetation cover, because the forests had thinned out greatly and evidently were insular.

The steppe formation of the northeastern part of the basin, which began at the close of Phase II, reached its maximum during Phase III. The extension of the steppes in combination with birch-pine forests or birch-pine-larch forests

TABLE 14

RESULT OF SPORE-POLLEN ANALYSIS OF SAMPLES FROM SECTION No. 21, THE 45-M
TERRACE OF THE PODKAMENNAYA TUNGUSKA RIVER NEAR THE VILLAGE OF PANOLIK

| | Sample no.: | 2 | 3 | 4 |
|---|---|---|---|---|
| | Number of grains counted: | 132 | 29 | 8 |
| SPORE AND POLLEN COMPOSITION (%) | | | | |
| GENERAL COMPOSITION | | | | |
| Tree pollen | | 48 | 48 | 2* |
| Grass and shrub pollen | | 50 | 52 | 6* |
| Spores | | 2 | — | — |
| POLLEN OF TREE SPECIES | | | | |
| Abies | | 3 | — | — |
| Picea | | 2 | 2* | — |
| Pinus | | 60 | 9* | 1* |
| Betula | | 35 | 3* | 1* |
| GRASS AND SHRUB POLLEN | | | | |
| Gramineae | | 5 | 3* | 1* |
| Artemisia | | 64 | 9* | 4* |
| Chenopodiaceae | | 13 | 1* | 1* |
| MOTLEY GRASSES | | | | |
| Caryophyllaceae | | 1 | — | — |
| Compositae | | 15 | — | — |
| Unidentified | | 2 | 2* | — |
| SPORES | | | | |
| Bryales | | 3 | — | — |

and the almost complete displacement of the dark-coniferous taiga were connected with the increased continentality and aridity of the climate.

The succession of the first, second, and third phases is clear, since Phases I and II or II and III are often found simultaneously in a section, and the gradual progression of changes from one phase to another is clear. The transitional period from Phase III to the subsequent Phase IV is less well defined. The fourth phase is characterized by mixed forests with broadleaf species, but Phases III and IV are not found anywhere in a single section. Study of the mode of occurrence of the deposits of Phases III and IV has shown that the deposits of Phase III are older. Through geomorphologic investigations, V. A. Rastvorova was able to distinguish terraces of different heights in a small segment of the lower Iya river. The pollen and spore composition indicated that the 45-m Iya terrace belongs to the end of Phase II and the beginning of Phase III. In a section of the younger, 19-m, terrace the pollen spectra correspond to two successive phases. The lower part of the terrace alluvium corresponds to Phase IV of mixed forests with broadleaf species; the upper part of the alluvium to Phase V of light-coniferous forests. Similar relationships between terrace deposits were found in the Angara valley as well: the 45-m terrace corresponds to Phase III, the 20–30-m terraces to Phases IV–V. We still do not have direct evidence of the transition from Phase III to Phase IV but, on the basis of the afore-mentioned modes of occurrence of the deposits, we can regard Phase IV as representing a later stage of development of the vegetation and as being subsequent to Phase III. Evidently the end of Phase III and the beginning of Phase IV correspond to the time of downcutting of the rivers and the interruption of the formation of terrace deposits.

The spread of broadleaf trees—the linden, elm, oak, and hazel—and, subsequently, some increase in the role of the dark-coniferous taiga and afforestation of the territory are characteristic of the fourth stage of development of the Angara vegetation. The pollen of the broadleaf species and of the spruce, fir, and Siberian pine are encountered constantly in the deposits of this phase. The role of club-moss and sphagnum spores increased. For a long time, the small amount of broadleaf pollen in the samples cast doubt upon the consistency of the finds. However, many sections have been taken which contain broadleaf pollens in the terrace deposits, whose mode of occurrence indicates that they are younger than the deposits of Phase III. Pollen of broadleaf species has also been found in the terrace deposits of the Yenisey and Lena rivers. By way of illustration, let us cite the results of analyses of several sections.

Section no. 22: the 25-m terrace of the Angara river at Narotay Island. The samples were collected by Ye. I. Sakharova in 1949 and by S. S. Voskresenskiy in 1951 from two pits. Table 15 shows the results of the spore-pollen analysis of four samples of light-gray, powdery, inequigranular sand and greenish-gray medium-grained sand. The samples still contain considerable quantities of grass pollens, as distinct from the pollen spectra of Phase III. The tree pollens include pine and birch, and the broadleaf species elm, linden, and hazel, which comprise 10 per cent of the total pollen. The grass pollen is diversified and contains a considerable amount of *Artemisia* and Chenopodiaceae.

Table 15 indicates that sparse pine and birch forests with an admixture of broadleaf species existed here, with a relatively rich grass cover, and a richer representation of spore plants. The spread of broadleaf species may have been associated with a considerable warming of the climate.

Section no. 23: the 19-m terrace of the Iya river at Kob. The alluvium, which rests on a sandstone socle, has the following stratification:

| | |
|---|---|
| 1. Sand with humus in the upper part | 0.65 m |
| 2. Loam, light gray, highly carbonated | 0.85* |
| 3. Loam, grayish-yellow, more friable and lighter in the lower part, becoming reddish-yellow loam | 1.2* |
| 4. Fine-grained, lilac-gray sand with a small amount of pebbles, becoming grayish-yellow loam farther down with interlayers of coarse-grained yellow sand 1–2 cm thick | 1.4 |
| 5. Medium-grained and coarse-grained light gray sand with a large quantity of pebbles | 1.63 |

Figure 10 and Table 15 indicate the results of the spore-pollen analysis of the deposits in this section. Conifer pollen predominates in the sandy deposits containing pebbles. Siberian pine pollen comprises 35 per cent and spruce 16 per cent of the total tree pollen. Fir, linden, and oak pollen also appears. The analysis indicates that during the accumulation of the sandy horizon the vegetation differed essentially from that of the third phase, i.e., the phase of birch and light-coniferous forests. The broadleaf species gained wide distribution and the role of the dark-coniferous taiga increased. The role of the dark-coniferous taiga in the forests was greater than at present in the Angara basin, but evidently the dark-coniferous taiga was not as highly developed as in the Early Quaternary period, during Phase II.

The pollen composition of the loam deposits is different. The broadleaf species

*[The original reads 0.25 and 2.1 m, respectively, here, but these must be misprints, as a downward progression of horizons is reported.—Translator.] [See also Table 15.—Editor.]

TABLE 15

RESULT OF SPORE-POLLEN ANALYSIS OF SAMPLES OF SECTION NO. 22, THE 23–M
ANGARA TERRACE AT NAROTAY ISLAND, AND SECTION NO. 23, THE ALLUVIUM
OF THE 19-M IYA TERRACE AT KOB

| Pit no.: | Pit I | | | | Pit II | | | |
|---|---|---|---|---|---|---|---|---|
| Sample no.: | 1 | 2 | 3 | 11 | 2 | 10 | 14 | 15 |
| Depth of sampling in m: | 0.7 | 0.9 | 1.10 | 2.9 | 0.25 | 4.0 | 6.45 | 7.03 |
| Number of grains counted: | 45 | 4 | 203 | 24 | 165 | 78 | 199 | 129 |

SPORE AND POLLEN COMPOSITION (%)

| | | | | | | | | |
|---|---|---|---|---|---|---|---|---|
| **GENERAL COMPOSITION** | | | | | | | | |
| Tree pollen | 67 | 1* | 48 | 7* | 92 | 55 | 98 | 97 |
| Grass and shrub pollen | 15 | — | 42 | 13* | 1 | 15 | <1 | 1 |
| Spores | 18 | 3* | 10 | 4* | 7 | 29 | 1 | 2 |
| Hydrophyte pollen | — | — | — | — | — | 1 | — | — |
| **POLLEN OF TREE SPECIES** | | | | | | | | |
| Abies | 3* | — | — | — | — | — | 2 | — |
| Picea | — | — | 1 | — | — | 16 | <1 | 2 |
| Pinus silvestris | 3* | — | 70 | 3* | 22 | 40 | 87 | 85 |
| P. sibirica | — | — | — | — | <1 | 35 | 5 | 10 |
| Pinus (deformed) | 19* | 1* | — | — | — | — | — | — |
| Betula | 5* | — | 13 | 4* | 78 | 7 | 5 | 2 |
| | | | | 2 forms | | | | |
| Alnus | — | — | 6 | — | — | — | — | — |
| Corylus | — | — | 8⎫ | — | — | — | — | — |
| Ulmus | — | — | 1⎬10 | — | — | — | — | — |
| Tilia | — | — | 1⎭ | — | — | — | — | 1 |
| Quercus | — | — | — | — | — | 2 | — | — |
| **POLLEN OF GRASSY PLANTS** | | | | | | | | |
| **AND SHRUBS** | | | | | | | | |
| Ericales | — | — | 1 | — | — | — | — | — |
| Gramineae | 3* | — | 8 | 1* | — | — | — | 1* |
| Chenopodiaceae | 3* | — | 8 | — | — | 1* | — | — |
| Artemisia | — | — | 50 | 9* | — | 1* | — | — |
| **MOTLEY GRASSES** | | | | | | | | |
| Rubiaceae | — | — | 1 | — | — | 1* | — | — |
| Compositae | — | — | 5 | 1* | — | — | — | 1* |
| Caryopylliceae | 1* | — | — | — | — | — | — | — |
| Umbelliferae | — | — | 1 | — | — | — | — | — |
| Polygonaceae | — | — | — | 1* | — | — | — | — |
| Unidentified | — | — | 12 | 1* | 2* | 9* | 1* | — |
| **SPORES** | | | | | | | | |
| Bryales | 5* | 3* | 70 | 3* | 10* | 1* | 2* | 1* |
| Sphagnum | — | — | 5 | — | — | 1* | — | — |
| Filicales | 3* | — | 15 | — | 1* | 7* | — | — |
| Lycopodium | — | — | 10 | — | — | — | — | — |
| L. clavatum | — | — | — | — | 1* | — | — | — |
| L. annotinum | — | — | — | — | + | — | — | — |
| Polypodiaceae | — | — | — | — | — | 4* | — | 1* |
| **HYDROPHYTE POLLEN** | | | | | | | | |
| Lentibulariaceae | — | — | — | — | — | 1* | — | — |

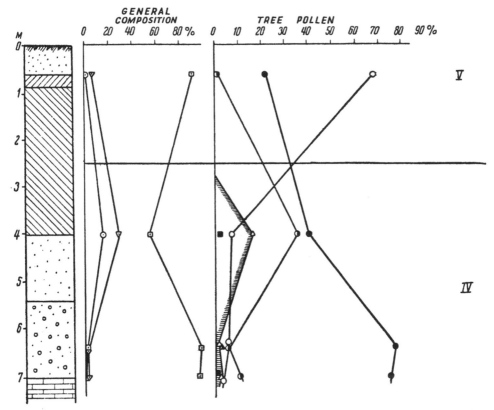

FIGURE 10. Spore-pollen diagram of the 19-m Iya river terrace at Kob (section no. 23).

disappear and spruce and Siberian pine pollen is found only occasionally. Birch occupies first place. The spectrum is analogous to Phase III. One may assume that two phases of development of vegetation are represented in this section: Phase IV, i.e., a phase of mixed-coniferous forests with broadleaf species, corresponding to the lower part of the section, and Phase V, i.e., a phase of light-coniferous and birch forests, corresponding to the upper part of the section. In the remaining sections of this complex of 20–30-m terraces, which is regarded as belonging to Phase IV, either there is little pollen or else a small time-segment of Phase IV is represented (Table 16).

Section no. 24: the 20-m Angara terrace at Moko Island. In the alluvium samples, grass pollen occupies a prominent place in the general pollen composition. Hazel pollen is found. The amount of Siberian pine pollen increases upward along the section. The deposits apparently correspond to the beginning of Phase IV.

Section no. 25: the 30-m Angara terrace below the mouth of the Dunayeva river. Hazel pollen is found among the tree pollen at a depth of 2.5 m.

Section no. 26: the 22-m Angara terrace above Nalyur. A large amount of grass pollen (spectra of the wooded-steppe type) appears in the alluvium. Hazel pollen appears; spruce pollen amounts to 6 per cent.

Section no. 27: the 30-m Angara terrace at Intey Island. Hazel, fir, and spruce pollen and sphagnum spores are found in the alluvium, along with pine and birch pollen.

RESULT OF SPORE-POLLEN ANALYSIS OF SAMPLES FROM THE

| Section no.: | 24 | | | | | | 25 | 26 | | | | 27 | 28 | | 29 | | | | |
|---|---|---|---|---|---|---|---|---|---|---|---|---|---|---|---|---|---|---|---|
| Sample no.: | 1 | 2 | 3 | 7 | 8 | 9 | 7 | 1 | 2 | 3 | 4 | 2 | 1 | 4 | 1 | 2 | 3 | 4 | 5 |
| Sampling depth in m: | — | — | — | — | — | — | 2.5 | 1.5 | 2.3 | 4.0 | 5.7 | 0.9 | 0.2 | 1.9 | 0.35 | 0.9 | 1.0 | 1.2 | 1.5 |
| Number of grains counted: | 11 | 10 | 124 | 40 | 43 | 12 | 43 | 36 | 26 | 107 | 201 | 310 | 105 | 264 | 62 | 14 | 43 | 110 | 215 |
| **GENERAL COMPOSITION** | | | | | | | | | | | | | | | SPORE AND POLLEN | | | | |
| Tree pollen | 6* | 4* | 71 | 36 | 37 | 4* | 84 | 19 | 5* | 29 | 39 | 85 | 70 | 90 | 53 | 6* | 56 | 58 | 96 |
| Grass and shrub pollen | 3* | 6* | 22 | 62 | 62 | 8* | 16 | 72 | 20* | 69 | 61 | 14 | — | 7 | 19 | 2* | 30 | 34 | 3 |
| Spores | 4* | — | 7 | 2 | 1 | — | — | 9 | 1* | 2 | — | 1 | 30 | 3 | 27 | 6* | 14 | 8 | 1 |
| **POLLEN OF TREE SPECIES** | | | | | | | | | | | | | | | | | | | |
| Larix | — | — | — | — | — | — | — | — | — | — | — | 1 | — | <1 | — | — | — | — | — |
| Abies | — | — | — | — | — | — | — | — | — | — | — | 1 | — | <1 | — | — | — | — | — |
| Picea | — | 1* | — | — | — | — | — | — | — | 6 | — | 2 | — | <1 | — | — | — | — | — |
| Pinus sibirica | — | — | 45 | 3* | 2* | 1* | — | — | — | — | — | — | ~40 | ~5 | — | — | — | — | + |
| P. silvestris | — | — | 2 | 4* | — | — | — | — | — | — | — | — | ~40 | ~79 | — | — | — | — | — |
| Pinus | 2* | 3* | 49 | 5* | 5* | 2* | 84 | 3* | 4* | 76 | 86 | 88 | — | — | 90 | 3* | 76 | 75 | 78 |
| Pinus subsp. Diploxylon | — | — | — | — | — | — | — | — | — | — | — | — | — | — | — | — | — | — | — |
| Betula | — | — | 4 | — | 4* | 1* | 14 | 4* | 1* | 16 | 13 | 9 | 17 | 4 | 10 | 3* | 17 | 25 | 22 |
| Alnus | 1* | — | — | — | 5* | — | — | — | — | — | 2 | 1 | 3 | 1 | — | — | 4 | — | — |
| Quercus | — | — | — | — | — | — | — | — | — | — | — | — | — | — | — | — | 1 | — | — |
| Tilia | — | — | — | — | — | — | — | — | — | — | — | — | — | — | — | — | — | — | — |
| Corylus | 2 | — | — | 2 | 2 | — | 2 | 1 | — | — | — | 1 | — | — | — | — | — | — | — |
| Ulmus | — | — | — | — | — | — | — | — | — | — | — | — | — | — | — | — | — | — | — |
| **GRASS AND SHRUB POLLEN** | | | | | | | | | | | | | | | | | | | |
| Gramineae | — | — | 0* | 9* | 10* | 6* | 3* | 4 | 2* | 7 | 21 | 2 | — | 3* | 1* | — | 2* | 6* | 1* |
| Cyperaceae | — | — | — | — | 1* | — | — | — | — | — | — | — | — | — | — | — | — | — | — |
| Chenopodiaceae | — | — | 4* | 7* | 3* | — | — | — | — | — | 2 | 7 | — | 1* | — | — | — | 6* | — |
| Ericales | — | — | — | — | — | — | — | — | — | — | — | 62 | — | — | — | — | — | 6* | — |
| Artemisia | 1* | 4* | 10* | 6* | 10* | 2* | 1* | 43 | 15* | 85 | 68 | 17 | — | 6* | 8* | 2* | 5* | 18* | 5* |
| **MOTLEY GRASSES** | | | | | | | | | | | | | | | | | | | |
| Compositae | — | — | 1* | — | 3* | — | — | 2 | 1* | — | 2 | 2 | — | — | 1* | — | 1* | 2* | 1* |
| Type Statice | — | — | — | — | — | — | — | — | — | — | — | — | — | — | — | 1* | — | — | — |
| Caryophyllaceae | — | — | — | — | — | — | — | 2 | — | — | 1 | — | — | — | 1* | — | — | — | — |
| Leguminosae | — | — | 1* | — | — | — | — | — | — | — | — | — | — | — | — | 1* | — | — | — |
| Rubiaceae | — | — | — | — | — | — | — | — | — | — | — | — | — | 1* | — | 1* | — | — | — |
| Polygonaceae | — | — | — | — | — | — | — | — | — | — | — | — | — | — | — | — | — | — | — |
| Umbelliferae | — | — | — | — | — | — | 3* | — | — | 1 | — | — | — | — | — | — | — | — | — |
| Cruciferae | — | — | 1* | — | — | — | — | — | — | — | — | — | — | — | — | — | — | — | — |
| Unidentified | — | 2* | 1* | 3* | — | — | — | 49 | 2* | 7 | 6 | 10 | — | 6* | 1* | — | — | — | — |
| Labiatae | — | — | — | — | — | — | — | — | — | — | — | — | — | — | — | — | — | — | — |
| Ranunculaceae | — | — | — | — | — | — | — | — | — | — | — | — | — | — | — | — | — | — | — |
| **SPORES** | | | | | | | | | | | | | | | | | | | |
| Bryales | 3* | — | 6* | 1* | 1* | — | — | — | — | — | — | — | 9 | 1* | 17* | 6* | 6* | 8* | — |
| Sphagnum | — | — | — | — | — | — | — | — | — | 1* | — | 2* | 3 | — | — | — | — | — | — |
| Lycopodium | 1* | — | — | — | — | — | — | 3* | 1* | 1* | — | — | 26 | 1* | — | — | — | 1* | — |
| L. clavatum | — | — | — | — | — | — | — | — | — | — | — | — | — | — | — | — | — | — | — |
| Polypodiaceae | — | — | — | — | — | — | — | — | — | — | — | — | — | — | — | — | — | — | — |
| Other Filicales | — | — | 2* | — | — | — | — | — | — | 1* | — | 2* | 62 | 6* | — | — | — | — | 2* |
| Unidentified | — | — | — | — | — | — | — | — | — | — | — | — | — | — | — | — | — | — | — |
| **HYDROPHYTE POLLEN** | | | | | | | | | | | | | | | | | | | |
| Myriophyllum | — | — | — | — | — | — | — | — | — | — | — | — | — | — | — | — | — | — | — |

Section no. 28: the 21-m Angara terrace at Srednekamennyy Island. The pollen of broadleaf trees does not occur in this section, but the pollen and spore composition of the section indicates that it belongs to Phase IV (perhaps the end of Phase IV), because the tree pollen includes much Siberian pine and because spruce and fir pollen is also present.

Section no. 29: the right bank of the Angara river at Shcherbakovo. An outcropping at the mouth of a ravine, 16-m high. The deposits consist of pebble gravel and sands. The lower part of the deposits contains pine, fir, spruce, and hazel pollen, and oak pollen is found higher up. Spruce and fir pollen does not appear in the upper part of the section; the role of birch and grasses increases. The deposits exposed by the ravine evidently represent the end of the fourth and the beginning of the fifth phase.

Section no. 30: the 20-m terrace of the Chadobets river (an outcropping on a cape at the river mouth) in the northern part of the Angara basin. Oak, linden, and hazel pollen is found in the samples. The pollen of broadleaf species amounts to 9 per cent of the total.

ALLUVIUM OF TERRACE DEPOSITS OF THE MIDDLE TERRACE COMPLEX, SECTIONS 24–36

| 30 | | | 31 | | | | | 32 | | | | 33 | | 34 | | | 35 | | 36 | | | |
|---|---|---|---|---|---|---|---|---|---|---|---|---|---|---|---|---|---|---|---|---|---|---|
| 6 | 7 | 266/4 | 1 | 2 | 3 | 4 | 5 | 1 | 2 | 3 | 4 | 2 | 4 | 1 | 4 | 7 | 3 | 4 | 1 | 2 | 3 | 4 |
| 1.8 | 2.0 | — | 0.25 | 0.55 | 0.75 | 1.5 | 2.0 | — | — | — | — | 0.2 | 0.5 | 0.5 | 0.9–1 | 2.75 | 0.5 | 0.75 | 0.75 | 0.95 | 1.2 | 1.4 |
| 10 | 169 | 185 | 68 | 28 | 25 | 29 | 28 | 10 | 153 | 43 | 37 | 176+ | 304 | 503 | 38 | 22 | 837 | 569 | 21 | 72 | 80 | 159 |
| **COMPOSITION (%)** | | | | | | | | | | | | | | | | | | | | | | |
| 7* | 98 | 25 | 35 | 75 | 72 | 72 | 59 | 1* | 65 | 19 | 30 | 75 | 90 | 79 | 53 | 14* | 51 | 47 | 13* | 96 | 95 | 97 |
| 2* | 2 | 29 | — | 11 | 12 | 14 | 39 | 4* | 23 | 65 | 58 | 15 | 4 | 12 | 18 | 6* | 7 | 6 | 4* | 3 | 4 | 1 |
| 1* | <1 | 46 | 65 | 14 | 16 | 14 | 4 | 5* | 12 | 16 | 12 | 10+ | 6 | 9 | 29 | 2* | 42 | 47 | 4* | 1 | 1 | 2 |
| — | 1 | — | — | — | — | — | — | — | 3 | — | 1* | — | — | — | — | — | <1 | + | — | — | — | — |
| — | <1 | — | 2* | 1* | 1* | 1* | — | — | 8 | — | — | <1 | — | <1 | — | — | 1 | 9 | — | — | — | — |
| + | + | 10 | — | — | — | — | — | — | — | — | — | — | — | — | 3* | — | + | + | — | — | — | — |
| + | + | 17 | — | — | — | — | — | — | — | — | — | — | — | — | — | — | + | + | — | — | — | — |
| 6* | 82 | 40 | 21* | 15 | 16* | 18* | 10* | 1* | 68 | 5* | 8* | 15 | 4 | 42 | — | 3* | 70 | 65 | 10* | 66 | 54 | 96 |
| — | — | 2 | — | — | — | — | — | — | — | — | — | — | — | — | 9* | — | — | — | — | — | — | — |
| 1* | 2 | 24 | 1* | 5* | 1* | 2* | 2* | — | 21 | 3* | 2* | 84 | 96 | 58 | 9* | 5* | 24 | 17 | 3* | 34 | 46 | 4 |
| — | — | 4 | — | — | — | — | 2* | — | — | — | — | 1 | — | — | 5* | — | 5 | — | — | — | — | — |
| — | — | 3 | — | — | — | — | 2* | — | — | — | — | — | — | — | — | — | — | — | — | — | — | — |
| — | 1 | 2 | — | — | — | — | — | — | — | — | — | — | — | — | — | 1* | — | — | — | — | — | — |
| — | — | 58 | — | 1* | — | 1* | 4* | 2* | 11 | 2* | 1* | — | — | 3 | — | — | 54 | 28 | — | 1* | — | — |
| — | — | 6 | — | — | — | — | — | — | — | — | — | 8 | 2* | 11 | 1* | 3* | — | — | — | — | — | — |
| — | 1* | 2 | — | 1* | 2* | — | 1* | 1* | 9 | — | 2* | — | — | — | — | — | 1 | 6 | — | 1* | — | — |
| — | — | — | — | — | — | — | — | — | 3 | 1* | — | — | — | — | — | — | — | — | — | — | — | — |
| — | 1* | 14 | — | 1* | 1* | 1* | 1* | 1* | 75 | 23* | 17* | 50 | 4 | 2 | — | 1* | 3 | 10 | 1* | 1* | 4* | — |
| — | — | 2 | — | — | — | 1* | — | — | 2 | 1* | — | 4 | — | 4 | — | — | 2 | 6 | — | — | — | — |
| 1* | — | — | — | — | — | — | — | — | 1* | 1* | — | — | — | 2 | — | 1* | <1 | — | — | — | — | — |
| — | — | — | — | — | — | — | — | — | — | — | — | — | — | 1 | — | — | 3 | 3 | — | — | — | — |
| — | — | — | — | — | — | 1* | — | — | — | — | — | — | — | 2 | — | — | 3 | 10 | — | — | — | — |
| 1* | — | 18 | — | — | — | 1* | 4* | — | — | — | — | 38 | 16* | 75 | 5* | 1* | 26 | 44 | 3* | 2* | 3* | 1* |
| — | — | — | — | — | — | — | — | — | — | — | — | — | — | — | — | — | <1 | 14 | — | — | — | — |
| — | — | — | — | — | — | — | — | — | — | — | — | — | — | — | — | — | 1 | — | — | — | — | — |
| — | — | 28 | — | 2* | — | 3* | 1* | 4* | 2* | 2* | 4* | ~ | 24 | 69 | 8* | ~ | 39 | 42 | 1,, | 1* | — | 1* |
| — | — | — | — | — | — | — | — | — | — | 1* | — | 1* | — | 8 | 2* | — | 2 | 12 | — | — | 1* | — |
| — | — | — | 36* | — | 1* | — | — | — | 2* | 1* | — | — | — | — | — | 1* | 5 | 8 | 2* | 1* | 1* | 2* |
| — | — | — | — | — | — | — | — | — | — | — | — | 3* | 46 | — | — | — | — | — | — | — | — | — |
| — | — | 72 | — | — | — | — | — | — | — | — | — | 9* | 22 | 17 | — | — | — | — | — | — | — | — |
| 2* | 1* | — | 8* | 2* | 3* | 1* | — | 1* | 14* | 3* | 1* | 3* | 3 | 6 | 1* | 1* | 54 | 38 | 1* | — | — | — |
| — | — | — | — | — | — | — | — | — | — | — | — | 1* | 5 | — | — | — | — | — | — | — | — | — |
| — | — | — | — | — | — | — | — | — | — | — | — | — | — | — | — | — | — | 1* | — | — | — | — |

Section no. 31: the 30-m terrace at Katanga (below the mouth of the Chamba river). Oak pollen was found and a large quantity of spruce pollen was detected in four samples higher up the section.

Section no. 32: the 12-m terrace of the Oskoba river (according to Yu. P. Parmuzin, this terrace becomes the 25-m terrace of the Katanga river). A considerable quantity of spruce and fir pollen was found here.

Let us examine three more sections simultaneously (nos. 33, 34, and 35). These sections have been described by Ye. I. Sakharova, who collected the samples for analysis.

Section no. 33: the 8–10-m terrace of the Oka river (right bank), 2.5 km northeast of Okinskiye Sachki. The following layers are visible in the terrace outcropping:

| | | |
|---|---|---|
| 1. Sod and humus horizon | 0.02 m | |
| 2. Mottled sandy loam | 0.02–0.50 | |
| 3. Clayey, fine-grained sand | 0.50–2.15 | |
| 4. Gray-yellow sandy loam | 2.5 –3.5 | |

Section no. 34: the 10-m terrace of the Oka river at Okinskiye Sachki. The terrace alluvium is represented by the following deposits:

| | | |
|---|---|---|
| 1. | Sandy loam with humus | 0.05–0.15 m |
| 2. | Dark-gray sandy loam | 0.15–0.38 |
| 3. | Mottled gray-yellow sandy loam | 0.38–0.70 |
| 4. | Grayish-green sandy loam | 0.70–1.60 |
| 5. | Fine-grained sand | 1.60–2.15 |
| 6. | Gray sandy loam | 2.15–3.35 |

Section no. 35: the 10-m Oka terrace at Verkhne-Okinskiy. The samples were taken from gray, clayey sandy loam in the lower-lying sands.

All three sections are from the same region and represent terraces of the same character and height. Spore-pollen analysis indicates they belong to Phases IV and V. Elm pollen was found in the lower part of section 34, while the lower part of section 35 contained fir, spruce, pine, and alder pollen along with birch and pine. Thus, we are dealing with the forest type of spore-pollen spectrum with considerable participation of dark-coniferous tree specimens. As already mentioned, these spectra correspond to Phase IV.

The upper part of the alluvium of sections 33 and 34 contains spore-pollen spectra characterized by a birch-pollen maximum, reaching 58–96 per cent of the total tree pollen. The percentage content of grass pollen, especially xerophytes, increases upward along the sections. Such spore-pollen spectra could have formed in alluvium deposits only during a change in the vegetation pattern, when the forests thinned out greatly and the main forest species became birch and, in places, pine and larch. The dark-coniferous taiga disappeared. This time segment may be regarded as an independent phase, Phase V, in the development of the vegetation.

The character of the spectra obtained from the lower part of the alluvium of the 10-m Oka terraces is similar to that of the lower part of the alluvium of the 19-m Iya terrace (section no. 23); the spectra of the upper parts of the alluvium of both terraces are also similar. For example, it was found that the 10-m Oka terraces in the valley section situated (according to Ye. I. Sakharova) in a marginal fault zone are of the same age as the 19-m terrace of the lower Iya river near Kob.

The deposits of the upper alluvium of the 20–25-m terrace of the Angara river at Tokarey Island (section no. 36) may be regarded as belonging to Phase V. Upward along the section, the birch pollen content increases to 46 per cent and 78 per cent in the spore-pollen spectra. Pollen of the broadleaf species and the representative tree species of the dark-coniferous taiga, viz., the spruce and the fir, is not encountered. The spectrum shows a preponderance of pine-birch forests and also, apparently, sparse larch forests. Unfortunately, the deposits of the upper alluvium of this entire complex of Angara terraces contain very few plant remains. Pollen finds are rare, and then only pine and birch are encountered.

The next two phases of the late Pleistocene, Phases VI and VII (Table 17), are associated with a younger complex of terraces.

Spore-pollen spectra of a similar nature are found in the lower alluvial horizons of the 14–16-m Angara terraces and the 6–12-m terraces of the Iya, lower Oka, Kezhma, Konda, and Irkineyeva rivers. This similarity consists chiefly in the appearance of fir and spruce pollen, which sometimes form a considerable maximum (35–58%) among the tree species. The percentage content of Siberian pine

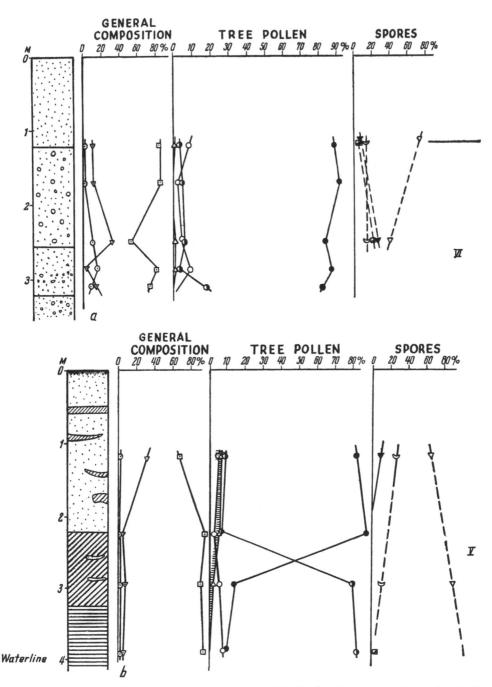

FIGURE 11. Spore-pollen diagrams: *a*, section no. 38, the first Iya terrace above the flood-plain, at Burkhun; *b*, section no. 39, the first Iya terrace above the floodplain, at Rodionovo.

## TABLE 17

### RESULT OF SPORE-POLLEN ANALYSIS OF SAMPLES OF TERRACE DEPOSITS OF THE ANGARA AND ITS TRIBUTARIES

SPORE AND POLLEN COMPOSITION (%)

| Section no.: | 37 | | | | 38 | | | | | 39 | | | | | 40 | | 41 | | | | 42 | | | | | 43 | 44 |
|---|---|---|---|---|---|---|---|---|---|---|---|---|---|---|---|---|---|---|---|---|---|---|---|---|---|---|---|
| Sample no.: | 1 | 3 | 9 | 11 | 4 | 5 | 6 | 7 | 8 | 2 | 4 | 6 | 8 | 10 | 7 | 8 | 2 | 3 | 4 | 5 | 5 | 7 | 8 | 9 | 10 | — | 2 |
| Sampling depth in m: | 0.12 | 1.0 | 4.33 | 5.0 | 1.2 | 1.7 | 2.5 | 2.85 | 3.1 | 0.42 | 1.18 | 2.26 | 2.96 | 3.9 | 6.0 | 10.0 | 3.0 | 6.0 | 9.0 | 11.0 | 2.05–3.0 | 4.25–4.6 | 4.65–5.5 | 5.5–7.0 | 7.0–8.5 | 6.0 | 0.5 |
| Number of grains counted: | 88 | 138 | 296 | 383 | 236 | 103 | 147 | 114 | 46 | 4 | 159 | 221 | 295 | 226 | 386 | 189 | 90 | 325 | 100 | 58 | 52 | 20 | 4 | 245 | 399 | 204 | 217 |
| **GENERAL COMPOSITION** | | | | | | | | | | | | | | | | | | | | | | | | | | | |
| Tree pollen | 38 | 5 | 80 | 63 | 85 | 86 | 54 | 81 | 74 | — | 68 | 95 | 90 | 93 | 6 | 79 | 25 | 85 | 81 | 53 | 19 | 2* | — | 18 | 29 | 95 | 90 |
| Grass and shrub pollen | 14 | 6 | 3 | 7 | 3 | 2 | 11 | 15 | 11 | — | 1 | <1 | 1 | 1 | 91 | 20 | 25 | 9 | 15 | 43 | 79 | 16* | — | 81 | 69 | 5 | 6 |
| Spores | 48 | 89 | 17 | 30 | 12 | 12 | 34 | 4 | 15 | — | 31 | 5 | 9 | 6 | 3 | 1 | 50 | 6 | 4 | 4 | 2 | 2* | — | 1 | 2 | 3 | 4 |
| Hydrophyte pollen | — | — | — | — | — | — | — | — | — | — | — | — | — | — | — | — | — | — | — | — | — | 3* | — | — | — | — | — |
| **POLLEN OF TREE SPECIES** | | | | | | | | | | | | | | | | | | | | | | | | | | | |
| Abies | 1* | — | — | — | — | — | — | — | — | — | 1* | — | — | — | 1* | 2 | — | — | — | — | — | — | — | — | — | — | — |
| Picea | — | — | 1 | <1 | 1 | 1 | 1 | 1 | — | — | 6 | 5 | 2 | 1 | 91 | 4 | <1 | 13 | 6 | — | — | — | — | <1 | 3 | 3 | 2 |
| Larix | — | — | — | <1 | <1 | 5 | — | — | — | — | — | — | — | — | — | — | — | 5* | 3 | — | — | — | — | — | — | — | — |
| Pinus sibirica | 9 | 9 | 9 | 14 | 4 | 5 | 6 | 3 | 18 | — | 8 | 6 | — | 7 | 6* | 54 | 32 | 71 | 31 | 45 | 4* | 11* | — | 7 | 15 | 28 | — |
| P. silvestris | 59 | 5* | 79 | 73 | 89 | 92 | 84 | 87 | 82 | — | 81 | 87 | 81 | 63 | — | 15 | 64 | 13 | 28 | 23 | 4* | 2* | — | 63 | 42 | 52 | — |
| Pinus | — | — | — | — | — | — | — | — | — | 1* | — | — | — | — | — | — | — | — | — | — | — | — | — | — | — | — | — |
| Betula | 20† | 1* | 10 | 12 | 9 | 3‡ | 6 | 9 | — | — | 5 | 2 | 29 | 29 | 1* | 21 | 32 | 13 | 38 | 22 | 2* | 3* | — | 29 | 19 | 11 | 89 |
| Alnus | — | 1* | <1 | 1 | — | 2 | 2 | — | 5 | 1* | — | 5 | 2 | 2 | 1* | 2 | 64 | 2 | 2 | — | 1 | — | — | 2 | 1 | 1 | 7 |
| Salix | — | — | — | — | — | — | — | — | — | — | 5 | — | — | — | 1* | — | 4 | — | — | — | 1* | — | — | — | — | — | 3 |
| **GRASS AND SHRUB POLLEN** | | | | | | | | | | | | | | | | | | | | | | | | | | | |
| Gramineae | — | — | — | 1* | — | 4* | 2* | — | 2* | — | 1* | 1* | — | — | 65 | 47 | 12* | 39 | 3* | 20 | 3 | 4* | — | 24 | 25 | 4 | 1* |
| Cyperaceae | — | — | 1* | 1* | — | 1* | — | — | — | — | 1* | — | — | — | 2 | 2 | 2* | 5 | 5* | 12 | 7 | — | — | <1 | — | 4 | — |
| Chenopodiaceae | — | — | 1* | 1* | — | — | — | — | — | — | — | — | — | — | 1 | 11 | 1* | 5* | — | — | 7 | — | — | 3 | 1 | 22 | 1* |
| Ericales | — | — | — | — | — | — | 2* | 5* | 3* | — | — | — | — | — | 7 | 26 | — | 13 | 3 | 44 | 63 | — | — | 13 | 13 | 13 | — |
| Artemisia | — | 1* | — | — | — | 2* | — | — | — | — | — | 1* | 1* | — | 5 | — | — | 7 | 7 | — | — | 11* | — | 45 | 55 | — | 5* |
| Caryophyllaceae | — | — | — | — | — | — | — | — | — | — | — | — | — | — | — | — | — | 8 | — | — | — | — | — | 2 | <1 | — | — |
| Polygonaceae | — | — | — | — | — | — | — | — | — | — | — | — | — | — | 1 | — | — | 1 | — | — | — | — | — | — | — | — | — |
| Onagraceae (Epilobium) | — | — | — | — | — | — | — | — | — | — | — | — | — | — | — | 4 | — | — | — | — | — | — | — | — | — | — | — |
| Umbelliferae | — | — | — | — | — | — | — | — | — | — | — | — | <1 | <1 | — | — | — | 2 | — | — | — | — | — | <1 | — | — | — |
| Rubiaceae | — | — | — | — | — | — | — | — | — | — | — | — | — | — | — | — | — | — | — | — | — | — | — | <1 | — | — | — |
| Plumbaginaceae (?) | 1* | — | — | — | — | — | — | — | — | — | — | — | — | — | — | — | 1 | — | — | — | — | — | — | 1 | — | — | — |
| Borraginaceae | — | — | — | — | — | — | — | — | — | — | — | 1* | — | — | — | — | — | 2 | — | — | — | — | — | — | — | — | — |
| Gentianaceae | — | — | — | — | — | — | — | — | — | — | — | — | — | — | — | — | — | — | — | — | — | — | — | 1 | — | — | — |
| Liliaceae | — | — | — | — | — | — | — | — | — | — | — | — | — | — | — | — | — | — | — | — | — | 1* | — | — | 8 | 5 | — |
| Labiatae | — | — | — | — | — | — | — | — | — | — | — | — | — | — | 1 | 3 | — | 2 | 12 | 20 | 22 | — | — | 9 | 11 | 39 | — |
| Other Compositae | 1* | 1* | 1* | 3* | 1* | 3* | — | 13* | 3* | 1* | 1* | 1* | <1 | <1 | 1 | 3 | 7* | 2 | — | — | 1* | — | — | — | — | 5 | 2* |
| Unidentified dicotyledons | 10* | 7* | 15* | 18* (8sp.) | 2* | 7* | — | — | — | 1* | 1* | 1* | <1 | <1 | 17 | 13 | — | 22 | 12 | 20 | 5 | — | — | 23 | 11 | 39 | 4* |
| Cruciferae | — | — | — | — | — | — | — | — | — | — | — | — | — | — | — | — | — | — | — | — | — | — | — | — | — | — | — |
| **SPORES** | | | | | | | | | | | | | | | | | | | | | | | | | | | |
| Bryales | 40 | 42 | 59 | 59 | 73 | 40 | 4* | 2* | 2* | 1* | 64 | 6* | 89 | 4* | 11* | — | 3 | 4 | 8* | 2* | 1* | 1* | — | 2* | 4* | 3* | 8* |
| Sphagnum | 7 | <1 | 10 | 7 | 7 | 26 | — | — | — | — | — | — | — | — | — | — | 3 | 5 | — | — | — | — | — | — | — | — | — |
| Lycopodium clavatum | 10 | 8 | 3 | 8 | — | 4 | — | — | 1* | — | — | 3 | — | — | — | 8 | — | 1 | — | — | — | 1* | — | — | — | — | — |
| L. pungens | — | — | — | 1 | — | — | — | 1* | — | — | — | — | — | — | — | — | — | — | — | — | — | — | — | — | — | — | — |
| L. annotinum | 5 | 19 | 6 | 3 | 3 | 6 | — | — | — | — | — | — | — | — | 7 | 3 | 6 | — | — | — | — | — | — | — | — | — | — |
| L. cf. alpinum | 5 | 15 | — | — | — | 6 | — | — | — | — | — | — | — | — | 5 | 8 | 8 | 5 | — | — | — | — | — | — | — | — | — |
| L. cf. appressum | — | <1 | — | — | — | — | — | — | — | — | — | — | — | — | 1 | 1 | 1 | 7 | — | — | — | — | — | — | — | — | — |
| L. typ. complanatum | 27 | — | 2 | 3 | — | — | — | 1* | 1* | 1* | — | — | <1 | 1 | 1 | — | 8 | 8 | — | — | — | — | — | — | — | — | — |
| Lycopodium | — | 2 | 18 | — | 3 | — | — | — | — | — | 26 | 4 | — | — | 1 | — | 2 | 51 | — | — | — | — | — | 5* | 5* | — | — |
| Polypodiaceae | 19 | 14 | 2 | 14 | — | 8 | — | — | — | — | — | — | — | — | 17 | 13 | 95 | 40 | 12 | 20 | 5 | — | — | 2* | 2* | 5* | 2* |
| Other Filicales | 12 | <1 | — | 5 | — | 6 | — | — | — | — | 2 | 11 | — | — | — | — | — | — | — | — | — | 1* | — | 3* | — | 2* | 1* |
| **POLLEN OF HYDROPHYTES** | | | | | | | | | | | | | | | | | | | | | | | | | | | |
| Hydrocharitaceae | — | — | — | — | — | — | — | — | — | — | — | — | 1* | — | — | — | — | — | — | — | — | — | — | — | — | 4* | — |
| Myriophyllum | — | — | 2* | — | — | — | 2* | — | — | — | — | — | — | 3* | — | — | — | — | — | — | — | — | — | — | — | — | — |

†Distorted spectrum. Birch pollen, stuck together in groups (not considered in calculating the percentages).

‡Pollen of only one species counted. The actual participation of birch pollen in the spectrum is greater than shown arbitrarily in the table.

# TABLE 17 (concluded)

| Section no.: | 45 | 45 | 46 | 47 | 47 | 47 | 47 | 48 | 48 | 48 | 48 | 49 | 49 | 50 | 50 | 50 | 50 | 51 | 51 | 52 | 52 | 52 | 52 | 52 | 52 | 53 | 53 |
|---|---|---|---|---|---|---|---|---|---|---|---|---|---|---|---|---|---|---|---|---|---|---|---|---|---|---|---|
| Sample no.: | 1 | 2 | — | 1 | 2 | 3 | 4 | 2 | 4 | 6 | 8 | 1 | 3 | 1 | 2 | 3 | 4 | 5 | 6 | 1 | 2 | 3 | 4 | 5 | 6 | 2 | 4 |
| Sampling depth in m: | 5.0 | 5.8 | 9.0 | 0.3–0.4 | 1.45–1.5 | 2.95 | 3.05–3.15 | 0.6 | 1.85 | 2.9 | 3.9 | 1.4–2.0 | 3.2–3.6 | 0.4 | 1.1 | 1.4 | 2.9 | 6.4 | 6.9 | 1.1 | 1.4 | 1.55 | 2.2 | 2.4 | 3.3 | 1.0 | 2.64 |
| Number of grains counted: | 284 | 234 | 34 | 79 | 65 | 24 | 122 | 24 | 29 | 5 | 12 | 184 | 31 | 188 | 59 | 43 | 32 | 214 | 265 | 36 | 45 | 24 | 7 | 17 | 12 | 43 | 16 |
| **SPORE AND POLLEN COMPOSITION (%)** | | | | | | | | | | | | | | | | | | | | | | | | | | | |
| **GENERAL COMPOSITION** | | | | | | | | | | | | | | | | | | | | | | | | | | | |
| Tree pollen | 53 | 63 | 84 | 41 | 23 | ~17 | 34 | 18* | 27* | — | 8* | 6 | 32 | 60 | 51 | 53 | 81 | 83 | 61 | 23 | 29 | 8* | 1* | 8* | 5* | 58 | 12* |
| Grass and shrub pollen | 2 | 3 | 7 | 58 | 74 | ~79 | 65 | 6* | 2* | 5* | 4* | 27 | 65 | 3 | 25 | 12 | — | 16 | 39 | 17 | 64 | 4* | 6* | 2* | 1* | 26 | 3* |
| Spores | 45 | 34 | 9 | 1 | 3 | 4 | 1 | — | — | — | — | 67 | 3 | 37 | 24 | 35 | 19 | 1 | <1 | 60 | 70 | 12* | — | 7* | 1* | 16 | 1* |
| Hydrophyte pollen | — | — | — | — | — | — | — | — | — | — | — | — | — | — | — | — | — | — | — | — | — | — | — | — | — | — | — |
| **POLLEN OF TREE SPECIES** | | | | | | | | | | | | | | | | | | | | | | | | | | | |
| Abies | 7 | 18 | 3 | — | — | — | — | — | — | — | — | — | — | — | — | — | — | — | — | — | — | — | — | — | — | — | — |
| Picea | 85 | 29 | 58 | — | — | — | — | — | — | — | — | — | — | — | — | — | — | — | — | — | — | — | — | — | — | — | — |
| Larix | — | — | — | — | — | — | <1 | — | — | — | — | — | — | — | — | — | — | — | — | — | — | — | — | — | — | — | — |
| Pinus sibirica | — | — | 38 | — | — | — | — | — | — | — | — | — | — | — | — | — | — | <1 | 2 | — | — | — | — | — | — | — | — |
| P. silvestris | — | — | — | — | — | — | — | 12* | 17* | — | 4* | 8* | 4* | + | 3 | 4 | 4 | — | — | — | — | 8* | — | 3* | 3* | — | — |
| Pinus | 50 | 40 | 1 | 97 | ~80 | ~2 | 83 | 3* | 9* | — | — | 3* | 5* | 51 | 70 | 74 | 70 | 97 | 85 | 9* | 5* | — | 1* | — | — | 92 | 10 |
| Betula | 5 | 13 | — | 3 | ~20 | 2* | 12 | 3* | 1* | — | 4* | 3* | 5* | 49 | 27 | 22 | 24 | 2 | 13 | — | — | — | 1* | 4* | — | — | 2 |
| Alnus | 3 | — | — | — | — | — | <1 | — | — | — | — | — | — | — | — | — | — | — | — | — | — | — | — | — | — | 8 | — |
| Salix | — | — | — | — | — | — | — | — | — | — | — | — | — | — | — | — | — | — | — | 8* | 8* | — | — | — | — | — | — |
| **GRASS AND SHRUB POLLEN** | | | | | | | | | | | | | | | | | | | | | | | | | | | |
| Gramineae | 5* | 2* | — | 37 | 6 | ~5 | 37 | — | — | — | — | 16 | 1* | — | — | — | — | 6* | 8 | — | 3 | — | — | — | — | — | — |
| Cyperaceae | — | — | — | — | — | — | — | — | — | — | — | 12 | — | — | — | — | — | 8 | — | 3 | — | — | — | — | — | — | — |
| Chenopodiaceae | — | — | — | 2 | 2 | 3 | 3 | — | — | — | — | — | — | — | — | — | — | 13* | 10 | — | — | — | — | — | — | — | — |
| Ericales | — | — | — | — | — | — | — | — | — | — | — | — | — | — | — | 1* | — | 16* | 53 | — | — | — | — | — | — | — | — |
| Artemisia | 1* | — | — | 46 | ~63 | ~60 | 36 | — | — | — | — | 62 | 13* | 4* | 7* | 4* | — | — | 5 | 87 | — | 4* | — | 1* | — | 3* | 1* |
| Caryophyllaceae | 1* | 2* | 2* | — | 2 | — | — | — | — | — | — | — | — | — | — | — | — | 5* | 5 | — | — | — | — | — | — | — | — |
| Polygonaceae | — | — | — | — | — | — | — | — | — | — | — | — | — | — | — | — | — | — | — | — | — | — | — | — | — | — | — |
| Onagraceae (Epilobium) | — | — | — | — | — | — | — | — | — | — | — | — | — | — | — | — | — | — | — | — | — | — | — | — | — | — | — |
| Umbelliferae | — | — | — | — | — | — | — | — | — | — | — | — | — | — | — | — | — | — | — | — | — | — | — | — | — | — | — |
| Rubiaceae | — | — | — | — | — | — | 4 | — | — | — | — | — | — | — | — | — | — | — | — | — | — | — | — | — | — | — | — |
| Plumbaginaceae | — | — | — | — | — | — | — | — | — | — | — | — | — | — | — | — | — | — | — | — | — | — | — | — | — | — | — |
| Borraginaceae | — | — | — | — | — | — | — | — | — | — | — | — | — | — | — | — | — | — | — | 1* | — | — | — | — | — | — | — |
| Gentianaceae | — | — | — | — | — | — | — | — | — | — | — | — | — | — | — | — | — | — | — | — | — | — | — | — | — | — | — |
| Liliaceae | — | — | — | — | — | — | — | — | — | — | — | 1* | — | 1* | 1* | — | — | — | — | — | — | — | — | — | — | 2* | — |
| Labiatae | — | 1* | — | — | — | — | — | — | — | — | — | 1* | — | 7* | 7* | — | — | 8* | — | 1* | 7 | — | — | — | — | — | — |
| Other Compositae | — | — | 1* | 8* | 14 | + | 10 | — | — | — | — | 8 | 6* | 1* | — | — | — | — | 1 | — | 3 | — | 3* | — | 1* | 2* | 1* |
| Unidentified dicotyledons | 2* | 1* | 1* | 7 | 14 | + | 10 | — | — | — | — | 1 | — | 7* | 7* | — | — | 8* | 13 | 1* | — | — | 2* | — | — | 6* | 1* |
| Cruciferae | — | — | — | — | — | — | — | — | — | — | — | 1* | — | — | — | — | — | — | — | — | — | — | — | — | — | — | — |
| **SPORES** | | | | | | | | | | | | | | | | | | | | | | | | | | | |
| Bryales | 65 | 90 | — | 1* | 1* | 1* | — | 6* | 2* | 3* | 4* | 9 | 9* | 71 | 2* | 9* | 4* | — | — | 20* | 3* | 10* | 3* | 4* | 6* | 6* | 1* |
| Sphagnum | 2 | 5 | 1* | — | — | — | — | — | — | 2* | 1* | 1 | 1 | 10 | 2* | — | 1* | — | — | 1* | — | 1* | 2* | — | — | — | — |
| Lycopodium clavatum | — | — | — | 1* | — | — | — | — | — | — | — | — | — | — | — | — | — | — | — | 3* | — | 1* | 2* | — | — | 6* | 1* |
| L. pungens | — | — | — | — | — | — | — | — | — | — | — | — | — | — | — | — | — | — | — | — | — | — | — | — | — | — | — |
| L. annotinum | — | — | — | — | — | — | — | — | — | — | — | — | — | — | — | — | — | — | — | — | — | — | — | — | — | — | — |
| L. cf. alpinum | — | — | — | — | — | — | — | — | — | — | — | — | — | — | — | — | — | — | — | — | — | — | — | — | — | — | — |
| L. cf. cupressum | — | — | — | — | — | — | — | — | — | — | — | — | — | — | — | — | — | — | — | — | — | — | — | — | — | — | — |
| L. typ. complanatum | — | — | — | 1* | 1* | — | — | — | — | — | — | 85 | 1* | 7 | — | — | — | * | — | — | — | — | — | — | — | — | — |
| Lycopodium | 2 | 4 | 2* | — | 1* | 1* | — | — | — | — | — | — | — | — | 10* | 6* | — | 3 | 1* | 1* | 1* | — | — | 1* | — | — | — |
| Polypodiaceae | 31 | 1 | — | — | — | — | — | — | — | — | — | 5 | — | 3 | — | — | — | — | — | 1* | 1* | — | — | 3 | — | 1* | — |
| Other Filicales | — | — | — | 1* | — | 1* | — | — | — | — | — | — | — | — | — | — | — | — | — | — | — | — | — | — | — | — | — |
| **POLLEN OF HYDROPHYTES** | | | | | | | | | | | | | | | | | | | | | | | | | | | |
| Hydrocharitaceae | — | — | 1* | — | — | — | — | — | — | — | — | — | — | — | — | — | — | — | — | — | — | — | — | — | — | — | — |
| Myriophyllum | — | — | — | — | — | — | — | — | — | — | — | — | — | — | — | — | — | — | — | — | — | — | — | — | — | — | — |

pollen increases. Birch pollen comprises only 1–12 per cent of the total tree pollen content and alder pollen is found frequently.

During the formation of the terraces of this complex, i.e., the deposits of the lower alluvium, natural conditions again favored the spread of dark-coniferous taiga and mixed-coniferous forests, with a corresponding reduction of the area occupied by sparse birch forests and wooded steppe. Our Phase VI corresponds to this last amelioration of the climate during the Pleistocene, which facilitated the growth of mixed-coniferous forests, in which the role of dark-coniferous taiga increased.

The deposits of the upper alluvium of this complex of terraces and the younger 6–9-m Angara terraces contain plant remains from the last cold and relatively dry period of the Pleistocene. During this period, the composition of the forests changed. The spruce, fir, and Siberian pine disappeared. Over broad expanses the steppes crowded out the forest and reached their maximum development in the Angara basin during this phase. The remaining forests consisted of sparse light-coniferous taiga and birch groves (possibly the dwarf birch).

Phases VI and VII can be illustrated by a number of cross-sections. The results of the spore-pollen analyses of these sections are summarized in Table 17.

Section no. 37: the 10-m Oka terrace near Shamanovo. The terrace alluvium, according to V. A. Rastvorova's description, consists of sands and loams. During the accumulation of the lower part of the alluvium (at a depth of 4–5 m in the section), forest conditions similar to the present conditions prevailed, judging by the participation of spruce, pine, birch, and some Siberian pine in the forest complex. The upper alluvium formed during more severe climatic conditions, as indicated by the gradual change in the spectra upward along the section. The quantity of grass pollen increased and the quantity of tree pollen decreased. This section represents Phase VI and the beginning of Phase VII.

Section no. 38: the first terrace above the floodplain of the Iya river near Burkhun. According to V. A. Rastvorova, the terrace alluvium here consists of pebble gravel, and higher up of sands and sandy loam. In general, the pollen and spore composition of the alluvium is similar to that of the present floodplain deposits in this same region and is also similar to that of the samples taken from section no. 37 at Shamanovo. This region was somewhat less forested at the time (Fig. 11a).

Section no. 48: the first terrace of the Iya river above the floodplain near the mouth of the Ilir river. A small amount of pollen was found in four samples. It is interesting that 17(!) of the 24 pollen grains found were spruce.

Section no. 39: the first terrace above the floodplain, the 4-m terrace of the Iya river near Rodionovo (Fig. 11b). The lower part of the section corresponds to the Siberian pine pollen maximum; spruce pollen runs as high as 6 per cent; birch pollen is infrequent. Upward along the section, pine pollen replaces the Siberian pine pollen. Here we see a picture of mixed-coniferous forests of spruce, Siberian pine, and pine gradually becoming light-coniferous forests.

Section no. 40: the 16-m Angara terrace at Prospikhino. At a depth of 10 m the sample contains fir, spruce, and Siberian pine pollen (54%). Higher, at a depth of 6 m, the grass-pollen content increases, and the percentage of pine pollen increases at the expense of Siberian pine. The analysis indicates that the deposits comprising the lower alluvium of the terrace belong to Phase VI.

Section no. 41: the 12-m terrace of the Koda river. According to S. S. Voskresenskiy, the alluvium of this terrace consists of the following horizons:

1. Loam                                          1–3 m
2. Sand                                          3–8
3. Sand, with considerable humus, with
    pebbles 3–5 cm in diameter                    8–11

As can be seen from Figure 12, the Siberian pine and spruce maximum corresponds to the two lower horizons of alluvium; the birch maximum (64%) corresponds to the upper horizon where the pollen of spruce and Siberian pine is minimum. The section is divided into two parts according to pollen composition: Phase VI, corresponding to the spread of the dark-coniferous taiga (expansion of its domain), and Phase VII, corresponding to the dominance of sparse light-coniferous and birch forests.

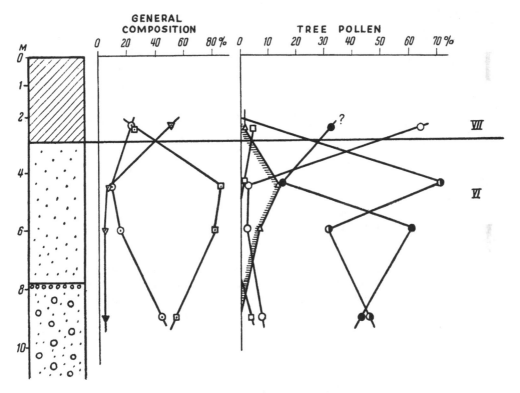

FIGURE 12. Spore-pollen diagram of section no. 41, the 12-m terrace of the Koda river.

Section no. 42: the 14-m Angara terrace at Kazachye. The samples were taken by Yu. P. Parmuzin from alluvium and show the following deposits:

1. Sandy loam of various hues, stratified and unstratified, grading downward into cross-bedded sandy      0–3 m
    loam with lenses of coarse-grained sand
2. Sandy loam with red-brown interlayers          3–4.5
3. Coarse-grained, cross-bedded sands           4.25–4.65
4. Sandy loam, fine, powdery                4.65–5.55
5. Pebble gravel with boulders of various composition    5.55–8.55

The pebble gravel horizon contains Siberian pine and spruce pollen, which decreases in amount upward while the amount of pine and birch pollen increases. Evidently the diagram (Fig. 13) reflects the end of Phase VI and the beginning of Phase VII. The pollen composition is distinctive in that it contains a large quantity of grass pollen, corresponding to wooded-steppe or even steppe conditions, which prevailed in this region during Phase VII.

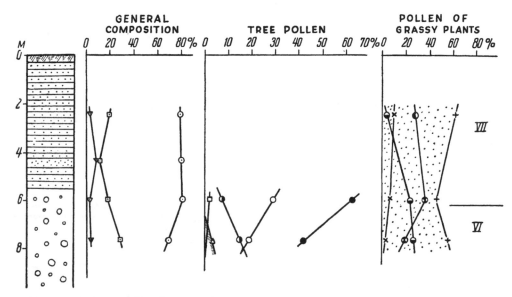

FIGURE 13. Spore-pollen diagram of section no. 42, the 14-m terrace of the Angara river at Kazachye.

Section no. 43: the 16-m terrace of the Kezhma river. A sample of close-grained sand was taken for analysis from the middle portion of the alluvium at a depth of 6 m. The spore-pollen analysis showed the spectra obtained here to be similar to the spectra of the lower parts of sections 37, 38, and 39. Conifer pollen predominated among the tree species. Siberian pine comprised 28 per cent, spruce 3 per cent, and pine 52 per cent. The pollen composition of this deposit also corresponds to Phase VI.

Section no. 44: the 16-m Angara terrace at Padun. The sample for analysis was taken from the upper alluvium horizon. Conifer pollen, including spruce, dominated the tree pollen. Birch pollen comprised only 7 per cent. Unfortunately, Siberian pine was not identified among the *Pinus* species, but the forest aspect of the spectrum, the presence of spruce and alder pollen, and the small amount of birch indicate that the deposits of the upper part of the 16-m Angara terrace in this region belong to Phase VI.

Section no. 45: the 12–15-m terrace of the Irkineyeva river near Bedoba. The samples were taken by Ye. A. Falkova at depths of 5.0 and 5.8 m, from a layer of peat and clay in the alluvium. Falkova mentioned a find of mammoth bones near the peat layer. The spore-pollen analysis revealed a spectrum of the forest type. The pollen of conifers dominated the tree species. The pollen counts of fir and spruce were very high (47% and 42%, respectively). Birch comprised only 3–13 per cent of the tree pollen. Sphagnum and club-moss spores were found. The high spruce and fir pollen content here, compared with the other sections, may be due to the swampy nature of the deposits; the spore-pollen spectra of swamp deposits are formed basically of sublocal components of the vegetation cover. However, a sample taken from clay higher in the section, and not from the peat, also contained considerable fir and spruce pollen. The lower part of the deposits, comprising the 12–15-m Irkineyeva terrace, definitely belongs to Phase VI.

Section no. 46: the 12–17-m terrace of the Irkineyeva river, near its mouth. The sample was taken by Ye. I. Sakharova at a depth of 9 m, from the pebble gravel horizon. It is particularly interesting to note the similarity between the spectrum taken from the pebble gravel stratum and that taken from the peat and clay of the 12–15-m terrace farther upstream, at Bedoba. The height of the deposits and the similarity of the spectra indicate that they come from the same terrace of the Irkineyeva river. The sample from the pebble gravel stratum contains tree pollen in the following proportions: spruce 58 per cent, Siberian pine 38 per cent, fir 3 per cent, and pine only 1 per cent; no birch pollen was found. Analysis of the samples from sections 45 and 46 indicates the existence of a continuous dark-coniferous taiga in the Irkineyeva basin during the formation of the lower deposits of the 12–17-m terrace.

Section no. 47: the 18-m Angara terrace at Zvezdochka. According to Yu. P. Parmuzin, the terrace alluvium contains the following deposits:

| | |
|---|---|
| 1. Sand | 0.02–0.4 m |
| 2. Sand, close-grained, interstratified, with inequigranular and coarse-grained sand | 0.4–2.93 |
| 3. Loess-like loam | 2.93–2.95 |
| 4. Coarse-grained sand | 2.95–3.5 |
| 5. Pebble gravel | 3.5–5.5 |
| 6. Limestone | To water level of river |

Samples of the upper alluvium were analyzed. The pollen content was similar to that of comparable samples from the upper alluvium of the 14-m Angara terrace at Kazachye (section no. 42). Here, too, grass pollens predominated in the spectra and pine and birch dominated the tree pollen (Fig. 14). The spectra correspond to Phase VII, i.e., to wooded steppe, sparse-birch, and light-coniferous forests. Phase VII continues in the alluvium of the younger terraces.

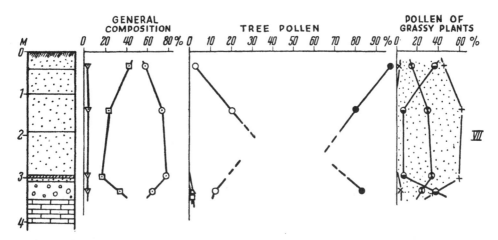

FIGURE 14. Spore-pollen diagram of section no. 47, the 18-m Angara terrace at Zvezdochka.

Section no. 49: the 9-m Angara terrace at Baygano. The spore-pollen spectra of alluvium samples from this section showed that grasses, particularly *Artemisia*, Gramineae, and orache (goosefoot species), played a large role. Pine and birch pollen was also found.

Section no. 50: the 10-m Angara terrace near Miloslavka. These deposits may be divided into two groups. The spore-pollen spectrum of the lower segment (clays and loams) corresponds to Phase VI and thus is synchronous with the lower alluvium of the 14–16-m Angara terrace in this same region. The upper segment corresponds to the spectra of the end of Phase VII, i.e., it is synchronous with the upper alluvium of the 6–10-m Angara terrace in this region. The spore-pollen analyses indicate that the accumulation of the deposits comprising the 10-m Angara terrace at Miloslavka was interrupted. At the same time, the lower portion (apparently the socle or base) of the terrace is of the same age as the lower alluvium of the older complex of 14–16-m Angara terraces in this part of the valley. This deduction is in keeping with the terrace structure. Ye. V. Koreneva showed that the clays and loams are separated from the higher-lying sandy loams by a layer of buried soil which is partially eroded; further, these two portions of the alluvium do not conform to one another (Fig. 15).

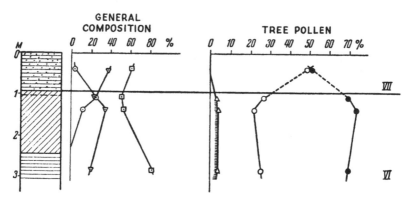

FIGURE 15. Spore-pollen diagram of section no. 50, the 10-m Angara terrace at Miloslavka.

Section no. 51: the 9-m Angara terrace near Tashlykova. This terrace consists of close-grained sand and porous loam with columnar jointings. The samples were taken from a layer of sand at depths of 6.4 and 6.9 m. The pollen composition corresponds to the end of Phase VII. There is a large quantity of grass pollen here, but the grass pollen content decreases upward along the section. *Artemisia* and Chenopodiaceae comprise 63 per cent of the grass pollens, while pine dominates the tree pollen.

Section no. 52: the 6–8-m Angara terrace at Bertsov Island. The deposits consist of loams with buried soil horizons; lower down there are clays with intercalations of sand; still lower, pebble gravel and sand. The samples contain few plant remains. The spore-pollen analysis showed large quantities of grass pollens, among which *Artemisia* predominated. This spore-pollen spectrum represents the wooded steppe. Pine and birch were encountered most frequently among the tree pollens.

Section no. 53: the 6-m Angara terrace at Pyana. The pollen and spore composition of the samples taken from a horizon of fine-grained sand at a depth of 1.0

m and from sands with pebbles at a depth of 2.64 m is similar to that of section 51. These deposits also belong to Phase VII.

Thus, the series of sections of alluvial deposits in the Angara basin indicate that the first and second terraces above the floodplain in the portions of the basin investigated formed during two distinct periods of physicogeographic development, corresponding to two successive phases of vegetation development, Phases VI and VII.

Analysis of the deposits of the floodplain terrace of the Angara and Iya rivers and analysis of peat from modern swamps give us some idea of the ensuing time-segment. Table 18 and the diagrams show the results of the analyses. The connection with Phase VII can still be seen in the sections which contain the older Holocene deposits. Phase VII is reflected in the considerable participation of birch pollen in the spectra, a typical feature of that phase. The later deposits reflect the gradual re-establishment of the forests, the increased role of dark-coniferous forests, and the reduction of the steppe areas. Only the uppermost samples show some evidence of a change in the spectra, apparently associated with a recent progression of steppes in the valleys and a reduction of the area of dark-coniferous taiga due to the expansion of birch-aspen forests. The disposition of the seven investigated sections is shown on the maps in Figure 20. We shall treat the entire Holocene as Phase VIII and not attempt a finer breakdown of the Holocene deposits, for want of sufficient evidence.

Section no. 54: the floodplain terrace of the Iya river, 1 km above Burkhun. The outcropping is located on an island. The deposits consist of sandy loam with lenses of laminated dark-gray sand and clay. The quantity of birch pollen decreases noticeably upward along the section and the percentage of fir and spruce pollen increases. The deposits correspond to the lower and middle Holocene (Table 18).

Section no. 55: a peat bog situated on the right slope of the Iya watershed north of the Bolshoy Korabl river. Figure 16a shows that the vegetation did not change substantially during the period when the 3-m peat deposit accumulated. Then, as now, there were light-coniferous pine forests with birch and larch in this region. Spruce was somewhat more widely distributed than now; its pollen amounts to 4–6 per cent in the spectra, while only occasional grains are found in the surface samples. These deposits correspond to the Upper Holocene.

Section no. 56: a peat bog on the left slope of the Angara watershed below Padun near Kadarinskaya Sopka [Hill]. The spore-pollen diagram (Fig. 16b) is similar to the diagram of section 55. The vegetation of the Middle and Late Holocene was similar to the present vegetation, as can be seen in the diagram. Upward along the section, spruce and fir decrease slightly in the pollen spectrum. The Siberian pine forms a slight maximum in the middle part of the section.

Sections nos. 57 and 58: the floodplain terrace of the lower Angara at Motygino. The samples were collected by Ye. I. Sakharova from the lower and middle parts of the floodplain alluvium. The lower part of the alluvium, apparently corresponding to the warm Middle Holocene, has a forest-type spore-pollen spectrum. In contrast to the spectra of the first Angara terrace above the floodplain, this contains a considerable amount (51%) of spruce, Siberian pine, and fir pollen, species of the dark-coniferous taiga. The dark-coniferous taiga of the northwestern part of the Angara basin was somewhat more widely distributed then than at present.

The same forest spectrum can be seen in the middle part of the alluvium of section 57 (Fig. 16c); the spruce pollen decreases somewhat and the birch pollen

# TABLE 18

## RESULT OF SPORE-POLLEN ANALYSIS OF SECTIONS 54–58 OF FLOODPLAIN AND MODERN SWAMP DEPOSITS OF THE ANGARA BASIN

| Section no. : Sample no. | 54:2 | 54:3 | 54:4 | 54:6 | 54:7 | 54:8 | 55:1 | 55:2 | 55:3 | 55:4 | 55:5 | 55:6 | 55:7 | 55:8 | 56:1 | 56:2 | 56:3 | 56:4 | 56:5 | 56:6 | 56:7 | 56:8 | 57:1 | 57:2 | 57:3 | 58:2 |
|---|---|---|---|---|---|---|---|---|---|---|---|---|---|---|---|---|---|---|---|---|---|---|---|---|---|---|
| Sampling depth in m | 0.29 | 0.45 | 0.95 | 1.97 | 2.39 | 2.67 | 0.5 | 1.0 | 1.5 | 2.13 | 2.5 | 2.6 | 2.7 | 2.9 | 0.15 | 0.2 | 0.4 | 0.5 | 0.7 | 0.9 | 1.1 | 1.25 | 0.7 | 1.5 | 1.5 | 3.6 |
| Number of grains counted | 190 | 7 | 46 | 52 | 47 | 50 | 216 | 84 | 43 | 43 | 43 | 36 | 62 | 92 | 483 | 386 | 444 | 483 | 417 | 512 | 499 | 373 | 661 | 192 | 285 | 342 |
| **GENERAL COMPOSITION** | | | | | | | | | | | | | | | | | | | | | | | | | | |
| Tree pollen | 68 | 4* | 74 | 13 | 20 | 16 | 63 | 80 | 54 | 61 | 49 | 69 | 48 | 58 | 62 | 75 | 74 | 62 | 90 | 72 | 70 | 87 | 41 | 35 | 36 | 58 |
| Grass and shrub pollen | 2 | — | 6 | — | 3 | — | 2 | 2 | 4 | 2 | — | 3 | 4 | 2 | 3 | 7 | <1 | 3 | 3 | 8 | 10 | 10 | 5 | 4 | 5 | 2 |
| Spores | 30 | 3* | 20 | 87 | 77 | 84 | 35 | 18 | 42 | 37 | 51 | 28 | 48 | 41 | 35 | 18 | 25 | 25 | 7 | 20 | 20 | 3 | 54 | 61 | 59 | 40 |
| | | | | | | | | | | | SPORE AND POLLEN COMPOSITION (%) | | | | | | | | | | | | | | | |
| **POLLEN OF TREE SPECIES** | | | | | | | | | | | | | | | | | | | | | | | | | | |
| *Abies* | <1 | — | — | — | — | — | 1 | 3 | — | — | — | — | — | 2 | 3 | 1 | 2 | 3 | 3 | 2 | <1 | <1 | <1 | 1 | — | 2 |
| *Picea* | 9 | 3* | — | — | — | — | + | 1 | 4 | 4 | — | 4 | — | 6 | <1 | 1 | 3 | 3 | 2 | 2 | 4 | 3 | 3 | 1 | 1 | 10 |
| *Larix* | <1 | — | — | — | — | — | — | — | — | — | 1* | — | — | — | — | — | — | — | — | — | — | — | — | — | — | — |
| *Pinus sibirica* | — | — | — | — | — | — | 7 | 9 | 4 | 4 | 16* | 72 | 87 | 83 | 5 | 6 | <1 | 8 | 13 | 8 | 4 | 2 | 31 | 14 | 34 | 39 |
| *P. silvestris* | 77 | — | 71 | — | — | — | 78 | 77 | 80 | 77 | — | — | — | — | 73 | 76 | 69 | 75 | 55 | 65 | 58 | 68 | 42 | 65 | 44 | 44 |
| *Pinus* | — | — | — | — | — | — | — | — | — | — | — | — | — | — | — | — | — | — | — | — | — | — | — | — | — | — |
| *Betula* | 17 | 1* | 29 | 1* | 6* | 7* | 14 | 10 | 15 | 15 | 4* | 24 | 13 | 9 | 18 | 15 | 26 | 12 | 27 | 20 | 34 | 26 | 21 | 10 | 15 | 2 |
| *Alnus* | — | — | — | — | 1* | 1* | — | — | — | — | — | — | — | — | <1 | 1 | <1 | <1 | — | 1 | 1 | 1 | 2 | 3 | 4 | 2 |
| *Salix* | — | — | — | — | — | — | — | — | — | — | — | — | — | — | — | — | — | — | — | — | — | — | — | 1 | 2 | 1 |
| **POLLEN OF GRASSY PLANTS AND SHRUBS** | | | | | | | | | | | | | | | | | | | | | | | | | | |
| Chenopodiaceae | — | — | — | — | — | — | — | — | — | — | — | — | — | — | — | — | — | — | — | — | — | — | — | — | — | — |
| Gramineae | — | — | — | — | — | — | — | — | — | — | — | — | — | — | 1* | — | — | — | — | 1 | 1* | 15 | 6 | 1* | 1* | — |
| Cyperaceae | — | — | — | — | — | — | — | — | 1* | — | — | — | 2* | 1* | 6* | 74 | 1* | 4* | 4* | 40 | 46 | 46 | 12 | — | 2* | — |
| Ericales | — | — | — | — | — | — | 2* | — | — | 1* | — | — | — | — | 6* | 22 | 2* | 4* | 1* | 46 | 1* | 18 | 3 | — | — | 5* |
| *Artemisia* | 5* | — | — | — | — | — | — | — | 1* | — | — | — | — | — | — | — | 2* | 4* | 4* | 12 | 1* | 13 | 15 | 1* | 4* | — |
| **MOTLEY GRASSES** | | | | | | | | | | | | | | | | | | | | | | | | | | |
| Compositae | — | — | — | — | — | — | — | — | — | — | — | 1* | — | — | — | 4 | — | 1* | — | — | — | — | 25 | 3* | 3* | — |
| Caryophyllaceae | — | — | — | — | — | — | — | — | — | — | — | — | — | — | — | — | — | — | — | 1 | 8 | — | — | — | — | — |
| Polygonaceae | — | — | — | — | — | — | — | — | — | — | — | — | — | — | — | — | — | — | — | — | — | — | 3 | 1* | 1* | — |
| *Polemonium* | — | — | — | — | — | — | — | 2* | 1* | 1* | — | — | — | — | — | — | — | — | — | — | — | — | — | 3 | 3 | — |
| Unidentified | 5* | 3* | 3* | — | 1* | — | 2* | — | — | 1* | 5* | — | — | — | 6* | — | — | 4* | 4* | — | — | — | 31 | 2 | 3* | 3* |
| **SPORES** | | | | | | | | | | | | | | | | | | | | | | | | | | |
| Bryales | 62 | — | 4* | 34 | 21 | 59 | 3 | 10* | 12* | 9* | 17* | 3* | 3 | 3 | 3 | — | — | — | — | — | — | 2* | 42 | 38 | 34 | 31 |
| *Sphagnum* | 18 | 1* | 2* | 2 | 10 | 7 | 65 | 65 | 6* | — | — | 3* | 54 | 55 | 86 | 100 | 100 | 99 | 93 | 99 | 100 | 3* | 7 | 10 | 11 | 4 |
| *Lycopodium clavatum* | 3 | — | 1* | 2 | 7 | 5 | — | — | — | — | — | — | — | — | — | — | — | — | — | — | — | — | — | — | — | — |
| *L. annotinum* | 9 | — | — | — | — | 7 | 7 | 1* | 1* | 2* | — | 10 | — | — | — | — | — | <1 | 3 | 1 | — | 3* | 8 | 14 | 5 | 7 |
| *Lycopodium* | — | — | — | — | — | — | — | — | — | — | — | — | — | — | — | — | — | — | — | — | — | — | — | — | — | — |
| Polypodiaceae | 9 | 2* | 2* | 62 | 31 | 20 | 25 | 4* | 1* | 5* | 5* | 7* | 33 | 42 | 11 | — | — | <1 | 4 | — | 2* | 2* | 40 | 37 | 50 | 56 |
| Filicales | — | — | — | — | — | — | — | — | 5* | — | — | — | — | — | — | — | — | — | — | — | — | — | 8 | — | — | 1 |
| Unidentified | — | — | — | — | — | 2 | — | — | — | — | — | — | — | — | — | — | — | — | — | — | — | — | 4 | — | 1 | — |
| *Selaginella selaginoides* | — | — | — | — | — | — | — | — | — | — | — | — | — | — | — | — | — | — | — | — | — | — | — | — | — | — |

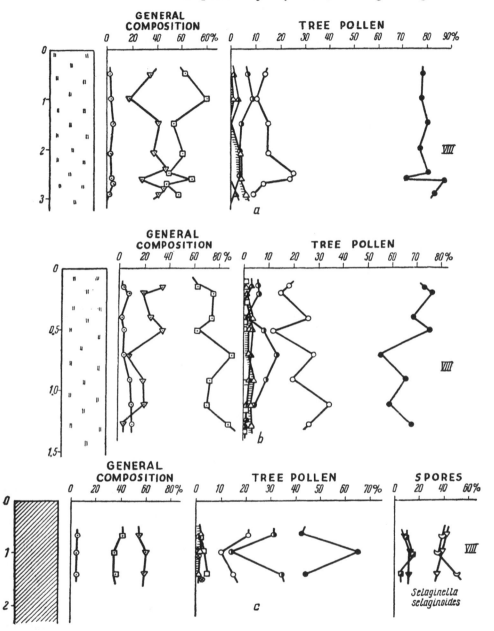

FIGURE 16. Spore-pollen diagrams: *a*, section no. 55, peat bog north of the Bolshoy Korabl river; *b*, section no. 56, peat bog below Padun; *c*, section no. 57, the Angara floodplain near Motygino.

increases to 21 per cent. The spectra are similar to the present-day spectra. Thus, it may be concluded that at the end of the Holocene the role of light-coniferous taiga and birch forests (more likely, birch-aspen) increased at the expense of the dark-coniferous taiga.

In 1955, M. I. Neyshtadt published a spore-pollen diagram of the lake-swamp deposits of the Umykey swamp in the Oka basin (Fig. 17), and M. N. Nikonov

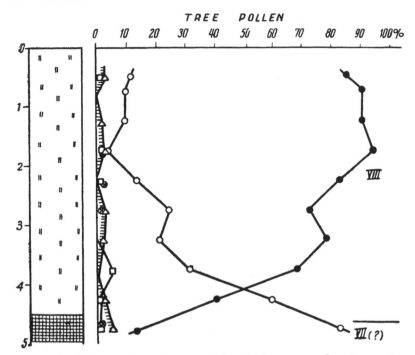

FIGURE 17. Spore-pollen diagram of the Umykey swamp deposits, section no. 59 (after M. N. Nikonov).

analyzed these deposits. Neyshtadt considers it possible to classify all the deposits found in this section as Holocene and to divide the time of their accumulation into Older, Early, Middle, and Late Holocene.

In our opinion, section no. 59 from the Umykey swamp in the Oka basin is truly typical of the Holocene, as indicated by the decrease of birch pollen from a maximum of 82 per cent to 20–30 per cent in the lower part of the diagram. This corresponds to the beginning of the Holocene. It is easy to picture the transition from Phase VI, characterized by the sections from the first terrace above the floodplain, to Phase VIII in the contemporary peat bogs, i.e., from the birch-pollen maximum to the much-reduced birch content. The pine-pollen curve shows a maximum higher along the section. Siberian pine is not distinguished from the *Pinus* pollen, but we may assume a Siberian pine maximum here in the Middle Holocene deposits. The Umykey swamp section corresponds well to sections 54, 55, and 56. During Phase VIII in the Angara basin, the steppe decreased and the role of the dark-coniferous taiga increased.

The vegetation assumed a modern appearance in the Late Holocene; the role of the birch-aspen and light-coniferous forests increased and the dark-coniferous taiga retreated somewhat.

*III*

The physical data we have examined indicate that the Angara flora did not experience profound changes during the Quaternary period, changes comparable with those of the period transitional between the Tertiary and the Quaternary. Very little change can be detected in the present flora, but changes in the vegetative cover have become more evident.

The many works which have been devoted to the history of the topography and the glaciation during the Quaternary indicate that the Angara vegetation developed under the influence of two basic natural factors: glaciation and orogenesis.

Drawing upon paleontologic data for all Eurasia (including our own conclusions), K. K. Markov (1955) stated the main features of natural development characteristic of the Quaternary period as a whole: the universality of the phenomena, the progressive nature of the changes in the natural conditions, the rhythmic nature of these changes, and the distinctive aspect of the natural development of the individual territories. The last three characteristics are especially evident in the material we have assembled.

The plant remains preserved in successive strata indicate that the changes in the vegetation correspond to definite climatic warnings and coolings.

Phase I: at first, mixed-coniferous forests with an admixture of broadleaf species, then birch and pine forests with rare relicts of exotic coniferous and broadleaf species, and with steppes in the young mountain region. Phase I corresponds to a climate which was cold and moist at first and then became cold and dry. Undoubtedly, the cooling was caused by the first glaciation during the Early Pleistocene.

Phase II: at first, dark-coniferous taiga with broadleaf species, then dark-coniferous taiga without the broadleafs. Phase II corresponds to a warm, moist climate which later, with the disappearance of the broadleaf species and the beginning of the degradation of the dark-coniferous taiga, became a cold and moist climate. We attribute the disappearance of the broadleaf species during the expansion of the dark-coniferous taiga to increased humidity and cooling, which also led to the formation and subsequent growth of glaciers. The taiga became degraded under the influence of progressive glaciation and permafrost. Therefore, we feel that the boundary of the second glaciation can be drawn above the spruce and Siberian pine maximum. [See Fig. 18.]

Phase III: light-coniferous pine and pine-larch taiga with birch and sparse birch woods and steppes (or tundra steppes) in places. Phase III corresponds to a cold dry climate (including xerophytation of the vegetation under conditions of physiological dryness). Evidently a glacial maximum occurred during this period, succeeded by glacier retreat and degradation of the permafrost.

Phase IV: mixed-coniferous forests and wooded steppes with broadleaf species, subsequent increased forestation and extension of the dark-coniferous taiga. Phase IV corresponds to an initially warm and relatively dry climate with complete disappearance of glaciers and permafrost, and then to a warm and moist climate, corresponding to the second interglacial period. Futhermore, by the end of this phase, the broadleaf species had disappeared and the dark-coniferous taiga reached its maximum development, which corresponds to cooling and to a moist climate, which brought about a new, a third glaciation. The glaciation itself caused further cooling, which, in turn, caused degradation of the taiga. We have drawn the boundary of the new glaciation along the time-line of degradation of the dark-coniferous taiga, i.e., above the spruce and Siberian pine maximum.

Phase V: light-coniferous pine-larch and sparse birch forests, succeeding the previous phase. Phase V corresponds to a cold and dry climate. The glaciation reached its maximum development during this period and then gradually retreated. The character of the vegetation was controlled completely by the existence of the ice sheet and permafrost.

Phase VI: mixed-coniferous forests and increased area of taiga. Phase VI corresponds to a climatic warning, to the retreat or complete disappearance of the glaciers. During this interglacial (?) period, the climate was moister but cooler than the climate of the previous interglacial period, Phase IV. At the end of Phase VI, under conditions of ever increasing humidity and renewed cooling, the fourth and last glaciation took place.

Phase VII: a continuation of the cold, but now dry climate of the last glaciation, a period of light-coniferous pine-larch and sparse birch forests and steppes.

Phase VIII: the last phase, light-coniferous, pine-larch taiga and, in places, dark-coniferous taiga. Phase VIII corresponds to a warm dry climate, which later, with an increased participation of dark-coniferous species, became warm and somewhat moister than at present, i.e., this phase corresponds to the climate of the postglacial period, the Holocene.

Thus, there was a climatic alternation: warm and dry, then warm and moist, cold and moist, cold and dry, then again warm and dry, and so on. The vegetation and, consequently, the climate changed rhythmically. The schematic diagram (Fig. 18), which is based on factual data, shows several such rhythms or cycles,

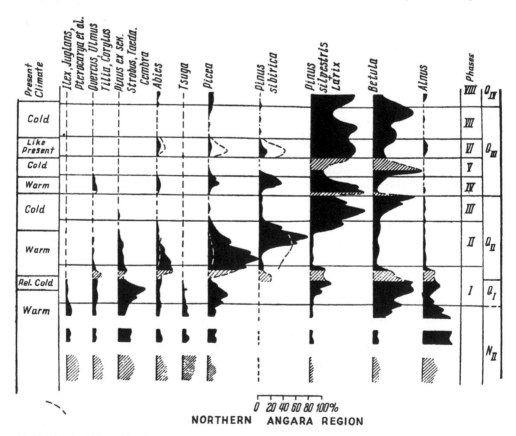

FIGURE 18. Schematic pollen diagram for the Angara basin (more accurate than the diagram published in 1955).

at least four. The cold periods were similar and still cannot be differentiated on the basis of spore-pollen spectra. The general features of the warm periods were similar, but showed several distinctive spectral components. The lower time-segment corresponding to a warm climate (first half of Phase II), has exotic conifers (*Pinus strobus*) and broadleafs (oak, elm, hazel, linden) in the spore-pollen spectra; higher up the section there is a pollen maximum of spruce and fir or Siberian pine and spruce; this segment also has a high species saturation.

The participation of broadleaf pollen (oak, hazel, linden, and elm) in the spectrum is also characteristic of the middle segment, which corresponds to a warm climate, but the exotic conifers are not found. The spruce and Siberian pine maximum is slight and there is a considerable quantity of grass pollen.

The Upper Interglacial (or Interstadial) shows only conifer pollens and, higher up the profile, dark-coniferous species. Broadleaf species are not found in the Angara basin, which may be explained by the nature of the preceding glaciation: evidently the third glaciation was maximum in the Pleistocene and forced the broadleaf species beyond the mountains on the southern margin of the area and led to their complete extinction in the Angara basin. As facts accumulate and more detailed investigations are made, undoubtedly many other differences will become apparent, from which the age of the deposits may be deduced. For example, such differences may consist in a gradual general impoverishment of the flora (reduction of the species saturation), a general increase in the role of the light-coniferous taiga, and a reduction of the dark-coniferous taiga in connection with the gradual general cooling and increased continetality of the climate during the Pleistocene. In a more detailed analysis, with identification of species, differences will also be discovered in the composition of the species which migrated into the Angara basin during the warm interglacial periods. Such differences have already provided a general orientation into the evolution of the vegetation of the Angara region and will lead to further clarifications and show the general direction of development of this flora.

The third characteristic feature, the "individuality (local aspect) of the natural conditions," becomes apparent upon comparing our system with that of the European part of the U.S.S.R. (V. P. Grichuk, 1950). For example, the broadleaf species of the Angara region, although they extended far to the north along the Angara, Yenisey, and Lena valleys, played a subordinate role in the vegetation cover and disappeared completely during the last interglacial period, when taiga or mixed forests of coniferous and small-leaf species prevailed. All this indicates that the amplitudes of climatic fluctuations were comparatively slight from the optimum of the glacial epochs to the optimum of the interglacial epochs, and also that the climate remained comparatively continental, preventing the development of broadleaf forests, while broadleaf forests prevailed in the European part of the U.S.S.R. during the interglacial epochs.

To get an idea, albeit a rough one, of the quantitative relationship between the expanses occupied by the different types of vegetation and how the vegetation was distributed in the Angara basin during the different phases, we have attempted to map the factual data (see Fig. 2, p. 385), and where we do not have such data we have assumed a distribution. Although a sample may have been taken from a point in a small basin, the spore-pollen spectrum represents the vegetation of the entire basin to a certain extent. For example, if the spectrum is a taiga spectrum, it can be assumed that taiga appeared throughout the basin, but if the spectrum is complex and contains pollen of vegetation characteristic of dark-coniferous taiga as well as steppe pollen in considerable quantity, it can be assumed that taiga and steppe coexisted in the basin. One may get a more accurate picture of the disposition of a particular type of vegetation by studying deposits from several sites. As a region of dark-coniferous taiga is approached, the amount of taiga pollen increases in the spectrum. On the other hand, the amount of steppe pollen increases in a region where steppe associations predominate. This change of pollen in the spectra as a function of areal distribution of various types of vegetation in the basin may be seen in surface samples of the present day by comparing the analyses given at the beginning of the article with the present vegetation.

In addition to the general similarity of the spectra, which has already been stressed, a small but regular increase of the percentage composition of the pollen of individual tree species as a function of their distribution may be observed. For example, in the surface samples, the amount of Siberian pine pollen increases upvalley, towards the mountains, because Siberian pine-larch forests grow in the mountains. The amount of birch pollen in the spectra increases northward in the basin, because towards the north, especially in the middle section of the basin, aspen-birch forests are encountered more and more frequently and birch pollen plays an ever increasing role in the floodplain spectra. The same type of situation is observed in the increased amount of grass pollen as the steppe portions of the valley are approached.

The mapping system is not equally valid everywhere, but depends on the number and position of the sites investigated. The vegetation is assumed for the places on the map where there are no investigated sites, on the basis of the nature of the vegetation of neighboring regions, the topography, and the transport of pollen by water and wind during the formation of the spectra.

Only four general types of vegetation are distinguished in our mapping systems. The role of broadleaf species is indicated by a special symbol. The composite type, called sparse light-coniferous and birch forests, includes several types, which appeared primarily in certain regions and predominated at various times. For example, there may have been birch forests or vegetation of the dwarf-birch type, which we were unable to distinguish by spore-pollen analysis, and there may have been sparse larch-birch or pine-larch forests. The spectra of these latter differ but little, because the larch pollen is not preserved.

The spruce-fir and fir taiga is similar to the Siberian pine–common pine and Siberian pine taiga with spruce and can be distinguished by the dominance of either spruce or Siberian pine. It has been expedient to differentiate these two types because the Siberian pine–common pine and Siberian pine taiga proved more stable as the climate deteriorated and yielded to light-coniferous taiga and birch at a later date than did the spruce taiga; possibly it tended to expand in area as the spruce degraded. The territorial distribution of these two types in the individual valleys is not always clear, because it was closely associated with the nature of the topography of the time. However, this distribution has been established in a number of cases and the qualitative relationship of the areas occupied by these types was similar to that shown in the maps.

The wooded-steppe and steppe type of vegetation is also composite and includes wooded-steppe, steppe, meadow-steppe, and tundra-steppe. They differed floristically at different times and in different regions. For example, the steppe landscapes with broadleaf species of Phase IV differed substantially from the species-poor cold steppes" of Phase VII, which extended far beyond the confines of the valleys. V. V. Reverdatto (1940), who studied the glacial relics in the Angara flora, pictures the vegetation of the periglacial regions as unique complexes" of bog-steppes "with a special combination of xerophytes, hydrophytes, and alpine plants. Scattered larch groves survived among these plant associations."

At the same time, the genesis of the flora on the steppes of the lower Angara during Phases II and III was different from that of the steppes in the Irkut valley during Phase III.

The maps on page 385 show clearly the differences between the vegetation of the different phases and the differences in the history of the vegetation of the

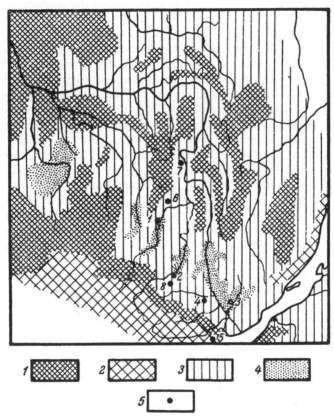

FIGURE 19. Map of the present vegetation of the Angara region (according to Ye. M. Lavrenko's map of 1939): 1, dark-coniferous taiga; 2, Siberian pine-larch forests (light-coniferous); 4, wooded steppe and steppe; 5, points at which surface samples were taken from the floodplains and peat bogs.

individual regions of the Angara basin. A comparison of the map of Phase II with the maps of Phases IV and VI and the present may (Fig. 19) show that, in general, during the Pleistocene dark-coniferous taiga yielded to light-coniferous. It must be remembered that the area occupied by dark-coniferous taiga on the map of contemporary vegetation published in 1939 (Ye. M. Lavrenko, *Karta rastitelnosti SSSR* [Vegetation map of the USSR]) is exaggerated, which may be seen by comparing it with the more detailed new map of 1954 (Ye. M. Lavrenko and V. R. Sochava (Editors), *Geobotanicheskaya karta SSSR* [Geobotanic map of the U.S.S.R.]. On the 1939 map, the dark-coniferous taiga is shown in places where now there are no birch-aspen, so-called second-growth forests. The map showing the area of dark-coniferous forests represents the vegetation of the era preceding the present era, rather than the present vegetation. Therefore, we have not shown the vegetation corresponding to the Holocene (Phase VIII) on our map. It is similar in general to that shown on the geobotanic map of 1939.

It is especially interesting to trace the distribution of the steppes (or wooded steppes) in the various phases. For Phase I, traces of steppe flora have been found only in the Baykal area. The existence of steppes during Phase II has definitely been established in the northeastern part of the Angara basin and traces of alpine steppes have been found in the Baykal region. During Phase III, the steppes expanded to the northeast, and to the south along the Angara valley; in the Baykal region, the Irkut valley had steppes and possibly steppes appeared northward along the Angara valley. Further, at the beginning of Phase IV, the Angara valley and its southern tributaries became steppes; steppes were preserved in the northeast. The situation during Phase V is still not clear because of the small number of sections studied. During this period of apparent maximum cooling, bog-steppes appeared among the larch-birch sparse forests. In Phase VI, during the considerable expansion of the taiga, the two steppe regions survived, in the northeast and in the south, but were reduced in area. During Phase VII, the steppes reached their maximum in the Angara basin not only in the river valleys but also on the watersheds. The northeastern and southern steppes merged. The steppes decreased in area during Phase VIII. In all probability the steppes were not completely crowded out by taiga in the northeastern part of the basin until Phase VIII, and taiga flourishes there at present. On the modern map we again see some expansion of the steppe domain in the southern part of the basin.

Many steppe relicts have been preserved in the present taiga region from Zayarsk to Bratsk and northward to 58° N. lat., in connection with the recent existence of steppes. For example, M. G. Popov (1953) points out the abundance of steppe plants encountered in the taiga "on steep southern slopes—rocky, stony, and even fine-grained earth—provided these slopes have a sparse forest cover or no forest cover at all."

It is also interesting to note that steppe orientation along the Sayan was first noted for the Middle Pleistocene, during Phase IV. Perhaps at that time the mountains were already high enough to exert an influence on the climate of the southwestern Angara basin. Thus, it becomes clear that the history of the Angara vegetation in the Quaternary period is more complex than we can imagine it at present. It would be difficult to form a clear concept of which appeared first in the Angara region, taiga or steppe (M. G. Popov, S. N. Korzhinskiy). This is a relative matter and may be posed specifically only for some specific region and specific time-interval. For example, the present-day taiga in the northeastern portion of the basin may be considered secondary, in view of the long existence of steppes there during the Pleistocene. However, the present steppes in the middle course of the Iya and Oka rivers formed during the second half of the Pleistocene in place of taiga, in connection with the uplifting of the Sayan mountains. Therefore, the taiga is primary here, having appeared during the Older Pleistocene, and the steppes are secondary.

During the Quaternary period the relationship of types of vegetation depended basically on the changes of climate and topography, which favored the development and spread of a particular type of vegetation.

According to the afore-mentioned history of vegetation, it may be assumed that at present in the Angara region the climate favors the spread of light-coniferous taiga and birch-aspen forests at the expense of dark-coniferous taiga, and also the spread of wooded steppe and steppes. Undoubtedly, the activity of man has wrought considerable changes in the natural process.

A number of botanists have noted that the present-day Angara steppes are floristically complex. They contain elements of the Mongolian steppe flora and of western steppe flora (V. V. Reverdatto, M. G. Popov, A. N. Krishtofovich, *et al.*). They also contain elements of high-alpine flora and a number of other glacial relicts. These facts have led to contradictory conclusions about the origin of the steppes. In our opinion, the spore-pollen method of analysis of the older deposits provides the proper solution of these problems. The results of spore-pollen analyses indicate that steppes have existed in the Angara basin since the Tertiary period. Further, the steppes in the northeastern part of the Angara basin apparently were closely associated with the steppes of the southern part of western Siberia, where a number of investigators have established their existence during the Pliocene (Bogolepov, 1955). In the south, the mountain steppes apparently first appeared at the end of the Tertiary and at that time were associated with the Mongolian steppes. In The Pleistocene, during the cold and dry period (end of the glaciation) and then during the warm and dry period (end of the inter-glacial stage), the two steppe floras combined (possibly there were several unions), leading to their mixing. During the periods of glacial maxima, "the mountain alpine vegetation descended" and spread out on the plains, as V. V. Reverdatto conceived it.

In conclusion, it must be mentioned that our conclusions are preliminary and undoubtedly require confirmation and supplementation. This will become possible as new data are gathered and the available data are subjected to a more detailed analysis.

N. S. CHEBOTAREVA, N. P. KUPRINA, AND I. M. KHOREVA

# STRATIGRAPHY OF THE QUATERNARY DEPOSITS OF THE MIDDLE LENA AND THE LOWER ALDAN RIVERS*

THE PRESENT ARTICLE is based on material assembled by the authors during three years of work with the Interdisciplinary Central Siberian Expedition of the Institute of Geological Sciences, Academy of Sciences of the U.S.S.R.,† in the middle course of the Lena river from the mouth of the Vitim river to the mouth of the Sinyaya, and on the lower Aldan river (from the mouth of the Aldan to the mouth of the Anga).

We cannot go into a detailed analysis of the numerous works on the Quaternary deposits of this region here, but will merely note that the foundations for the present knowledge of the geology and geomorphology of this region were laid by the works of V. A. Obruchev, A. G. Rzhonsnitskiy, A. A. Grigoreva, and others. In recent years, work has been conducted which has provided more exact information and detail on a number of problems of the Quaternary geology of the deposits of this region. However, to date, no stratigraphic scheme based on paleontology has been offered.

In developing our proposed system, we have depended chiefly on finds of fauna and on the data of spore-pollen analysis. We have stressed the lithologic characteristics of the deposits. The accuracy of our scheme has been confirmed by comparison with the systems devised in 1954 by E. I. Ravskiy for the southern edge of the Tunguska basin and by M. N. Alekseyev for the southern part of the Vilyuy basin; both these systems were based on biostratigraphy.

The data on the mineralogic composition of the deposits are based, in part, on analyses conducted in the laboratory of the Scientific Research Institute for the Geologic Prospecting of Gold‡ and on analyses made by I. A. Yefimov, V. I. Muravyev, and Yu. V. Bulava, colleagues of the Institute of Geologic Sciences, Academy of Sciences of the U.S.S.R.§ The spore-pollen analyses were conducted by R. E. Giterman, G. M. Brattseva, and I. Z. Kotova at the laboratory of the Institute of Geologic Sciences. The mammal bone remains were identified by E. A. Vangengeym, a colleague at the Institute of Geologic Sciences.

The portion of the Lena valley investigated may be divided into three sections. The first section, from Vitim to Olekminsk, has the most complex structure. Here the valley intersects the Angara-Lena Lower Paleozoic marginal fault, within which one may distinguish the Lena folded zone and the Berezovskaya depression, which have Cambrian and Silurian deposits crumpled into folds.

*Translated from K. K. Markov and A. I. Popov (Editors), *Lednikovyy period na territorii yevropeyskoy chasti SSSR i Sibiri*, Moskovskiy gosudarstvenny Universitet, Geograficheskiy Fakultet i Muzey Zemlevedeniya, 1959, pp. 498–509.

†[Kompleksnaya Tsentralno-Sibirskaya ekspeditsiya Instituta geologicheskikh nauk, Akademii nauk SSSR.]

‡[Nauchno-issledovatelskiy geologo-razvedochnyy institut Zolota (NIGRI Zoloto).]

§[Institut geologicheskikh nauk Akademii nauk SSSR (IGN, AN SSSR).]

These structures are separated by the Urin anticlinorium, created during the Proterozoic and comprising a branch of the Baykal folded zone. This part of the valley differs geomorphologically from the other parts. The valley is longitudinal in the Lena folded zone and in the Berezovskaya depression. Usually it is narrow and deeply incised. The riverbed is no more than 600–700 m wide and has practically no islands. Terraces and floodplains stretch along the valley in narrow bands rarely more than 300–800 m wide. The high terraces, which are widely developed in this region, are an exception (terrace VII is 150–170 m wide; terrace VIII, 200–250 m wide). Morphologically, they are well defined as broad level surfaces sloping slightly towards the river. They reach a width of 3–4 km. The rear suture of the terraces is almost indistinguishable, and the terraces gradually merge with the watershed (Fig. 1).

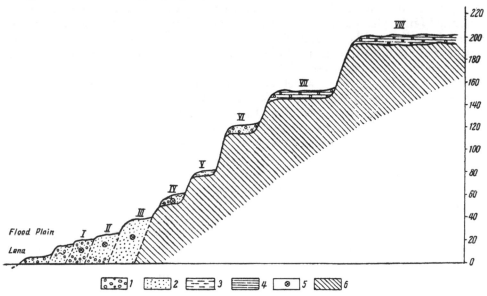

FIGURE 1. Relationship of the terraces of the middle course of the Lena river: 1, pebble gravel; 2, sand; 3, loams; 4, clays; 5, sites of faunistic finds; 6, socle.

Terraces VI (100–120 m), V (70–80 m), IV (50–60 m), and III (35–40 m) are, basically, compound terraces, but sometimes are wholly aggradational terraces; further, the lower the terrace is, the more widely is it distributed and the more frequently is it an aggradational terrace. The cusp and the surface of these terraces are well defined. Often the surface is mounded, in connection with the eolation of the sands.

The first and second terraces occur almost everywhere. They are very clearly defined in the relief and usually are aggraded. Areas where these terraces have a socle as much as 18–20 m high are the exception.

The floodplain, usually multistage, is very widely developed. The high floodplain (as much as 10–12 m high) has a clearly defined floodplain relief. Where the floodplain is comprised of sandy material, mounds are observed on its surface and it reaches heights of 15–17 m. In addition to the high floodplain, there are several projections (cusps) and a well-defined river shoal (sand bar), consisting of pebble gravel. Sometimes one encounters sectors where the floodplain has a low socle, usually not more than 2–3 m high.

The Lena valley becomes transverse upon intersecting the Urin anticlinorium. The alteration of wide and narrow sectors, characteristic of this section of the valley, is associated with its lithology.

The second section of the valley, from Olekminsk to Pokrovsk, lies within the limits of the northern slope of the Aldan shield, which is comprised of uniform, almost horizontal Cambrian strata. The valley structure is extremely simple in this sector. The same complex of terraces occurs here, but aggradational terraces are rare. The valley is very narrow here.

The third section of the Lena valley (from Pokrovsk to the mouth of the Aldan) together with the lower Aldan lies within the southern limb of the Vilyuy basin and the Verkhoyansk marginal fault, which are occupied by Mesozoic rocks (Vakhrameyev and Pushcharovskiy, 1954). The latest data indicate that Tertiary deposits up to 700 m thick also occur extensively in the Verkhoyansk marginal fault. This portion of the valley differs sharply from the preceding ones. Here the valley reaches widths of tens of kilometers, the riverbed is 6–8 km wide (near the mouth of the Aldan, 10 km wide), and there are many islands. There is no sandbar. A narrow, essentially sandy floodplain 4–5 m high stretches along the riverbed. The terraces, which are well defined in this portion of the valley, differ considerably in width. The 100–120-m terrace is the highest terrace here.

The valley is well developed along the entire lower Aldan. At the mouth of the Aldan, where the riverbed reaches a width of 4 km, the valley reaches widths of several tens of kilometers. The difference in structure of the terraces on the right and left banks is immediately evident. Cobbles form part of the terrace structure on the right bank, but do not occur on the left bank. Further, the terraces on the right bank are more clearly defined morphologically.

The Aldan valley has four terraces (Fig. 2). Terraces II, III, and IV are compound terraces. The socle consists of Tertiary deposits. The highest terrace (IV), which reaches heights of 70–80 m, appears on both sides of the Aldan valley and also in the lower reaches of the Amga. The outcropping at Mamontova Gora [Mammoth Mountain] reveals the structure of this terrace. The next terrace (III) is not widely developed and reaches heights of 40–50 m.

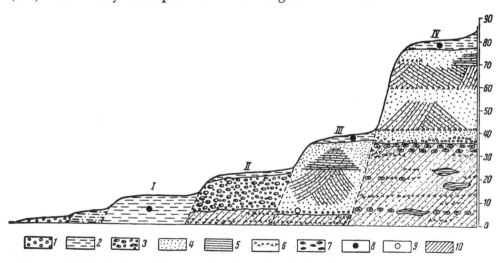

FIGURE 2. Relationship of the terraces of the lower reaches of the Aldan river: 1, pebble gravel; 2, loams; 3, cobbles; 4, sands; 5, clays; 6, plant remains; 7, concretions; 8, Upper Paleolithic fauna complex $Q_{III}$; 9, site of the discovery of ancient $Q_I$ fauna; 10, socle.

Terrace II is comparatively well developed in the valley and is 20–30 m high. Even now there is some evidence that this terrace should be regarded as two terraces. Terrace I appears extensively in the Aldan valley. It is 11–15 m high, with a maximum width of several kilometers. A floodplain 5–6 m high and several hundreds of meters to several kilometers wide is clearly distinguishable here. The lower floodplain cannot be distinguished easily, because its cusp is not always well defined, but usually it rises gradually from the channel shoal, which appears along the riverbed over all of the investigated sector and is comprised basically of pebble gravel.

Now let us examine the characteristics of the deposits which comprise the terraces.

## Tertiary Deposits

Thin (6–7 m) reddish alluvial deposits appear on the high terraces (150–250 m) of the Lena valley between Vitim and Pokrovsk. Usually these deposits are brown or reddish-brown clays, loams, and sandy loams with a small admixture of pebbles of highly uniform petrographic composition. Quartz pebbles comprise more than 50 per cent of the pebble deposits and considerable quartzite, often ferruginous, is present, in addition to sandstone and porphyry. The latter are very weathered. The mineralogic composition of these deposits is also uniform. Quartz prevails, comprising about 60 per cent of the material, feldspar comprises 12–15 per cent, and hydromica appears in considerable quantities (29% to 0.72%) in almost all samples. Typical weathered minerals, vermiculite and montmorillonite, are found in some samples. Ilmenite clearly predominates in the heavy fraction, comprising 40–50 per cent; hornblende and epidote are found in smaller quantities. Zircon, garnet, sphene, and magnetite almost always appear, and sometimes tourmaline, staurolite, and disthene are found. No pyroxenes are present.

Thus, the preservation of pebbles and the mineralogic composition of the deposits indicate that they are quite old. These deposits do not contain either spores or pollen. No paleontologic finds of these deposits have been made in this region. Therefore, age can be determined only by comparing the deposits found here with Tertiary deposits found in other regions (Velikovskaya, 1955).

Tertiary deposits appear extensively in the Lena valley in the section south of Pokrovsk. They can be characterized by a cross-section which has been studied in detail, namely, the section known in the literature as Kangalasskiy Mys [Cape Kangalasy], situated 30 km upriver from Yakutsk.

As early as 1928, G. A. Ivanov differentiated three horizons here: A, B, and C. His horizon A consists of coal-bearing deposits, classified by later investigators as Lower Cretaceous, on the basis of spore-pollen analysis.

Horizon B consists of lilac-gray or brownish-gray, massive, inequigranular sandy loams enriched with plant remains. Scattered pebbles of quartz, silicon, quartzite, porphyrite, chalcedony, shales, and, in places, intercalations of conglomerate were found at the base of these deposits. This horizon is 8–12 m thick. According to V. A. Kakhrameyev, a diversified pollen of angiosperms, indicating that the deposits are Upper Cretaceous or Tertiary, was encountered in the lower part of horizon B at a depth of 67–69 m along with pollen which appeared to be Jurassic and Lower Cretaceous. In Vakhrameyev's opinion, the paleogenic aspect of the pines and the presence of linden indicate a Tertiary age.

Horizon C consists of yellow, white, and light-gray cross-bedded quartz sands, principally coarse-grained, with an admixture of gravel. Downward in the section the sands assume a uniform ochre color, and lenses of yellowish gray-brown sandstone appear in them. Near the roof, amid the sands, there is a layer of reddish conglomerate with quartz, quartzite, and quartz porphyry pebbles. The conglomerate band is 10–12 m thick. Horizon C is 50–60 m thick and consists of two independent subhorizons: the lower subhorizon, consisting of gray sands, and the upper one, consisting of highly ferruginous sands with pebbles and intercalations of conglomerates.

Spore-pollen analysis of a sample taken at the contact of the gray sands and the ferruginous sands showed that the vegetation which existed at the time these deposits were laid down was highly diversified and included Tertiary species. Analysis of the upper band of of horizon C, 11 m below the surface, indicated the presence of a large quantity of redeposited pollen of the Tertiary type. According to A. N. Sladkov (personal communication), only Quaternary pollen is found in the uppermost part of the ferruginous sands; therefore, he considers the horizon of ferruginous sands to be transitional between the Neogene and the Quaternary.

Tertiary deposits occur extensively and are very thick within the Verkhoyansk marginal fault. Tertiary deposits emerge in the socles of all terraces of the lower Aldan. The most complete section is found in the outcropping at Mamontova Gora, where we detected three layers: the lower, the middle, and the upper.

The lower layer consists of brownish, ferruginous, quite well cemented sands with a large quantity of siderite concretions; pebbles are encountered, in places the stratification is apparent, and accumulations of plant remains appear, from which large quantities of *Juglans cinerea* fruit and cones have been collected. The heavy fraction of the concentrate of these sands contains 0.4 per cent magnetite, 45.3 per cent ilmenite, 52 per cent hornblende, 0.2 per cent sphene, 0.4 per cent rutile, and 1.7 per cent zircon.

A. N. Krishtofovich and A. F. Yefimova established the following species among the numerous plant remains in the clayey-siderite concretions of the lower layer: *Magnolia* sp., *Juglans cinerea, Corylus, Acer, Salix, Betula, Equisetum*, etc. (in I. I. Tuchkov's collections).

In the lower stratum, A. P. Vaskovskiy detected *Pinus radiata* Douge and *P. Monticola* Douge, which he considered to be Pliocene; the first of these grows at present in California, the second in the Rocky Mountains (Vaskovskiy and Tuchkov, 1953).

Spore-pollen analysis indicates that the vegetation which existed during the period when the lower layer was deposited was highly diverse and contained Tertiary forms as well. The gymnosperms are represented by various pines, and the angiosperms by various broadleaf trees, *Juglans* sp., *Ostrea*, etc. However, the species saturation is lower than for the vegetation of the European Neogene.

Various opinions exist as to the age of the Mamontova Gora deposits. The disagreement centers about the age of the middle and upper strata. It is agreed that the lower stratum is Tertiary.

The following conclusions may be drawn from an analysis of the available data on the Tertiary deposits. The Tertiary deposits on the southern limb (wall) of the Vilyuy basin and the Verkhoyansk marginal fault occupy a lower hypsometric position than those of the upper Lena valley. This is associated with differences in the tectonic regimes of these territories, as indicated by the change in the thickness and the character of the deposits in the various sectors. The upper portion is thin (5–6 m) and its deposits consist of coarse clastic material, while the deposits of the southern limb of the Vilyuy basin are much thicker (50–60 m) and contain fine sediments in addition to the coarse clastic material. The Tertiary deposits of the Verkhoyansk marginal fault are of the order of 700 m thick. All this indicates a very intensive downwarping of the Verkhoyansk fault and a considerably less intense downwarping of the southern slope of the Vilyuy basin, while the territory which confines the upper Lena valley was uplifted in comparison with the downwarped portions.

## Deposits Transitional between Tertiary and Quaternary

In the middle course of the Lena river such deposits are confined to the 100–120-m terrace. In the section from Vitim to Pokrovsk, they appear as individual small areas and are represented by sandy-pebble gravel or sandy deposits. For example, these deposits lie on the 110-m terrace along the left bank of the Lena below the mouth of the Solyanka. The alluvial layer is about 6 m thick. It is a yellow-brown sandy loam with a large quantity of mica, containing many pebbles which differ in granulometric and petrographic composition.

No fauna restricted to these deposits has been discovered in this region. The very limited results of spore-pollen analysis for these deposits indicate that two types of forests prevailed during the period when these deposits were laid down: spruce-larch taiga with a small admixture of broadleaf species (individual grains of *Tilia, Ulmus,* and *Quercus* have been found), and pine forests on the sandy expanses.

Comparison of the transitional deposits and the Tertiary deposits found in this region points up a certain difference in the petrographic composition of the pebble gravel and the mineralogic composition of the deposits: in the transitional deposits there are many pebbles of igneous rock, which have disappeared almost completely in the Tertiary deposits; there are practically no weathered minerals in the transitional deposits, while such are characteristic of the Tertiary deposits.

As already noted, A. N. Sladkov, on the basis of spore-pollen analysis, classified the highest stratum of the C horizon (noted by G. A. Ivanov in the outcropping of Cape Kangalasy) as transitional between Neogene and Quaternary. Tracing the sequence of deposits of this high, 100-m terrace, up the Lena, one observes that the ferruginous sands of Cape Kangalasy are widely distributed and occur in many outcroppings as far as Tabaginskiy Mys [Cape Tabaga]. The above data support our conclusion that transitional deposits occur in the 100–120-m terrace.

The middle stratum of Mamontova Gora consists of these same deposits, i.e., of inequigranular gray and yellow friable sands with well-defined cross-bedding. Plant remains appear as thin intercalations. Downstream in this outcropping there is a facies transition from sands to pebble gravel, consisting of pebbles of very different size and degree of roundness. This layer is 10–12 m thick. The sands and pebbles are similar in mineralogic composition to the heavy fraction. The main mass of the heavy concentrate consists of garnet and ilmenite; magnetite, rutile, and zircon are present in small amounts.

In contrast to the lower layer, only individual specimens of the cones and fruit of the *Juglans cinerea* have been found in the middle layer.

Spore-pollen analysis of the samples from this layer produced results similar to those for the lower layer. This is probably due to littering of the middle layer with Tertiary pollen. It is firmly established fact that all the post-Tertiary deposits of the Aldan river valley are littered with Tertiary pollen. One must keep in mind that the method of distinguishing between pollen that has and that has not been redeposited is poorly conceived and that the data of spore-pollen analysis cannot be relied upon completely in this case, in particular because they conflict with the geologic data.

## Lower Pleistocene Deposits

These are poorly represented and on the Lena are confined to the 70–80-m terrace. Usually they are loamy or clayey deposits with pebbles, and are thin. At the settlement of Delgeyskoye, the alluvium of the 80-m terrace consists of pale yellow clays with a slight admixture of fine-grained sands and an occasional pebble. This layer is 0.2 m thick. Below it are dark-violet clays with marl fragments, representing the eluvium of indigenous deposits.

At present no paleontologic and paleofloristic finds of the Lower Pleistocene have been recorded. Therefore, the age of these deposits is judged by the following information. The terrace rests directly against the 100–120-m terrace, whose alluvium has been established as transitional between Tertiary and Quaternary. Further, the 50–60-m terrace rests directly against the 80-m terrace and belongs to the upper part of the Lower Pleistocene or the lower part of the Middle Pleistocene. Thus, there is some basis for classifying the deposits found on the 70–80-m terrace as Lower Pleistocene. No such well-substantiated Lower Pleistocene deposits are known in other sections of the Lena valley with which one might compare the deposits described above.

The deposits of the 40–50-m terrace in the lower course of the Aldan can be classified as Lower Pleistocene. On its base, on the Tertiary sands, there is a thin layer of highly ferruginous friable pebble gravel, which is superficially quite similar to the pebble gravel of the "middle" layer of Mamontova Gora. Above the pebble gravels there are rose-colored, medium-grained, cross-bedded sands with occasional pebbles. This layer is 10–12 m thick. The mineralogic composition of these deposits differs in several ways from the heavy fraction of the Tertiary deposits and the deposits transitional between Tertiary and Lower Pleistocene: there is a different ratio of magnetite and ilmenite, magnetite predominating but ilmenite comprising about 30 per cent, and the remaining minerals appear in about the same percentage as indicated previously.

Upward along the section the rosy sands grade into gray fine-grained sands without apparent interruption. Their stratification is well defined, as in the rosy sands. This layer is 15–20 m thick.

An *Equus* cf. *senmeniensis* bone was found in the pebble gravel. Further, bones of other ancient mammals, viz. *Elephas* cf. *nomadicus*,* *Trogontherium* cf. *cuvieri*, a fragment of the lower jaw of *Alces latifrons*, and a horn fragment, were found at the base of the outcropping of this terrace. The bones were only slightly rounded and were in a state of preservation nearly equal to that of *Equus* cf. *senmeniensis*; because of this E. A. Vangengeym considered them as coming from the same pebble gravel.

*Also, a tooth of the same species of elephant was discovered at the Ytyk-Kel Museum in a sample which had been taken from the Amga river.

Older Quaternary and Tertiary faunistic complexes have not yet been found in Yakutia, so the afore-mentioned finds must be compared with those of rather remote territories. *Alces latifrons* has been found in the U.S.S.R. in the Tiraspol gravel complex (described by M. V. Pavlova) and in the older Quaternary pebble gravels of the Ishim river (determined by V. I. Gromov). *Trogontherium cuvieri* has been identified in the pebble gravels of a tributary of the Ishim, where it appeared together with Pliocene fauna. *Elephas nomadicus*, *Trogontherium cuvieri*, and *Equus senmeniensis* have been found in the Nihowan fauna of northern China (northern Hopeh, about 40° N. lat.), which Chinese geologists classify as Eopleistocene and equate with the Villafranca of Europe. The Chinese geologists place the formation of the pebble gravels of Ordos and Inner Mongolia in this same period.

All representatives of this fauna existed to the end of the Mindel glaciation. The *Alces latifrons* is not found in China. To clarify the exact stratigraphic position of the finds, one must have additional material. Perhaps all these animals are representatives of a single complex, characteristic of Siberia.

Thus, the afore-named finds provide a basis for dating the deposits of the 40–50-m terrace of the Aldan river as Lower Pleistocene.

The deposits which we have attributed to the upper part of the Lower Pleistocene or the lower part of the Middle Pleistocene are very widely distributed. They are confined to the 50–60-m terrace of the Lena and usually consist of sandy-pebble gravel or loamy deposits. They vary greatly in thickness, from 1–2 m to 60 m.

There is an interesting section of the 60-m terrace at Olekminsk. The socle of the terrace, which is about 50 m high, is composed of the upper Lena suite of the Upper Cambrian. The alluvium consists of a 10-m layer of fine yellowish-brown compact sandy loam; at the contact with the bedrock there is a thin, 10–20-cm, layer of pebble gravel. The mineralogic composition of these deposits differs somewhat from that of the earlier deposits. The heavy fraction consists essentially of hornblende, and there is a considerable quantity of epidote, pyroxenes, colored mica, magnetite, ilmenite, and garnet. It should be noted that these deposits contain pyroxenes, which were absent almost entirely in the older deposits. Further, the older deposits did not have the fairly high mica content of these deposits.

In the pebble-gravel layer on the 60-m terrace at Olekminsk, we found the radial bone of a *Rhinoceros antiquitatis*. According to V. I. Gromov (1948), the *Rhinoceros antiquitatis* appeared in Siberia in the Mindel-Riss interglacial and continued to exist there until the Riss-Würm period, and until the Würm in the north.

Comparison of the rhinoceros bone found in the deposits of the 60-m Lena terrace with numerous remains of this animal which have been attributed to the second half of the Siberian Pleistocene shows that the Lena remnant has been heavily mineralized. This led E. A. Vangengeym to classify the Lena fragment as Lower Pleistocene, in analogy with finds in northern China. The deposits containing this bone evidently belong to the upper part of the Lower Pleistocene or the lower part of the Middle Pleistocene. This definition also agrees with the spore-pollen analysis, which shows that tree pollens clearly predominated and included considerable *Pinus* (20–80%) and *Picea* (up to 55%). Since *Pinus* pollen is airborne, the *Picea* must have played a large role in the plant composition; however, this does not exclude the possibility that *Pinus* predominated on the sandy expanses. *Abies* pollen was found in considerable quantities, up to 20 per cent. It is interesting that pollen grains of the broadleaf species *Ulmus* and *Tilia* were present in all the samples analyzed.

## Middle Pleistocene Deposits

These occur widely in the Lena valley. The deposits which we have classified as lower Middle Pleistocene appear on the 35–40-m terrace. In the region we investigated, they appear as small isolated sections. As a rule, they are yellow-brown or yellowish-gray sands, sometimes cross-bedded. In the section from Olekminsk to Pokrovsk, these deposits are usually sandy loams or loams 1–3 m thick. They have a large quantity of magnetite in the heavy fraction, often more than 50 per cent. There is considerable ilmenite (up to 30%), and almost equal amounts of epidote, hornblende, garnet, and zircon. Individual grains of limonite, staurolite, tourmaline, rutile, and kyanite are found.

In the section from Pokrovsk to the mouth of the Aldan, these deposits comprise the very widely developed "Bestyakh" terrace, which is 35–40 m high.

These deposits have been classified as Middle Pleistocene on the basis of the discovery of a *Bison priscus* aff. *longicornis* bone (according to Gromov) in them (V. P. Chernyshkov and B. N. Mozhayev). This bison is an indicator of the Khazar faunistic complex and, in Europe, characterizes the epoch from the Middle Pleistocene to the glacial maximum (Gromov, 1948). The Middle Pleistocene age of these deposits is also confirmed by the paleontologic finds in the 12-m terrace of the Suola river, a left tributary of the Lena. This terrace rests directly against the "Bestyakh" terrace. Many bone remains of *Elephas primigenius* of the early (Riss) type were found in the deposits of this terrace. The very widely distributed deposits of the 25–30-m Lena terrace have been dated upper Middle Pleistocene. They consist primarily of sandy and sandy-pebble gravel formations, the sands being of a light yellow and a brownish-yellow hue.

The mineralogic composition of the heavy fraction of these deposits shows that the main mass of the heavy concentrate consists of hornblende, epidote, and zoisite, with considerably smaller quantities of magnetite and ilmenite.

The age of these deposits was determined on the basis of mammal bones. S. S. Korzhuyev found *Elephas primigenius* bones (two pelvic bones of an adult specimen, one pelvic bone of a young specimen, and one vertebra) in the 25–30-m terrace on the left bank of the Lena below the mouth of the Dobraya river about 6 m below the terrace brow. In addition a tooth,[4]M, was discovered. The tooth had 11 laminae per 10 cm, which exceeds the number of laminae of the typical earlier form of *Elephas primigenius* belonging to the Riss, and is fewer than in the typical later form. Thus, this dscovery indicates that the deposits are upper Middle Pleistocene. Furthermore, in 1952, G. I. Bushinskiy found an *Elephas primigenius* tooth near Olekminsk in a small gully which washes the 30-m terrace. The tooth was considerably abraded and smoothed and cannot be identified as either an early or a late form.

The Middle Pleistocene deposits of the lower Aldan are confined to the terrace which has a relative height of 20–30 m. For example, on the 30-m terrace, 20 km below the mouth of the Tatta river, the Middle Pleistocene deposits consist of gray medium-grained sands with lenses of coarse-grained sands having clearly defined cross-bedding. The transition to the overlying loams is abrupt. The contact with the underlying Tertiary deposits comprising the socle of the terrace could not be found and the structure of the lower part of these deposits could not be determined, so their exact thickness could not be established, but it is approximately 20 m.

The inclination of this terrace against the higher terrace, whose alluvium is Lower Pleistocene, and the inclination of this terrace towards the lower (15-m) terrace, which has been identified as Upper Pleistocene on the basis of numerous Upper Paleolithic mammal finds, give indirect evidence that these deposits belong to the Middle Pleistocene.

## Upper Pleistocene Deposits

The Upper Pleistocene deposits comprise the first Lena terrace (18–20 m) above the floodplain and consist of sandy material with a small number of pebbles.

The mineralogic composition of these deposits is almost identical with that of the Middle Pleistocene deposits. Their age has been determined from the discoveries of fauna in them. A. P. Okladnikov (1948) reported a find of bones in the 15–20-m terrace of the left bank of the Lena, 1 km above Mironovo and 5 m beneath the surface. According to V. I. Gromov, these are the bones of a mammoth (lower jawbone, tusk, scapula, and other parts of the skeleton).

S. S. Korzhuyev made a very interesting find near Pokrovsk. On the socle of the 20-m terrace, he found a very smooth *Alces latifrons* horn and an *Elephas primigenius* tooth ($_6$m). The tooth was rounded and abraded almost to the base. Thus, all these discoveries indicate that the deposits of this terrace were formed during the post-Riss era, i.e., towards the Upper Pleistocene.

The Upper Pleistocene deposits of the Aldan valley are confined to the first terrace above the floodplain (terrace I), which is 11–15 m high; further, all Aldan terraces, including the 70–80-m terrace, and even the Lena-Amga watershed, are definitely overlain with deposits whose upper strata contain Upper Pleistocene fauna. These deposits are known in the literature as "covering loams." Often the first terrace above the floodplain consists entirely of dark-gray loams with plant remains, occasional pebbles, and a fairly thick stratum of fossil ice. Stratification is not always observed. The bones of various mammals are found in the loams: *Rhinoceros antiquitatis, Elephas primigenius, Boss* sp., *Equus caballus, Cervus* sp. (*Alces*), *Rangifer tarandus, Cervus elaphus*—all of which belong to the Upper Paleolithic complex. Sometimes terrace I contains sands.

The cover deposits consist of gray or dark-gray loams and sandy loams with poorly defined stratification, with sand lenses, sometimes coarse-grained with pebbles, with wood, and other plant remains; often they contain mammal bones. The cover deposits vary greatly in thickness; on the terrace surfaces they are usually about 5 m thick, rarely more. On the Lena-Amga interfluve they become several tens of meters thick.

Almost no data are available for determining the lower limit of loam accumulation. The cover deposits may be assumed to be of different ages, as indicated by the presence of pollens of the broadleaf species in the lower horizons of these deposits, according to A. I. Popova's data (1955) taken from a borehole in the Churpacha and Chuya outcroppings, and also by the find of an *Elephas* sp. tooth with primitive features (which Gromov believes places it at the end of the Riss, the Riss-Würm period).

TABLE 1

STRATIGRAPHY OF THE QUATERNARY DEPOSITS OF THE MIDDLE LENA AND THE LOWER ALDAN

| Era | System | Epoch | Stage | Upper Lena, Vitim to Olekminsk | Middle Lena, Olekminsk to Pokrovsk | Lower Lena, Pokrovsk to mouth of Aldan | Lower course of Aldan river | Type | Composition of the heavy fraction | Mammals | Spore-pollen complex |
|---|---|---|---|---|---|---|---|---|---|---|---|
| CENOZOIC | QUATERNARY PERIOD | Holocene | Upper | Floodplain | Floodplain | Floodplain | | Loams, sandy loams, and sands of floodplain | Dominance of magnetite (up to 40%), garnet, sphene, rutile | *Elephas primigenius* (late form), *Rhinoceros antiquitatis, Equus caballus,* and others | Taiga similar to the present taiga |
| | | | | I 18–20 | I 18–20 | I up to 20 | I 11–15 | Sandy-pebble gravel deposits and loams of Aldan terrace I. Upper part of covering loams of Lena-Amga watershed and high Aldan terraces | | | |
| | | PLEISTOCENE | Middle | II 25–30 | II 25–30 | II 25–30 | | Sandy-pebble gravel deposits | Predominance of hornblende | *Elephas primigenius* (early form), *Lemmus obensis, Ochotona* cf. *hyperborea, Dicrostonyx torquatus* | |
| | | | | III 35–40 | III 35–40 | III 35–40 | II 22–30 | Sandy deposits (Lena river); cobbles and sandy loam (Aldan) | Magnetite, ilmenite, epidote, hornblende | *Bison priscus* aff. *longicornis* | |
| | | | Lower | IV 50–60 | IV 50–60 | IV 50–60 | | Sandy-pebble gravel deposits and loams | Pyroxenes, much mica | *Rhinoceros* cf. *antiquitatis* | Dark coniferous taiga (spruce, fir), pines predominating on sandy expanses with small admixture of broadleafs |
| | | | | V 70–80 | V 70–80 | III(?) 70–80 | III 40–50 | Sandy and loamy deposits with pebbles (Lena); sands and pebble gravels (Aldan) | | *Elephas* cf. *nomadicus, Equus* cf. *seramsniensis, Alces* cf. *latifrons, Trogontherium cuvieri* | |
| | | EOPLEISTOCENE | | VI 100–120 | VI 100–120 | IV(?) 100–120 | IV 70–80 | Sandy-pebble gravel deposits and sands (Lena), few weathered minerals. Sands (middle layer of Mamontova Gora) (Aldan) | Ilmenite dominates (16–20%), little magnetite, garnet (68–69%), sphene, rutile, zircon | | Taiga with admixture of broadleafs (*Tilia, Quercus, Ulmus*) |
| | TERTIARY | PLIOCENE | | VII 150–170 | VII 150–170 | | | Red-bed deposits (clays, loams with dominance of stable rocks in pebbles and presence of minerals of weathering crust; vermiculite and monmorillonite) (Lena). | Ilmenite dominates (up to 46.3%), magnetite (0.4%), garnet (52%), sphene, rutile | | Tertiary flora: mixed forest with various pines and spruces and broadleaf (*Tilia, Quercus, Ulmus*) and thermophilic Tertiary forms (*Juglans* sp., *Ostrea* sp, *Engelhardtia, Carya* sp. |
| | | | | VIII 200–250 | VIII 200–250 | | | Sands of lower layer of Mamontova Gora | | | |

The following conclusion may be drawn from an analysis of the thicknesses of the Quaternary deposits over the entire investigated territory. As a whole the Quaternary deposits are not thick; however, deposits several tens of meters thick have been found in the uppera part of the middle course of the Lena river. Thus, there were differential movements embracing small regions on a background of general stability in the territory.

The floodplain deposits are Holocene. They are found everywhere and consist of clayey or sandy material.

In addition to the floodplain, there is a clearly defined channel shoal consisting of pebble gravel. The pebble gravels contain rocks essentially from three source areas: the Vitim-Patom upland, the Lena Plateau, and the Upper Yana region.

Table 1 presents a preliminary stratigraphic scheme of the Quaternary deposits of the middle Lena and lower Aldan region.

A. P. VASKOVSKIY

# A BRIEF OUTLINE OF THE VEGETATION, CLIMATE, AND CHRONOLOGY OF THE QUATERNARY PERIOD IN THE UPPER REACHES OF THE KOLYMA AND INDIGIRKA RIVERS AND ON THE NORTHERN COAST OF THE SEA OF OKHOTSK*†

*Introduction*

IN THIS ARTICLE we have attempted to outline the basic features of the paleogeography and the chronologic stages of the Quaternary period in the southern part of that remote region of the U.S.S.R. called the Far Northeast by those who have studied it. This name is usually applied to the territory east of the lower Lena and Aldan rivers and the line connecting the Aldan valley at the mouth of the Maya river with the [river] port of Ayan.

The Quaternary deposits are enormously important for the Far Northeast, because numerous deposits of useful minerals are Quaternary and the solution of a number of important economic problems is associated with this period. Naturally, the proper solution of these problems requires the development of the principles of Quaternary stratigraphy and paleogeography in this region. However, these principles have not yet been established and the attempts made thus far to classify the Quaternary deposits, especially the deposits of the southern part of the territory, are based on indirect evidence and not on paleontologic data, which are decisive for stratigraphic syntheses. The paleontologic evidence has been scarce and the conditions under which it has been gathered have often been so vague that in many cases not only is it of no use for developing a scheme of Quaternary chronology and paleogeography, but also it has given rise to misunderstanding and error.

It suffices to recall the history of the simultaneous find of remains of the butternut and the mammoth in V. N. Zverev's collections from the Aldan which caused A. N. Krishtofovich to hesitate, till the last years of his life, between classifying the Aldan finds as Pliocene or Pleistocene, while in his earlier works he upheld I. D. Cherskiy's erroneous hypothesis that the Pleistocene climate of eastern Siberia had been mild.

This discrepancy between paleontologic data and the events of the Quaternary

*Translated from K. K. Markov and A. I. Popov (Editors), *Lednikovyy period na territorii yevropeyskoy chasti SSSR i Sibiri,* Moskovskiy gosudarstvennyy Universitet, Geograficheskiy Fakultet i Muzey Zemlevedeniya, 1959, pp. 510–556.

†Presented at the Seminar on the Study of the Quaternary Period, Moscow State University, May, 1955.

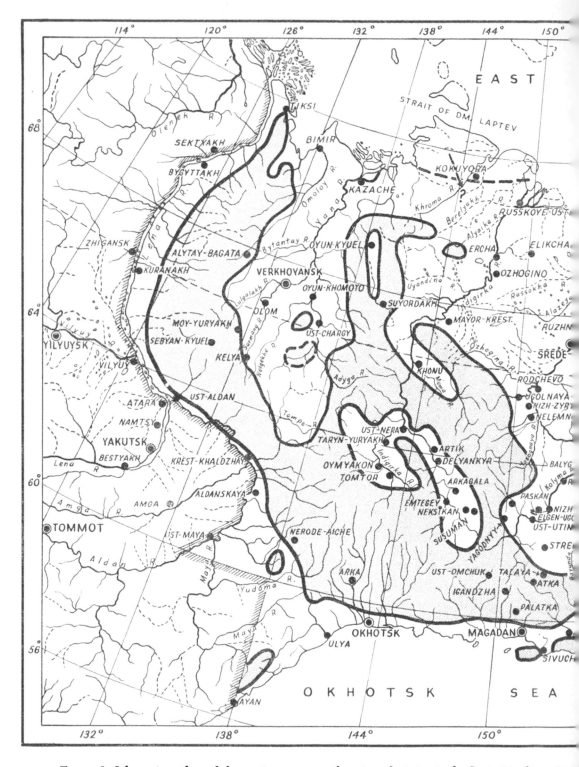

FIGURE 1. Schematic outline of the maximum extent of ancient glaciation in the Soviet Northeast (cor... in the hatched area. Only the valleys were occupied by glaciers and only here and there did the glaciers... The maximum advance of the ice was very similar during the two glacial epochs which left clear traces... to the south and west of the hatched area indicates the limit of the area to which this scheme applies.

by A. P. Vaskovskiy.) This scheme reflects the maximum advance of the ice. The glaciation was not continuous
emerging onto the piedmont and in the lowland of the Lower Anadyr river, leave trails of the Malaspina type.
ashed line in the north indicates the hypothetical limit of glacier advance, as indicated by Saks. The heavy line

epoch and the incorrect interpretation of pollen analysis data led Yu. N. Popov (1947) to the erroneous conclusion that the Pleistocene climate was moist and mild, that it changed little during the Quaternary period, and, consequently, that no alternation of glacial and interglacial epochs occurred. If this had been true, it would have been very difficult to distinguish the Quaternary deposits in the Northeast, but fortunately this is not the case.

V. N. Saks, the compiler of the first and only survey of Quaternary deposits of the Northeast, had to contend with a paucity of paleontologic finds and the lack of a strict correlation between the finds and the specific stratigraphic horizons. His regional sketches, which generalized all that was known of the Quaternary geology of the northern half of the Far Northeast at the time, were printed in various editions from 1936 to 1946, and conclusions based on these sketches have been included in a number of his general works (Saks, 1947a, 1947b).

In 1937–1938, Saks himself worked in the Far Northeasten U.S.S.R. and made small but valuable paleontologic collections. Subsequently, he investigated Quaternary deposits in other regions of the Soviet Arctic and carefully studied all the special literature. Saks' surveys reflect the present state of knowledge of the Quaternary deposits of the Soviet Arctic with almost exhaustive thoroughness. One of the characteristic features of the state of knowledge in Saks' day was the scarcity of geologic and paleontologic data on the Soviet Far Northeast. Therefore, Saks had to solve many problems by drawing on material from other, distant, regions; he solved many problems arbitrarily, and some he left completely unsolved, but through his great experience in the study of Quaternary deposits and his masterful analysis of the scanty published data, he arrived at a proper solution of many important problems. For example, following Ya. S. Edelshteyn, he portrayed the Pleistocene glaciation of the Northeast in almost its true dimensions, while I. D. Cherskiy assumed that only isolated glaciers existed at that time and V. A. Obruchev presumed sheet glaciation.

In opposition to Cherskiy, Saks proposed that the ancient glaciation of the Northeast was associated with cold periods and not with a mild, moist climate. Our analysis, based on additional data, upholds Saks' opinion, and merely adds some details to Saks' conclusions.

Using the data of S. V. Obruchev and the geologists of Dalstroy* and the Arctic Institute [in Leningrad] as his basis, Saks defended his view of the multiplicity of glacial epochs in the Soviet Northeast and specified three epochs of ice expansion, although he was not able to state whether the last of them, the Sartan,† was an independent glaciation or a stage of the preceding glaciation.

Saks' summaries pertain chiefly to the part of the Far Northeast which lies beyond the Arctic Circle, and only some scattered information (published by various authors) is included on the Quaternary deposits of the southern part of the territory. However, it is the southern part of the Soviet Far Northeast which has the valuable mineral deposits associated with the Quaternary, and the composition of these deposits and their relationship to earlier marine deposits help solve many questions of Quaternary geology, the significance of which goes far beyond the Northeast.

These circumstances led the Geologic Prospecting Administration of Dalstroy‡ to summarize the evidence on the stratigraphy and paleogeography of the Quaternary period in the southern part of the territory. However, this attempt encountered difficulties from the start, owing to the contradictions in the many

---

*[Administration for construction in the (Soviet) Far East.—Editor.]

†[Named for the Sartan river, in the upper Yana basin—Translator.]

‡[Geologorazvedochnoye Upravleniye Dalstroya.]

basic postulates of Quaternary paleogeography proclaimed in the works of investigators who dealt with the Quaternary deposits of the Northeast. Further, the reports of geologists who participated in the regional geologic survey contain few factual data, even for places extremely favorable for the collection of paleontologic Quaternary remains. In most cases, generalizations or references to remote regions were substituted for direct evidence. Therefore, the paleontologic data required for documentation of the main stages of the Quaternary history of the region described had yet to be collected.

I participated in the planning and execution of the considerable program of Quaternary investigations in the southern part of the Soviet Northeast, and an important part of that program consisted in the collection of paleontologic remains.

Specially organized teams collected samples for spore-pollen analysis (S. L. Khaykina, Ye. M. Voyevodova, Ye. D. Vlasova, M. V. Agranovich, A. P. Vaskovskiy) and the macroremains of flora (I. I. Tuchkov, A. P. Vaskovskiy, S. L. Khaykina, A. D. Kochetkova, *et al.*). Further, collections of fauna and flora made by the prospector geologists were examined and, what is especially important, collections by geologists who investigated the high terraces of the Kolyma-Indigirka region were examined. Extremely valuable paleontologic data were obtained during these prospecting ventures. The prospectors not only sent their finds to Magadan, but often reported immediately on discoveries of fossil animals and plants, and thus fresh finds were studied on the spot or delivered to the laboratory in an optimum state of preservation.

For example, mummified rhinoceros, bison, and suslik remains were found in the Elgi basin by prospectors and delivered to scientific research organizations by Yu. N. Popov, A. P. Vaskovskiy, M. M. Oradovskaya, and others. Similarly, some valuable finds of animals of the "mammoth complex" were made in the Berelyakh basin; the carcass of a horse was found in the Bolshoy Taryn basin, cones of the Anadyr spruce were discovered in deposits crowning the high terraces of the Berelyakh river and were delivered to scientific institutes.

The remains were classified in part at the Geologic Prospecting Administration of Dalstroy and in part at scientific research institutes in Moscow and Leningrad. The pollen and spores were studied by R. A. Baskovich, A. P. Vaskovskiy, and Ye. V. Dudkevich; the vertebrate remains were studied by Yu. N. Nopov, N. K. Vereshchagin, B. S. Vinogradov, V. Ye. Garutt, and others; the mollusks were studied by A. P. Vaskovskiy, A. F. Yefimova, and O. A. Skarlatto. An atlas of the morphology of the pollens of plants comprising the present flora of the Northeast was compiled by A. P. Vaskovskiy, S. L. Khaykina, Ye. M. Voeyvodova, and R. A. Baskovich to ensure reliable identification of the pollen grains. In addition, present-day pollen spectra of the soils beneath the most important plant groupings of the Northeast were taken and investigated along a meridional profile from Pevek in the north to Magadan in the south (A. P. Vaskovskiy, R. A. Baskovich, and S. L. Khaykina).

Comparison of the results of these works has shown the advantages and disadvantages of the methods employed. For example, some generally known inadequacies of the pollen method became especially evident under the peculiar floristic conditions of the Pleistocene and the present era in the Northeast. In particular, the poor state of preservation of the larch pollen, both in the fossil and in the modern spectra, which does not lead to major distortions of our conception of the forest composition in other localities, makes it very difficult to interpret the spectra of the Northeast, where various species of the larch played an enormous role in the Quaternary period and where one of them, *Larix dahurica* Turcz., predominates even today.

The pollen grains of the "dwarf Siberian pine"[*] and the Siberian ("cedar") pine[†] (belonging to one section of the species *Pinus*) are so much alike in the fossil spectra that they are either inaccurately identified in mass analyses or combined by palynologists under the heading *Pinus* e subgen. *Haploxylon*. This uncertainty has led to confusion in determining climatic conditions, since the dwarf Siberian pine is an indicator of the severe climate of the *uryema* [bottomland] tundra[‡][§] and the subalpine belt of mountains, while the Siberian pine is an indicator of a much warmer climate than the present climate of the Northeast.

The impossibility of differentiating between the species of spruce and pine by the pollen methods causes great difficulty in determining the appearance of the vegetation and thus the climatic conditions of the past. Further, during various epochs of the Quaternary period in the Northeast there were dark-coniferous forests of very similar genus composition but completely different species composition, indicating different ecologic conditions and different geographic affinities.

For these and other reasons, pollen analysis data should be used only in combination with a study of the macroremains of plants, chiefly pine cones, which are well preserved in friable deposits and which give direct indications of the species composition of the "builders" of the Quaternary plant communities. A. P. Vaskovskiy, R. A. Baskovich, and S. L. Khaykina correlated data obtained by the pollen analysis method with other methods. As a result, it became clear that plant remains are a reliable basis for constructing a scheme of Quaternary stratigraphy and establishing the paleoclimatic conditions of the Quaternary period.

## *Vegetation and Climate of the Late Pliocene Period*

To get an idea of the important events that took place in the Northeast during the Quaternary period, we shall begin our analysis of the paleogeographic conditions with the second half of the Pliocene, when this territory was still unaffected by the abrupt climatic changes which are a basic feature of the Quaternary. This is also essential because in recent years joint finds of marine Pliocene fauna and macroremains of plants and pollen have made it possible to associate the age of the lower limit of the glaciation of the Northeast (Fig. 1) and the age of many deposits of useful minerals with the stratigraphic scale constructed long ago on the basis of marine fauna studies.

### PLIOCENE DEPOSITS OF WESTERN KAMCHATKA

The so-called Etolon‖ suite in the Kavran stratum (Dyakov, 1935, 1936) is the youngest marine suite in the Tigil deposits of western Kamchatka, which are the most studied Cenozoic deposits and have the most complete profile. This marine suite is composed of dove-colored, gray, soft, interbedded sandstones with abundant and well-preserved marine fauna among which various species of sea scallops (*Pecten*) can be distinguished by their large and beautiful shells. The general structure of the fauna is as follows: *Acila* (*Truncacila*) Cobboldiae Sow.,

[*][*Pinus pumila.*—Translator.]
[†][*Pinus sibirica.*—Translator.]
[‡][Uryema—a floodplain or riverbank area covered with a low forest or shrub cover of willow.—Translator.]
[§]I proposed this name for the unusual type of wooded tundra in the Anadyr and Penzhina river basins (Vaskovskiy, 1950).
‖[Named for the Etolon river on the West Coast of Kamchatka.—Translator.]

*Arca* (*Anabara*) *trilineata* Conr., *Cardita etolonensis* Slodk., *C. kamtschatica* Slodk., *C. kavranensis* Slodk., *C. beringiana* Slodk., *C. monilicostata* Gabb. var. *ochotica* Slodk., *Glycymeris jessoensis* Sow., *Lyocima fluctuosa* Gid., *Modiolus tenuistriatus* Slodk., *Mactra* (*Spisula*) *polynyma* Stump., *Mya arenaria* L. var. *japonica* Jay, *Mytilus kamtschaticus* Slodk., *Pitaria Gretschischkini* Slodk., *P. kavranensis* Slodk., *Pecten* (*Patiopecten*) *caurinus* Gold., *P.* (*Pallium*) *swiftii* Bern. var. *nutteri* Arn. et var. *etchegoini* And., *Crepidula kamtschatica* Ilyina, *Neptunea lirata* Mart., *Turritella Gretschischkani* Ilyina, *Coptothyris ovalis* L. Kryscht., *Diestothyis ochotica* L. Kryscht., *Terobratula transversa* Sow. var. *elegans* L. Kryscht., and many other forms. The faunal composition leaves no doubt that the sea water which nourished these forms of life was comparatively warm.

The age of the Etolon suite has been established as Middle or Upper Pliocene.[*] Above the Etolon suite is the so-called Ermanovka[†] continental, lignite-bearing stratum, in which A. I. Poyarkova, on the basis of collections made by the geologists A. I. Shcherbakov, B. V. Nalivkin, B. F. Dyakov, and others, described the following plants: *Osmunda* sp., *Equisetum heleocharis* Eurh., *Struthiopteris* cf. *germanica* Willd., *Tsuga kamtschatica* A. Pojark., *Phragmites* cf. *communis* Trin., *Digraphis arundinacea* (L.) Trin., *Salix* cf. *dayana* Knowlt., *S. coalingensis* Dort., *S. amygdaloides* A. Pojark., *S. serratifolia* A. Pojark., *S.* cf. *remotidens* Knowlt., *S. kamtschatica* A. Pojark., *S. multinervis* A. Pojark., *S. minima* A. Pojark., *S. inquirenda* Knowlt., *S. glauca* L., *Betula* cf. *Benderei* Knowlt., *Alnus cuneata* A. Pojark., *A.* cf. *americana* L., *A. hirsutifolia* A. Pojark., *Corylus McQuarrii* Forb, *Ilex integerrimum* A. Pojark., *Acer* sp., *Quercus* sp., *Juglans* sp., *Phyllites* sp., *Vaccinim* sp. Many forms in this list are indicators (markers) of the American Pliocene, and A. I. Poyarkova established the age of the Ermanovka series as Late Pliocene–Early Pleistocene on the basis of these marker species.

In the northernmost parts of Kamchatka (near the mouth of the Gusina river and Rekinniki Bay), A. G. Pogozhev and later S. L. Khaykina and A. D. Kochetkova studied analogues of the Etolon and Ermanovka suites of the Tigil section. They established the presence of characteristic scallops of the Etolon series in the lower of these analogous suites: *Pecten swiftii* var. *etchegoini*, *P. swiftii* var. *nutteri*, and others and also *P. jessoensis* and other forms indicating it is of the same age as the Etolon suite of the Tigil deposits. S. L. Khaykina collected pine cones and nuts from the continental deposits lying above these strata, the continental deposits being analogues of the Ermanovka suite. A. P. Vaskovskiy (1954) classified these cones and nuts as *Metasequoia disticha* (Heer) Miki, *Juglans cinerea* L., and a new species of pine, the *Pinus itelmenorum* Vassk., belonging to the section *Strobus* (from Rekinniki Bay).

Earlier, I identified coniferous remains of the family Taxodiaceae in Pogozhev's collections from the Ermanovka strata. Pollen analysis made by Ye. M. Voyevodova from the same sample of the Pogozhev collection in which a twig of Taxodiaceae was found indicated the presence of Taxodiaceae pollen, several types of pines (including stroboids), spruces (from both sections comprising the genus *Picea*: *Eupicea* and *Omorica*), and representatives of the birch (*Betula*, *Alnus*, *Corylus*) and willow families.

The lists of flora remains reported by the above investigators have many features in common.

For example, the presence of Taxodiaceae remains was established by A. P. Vaskovskiy, who described the cones, and by Ye. M. Voyevodova, who analyzed

---

[*]In our opinion, the Etolon suite probably is Upper Pliocene.

[†][Named for the village of Ermanovka on the west coast of Kamchatka.—Translator.]

the pollen. The presence of remains of relatives of the walnut (*Juglans*) was established by all authors, although A. I. Poyarkova was not very certain of this species because of its poor state of preservation. However, Khaykina's find of *Juglans cinerea* fruit in a good state of preservation (identified by Vaskovskiy) fully confirms the genus identifications of the two other authors. Furthermore, she showed the possibility of applying the principle of actualism for revealing the ecologic conditions of the period during which these fruits existed, inasmuch as the American butternut has survived to our era in the eastern part of North America.

One must agree with Poyarkova that the Ermanovka suite belongs in part to the Late Pliocene and in part to the Early Pleistocene. It seems probable to us that the line of demarcation between the Pliocene and the Quaternary lies within the Ermanovka suite.

Now let us attempt to picture the ecologic conditions that existed on the west coast of Kamchatka during the Upper Pliocene. Both the fauna of the Etolon suite and the flora of the Ermanovka stratum indicate that the climate of the west coast of Kamchatka and all the neighboring areas was warmer than at present. At present, the scallops *Pecten swiftii* and *P. jessoensis*, which are characteristic of the Etolon suite, are not found north of Aniv Bay in southern Sakhalin Island, and for the most part they are found in the more southerly parts of the Sea of Japan. The northern limit of these marine fauna corresponds to the northern limit of the characteristic representatives of the so-called Manchurian flora on the shores of these seas, and their domains adjoin the region of broadleaf and coniferous-broadleaf forests in the lowlands of the Far East.

The fruit of *Juglans cinerea* and the cones of *Pinus itelmenorum* (which is similar to *Pinus strobus*) are highly important for determining the climate of the period when the lower part of the Ermanovka stratum formed (the Late Pliocene). At present, *Juglans cinerea* does not grow north of the +5° C annual isotherm and there is no reason to assume that it grew at much colder temperatures.

The climatic conditions of Green Bay, Wisconsin, which is within the domain of the butternut and *Pinus strobus*, may be regarded as characteristic of the region of *Juglans cinerea* distribution.

| Green Bay, Wis. | Temperature, °C | Precipitation, mm |
|---|---|---|
| January | −9.0 | 38 |
| July | +21.0 | 86 |
| Annual | +6.6 | 790 |

The climate of Ridgeway, Pennsylvania, is also characteristic for *Pinus strobus*. Ridgeway lies 200 m below a forest of *Pinus strobus* and *Tsuga canadensis*. The mean January temperature of Ridgeway is −4.1° C, the July temperature +19.0° C, the annual temperature +8.3° C, and the annual precipitation is 982 mm. Let us note that the mean annual air temperature on the shores of the seas in which the *Pecten swiftii* and *P. jessoensis* occur does not fall below +4° C, and, thus, the faunistic and floristic data confirm one another.

Consequently, the climate of the Kamchatka lowlands at the end of the Pliocene must have been characteristic of the climate of deciduous forests of the temperate zone (according to L. S. Berg, 1938)[*] and their vegetation must have been characteristic of the coniferous-broadleaf forests.

[*]According to Berg's classification of climates, coniferous-broadleaf forests also belong in this climatic region.

Judging by the find of cones of a stroboid pine and the pollen of hemlock and many other representatives of the pine family, a vertical belt of pure coniferous forests similar to the Laurentian forests of eastern North America must have grown above the belt of coniferous-broadleaf forests. The Laurentian pine forest grows along the Appalachians far to the south, above the belt of Appalachian broadleaf forests. Perhaps even less thermophilic forests of the mountain-taiga type existed still higher up the mountains. The presence of a whole series of oak-association plants in the lists of fossil flora (maple, oak, hazel, butternut, holly, and ostrich fern) and the presence of macroremains and pollen of the Taxodiaceae family also indicate that the basic type of vegetation of the Kamchatka lowlands was the coniferous-broadleaf forest. Thus, at the end of the Pliocene, coniferous-broadleaf forests, being a late derivative of the Turgay flora, grew in the lowlands of the northwestern part of Kamchatka.

Naturally, at that time there could not have been either a tundra zone within the limits of the present land area of the Northeast or any substantial glaciation of its mountains.

PLIOCENE DEPOSITS OF THE LOWER ALDAN

The relationship between layers with butternut and layers with pectenides on Kamchatka permits us to classify, with confidence, the well-known strata with *Juglans cinerea* in the deposits of Mamontova Gora on the Aldan as Upper Pliocene.

From the collections of I. I. Tuchkov, I have described a cone of the Monterey pine, *Pinus radiata* D. Don., and a fragment of a pine cone from the section *Strobus*—identified first as *Pinus monticola* though it may actually belong to the new type *Pinus itelmenorum* Vassk., belonging to the same section—taken from these deposits. M. N. Karavayev (1948) found cones of conifers of the family Taxodiaceae, cones of the *Pinus attenuata* Lemm, and a number of other southern forms in the Mamontova Gora deposits. A. N. Krishtofovich (1915) identified remains of *Picea Wollosowiczii* Suk. in the nut-bearing layers; however, the pictures of these cones in his work indicate that it would be impossible to identify their species.

Pollen analysis of samples from these same layers indicated the presence of the pollen of oak, hornbeam, birch, alder, and many representatives of the pine family. This floristic complex also indicates a moderately warm and quite moist climate; California forms are present and, what is more, they are not alpine types but vegetation of California lowlands.

Study of this outcropping by the geologists of Dalstroy shed new light on the ancient problem of the coexistence of mammoth bones and thermophilic flora in this deposit, indicating that the bones of the mammoth and other vertebrates of the Upper Paleolithic complex lie higher stratigraphically than the remains of the butternut and the exotic pines and are associated with cones of typical Siberian taiga spruce. However, the find of *Elephas meridionalis* Nesti at Vilyuy, described by I. A. Dubrovo (1954),[*] is probably associated with Pliocene (or Early Quaternary) deposits.

[*]See also the article "Stratigraphy of the Quaternary deposits of the middle Lena and the lower Aldan rivers," by N. S. Chebotareva, N. P. Kuprina, and I. M. Khoreva in the same volume [Dubrovo, 1954].

## Vegetation and Climate of the Early Quaternary Period

In 1953, S. L. Khaykina discovered a large number of conifer cones exclusively of the pine family at the mouth of the Gusina river in western Kamchatka above the Upper Pliocene deposits containing *Juglans cinerea* and *Metasequoia disticha*. A. P. Vaskovskiy, who studied these collections, identified the following forms: *Picea Bilibinii* Vassk., *P. hondoensis* Mayr. var. *tripartita* Vassk., *P. vitjasii* Vassk., *P. antiqua* Vassk., *Pinus* cf. *Nagajevii* Vassk., *Tsuga minuta* Vassk.

This list includes chiefly new and extinct species, which makes it difficult to judge the ecologic conditions of the flora and the correlation with other sections. However, there are many forms here that are found in another deposit of fossil flora, viz., the friable, sandy-pebble gravel layer of Nagayevo Bay. Vaskovskiy identified the following from the abundant collections of Ye. M. Voyevodova, S. L. Khaykina, and himself (collections made in 1952): *Picea Bilibinii* Vassk., *Picea rubra* Link., *Pinus monticola* Dougl., *Pinus Nagajevii* Vassk., *Tsuga minuta* Vassk., *Tsuga* cf. *heterophylla* (Rasin) Sarg., *Pseudotsuga magadanica* Vassk., and *Larix occidentalis* Nutt. Despite special, painstaking search, neither nuts of the Juglandaceae family nor cones of the Taxodiaceae family were found. No traces of these two families were found in the pollen analysis, except for one grain of *Pterocarya*, undoubtedly found in the secondary bedding.

Further, 323 grains of birch and pine pollen were found in the spore-pollen spectrum: *Picea* (both sections), *Pinus* (chiefly from the sections *Strobus* and *Eupitys*), *Tsuga*, *Alnus*, *Betula*, and *Corylus*. Spruce and pine pollen comprised 50 per cent of the tree portion of the spectrum, which corresponds approximately to their role in the cone collections.

In reconstructing the climatic conditions under which this forest grew, first we must keep in mind that no traces of trees of the families Juglandaceae and Taxodiaceae were found in the collections. This indicates clearly that the climate was colder than the climate at the time of formation of the Upper Pliocene deposits of Kamchatka. The cones found in the Nagayevo layer belong to the pine family, but they do not contain a single species that exists today, not only in the vicinity of Magadan, but generally on the Eurasian continent. Further, they contain several forms that have survived to our day in the North American Cordillera and in the mountains of Japan. The western American mountain species are: *Pinus monticola* Dougl., *Larix occidentalis* Nutt., and *Tsuga* cf. *heterophylla* (Rasin) Sarg.; the Japanese species is *Picea hondoensis* var. *tripartita*, which is often found in Kamchatka deposits of the same age and differs from the nominal subspecies only in structural details of the bract scale.

It must be remembered that the western American mountain pine, *Pinus monticola* Dougl., was found on the lower Omoloy by K. A. Vollosovich and was identified by V. N. Sukachev (1910*b*), who also identified a spruce, *Picea Wollosowiczii* Suk., which is similar to the Cordilleran Brewer spruce (*Picea Breweriana* Wats.). Sukachev assumed that the Northeast had a comparatively severe climate at the time these two trees flourished, because both trees reach the upper tree limit in the Cordillera. However, in Oregon, where the tree line is at an elevation of about 2500 m, the upper forest belt (1400–2500 m) consists of: *Picea Engelmannii* Eng., *Abies lasiocarpa* Nutt., *Tsuga mertensiana* Sarg., *Chamaecyparis nootkaensis* (Lamb.) Spach., and, often, a relative of our dwarf Siberian pine,

*Pinus albicaulis* Eng. At the same time, the species found in the Nagayevo layer —*Larix occidentalis*, *Tsuga heterophylla*, and *Pinus monticola*, along with other conifers—form the middle belt of vegetation (500–1800 m), changing downslope into a belt where *Pinus ponderosa* predominates. In the mountains on the Oregon-California line, where the tree line is at an elevation of about 3000 m, the *Picea Breweriana* reaches an average elevation of 1800 m and only individual trees grow at elevations as great as 2500 m.

Thus, none of these species, generally speaking, reaches the upper tree line, which indicates a mild, cool climate with positive annual temperatures, rather than a severe climate.

A *Pinus monticola* forest in St. Joe National Forest also contains: *Larix occidentalis*, *Pseudotsuga caesia* (apparently, a close relative of *Ps. magadannica* m.), *Thuja plicata*, *Tsuga heterophylla*, *Abies grandis*. Judging by the numerous descriptions and photographs of other places in the Cordillera, this is a standard environment (Schenk, 1939).

Just 200 m below this forest (three inhabitants of which existed near Magadan during the period of deposition of the Nagayevo layer), there is the Potlach (Idaho) Meteorological Station (elev. 770 m) which reports a January temperature of $-3.2°$ C, a July temperature of $+18.3°$ C, an annual temperature of $+7.5°$ C, and an annual precipitation of 580 mm with a maximum of 71 mm in December. The data of Polson Meteorological Station at Flathead Lake, Montana ($47°$ 45′ N., elev. 886 m) are indicative of the *Larix occidentalis* and the *Pseudotsuga caesia*, which are found below the *Pinus monticola* in *Pinus ponderosa* forests. The mean January temperature at Polson is $-5.0°$ C, July $+19.8°$ C, annual $+7.0°$ C, annual precipitation 463 mm with a maximum of 61 mm in November. Data of meteorological stations in the *Picea rubra* belt of the eastern American mountains indicate approximately the same climatic conditions (Dice, 1946; Schenk, 1939) as do the data for the *Picea hondoensis* in the mountains of Honshu, although the annual precipitation is greater (900–1700 mm).

Keeping in mind that the data cited here refer to the lower parts of this belt, the average climatic conditions for the growth of the Magayevo fossil forest may be estimated approximately as follows: mean annual temperature $+4°$ C, July temperature $+16°$ C, January $-7°$ C, precipitation 500–600 mm per year. These conditions are only slightly more severe than those of Leningrad today.

Like the Ermanovka Upper Pliocene suite of Kamchatka, evidently the Nagayevo layer was deposited near sea level. This permits us to trace the change of climate of the lowlands at the 60th parallel (which connects northern Kamchatka with Magadan) from the Late Pliocene to the deposition of the Nagayevo layer.

Judging by the change of flora, the mean annual temperature decreased by $2.5°$ C, the July temperature by $6.5°$ C, during that period. It is difficult to establish the change of precipitation, since these mountain forests are adaptable to a quite broad amplitude of moisture fluctuations. However, the plant communities are mesophytic, and lack the more hygrophilic spruces (*Picea sitchensis*, *P. Engelmannii*) which are so typical of the moisture-rich coastal ranges of the Cordillera. This rules out the possibility of a very great increase of precipitation.

It should be noted that there are small mountain glaciers, which constitute an attraction of Glacier National Park, near the Flathead Lake and Potlach Meteorological Stations, whose data were examined in connection with this forest belt. Although the glaciers here are quite numerous (90), the area they occupy is negligible and they do not extend below 1800 m (as in the southern glacier regions of the present Soviet Northeast). However, 3–4° farther north, in the Selkirk Mountains, 150 glaciers, occupying an area of 290 km², have been counted; the largest of these is 16 km long (Boli, 1948). The firn line here is at 2300 m, i.e., at the same height as the present firn line of the Suntar-Khayata range.

A forest belt, similar to that represented by the Nagayevo finds, is situated in the Rocky Mountains at an elevation of about 1000–1800 m and lies just 1000 m below the firn line and 500 m below the glacier termini. The flora finds in Nagayevo and on the Omoloy were made near sea level, and the excellent state of preservation of fragile cones rules out the possibility that they were transported from some distance either horizontally or vertically. Therefore, it should be assumed that the firn line of that time was much lower than at present in Idaho and British Columbia.

It seems probable that the first contact of the mountains of the Northeast with the chionosphere during the Cenozoic took place during the deposition of the Nagayevo layer and that the first glaciation (whatever its scale) was associated with a further deterioration of the climate, leading to the development of the Nagayevo flora, although the glacial maximum could have taken place considerably later. However this may be, there is not the slightest possibility that a Cenozoic glaciation could have occurred in the Northeast before the beginning of the Quaternary.

The most important feature of the early Quaternary flora represented by the vegetation of the Nagayevo layer is the presence of the western American and Japanese alpine forms, indicating a direct continental tie with Japan and the Cordillera across the Northeast at that time or shortly before it. The existence of this unique mixed flora, which may be called Nippono-Cordilleran, explains many features of the present flora (and fauna) of the Northeast, in which western American and Japanese species have been preserved.

Early Quaternary flora is known in the central and northern regions of the Soviet Northeast, as well as on the shores of the Sea of Okhotsk. The first find of Nippono-Cordilleran flora on the lower Omoloy, by K. A. Vollosovich in 1909 in deposits underlying the 80-m level of erosion, belongs to this group. The collections of G. P. Yefimov at the mouth of the Kegyulyur (a tributary of the Omoloy) were taken from these same deposits. Vaskovskiy identified *Pinus* e sect. *Strobus* (cf. *Pinus monticola* Dougl.) in the Yefimov collections.

The upper part of the Dzhelkan lignite layer of the Upper Nera depression and a number of other depressions of the Nera basin belongs to the Early Quaternary; Vaskovskiy identified *Pinus monticola* Dougl., *P. Nagajevii* Vassk., *Picea anadyrensis* Kryscht., and *Tsuga minuta* Vassk. in the collections of N. I. Orlov and others taken from the Dzhelkan lignite-bearing layer. The first three species are characteristic of all the Early Quaternary deposits of the Northeast and are also found in the Nagayevo layer. *Picea anadyrensis* appears here for the first time and allows us to determine the lower age limit of this species of spruce in the Soviet Northeast.

The lower part of the Dzhelkan layer, containing, according to the data of S. L. Khaykina, pollen of the *Juglans* and conifers of the Taxodiaceae family, is probably Pliocene. The conifer cones, which Yu. N. Popov (1947) classified as *Picea Wollosowiczii* on the assumption that they were found on the "150-m terraces of the Khudzhakh and the Taryn-Yuryakh," belong to these deposits. I had the opportunity to examine these cones (collected by A. I. Popov) and am convinced that their species cannot be identified because they are limonitized, rolled, and without scales. Furthermore, they were not found on the 150-m terrace, but in Pliocene deposits comprising the base of the 200-m aggradation terrace and containing *Juglans* pollen and representatives of the Taxodiaceae family. Unfortunately, no one has yet found cones of the *Picea Wollosowiczii* or other spruces on the 150-m terraces of the Taryn-Yuryakh.

Yu. N. Popov's assertion that "Zverev's find of conifer cones, the American butternut, and mammoth bones establishes the annual isotherm of the Age of the Mammoth as $+5°$ C" and that these fossils are early Pleistocene is also erroneous. As shown above, the annual isotherm of the butternut was probably even higher than $+5°$ C; this find should be classified as Tertiary and gives no indication of the climate during the Age of the Mammoth and the Pleistocene in general.

Unfortunately, there is a large gap in the paleontologic chronicle of the Soviet Northeast after the deposition of the layers with the Nippono-Cordilleran flora described. The following evidence pertains to later times and was found in another region, namely, in the Elgi river basin, a left tributary of the Indigirka, and in a number of neighboring valleys. We shall describe them presently.

# The Quaternary Geology of the Elgi, Bolshoy Taryn, and Berelyakh Basins and their Significance for Determining the Middle Stages of the Quaternary History of the Soviet Northeast

## TRACES OF THREE GLACIATIONS

Mining operations have been carried out in the floodplain of the Elgi river and on the high terraces cut by its tributaries. Such operations were also conducted in the Bolshoy Taryn valley. The Bolshoy Taryn is a right tributary of the Indigirka, entering the Indigirka almost directly opposite the Elgi river [64° 30′ N.; 143° 00′ E.].

Abundant and well-preserved floristic and faunistic remains, which shed light on the Quaternary history of the Northeast, have been found in the friable deposits which composed the terraces of these rivers to a greater or lesser degree.

It is fortunate for the history of the Quaternary period that plant remains were found in the deposits of high terraces of various ages in the Elgi valley near the Tobychan in conjunction with clear and sharp traces of ancient glaciation.

Traces of three glaciations have been observed in the Tobychan basin. The oldest of these left only uncoordinated erratic boulders on the mountain slopes at heights of 1200–1750 m above sea level. These boulders, which have been described by many authors, are found near the mountain installations along the left bank of the Elgi and did not cross over, as it were, to the right bank. They are not associated with the ancient river terraces; the glaciers which left them probably rose above the level of the peneplane and adjoined the high mountain chains.

Most authors, in writing of this glaciation, have been inclined to think it was of the ice-sheet type. However, traces of this glaciation are confined to the immediate environs of the high mountains, suggesting that it produced glaciers whose

termini fused into broad piedmont glaciers. Hence, this glaciation should be classified as alpine.

The second glaciation left its traces on the ancient leveling planes (peneplana-
tion planes) which were 150–220 m higher than the present level of the Tobychan river. Fragments of these leveling planes, bearing traces of mechanical
working by glaciers, and erratic boulders are arranged in strips which run across
the valleys of the large tributaries of the Elgi. Often these planes are intersected
by young troughs, whose floors are situated at the level of the thalwegs of the
river valleys or somewhat above them.

Of the young troughs worked by the third glaciation, the ancient valley of the
Tobychan river, from which the Tobychan was dislodged by the glacier and
which is now occupied by moraine deposits and Lake Chernyay, has been well
known since the expedition of S. V. Obruchev (Obruchev, 1932). It emerges into
the Elgi valley at the floodplain level and has a very thick moraine cover. Below,
we shall show that the third glaciation, which left its traces near the Tobychan
valley, was probably synchronous with the Würm glaciation of the Alps or the
Valday glaciation of the Russian Plain and that the interglacial epoch preceding
it corresponds to the Riss-Würm alpine system. There is also reason to believe
that the second glaciation of the Elgi basin was synchronous with the Riss
(Dnieper) glaciation.*

It is still impossible to draw a European parallel for the first glaciation, which
left erratic boulders on the ancient leveling planes, because of the lack of paleontologic evidence. Therefore, we shall establish a separate name for it and call it
the "Tobychan" glaciation.

### INTERGLACIAL FLORA AND THE DRAINAGE PATTERN OF BAZOVO CREEK

The highest of the erosion levels in the Elgi basin on which flora was found is
the remaining fragment of the ancient valley between the streamlets Bazovskiy
and Promezhutochnyy [*Bazovskiy ruchey* and *Promezhutochnyy ruchey*].

This ancient valley, which lies 370 m above the Elgi floodplain, consists of
alluvial deposits which conceal a buried canyon 40 m deep. Plant remains delivered to Magadan by V. M. Rodionov and G. P. Doroshenko and studied by A. P.
Vaskovskiy were found in these deposits. Among these remains, the following
were identified: *Tsuga* sp. (wood) and excellently preserved cones of *Picea
anadyrensis* Kryscht., *Larix sibirica* Ldb., *L. dahurica* Turcz., fragments of cones
of the *Picea* e sect. *omorica* and *Pinus* e sect. *eupitus*, needles of the *Picea* e sect.
*eupicea*, and inflorescence of *Betula* sp. The pollen spectra studied by R. A.
Baskovich contained pollen of all species and sections for which there are corresponding macroremains of plants and, further, much pine pollen of the subgenus *Haploxylon* (including *Pinus pumila*), *Alnus*, *Salix*, and *Corylus*.

Heather-like shrubs (evidently, representatives of the genus *Vaccinium* for the
most part) predominate in the non-arboreal part of the spectrum and the sphagnum mosses dominate the spores; probably they played a major role in the vegetation of the period. The Polypodiaceae were also abundant.

Thus, a general list of plants obtained by various methods for this deposit is
as follows: Lycopodiaceae, Polypodiaceae, *Sphagnum*, *Tsuga*, *Picea anadyrensis*,
*P.* e sect. *eupicea*, *P.* e sect. *omorica*, *Larix sibirica*, *L. dahurica*, *Pinus* e sect.

---

*In developing a unified scheme of Quaternary stratigraphy for the deposits of the Soviet
Northeast, the local name "Elgi glaciation" was given to the Second Glaciation of the Northeast, and the name "Bokhapcha" to the Third Glaciation.

*eupitus, Pinus* e subgen. *Haploxylon, Pinus pumila, Betula* sp., *Alnus* sp., *Corylus* sp., *Salix* sp., Ericaceae, and Vaccinaceae.

The above enumeration shows that during the formation of this ancient valley, the Elgi valley belonged to the area of taiga development with a fairly diverse complex of conifers, serving as "builders" of its plant communities. However, the composition of the flora was now much less exotic than that of the early Quaternary of the Nippono-Cordilleran coniferous forests of the Northeast, since it lacked broadleaf and coniferous trees, except for the pine family.

*Tsuga* and *Picea anadyrensis*, which are characteristic of the early Quaternary flora, still grew at that time but the typical representatives of Siberian flora, *Larix sibirica* and even *L. dahurica*, were making their appearance; these latter two have now gained complete supremacy in the cold pre-tundra forests and the forested tundras of the Northeast. These floristic features indicate a substantial change of climate, which took place during the period between the deposition of the Nagayevo layer and its analogues and the time these flora appeared.

The presence of *Larix dahurica* and *Pinus pumila* in the afore-named flora indicates a low annual temperature, inasmuch as these two species are not found south (and lower in altitude) of the mean annual zero isotherm and have a number of devices for protecting themselves against intense cold. The presence of the Anadyr spruce and the hazel, on the other hand, indicates that the mean annual temperature did not drop much below zero, and perhaps did not even drop below zero.

Keeping in mind the vertical zonality of the forestation, it would seem probable that the mean annual temperature in the lowlands of the Northeast was close to zero at that time and that soon afterward negative annual temperatures became established in the Northeast and with them permafrost, which varied in thickness from time to time, and which has not completely disappeared today.

Probably permafrost existed in the Northeast even earlier, during the first glaciation. At present we cannot say whether it disappeared completely during the subsequent interglacial epoch or merely became less extensive.

Let us note that flora was found at a height of 370 m above the level of the Elgi river at the source of Bazovskiy Ruchey [Bazovo creek] and, as a comparison of this flora with the pollen spectra taken by M. D. Yelyanov from the friable deposits of the 400-m Indigirka terrace show, this flora corresponds to the time of formation of the 400-m Indigirka terrace or of levels close to this terrace.

This flora is younger than the erratic boulders described earlier and is older than the traces of the second glaciation in the Elgi basin, traces found on the ancient leveling planes at elevations 150–200 m higher than the present level of the Tobychan river. Thus, it belongs to the first interglacial epoch in the Northeast.

Apparently, the finds on the 110-m terrace of the right bank of the Susuman, which fuses with the Berelyakh terrace at a similar height, represent one of the last stages of development of the flora of the first interglacial period. S. L. Khaykina and M. M. Oradovskaya, and A. P. Vaskovskiy, identified the following in the deposits of this terrace, on the basis of collections made by the geologist-prospectors and later the collections of A. I. Sudakov: cones of *Picea canadensis* B.S.P., *Picea obovata* Ldb., *P. anadryensis* Kryscht., *P. praeajanensis* Vassk. (a new species, similar to *P. ajanensis* Fisch.), *Larix dahurica* Turcz., and the wood of *Populus* sp.

The pieces of *Populus* trunks bore traces of working, which Vaskovskiy and A. P. Okladnikov (1948) ascribed to ancient man, but which I. G. Pidoplichko

(1950) considers the work of beavers. The pollen analysis, which S. L. Khaykina made from her own samples, revealed pollen of all genuses of coniferous plants whose macroremains were found on the Susuman terrace and, further, numerous grains of both subspieces of *Pinus* (in particular *Pinus pumila*), alder, willow, birch, and poplar.

Khaykina identified individual pollen grains of oak and a few hazel pollen grains in the samples, which led Vaskovskiy to conclude that the flora in these samples was characteristic of the southernmost taiga regions, which are still reached by occasional representatives of the broadleaf species Vaskovskiy and Okladnikov, 1948). However, additional pollen collections made by M. M. Ora-dovskaya* at my request and analyzed by Ye. M. Voyevodova showed no oak at all and just a few grains of hazel, probably transported thither. Furthermore, fir pollen was discovered, although in small quantities (up to 4%). Thus, this flora has a more northerly aspect than had previously been supposed. However, it has analogues in parts of the present Siberian and Hudson taiga zones of both northern continents where the winters are less severe than in the Berelyakh basin, although the summers of the period during which this flora existed probably were slightly warmer than the present.

This supposition is based on the data of the present climatic conditions for the growth of the Canadian spruce, which reaches the northern limit of the forests in America and which is the most representative member of the Hudson taiga. The climate of Fairbanks and Dawson represents the average climatic conditions of this flora, since the presence of *Abies, Picea praeajanensis*, and other representatives of the more southerly taiga in the samples rules out an analogy with the northernmost part of the Canadian spruce habitat. At Fairbanks, the mean annual temperature is −3.2° C, July +15.7° C, January −24.8° C; at Dawson the mean annual temperature is −5.2° C, the July temperature +15.2° C, January −31.4° C. In this region the annual precipitation amounts to 300–400 mm. Evidently, the Susuman flora records a stage when the climate deteriorated after the interglacial optimum and this deterioration, progressing, abetted a new advance of the glaciers.

Most likely, I. I. Tuchkov's samples from the Viliga river, which I investigated and which included cones of the American Engelmann spruce (*Picea Engel-manni* Eng.) and the Olga larch (*Larix olgensis* Henry),† belong to this interglacial period. The discovery of these two species indicates a moist, cool, and probably more maritime climate, which indicates that the Sea of Okhotsk existed during the first interglacial period.

## *Vegetation of the First Glacial Epoch*

The fine-grained and sandy material has been thoroughly washed out of the deposits of the second glaciation and thus far no paleontologic remains have been found in them.

*These samplings were taken in an attempt to resolve the contradiction between the species composition of the conifers described after the publication of an article by Vaskovskiy and Okladnikov (1948), which indicated a comparatively severe climate and the presence of oak and hazel in the pollen spectrum.

†First, I assumed that the cone belonged to a new species and called it *Larix viligensis*, but later I had the opportunity to acquaint myself with extensive collections of Far Eastern larch and I discovered that it fell within the species framework of the Olga larch.

Only in the upper reaches of the Kolyma, on the 125-m mixed river terrace near the mouth of the Bolshoy Khatynnakh river, have traces of intense cooling been established in the pollen spectrum (O. V. Kashmenskaya), while the spectra of the higher and lower terraces of this region indicate a richer vegetation and a warmer climate than the present.

The pollen spectrum of the deposits of the 125-m terrace do not include either spruce,* or relatives of the common pine, or larch, but it does contain a considerable amount of pollen of the dwarf Siberian pine, the birch, and the alder. Thus, this spectrum indicates a climate at least as severe as the present climate. O. V. Kashmenskaya concluded that it corresponds to an epoch of glacial advance. The climatic situation during this epoch, as far as can be ascertained from the pollen spectrum, was similar to the conditions under which the last glaciation occurred.

## The Flora of the Second Interglacial Period in the Elgi Basin and in Other Regions of the Northeast

As we shall see, remains of flora (and fauna) associated with the epoch of the third glaciation indicate a severe, harsh climate, in many respects similar to the present climate of the Northeast. The complete absence of even cold-resistant trees like the spruce, the common pine, and the Siberian larch indicates the severity of the climate.

In the Elgi valley and the lower reaches of its left tributaries, the moraines of the third glaciation rest on the present valley floors in places. Therefore, the plant remains found by G. P. Doroshenko in the friable deposits of the 80–100-m Elgi terrace at the point where it is intercepted by the Levyy Promezhutochnyy [creek] is particularly interesting. The floristic remains are, chiefly, well-preserved conifer cones, found at a depth of 15.8 m in a layer of gray loam with interlayers of sand and peat, covered with a 13-m layer of well-rounded pebble gravels. These remains were studied by Vaskovskiy, who identified the following among them: *Picea obovata* Ldb., *Larix sukaczewii* Djil., *L. sibirica* Ldb., *L. dahurica* Turcz. A pollen analysis made by R. A. Baskovich revealed *Betula* sp., *Alnus* sp., and *Salix* sp., in addition to the above.

Thus, this flora, which is situated between terraces that preserve clear traces of the next to the last and the last glaciation, belongs to the last interglacial period, as indicated by the three species of conifers that do not grow in the Soviet Northeast at present and which did not grow there during the last and next to the last glaciations. However, this flora does not contain any species alien to either the U.S.S.R. or the southern parts of Siberia, although it indicates a much warmer climate than the present climate of the Soviet Northeast.

This find of flora of the last interglacial period is not the only find in the central part of the Northeast and not even the richest, we have begun our description of the flora layer with this find simply because its position between adjacent well-authenticated glacial terraces is perfectly clear.

A very rich deposit of interglacial flora of almost the same composition was discovered by Vaskovskiy in the 40-m aggradation terrace of the Indigirka near

*Except for the lowest layers of friable deposits at their contact with the deposits of the terrace socle. Perhaps this layer was deposited during a warmer period.

Ust-Nera in 1953. Here, amid the thick fluvial pebble gravels is a layer of fine-grained sands and loams containing many trunks and cones of coniferous trees. The latter include: *Larix sibirica* Ldb. (more than 200 cones) and *Picea obovata* Ldb. (more than 30 cones). The following were also found: *Pinus silvestris* L. (15 cones) and *Larix dahurica* Turcz. (only 1 cone). A pollen analysis made by R. A. Baskovich revealed the presence of *Pinus pumila, P.* e sect. *cembra* (non *P. pumila*), *Picea* e sect. *omorica, Betula,* and *Alnus.*

This deposit of flora in which larch cones predominate shows very convincingly how poorly the larch pollen is preserved in modern and fossil spectra. The larch pollen comprises only 4 per cent of the total tree pollen in this deposit, while *Pinus pumila* pollen, which was not found among the macroremains, comprised 49 per cent.

A third deposit of similar flora was discovered by the geologist Shaposhnikov in the deposits of the 40-m compound terrace of the Berelyakh river above the mouth of the Susuman. Vaskovskiy identified *Picea obovata* Ldb. and *Larix dahurica* Turcz. in this sample. The pollen spectrum of this deposit differed only in details from the spectrum of the Ust-Nera terrace.

This flora is alien to the Northeast, but includes species whose ecologic requirements are well known and which are common to the more southerly part of Siberia. Therefore, we can define the climatic conditions under which that taiga existed with fair certainty. The presence of the spruce and pine in the floral register, the spruce now forming only insignificant islands in the southern part of the Northeast and the pine not occurring at all, indicates a less severe climate than the present, while the simultaneous find of *Pinus pumila* and *Pinus sibirica,* the *Larix sibirica* and the *Larix dahurica,* and omoricoid spruces causes us to seek climatic analogues of the Ust-Nera terrace in the limited region of south-eastern Siberia where the habitats of these species come into contact with each other.

The habitats of *Larix sibirica* and *L. dahurica* and *Pinus sibirica* converge at Baykal and the upper Lena. Here, other species of trees and shrubs that participated in the Ust-Nera flora are also widely distributed: *Pinus pumila,* the Siberian spruce, and the common pine. Since it cannot be assumed that a water basin even remotely the size of Lake Baykal, which would have ameliorated the climate, existed at that time [of less severe climate] in the vicinity of modern Ust-Nera, we can get an approximate idea of the climate at the time the Ust-Nera flora existed from the climate of Kirensk and Uchur. Uchur may be regarded as the western limit of spruces of the section *Omorica,* whose pollen was found in the Ust-Nera spectrum. Therefore, we shall call the above-described floristic complex the Uchur-Kirensk flora.

The climatic data of these two points are very similar; averaging theme we may regard the following conditions as being approximately those under which the Ust-Nera interglacial flora grew: average annual temperature −6° C, average January temperature −30° C, average July temperature +18° C, annual precipitation 350 mm. This severe and sharply continental climate is still almost 10° warmer in annual temperature and 3.5° warmer in the July temperature at Ust-Nera than at present. However, the average annual temperatures were lower than in Magadan at present, and thus the climate of the interglacial epoch was not mild and warm.

Numerous reliable finds of fauna of the "mammoth" complex are associated with the Uchur-Kirensk floristic finds. In particular, the prospectors found bones of the mammoth (*Mammonteus primigenius* Blum.) and the hairy rhinoceros

(*Rhinoceros antiquitatis* Blum.) together with the Berelyakh finds of Siberian spruce and both larches now growing in Siberia.

Yu. N. Popov (1948) described these finds, considering them comparable with and of the same age as deposits containing very rich Late Pleistocene fauna, which occur widely in the valleys of the Elgi tributaries. However, the two deposits are not of the same age and are not homotaxic and Popov's incorrect identification of these deposits led him to incorrect paleoclimatic conclusions.

Many representatives of this flora are found in the various deposits of the last interglacial flora of the U.S.S.R., including the classical ones. However, even in Povenets, which is just a little to the south of Ust-Nera, the deposits of the last interglacial period contain pollen not only of coniferous but also of broadleaf trees. Oak and hazel pollen are found at Povenets and the hazel pollen comprises more than half the tree pollen of the lower portions of the section.

There is nothing similar in the interglacial flora of the Ust-Nera, Berelyakh, and Elgi deposits, where the paleontologic data indicate the complete absence of broadleaf species in the vegetation and the complete supremacy of taiga, albeit a comparatively southern taiga.

Consequently, although the physicogeographic zones which passed through the U.S.S.R. in the last interglacial period were considerably displaced to the north compared with the present zones, there was an abrupt southward shift of the geographic zones in the Northeast then as today. Only the existence of a cold sea at the southern edge of the country in those days could explain this sharp distortion of the outline of natural zones; today the cold breath of the Sea of Okhotsk shifts the tundra zone along its shores southward to 59° N. lat.

One of the stages of development of the interglacial climate and vegetation is indicated by finds in the deposits of the 60-m terrace of the Berelyakh, rising between the mouths of its tributaries, the Sennyy and the Tengkelyakh, where A. S. Galun and A. I. Popov found a large number of conifer cones identified by Vaskovskiy as *Picea anadyrensis* Kryscht. and *Picea obovata* Ldb. This flora is older than the flora of the 40-m Berelyakh terrace and does not contain any vegetation foreign to modern Siberia.

We also classify the fossil peat containing fir trunks and pollen of *Pinus sibirica*,* *P. silvestris*, *Picea*, *Larix*, *Alnus*, and *Ulmus*, described by Yu. A. Bilibin, in the Allakh-Yun region as the last interglacial. Bilibin also classifies the remains of *Equus caballus fossils* found there as interglacial. Undoubtedly, V. N. Saks' find of conifer cones in the powdery loams of the 25–40-m Seledema terrace can be attributed to the early stages of development of this interglacial flora. He classifies these cones as *Picea Wollosowiczii*, following the somewhat uncertain identification by A. V. Yarmolenko, but probably they are *Picea anadyrensis*.

Wood of *Betula* sp., *Picea* sp., and *Corylus* sp. was found together with the cones, although *Corylus* sp. perhaps was identified incorrectly, inasmuch as the pollen spectrum does not contain hazel pollen, which would seem improbable, considering its collosal pollen productivity (according to the investigations of Pol, it is even more productive than birch), if it actually grew here. The pollen spectrum contained willow (54%), birch (23%), spruce (9%), Siberian pine,† (9%), and alder (9%). Bones of *Mammonteus primigenius*, *Equus caballus*, and *Bison priscus* were also found here.

If we enter the corrections, which seem needed in this list of flora, it would be typical of the middle stages of the last interglacial period.

*In our opinion, part of this is *Pinus pumila*.

†As V. N. Saks wrote, "obviously this is pine pollen of the section *Cembra*."

All the data help confirm an old report by Benkendorf that he found a cone and branch of the spruce in the stomach of a mammoth taken from the Indigirka in 1846 and lost again to the river. Evidently, this lost discovery of a mammoth belongs to the last interglacial period.

In re-creating the nature of the Northeast landscapes during the interglacial period on the basis of these data, we may conclude that the central and southern parts of the Northeast were forested with the south Siberian type of taiga at that time, while groves of alder (for the most part, probably shrubby), white birch, and willow grew in the northern parts of the country now occupied by tundra. This picture is now observed in southern Greenland and the lower Anadyr, and larch forests (or forested tundra) grew in the southern part of the modern tundra zone. The firn line in the southern part of eastern Siberia and in the Altay mountains, where forests now grow that are similar to the interglacial forests of the mountain regions of the Northeast, is higher than in the mountains of the present Northeast; thus we feel we can say that there were no glaciers in these mountains, at least during the period of the climatic optimum of the last interglacial period.

## The Paleogeography of the Last Glacial Period

I. D. Cherskiy, and recently Yu. N. Popov, held to the view that the climate of the entire Pleistocene era (and, consequently, of the glacial epochs) was comparatively mild and moist in the Soviet Northeast. Others (S. N. Obruchev, V. N. Saks, N. V. Tupitsyn, P. I. Skornyakov, L. A. Snyatkov, A. P. Vaskovskiy, Ye. T. Shatalov, P. N. Kropotkin, *et al.*) have held that the epochs of glacier advance in the Northeast as in other places were associated with the onset of a cold and severe climate. Much indirect evidence drawn from the arsenals of various sciences has been used by both parties, but no paleontologic data connected with the epochs of glacier advance were available. The adherents of the first point of view often used data of paleobotany, but as we have already seen from the collections made at Mamontova Gora, these researchers often confused finds of different age or arbitrarily assigned them to specific strata.

In recent works, the incorrect interpretation of pollen spectra has compounded these errors. The pollen spectra have been interpreted in an extremely linear manner, without consideration of the possibility that the older pollen of some trees (particularly the pine) could have been transported by the wind, even though the palynologists who developed the pollen-analysis method insisted that this be considered.

To solve this problem one must get paleontologic evidence from the deposits which formed during the epoch of ice expansion. This has been done by special field and laboratory research.

The deposits of the early epochs of glacial advance in the most accessible places had been thoroughly washed free of the fine fractions which retain pollen and other plant fragments; thus paleontologic samples were taken from deposits associated with the last glacial epoch, which we consider parallel with the Würm glaciation of the Alps or the Valday glaciation of the Russian Plain. Paleobotanic remains from the morainal and fluvioglacial deposits of that epoch were extracted in the following places: the middle course of the Tama river in the Chibagalakh basin, a left tributary of the Indigirka, by A. P. Vaskovskiy in 1939; and on the lower Taklaun river, a tributary of the Bolshoy Taryn, by Vaskovskiy in 1953; in the Dyadya Vanya valley, a left tributary of the Bokhapcha river in the Kolyma basin, by A. P. Vaskovskiy and S. L. Khaykina in 1950; near Lake Jack London in the Kolyma basin, by V. I. Safronov in 1951; at the head of the Bukesendzha river, which empties into Srednyaya Bay on the Sea of Okhotsk, by Ye. D. Vlasova in 1950; in the morainal deposits of the Berelyakh valley, at the stream Severnyy Rog, by Vaskovskiy in 1954; and on the shores of Egvekinot Bay in Zaliv Kresta [Cross Gulf] of the Bering Sea by Vaskovskiy in 1954.

Fragments of thin tree trunks, which I classified anatomically as *Larix* sp. and *Betula* e sect. *alba*, were found in the morainal deposits of Lake Jack London. The annual rings of the larch showed that it grew under extremely unfavorable conditions, apparently near the upper tree limit. K. I. Solonevich identified remains of *Pinus* e sect. *Cembra* (probably, *P. pumila*) and *Salix* sp. in the fluvioglacial deposits of the Tama river.

In V. G. Bulychev's samples from the deposits of the 10–12-m terrace of the Kolyma river, in the continuation of which, in the Bokhapcha valley, there are splendidly preserved moraines of the last glaciation in which Vaskovskiy found leaf remnants of *Betula exilis* Suk. and *Vaccinium uliginosum* L.

Pollen analysis yielded very important evidence. The absence of spruce and pine pollen of the subgenus *Diploxylon* and of the pollen of the more thermophilic coniferous and deciduous trees is a characteristic feature of the spectra taken from the morainal and fluvioglacial deposits left by the last glacial epoch. There are a few grains of the enumerated trees in some spectra (2 or 3 grains) but the number of these does not exceed that of grains of these same trees found in present-day spectra of the deposits of the Kolyma floodplain and some contemporary forest spectra of the Northeast, where neither spruce, pine, nor the more thermophilic trees grow.

Thus, pollen analyses indicate the absence of pine, spruce, and other thermophilic trees in the neighborhood of all glaciers of the last glaciation whose deposits have been studied. Morainal and fluvioglacial deposits contain the pollen of alder, birch, willow, and dwarf Siberian pine, which, together, predominate over the other components of the spectra. Samples from the forest regions of the territory also contain occasional larch grains. The study of modern spectra from the Northeast shows definitely that most of the birch, alder, and willow pollen found in the spectra of the glacial deposits is of the shrub type, not representative of trees of these species. This specific feature of the Late Quaternary and modern pollen spectra of the Northeast makes them generally incomparable with the scale of modern zonal spectra developed by V. P. Grichuk (1950) for the European U.S.S.R. Further, many researchers mechanically compare spectra of the Northeast with the European standard, without considering the peculiar and, in general, "non-arboreal" nature of the so-called "arboreal" portion of the spectrum in the Northeast.

The essential features of the pollen spectra of the morainal and fluvioglacial deposits are: the small sphagnum spore content compared with the spectra of the epoch when the Uchur-Kirensk floristic complex developed, the lesser development of club mosses (Lycopodiaceae), and the appearance, sometimes in considerable amounts, of spores of the Siberian resurrection plant, *Selaginella sibirica* (Milde) Hieron, and pollen grains of the wormwoods. The appearance of pollen grains of *Pinus pumila*, which were of much smaller size than at present, is also noteworthy.

Analysis of these remains invites the conclusion that the plant communities which existed in the Northeast during the last glaciation may have analogues only in the present vegetation of the Northeast, if it is examined in three dimensions. Then, as now, even such hardy conifers as the spruce and the common pine did not grow. Larch, dwarf Siberian pine, birch, willow, and alder (mostly of the shrub type) grew in the central part of the region, i.e., it was approximately the same plant complex which is characteristic of the southern and central parts of the Northeast today.

However, judging by the extremely stunted nature of a larch trunk taken from the morainal deposits near Lake Jack London, the tree line was lower then than now (it ran somewhat above 800 m, while now the larch reaches an absolute elevation of 1100 m above the lake), and the northern limit of the larch passed somewhat south of the present limit, although the poor state of preservation of the larch pollen (larch formed the forest boundary) makes any determination of the exact position of this boundary very difficult.

This similarity between the genus composition of the flora of the last glaciation and the modern flora of the Northeast led me to seek an explanation of these features of glacial deposits in the presence of *Selaginella sibirica* spores and pollen grains of *Artemisia* in the present plant associations of this territory.

The *Selaginella sibirica* and various species of *Artemisia* are characteristic components of the steppe associations discovered by V. A. Sheludyakova (1938) on the southern slopes of the upper Indigirka and by M. I. Yarovoy (1939) on the upper Yana. I found them widely developed in the basin of the Berelyakh and other tributaries of the upper Kolyma up to Seymchan and Buyunda in the east. Thus, the development of these steppe groups in conjunction with pre-tundra forests is a characteristic feature of the vegetation of this part of the forest zone of the Northeast within the Lena-Chaun mountain arc. These groups pass along the southern slopes of the mountains, gradually losing part of their steppe components, but preserving the wormwood and being enriched with club mosses nearly to the upper limit of the forest. Modern spectra taken from the soils of this region contain a considerable quantity of club moss spores and pollen grains of wormwood. Furthermore, near the upper limit of the forest they (like the spectra of other plant groups of the sparse larch forest zone and the belt of dwarf Siberian pine) contain the same kind of small *Pinus pumila* grains that are characteristic of the morainal deposits of the last glaciation. Let us note further that modern subalpine and alpine spectra, taken from the basin of the Sea of Okhotsk where a moister and foggier climate prevails, do not contain either club mosses or wormwood. Taking all these data into account, we may say that in the central part of the region the vegetation of the terminal moraine belt of the last glaciation, which lies at elevations of 700–800 m here, was similar to the present vegetation of the subalpine belt of the southern part of the Soviet Northeast. In places the moraines evidently descended into the sparse larch forest belt and in places the glaciers stopped in the *Pinus pumila* belt. Apparently, sparse stunted forests of *Larix dahurica* with underbrush of *Betula Middendorfii*, *B. exilis*, *Alnus fruticosa*, and willow bushes grew below the terminal moraine belt.

The belt of terminal moraines was approximately at the boundary of the *Pinus pumila* belt and the larch forest belt and we can get an approximate idea of the conditions during the glacial epoch from the present climate at this boundary.

The July isotherm, $+11°$ C, is now the temperature limit of *Larix dahurica* advance upward and northward. Therefore, the mean July temperature of the central part of this region in the terminal moraine belt during the last glaciation was probably $+10°$ to $+11°$ C and thus July was probably 2–2.5° colder than now. The winter was somewhat milder than at present, but was still very severe. Data from stations within or near the vertical *Pinus pumila* belt indicate that the mean January temperature of the terminal moraine belt during the last glaciation was approximately $-36°$ C and the mean annual temperature $-14°$ C. The amount of precipitation must have increased somewhat, because one of the precipitation maxima (in vertical cross-section) of the Soviet Northeast occurs in the present *Pinus pumila* belt. However, the precipitation increase apparently was not significant and probably was due chiefly to the increase of winter precipitation, because the presence of such heliophiles as *Artemisia* and *Selaginella sibirica* in the pollen spectra leads us to assume that the summer climate of the glacial era was continental to a certain extent, similar to that which now exists in the subalpine belt of the upper Indigirka and Kolyma. Considering this, we may assume that the total precipitation at the terminal moraine level (700–850 m absolute elevation) was no more than 400–350 mm and still less in the lower parts of the region.

Thus far we have treated the climatic conditions and the composition of the vegetation of the last glaciation of the central part of the Soviet Far Northeast. Only meager data are available on the parts of this region now occupied by tundra, which allow us to establish only the most general features of the climate of that era. The pollen samples, taken by Vaskovskiy from the marvellously preserved moraines of Egvekinot Bay and studied by R. A. Baskovich, present a picture qualitatively identical with modern spectra of the Chukchi tundra taken and analyzed by these same persons. These samples contain a very small amount of *Pinus pumula*, *Alnus*, and *Betula* grains, completely commensurable with the amount of pollen in the modern tundra spectra. No sphagnum spores were found, but spores of the *Selaginella* sp. and *Lycopodium* sp. were found.

Generally speaking, this is definitely a tundra spectrum and, thus, during the last glaciation the northern limit of the forest ran much farther south, not farther north, than at present.

Summarizing our knowledge of the vegetation and climate of the last glacial maximum, we may say that the vegetation of the Northeast at that time differed from the present vegetation quantitatively rather than qualitatively. The same sparse larch forests that are characteristic of the upper Kolyma and Indigirka today grew in the lowlands, but the trees were more stunted then than today. For convenience of presentation, we shall call this the "Kolyma floristic complex," since the upper and often the middle course of the modern Kolyma river and also the middle zone of mountain structures within this basin are in the region of development of similar plant associations.

The general picture of horizontal and vertical zonality was approximately the same as today, although the tundra zone was more extensive and the upper limit of the forest in the upper Indigirka basin was 400–500 m lower and in the upper Kolyma 300 m lower than at present. The temperature indices in the southern part of the Kolyma lowland evidently were similar to those of modern Omsukchan and Kresty Kolymskiye [on the lower Kolyma river] but in those days Omsukchan and Kresty had a climate corresponding to the present tundra climate. Thus, the mean annual temperatures of that time were somewhat higher than at present, because of the warmer winter temperatures, which were nearly compensated by lower summer temperatures. As a whole, this was a severe climate similar to the present climate of the Northeast, but from the ecologic point of view it was even more severe than the present climate, because the slight decrease of summer temperatures exerted a greater influence on the plant and animal world than a lowering of the winter temperatures.

We have described the characteristic features of the Kolyma complex of fossil flora, showing that this complex (and the severe climate which gave rise to it) existed during the maximum advance of the last glaciation. However, this complex appeared in the lowlands of the Northeast earlier than the apogee of the last glaciation. For example, A. P. Vaskovskiy, S. L. Khaykina, and Ye. D. Vlasova established that the pollen spectra of all the numerous terraces that rise 2–100 m above the thalweg in the portion of the Kolyma valley between the mouths of the Bokhapcha and Ortukan rivers have all the features that are typical of the Kolyma floristic complex. These spectra lack the spruce and pine (except *Pinus pumila*) that are so characteristic of the Uchur-Kirensk complex. When spruce and pine pollen is present it is found in negligible amounts so that it was probably transported there by various means. All finds of macroremains of plants associated with this series of terraces indicate that the flora of that time was similar to the present Kolyma flora.

However, the maximum advance of the last glaciation is associated with the formation of the deposits which crown the 10–12-m (and probably the 30-m) terrace of this part of the Kolyma valley.

In the upper Berelyakh basin, the characteristic features of the Kolyma flora are peculiar to the pollen spectra of the comparatively thick friable deposits that are covered by moraines of the most recent glaciations and which lie directly at water level on the stream Severnyy Rog; but in the middle course of the Berelyakh, which experienced a comparatively intensive uplifting, spruce and pine remains are not found below the 40-m level.*

Spruce and ordinary pine are not found below 60–80 m in the Elgi valley and here the moraines of the last glaciation are associated with the terraces at 20 m and lower.

Thus, the Uchur-Kirensk floristic complex was replaced by the Kolyma complex long before the glacial maximum, but after the climatic optimum of the last interglacial. The beginning of the reign of the Kolyma flora, which is well documented by the pollen spectra and finds of the macroremains of plants, serves as a clear chronologic guide by which one can establish the age relationships between the terraces of various regions of the Northeast. A considerable deterioration of the climate compared with the period during which the Uchur-Kirensk flora dominated is indicated and the establishment of a climate which differed only in details from the present climate of the Northeast.

The abundant and rich finds of the "mammoth" fauna complex are associated with the deposits of this period; as we saw earlier, finds of "mammoth" fauna are also connected with finds of Uchur-Kirensk flora, reflecting the conditions of the last interglacial period.

Thus, the "mammoth" complex in the Northeast has the same chronologic relationships with the glaciers of the last glaciation and the preceding interglacial stage as it does, judging by the works of V. I. Gromov (1948) and other authors, in the European U.S.S.R. and western Siberia. Therefore, we have reason to believe that this glaciation and the preceding interglacial period were synchronous with the last glaciation in Europe and its last interglacial.

The general composition of the Kolyma floristic complex reveals only very

*Counting from the upper edge of the alluvial deposits which crown the terrace to the summer water level of the river. Possibly, a somewhat lower terrace, not yet investigated paleontologically, will serve as this boundary line in the future.

small fluctuations in the various taiga regions of the Northeast. One of the best-known finds of this complex is the stomach and buccal cavity of the famous Berezovka mammoth, whose food remains were studied by V. N. Sukachev.

The following plants were found in the digestive tract of the mammoth and beneath its body (the species names are given in agreement with the systematic interpretations of the authors of *Flora SSSR* [Flora of the U.S.S.R.], 1934–1955). Mosses: *Hypnum fiuitans* Dill., *Aulacomnium turgidum* (Whlb.) Schwager; conifers: *Larix* sp.; Gramineae: *Alopecurus alpinus* Sm., *Beckmania syzigachne* (Stend.) Fern., *Agropyrum cristatum* (L.) Gaertn., *Hordeum brevisubulatum* (Trin.) Link., *Agrostis* sp.; Cyperaceae: *Carex tripartita* All., *Carex* sp.; birches: *Betula* sp., *Alnus* sp.; motley grasses: *Ranunculus acris* L., *Oxytropis sordida* (Will.) Pers., *Papaver nudicaule* L. (?), *Thymus serpillum* L.

Sukachev pointed out that most of the plants comprising this list also appear in the present vegetation of the Kolyma basin, but he assumed that three of the enumerated plants, namely *Agropyrum cristatum*, *Hordeum violaceum* (he considered *H. brevisubulatum* to be a variety of this), and possibly *Oxytropis sordida*, are not found at present in the Kolyma basin (Sukachev, 1914). However, subsequent work has shown that these three plants do grow in the forest belt of the Kolyma basin and thus all types found with the Berezovka mammoth inhabit the present Kolyma taiga. The combination of arcto-alpine plants (*Carex tripartita*, *Papaver nudicaule*, *Aulacomnium turgidum*) and steppe plants (*Thymus serpillum*, *Agropyrum cristatum*), which appears strange at first glance, reflects the coexistence of steppe associations on southern slopes and forested tundra communities on northern slopes, which is also a highly characteristic feature of the present vegetation of the Northeast. Therefore, it may be said that the Berezovka mammoth lived in a geographic situation similar to the present one in the place where it was found.

The plant remains found in the deposits which form a fairly thick stratum in the upper Indigirka basin beneath the valley floors of the Elgi, the Bolshoy Taryn, and its tributaries also indicate that the geographic medium of that era was similar to the present. A considerable subsidence took place here in the Late Quaternary, as a result of which the low mixed terraces of the Elgi and the Bolshoy Taryn and the floors of the valleys incised in them were buried beneath a layer of pebble gravels and loams. The lower parts of this layer contain abundant remains and sometimes entire bodies of vertebrates of the "mammoth" faunistic complex. Many other finds, in particular the body of a horse discovered in the Bolshoy Taryn basin, belong to this same formation. Vaskovskiy investigated the mode of occurrence of this find.

The following is a list of vertebrates whose remains have been extracted thus far from the buried canyons and valleys of the Elgi-Taryn region: *Mammonteus primigenius* Blum., *Rhinoceros antiquitatis* Blum., *Bison priscus* Boj., *Ovis nivicola* Esch., *Rangifer tarandus* L., *Equus caballus fossilis* L., *Citellus glacialis* B. Vinogr., *Lepus timidus* L., *Vulpes vulpes* L., and *Canis* sp. This list contains all species characteristic of the mammoth complex of northern Europe. The only exception is a recently described species of suslik, which has not been found elsewhere, and which may be endemic of eastern Siberia.

Yu. N. Popov, who studied the mode of occurrence of several finds in the Elgi basin in 1947, took samples for pollen analysis from the Struyka valley from strata containing remains of various animals of the mammoth complex. S. L. Khaykina discovered the pollen of birch, willow, alder, and grassy plants in these samples, among which were Gramineae, Polygonaceae, and Dipsacaceae. Individual pollen grains of the linden and hazel were found in various parts of the section, and, on the basis of this, Khaykina (1948) and Popov (1947) considered the climate at the time of the deposition of these layers to be moist and warm.

This statement raised a number of objections. First of all, no conclusions about climate can be drawn from a find of individual pollen grains of the linden and hazel, and even if this were possible, the pollen grain of linden, which Popov considered proof of a mild and moist climate, was discovered in the same layer as animal remains which indicated the presence of permafrost and Dipsacaceae pollen, which Popov asserts is evidence of a dry climate. Furthermore, the hazel pollen was found immediately above a layer containing the soft parts of animals and, if we follow Popov, this is proof of the return of a "warm and moist climate" after the deposition of the bone-bearing layer. This should have destroyed the permafrost which preserved the animal tissues; however, this did not occur. Neither the Dipsacaceae pollen nor the Polygonaceae pollen in themselves indicate a warm or a dry climate, because the representative of the Dipsacaceae, *Drosera rotundifolia* L., is still abundant in the sphagnum swamps of the Northeast, and the Polygonaceae family, which is very widely distributed over the entire Northeast, contains a multitude of arcto-alpine species both here and elsewhere.

Plant remains, accurately identified by V. Petrov as of the tuber *Polygonum viviparum* L., were found in the cheek pouches of a fossil suslik from the Dirin-Yuryakh deposits (Popov extends all conclusions which follow, in his opinion, from his study of the Struyka cross-sections to the Dirin-Yuryakh deposits). This jointweed is an arcto-alpine representative of the Polygonaceae and, as a result of the very severe climatic conditions, descended onto the meadows of the vertical forest belt in the Kolyma-Indigirka region, though it is much more abundant in the vertical belt of the forested tundras.

The contradictions in Popov's construction led me to visit these valleys of the Elgi basin in 1953, where remains of the bodies of Pleistocene animals had been found, and to collect new material from the bone-bearing deposits of the Dirin-Yuryakh and Struyka rivers for pollen analysis. These collections were analyzed by R. A. Baskovich and the spectra discovered in them revealed the following characteristic features:

1. Neither the Dirin-Yuryakh nor the Struyka deposits contained pollen of plants alien to the present Elgi basin.

2. Either birch and alder (probably shrub species, for the most part) or dwarf Siberian pine dominate the "tree components" of the spectrum. Larch is found consistently, but in smaller amounts. Pollen of heather-like shrubs (especially *Vaccinium*) is abundant. The motley grasses are represented by Gramineae, Cyperaceae, Urticaceae, Caryophyllaceae, Cruciferae, Umbelliferae, and, to a lesser extent, other families. Sphagnum predominates among the spores, besides which there are Lycopodiaceae, Polypodiaceae, Equisetaceae, Bryales, and *Selaginella sibirica* (Milde) Hieron. A single pollen grain of common pine was found in the Dirin-Yuryakh deposit.

3. Neither linden nor hazel pollen was found, which proves the fortuitous nature of its appearance in the samples collected by Yu. N. Popov.

Let us add that remains of tree trunks, which Ye. V. Budkevich has identified as larch, were found in the bone-bearing deposits beneath the floodplain of Nochnoy Ruchey [Nochnoy Creek] in the Elgi basin. The composition of the pollen spectra taken by me in the layer of loams which contained the body of a fossil horse in the Bolshoy Taryn valley (Sana Creek) 16.5 m below the floodplain of the Sana was complementary [*vide infra*]. The presence of *Selaginella sibirica* in the spectra and the larger quantities of *Artemisia* here than in the spectra of the Dirin-Yuryakh and Struyka deposits are noteworthy. As indicated previously, this combination is a characteristic feature of the spectra of the glacial maximum during the last glaciation and the spectra of the present xerophytic meadows which cover the southern slopes of the upper Indigirka basin. I was fortunate enough to find macroremains of flora in the deposits of the 8-m aggradation terrace of the Bolshoy Taryn (which covered the body of the fossil horse), viz.: cones of the *Larix dahurica* Turcz., leaves of the *Betula Middendorfii* Tautv. et M., *B. exilis* Suk., *Salix* sp., and *Vaccinium uliginosum* L. Undoubtedly, the body of the horse found in the Bolshoy Taryn basin was similar in age to the Elgi graveyards of Pleistocene animals; in fact there is so little doubt of this that the finds have been classified as belonging to the period when the Kolyma floristic complex predominated and, consequently, to a period of severe climate similar to the present climate of the upper Kolyma and Indigirka rivers.

On the basis of direct field observations, Vaskovskiy also established that the deposits which contained the body of the horse and the moraines of the last glaciation are related chronologically, because the 8-m terrace, beneath whose deposits the fossil horse was found, slopes towards the moraines of the maximum stage of this glaciation on the lower Taklaun (right tributary of the Bolshoy Taryn), while in the upper reaches of this river, the 8-m terrace bears the moraine of one of the stages of glacier retreat. Thus, the fossil horse probably belongs to the maximum stage of glaciation or the initial stages of glacier retreat.

These data are convincing evidence that a cold and comparatively dry climate prevailed during the period of formation of the thick layer of friable deposits now buried beneath the floodplains of the streams which flow into the lower Elgi and the Bolshoy Taryn and which contain abundant remains of animals of the mammoth faunistic complex. The climate was neither mild nor moist at any time during the formation of this layer, and varied but slightly. Further, the composition of the vegetation indicates that the climate at the time the Elgi-Taryn bone-bearing layer was deposited was much more severe than the climate of the period when the "25-m" (actually the 40-m)[*] Berelyakh terrace formed, which Popov compared with the Struyka section and which actually formed during an earlier period, namely, during the last interglacial period.

Although the climate of this period was not as cold as at present, it was severe, similar to the present climate of southern Yakutia. Thus, in the Soviet Northeast the finds of mammoth fauna are associated with deposits containing the remains of plants belonging to the Uchur-Kirensk and the Kolyma floristic complexes. These complexes belong, correspondingly, to the last interglacial and last glacial periods, but the mammoth faunistic complex apparently appeared here before the interglacial climatic optimum, possibly during one of the stages of the last glaciation. In any event, it formed during the late stages of the Pleistocene, as evidenced by A. A. Volosatov's find on the 80–100-m terrace of the Talbykchan river (Selennyakh basin) of an elephant tooth which, according to V. V. Menner, represents a primitive type of mammoth with a number of features similar to *Elephas trongontherii* Pohl.

Thus, the mammoth faunistic complex falls within the same chronologic framework in the Northeast as in Europe and, as in Europe, apparently it survived the

[*]Counting from the upper rim of the alluvial deposits which cover the terrace to its base.

alternation of cold–somewhat warmer–cold climate, which in all cases was a very severe continental climate. Therefore, let us reject once and for all I. D. Cherskiy's idea (1891), supported by M. Yokoyama and Yu. N. Popov, that eastern Siberia had a moist and warm climate during the entire Pleistocene.* The severity and continentality of the Upper Paleolithic climate explain the combination of steppe and arctic elements in the mammoth floristic complex far better than Cherskiy's hypothesis† that the Pleistocene climate of the Northeast was moist and mild.

These contrasting groups of animals usually do not come into contact with each other in the moderate and moist climate of the western parts of the present forest zone, being separated by a habitat of forest fauna adapted to these conditions. However, in the present severe and dry climate of the upper Indigirka and Yana rivers, the steppe and arcto-alpine plants and animals do come into direct contact with each other and live side by side. Here, in the river valleys of the lower vertical zone, grow such typical representatives of the Mongolian steppe as the fescue (*Festuca lenensis* Drob.), the koeleria (*Koeleria gracilis* Pers.), the ephedra (*Ephedra monosperma* C. A. M.), and on the Yana feather grass as well (*Stipa decipiens* Sm.); together with these one finds arctic sedges (*Carex stans* Dr.). On the upper Indigirka, steppe representatives of the locust family (*Bryodema tuberculatum dilutum* Stoll.) and leaf beetles (*Entomoscelus adonidis* Pall.) are found together with arcto-alpine white partridges (*Lagopus albus* L., *L. mutus* Mont.), arctic locusts (*Tetrix fuliginosa* Zett., *Melanopus frigidus* Boh.), and the Siberian pony (*Anthus gustavi* Sw.), an inhabitant of the tundra and forested tundra. In addition, such typical alpine animals as the snow sheep and the pika (*Ochotona hyperborea* Pall.) often appear, sometimes descending to the Indigirka itself. Taiga animals and birds dominate the fauna register of the upper Indigirka and its neighboring rivers and this distinguishes the present fauna from the glacial fauna. However, the mixture of animals of diverse ecologic composition—alpine, steppe, tundra, and taiga—is a characteristic feature of these fauna, caused by the severity and continentality of the climate.

The vegetation of the upper Indigirka combines plant goupings of very different composition: steppe fragments on the southern slopes, tundra sedge meadows and larch forests in the extra-floodplain portions of the valleys, and so on. Furthermore, in the lower strata of the extra-floodplain areas, larch forests of the Yana-Chukchi mountain tundra vegetation or of vegetation generally associated with the Arctic prevail, however paradoxical this may appear. For example, the prevailing moss in the moss larch forests which dominate the mountain valleys of the lower forest belt is the arcto-alpine *Aulacomnium turgidum* (Whlb.) Schwager; thus one can speak of "forested tundra moss beds." In their shrub layer, birches of the arctic section *nana* and a number of willows, belonging to sections spatially or genetically associated with other arctic regions, play a large role. Thus, the term mountain "tundra-forest-steppe" would be a good, brief characterization of the general aspect of the vegetation of the central part of the

---

*Yokoyama (1910), on the basis of an erroneous classificaiton of the "coral" layers of Japan as Pleistocene, assumed the Pleistocene climate to have been warmer than the present climate. However, Shigeru Miki proved that the Ekoda strata near Tokyo, in which there are abundant representatives of conifers now growing in the mountains 1500–2000 m above the Ekoda finds, belong to the last glacial era. This indicates that the summer temperatures were 7–10° colder than at present (Miki, 1938).

†Cherskiy speaks of the entire Quaternary period, but since he bases his judgments on the ecologic features of the mammoth faunistic complex, he was really thinking principally about the time during which this complex existed.

region, in which forest elements prevail. The material described above allows us to assume that the vegetation of the central part of the Northeast during the Late Quaternary epoch of glacier expansion was also "tundra-forest-steppe," in which tundra and probably steppe components played a larger role than at present. Having penetrated into the mountains, where the variety of natural conditions permits organisms with highly diverse ecologic requirements to exist in close proximity to each other, these mixed flora and fauna subsequently descended into the lowlands as well.

Thus, we conclude that the present geographic situation in the Soviet Northeast differs quantitatively rather than qualitatively from the geographic situation during the last glacial expansion and that the situation during the glacial era can be pictured easily on the basis of the main features of the present aspect of the country. Therefore, although we agree with D. M. Kolosov (1952) that the landscapes of the lowlands of the Northeast have many features of the Late Glacial epoch, we cannot agree with his conclusion that the present development of the mountains and lowlands of the Northeast took place in several stages. Both the mountains and the valleys have preserved many features of the Late Glacial geographic situation.

The buried ice of the East Siberian Lowland, which Kolosov considers proof that the lowlands belonged to a different stage of development because the buried ice did not melt at the same time as the mountain glaciers, is not of glacial origin. The conditions under which this buried ice formed and melted were different than those of the glaciers and, generally speaking, one cannot compare their stages of degradation.

The prevalence of a cold and comparatively dry climate during the glacial epochs offers a good explanation for the existence of species in the present flora and fauna of the Soviet Far Northeast which are identical with or related to species which occupy the highlands of southern Asia. For example, they include *Caragana jubata* (Pall.) Poir, a relative of the widely known yellow "acacia,"* found some time ago beyond the Arctic Circle in Yakutia (Komarov, 1926; Karavayev, 1948) and found recently by Vaskovskiy in the alpine belt of the Olkha Basalt Plateau on the watershed of the Kolyma river and Sea of Okhotsk. This group includes the Kurilean tea plant (*Dasyphora fruticosa* (L.) Rydb.), the hawk's beard (*Crepis nana* L.), the arcto-alpine wormwood (*Artemisia subarctica* Kraschen.), which is similar to the Tien-shan *Artemisia disjuncta* Kraschen., and a number of other plants found widely in the Soviet Northeast, but associated genetically with southern Asia. In southern Asia these plants or their close relatives are adapted to the subalpine and alpine belts of mountains which join southern and northeastern Asia. Naturally, this could have developed only by a considerable depression of the tree line in these mountains and this, in turn, required considerable cooling. These same conclusions can be drawn from an analysis of the species composition of the present fauna of the Soviet Northeast, because these are the very lines along which many animals genetically associated with the South Asian highlands reach the tundra zone. These include representatives of the marmot (genus *Marmota*), the mountain sheep (genus *Ovis*), the pika (genus *Ochotona*), the suslik (genus *Citellus*), and a number of invertebrates.

Probably the broad development of geographic ties which led to the penetration of southern Asian alpine plants into northern Asia and their development

---

*[*Caragana arborescens* Lam.—Translator.]

into arcto-alpine species was brought about by the cold climate of the glacial epochs, during which these geographic communication lines existed, apparently during both late epochs of ice expansion.

The position of the firn line during the last glaciation may be traced quite accurately from the remains of cirques in an excellent state of preservation on the slopes of all the uplands where glaciers formed.

Researchers who have indicated the height of the "snow line" in various regions of the Northeast have not defined exactly what they mean by snow line and have drawn it along a line connecting the lower ends of the cirques. In one of my works devoted to the present glaciation of the Northeast (Vaskovskiy, 1955), I indicated that many authors had used the term "snow line" to mean the firn line and not the climatic snow line.

This also pertains to the "snow line" of ancient glaciers. Many authors (e.g., M. Ye. Melnik) held that this line lies below the lower boundary of cirque distribution; however, this is not true, because the lower parts of the cirques of the ancient glaciation bear traces of working by moving ice. Further, in many places modern cirque glaciers form a terminal moraine within the cirque they occupy, which may take place only below the firm line (and climatic snow line).

Thus, the firn line of both present-day glaciers and glaciers of the last glaciation lie and lay above the lower ends of the cirques of the corresponding epochs. As our observations at the upper boundary of the glacial (in the narrow sense of the word) working of the floors of these cirques indicate, it lay, on an average, 100 m higher than the lower edges of the cirques.

Taking this into consideration, we may define the position of the firn line of the last glaciation as follows. The highest position it occupied in the Suntar-Khayata range was 1900 m* on the northern slopes, and about 1800 m on the southern slopes, which indicates a stronger influence of the maritime climate than at present.

This relatively small (400–600 m) depression of the snow line entailed a considerable development of mountain glaciers. Suffice it to say that the ice sheet of the last glacial epoch, descending from the heights of the Suntar-Khayata range southward and occupying the present upper reaches of the Yudoma river, was more than 300 km long, while the glacier flowing northward along the Suntar valley became 150 km long. The present glaciers in the Northeast are not even as much as 8 km long.

From the Suntar-Khayata range, the firn line descended in all directions but the west, and it was lowest in the coastal regions adjacent to the seas of the Pacific Ocean, while the Arctic Ocean exerted a much weaker influence on its position. In the central part of the Cherskiy mountain system the firn line dropped to 1700–1800 m, in the northern part of the system to 1300–1400 m, and in the southern part to 1500–1700 m. In the northern part of the Verkhoyansk range, at 68° N. lat., the firn line on the northern slopes lay at 1400 m, while on the southern and western slopes it was at 1500 m. In the mountains of the central part of the Chukchi peninsula (Amguyemo-Kuvet Massif), it dropped to 1100 m, and towards the shores of Krest Gulf, the snow line was only 200–300 m above sea level, and the glaciers descended into the sea.

*The present firn line is at an absolute height of 2300 m here.

Of course, far fewer data are available on the firn line of the last glaciation. In the central regions of the Northeast (the upper Indigirka), traces of the firn line were discovered at elevations of 1300–1400 m, but during the period of ice expansion it must have been 100–200 m lower, because the country has risen approximately 100–200 m since that era, as recorded by a series of terraces.

A. P. Vaskovskiy and M. V. Agranovich observed cirques and remnants of the moraines of this glaciation at about 1000 m above sea level along the northern coast of the Sea of Okhotsk. These splendidly preserved cirques of the last glaciation lay below the surface, which bore older moraines, clearly cutting into its edges.

Thus, the firn line of the next to the last glaciation was at an elevation of about 1100 m in the region of the present coast of the Sea of Okhotsk, i.e., it was 200–400 m lower than in the central part of the country. Consequently, the Sea of Okhotsk existed then, although it did not exert as great an influence on the position of the firn line as it did during the last glaciation.

## The Postglacial Period

Quite numerous finds of fossil flora have been made in the peat bogs of the tundra zone of the Northeast and a considerable number of pollen analyses from the southern part of the country allow us to establish the general features of the vegetation and climate of the postglacial period as well.

The observations of geobotanists (V. A. Sheludyakova, L. N. Tyulina, B. A. Tikhomirov, B. N. Gorodkov, and V. B. Sochava), summarized in a very valuable review by Tikhomirov (1941), have established a stage of advance of the northern tree limit which took it farther north than at present. In the extreme western part of the territory examined, in the Lena delta, this was proved by Tikhomirov, who found willow trunks 10–12 cm in diameter and larch wood in the postglacial peat bogs. These species now grow much farther south and are not found here at all. However, the grassy vegetation and mosses found in that peat bog still grow there. Further, the peat was investigated by the pollen method and the analysis showed the presence of a very small quantity of tree pollen: *Pinus* sp. (*P. silvestris*), *P. pumila*, *Picea* (1–4 grains), *Salix, Betula,* and *Alnus*. Tikhomirov also found wood remains in peat bogs near Tiksi Bay. Here they were represented by the wood and seeds of the *Larix dahurica* and by the wood of the white birch, spruce, and willow. Tikhomirov assumes the existence of sparse forests here during the deposition of the peat bog, obviously forming part of the forested tundra. It may be noted that single grains of airborne spruce and common-pine pollen also appear in the present forest and even forested tundra spectra of the European North in much greater quantities, while some pollen grains of spruce and a considerable quantity of pine pollen are also found in the spectra of the present tundra. Therefore, it would seem doubtful that the spruce and pine grew in the Lena delta and near Tiksi Bay in the postglacial epoch. It seems to me that the identification of spruce wood in the Tiksi peat bogs was erroneous or that its stratigraphic position was determined incorrectly. However, undoubtedly the Dahurian larch, the white birches, and the arborescent willows pushed northward from their present limit during one of the moments in postglacial history.

However, the larch did not push more than 100–200 km north of its present limit. As near as Cape Svyatoy Nos, B. V. Pepelyayev found only fragments of the trunks of large bushes and no tree trunks in the peat bogs covering the upper layers of ice. Thus, even in the period of maximum northward postglacial advance of forest vegetation, Svyatoy Nos was beyond the zone of coniferous forests and forested tundra although, judging by K. A. Vollosovich's observations, birch of the section *Alba* reached this far.

The data of Tikhomirov (1941) and S. G. Pavlov indicate that the peat bogs of the northern coast of the Chukchi peninsula do not contain tree fragments and, consequently, during the postglacial climatic optimum the larch limit, as now, passed much farther south in the east than in the west.

In the central and southern parts of the territory described, the vegetation evidently differed little from the present vegetation. At any rate, the works of M. V. Agranovich, A. P. Vaskovskiy, S. L. Khaykina, and Ye. D. Vlasova in the post-glacial terraces of the Kolyma and the northern coast of the Sea of Okhotsk have revealed the pollen of only those plants which now grow in the area. The presence of a small amount of pollen of the spruce and common pine, discovered in the deposits of these terraces (and even in the deposits of the most recent alluvial islands), can be explained entirely by transport from a remote area. If the spruce and common pine had penertated somewhat farther into the Northeast than at present, this would have occurred only in the western part of the country and, what is more, south of the main ridge of the Lena-Chaun mountain arc. Small groves of spruce (*Picea obovata* Ldb.) found today on the western shores of Shelekhov Gulf undoubtedly increased the area occupied by these trees.

Thus, a correction should be entered in the system of regional division of the types of pollen spectra of the U.S.S.R. introduced in M. I. Neyshtadt's work (1955). The Holocene pollen spectra of the forest region of the present Far Northeast, classified by Neyshtadt as the East Siberian type in which spruce and common pine are characteristic, should be classified as a special Yana-Kolyma type, identified by the complete absence of autochthonic pollen of spruce and all pines, except the dwarf Siberian pine. The western boundary of this type of spectrum should be drawn along the eastern water divide of the Lena basin. We do not yet have enough data on the region of development of the present Bering Sea *uryema* [bottomland] tundra (the Anadyr river, Penzhina basin, and south-ern part of the Koryak upland).

However, the results of pollen analyses conducted by Neyshtadt in L. N. Tyulina's samples from Holocene peat bogs on the Mayn river (Neyshtadt and Tyulina, 1936) indicate that the vegetation there during the postglacial period was very similar to the present vegetation. At any rate, the same trees and bushes grew there then that exist now above the section investigated (larch, birch, alder, dwarf Siberian pine). Further, the pollen diagrams of the Holocene peat bogs of western Kamchatka, shown in Neyshtadt's work (1955), are identical qualita-tively with the present spectra. The "tree" pollen in these spectra are represented by birch, alder, dwarf Siberian pine, and willow, i.e., species which play the largest role in the vegetation of this region today. Consequently, conifers did not penetrate to that region in the Holocene, just as they do not penetrate as far as the western coast of Kamchatka today.

Therefore, we cannot agree with V. B. Sochava and B. A. Tikhomirov that arctic elements penetrated into the flora of the Asiatic regions adjacent to the Bering Sea in most recent times. Tundra and forested tundra definitely occupied a considerable portion of that region during the last glaciation and have not yet yielded the most southeasterly portions of their domain to the forests.

Thus, the changes which occurred in the vegetation of the Soviet Northeast during the postglacial thermal optimum were not very large and were reflected in a relatively slight (100–200 km) northward and eastward expansion of a forest consisting of the same trees as at present and in a 100–200 m upward expansion of the forest limit. Apparently, the temperature increase was also slight, since the July isotherm of $+11°$ C, which marks the northern limit of the *Larix dahurica*, was only slightly to the north and higher than its present position. Despite this, the cirques of the last glaciation and the present glaciers often differ from each other spatially and are a considerable vertical distance apart (about 300–400 m); hence part of the ancient glaciers melted, at least in the central part of the territory, perhaps surviving on the Koryak upland.

Let us add that man's postglacial camps on the northern coast of the Sea of Okhotsk indicate a climate similar to the present. For example, the culture-yielding stratum of the campsite of ancient people discovered in 1951 by M. V. Agranovich and A. P. Vaskovskiy on the 10-m terrace of Srednyaya Bay in Babushkin Gulf included sea and land fauna which exist there now. Here, among the kitchen wastes were enormous numbers of mollusk shells of the types *Mytilus edulis* L., *Thais lima* Martyn., *Littorina squalida* Brod. et Sow., which constitute the bulk of the fauna of the present littoral of Srednyaya Bay. Reindeer (*Rangifer tarandus* L.) and seal (*Phoca hispida* Schr.) bones were also found. Pollen analysis showed that the general appearance of the vegetation at the time the camp existed was the same as now. In A. P. Okladnikov's opinion, the tools and utensils indicate that this camp belonged to the culture which he calls "Ancient Koryak" and which corresponds to the Metal Age in Europe.

*Conclusions*

In concluding, we shall outline the deduction we have drawn from our analysis of the material presented above.

1. A comparative study of the plant remains from strata covering the Upper Pliocene marine deposits of northern Kamchatka and from the continental friable deposits of the southern part of the Soviet Far Northeast has allowed us to establish the lower limit of the geologic age of these deposits.

2. On the basis of stratigraphic relations between the continental and the marine deposits, it may be stated that the strata which contain fruit of the American butternut and Metasequoia cones are Upper Pliocene in northern Kamchatka and indicate a comparatively mild and warm climate and the existence of coniferous-broadleaf forests at that time. A comparison of the composition of the flora of these strata with the strata containing the butternut, which comprise the lower part of the famous Mamontova Gora outcropping on the Aldan river, indicates that the Aldan deposits are also Upper Pliocene. At the same time, study of the plant remains in strata containing mammoth remains, which crown the Mamontova Gora section, shows that they formed later than the strata containing the butternut and under much more severe climatic conditions, similar to the present climate of the Siberian taiga.

3. The flora of the Kamchatka strata which were deposited somewhat later than the nut-bearing strata record a considerable deterioration of the climate and a change in the composition of the vegetation. It does not contain remains of broadleaf trees and the abundant conifer remains in it belong exclusively to the pine family. However, no representatives of pines characteristic of the modern flora of the Northeast and even of Siberia as a whole are found in this flora. On the contrary, it contains species identical with or very similar to the conifers which in part make up the forests that cover the central mountain slopes of the North American Cordillera and the mountains of central Japan. The author calls

this floristic complex the Nippono-Cordilleran and classifies it as Early Quaternary. Analysis of the spatial relationships between the snow line and the present mountain forests, in which representatives of the Nippono-Cordilleran fossil flora have been preserved, indicates that the chionosphere first came into contact with the mountain summits of the Northeast at the time this flora developed.

4. The great gap in paleontologic documentation prevents reconstruction of the stages of the Quaternary history of the Northeast immediately following the deposition of the strata containing the Nippono-Cordilleran flora. The stages of flora development closest in time are represented by finds of conifer remains in the ancient valley of the right bank of the Elgi river, which indicate a further cooling of the climate. The position of these finds, between two erosion levels at which traces of glacial activity are found, indicates that this flora belongs to the first* interglacial epoch. One of its characteristic features is the mixture of alpine North Pacific types (*Tsuga*) and Siberian types of conifers. The *Larix dahurica*, which later reigned supreme in the coniferous forests of the Northeast, first appeared in this flora.

Later stages of development of this flora are represented by finds on the 110-m Susuman river terrace which indicate the existence of taiga vegetation containing, as before, mixed Siberian and North American conifers. However, the Susuman finds do not represent the Cordilleran mountain types, but conifer species which form the so-called Hudson coniferous forests, i.e., the North American taiga. The name "Hudson-Siberian flora," which we have assigned to this group, reflects its geographic ties with the present vegetation and its ecologic features.

5. Thus far, no paleontologic documents have been found that are synchronous with the glaciation preceding the deposition of the strata that contain the Hudson-Siberian flora or with the glaciation which succeeded the deposits containing the Hudson-Siberian flora and which the author has somewhat arbitrarily identified with the Riss glaciation of the Alps.†

Making up for this lack are the finds of a flora which is associated with the last interglacial period; these finds have been made in many regions and indicate the existence of taiga in the southern part of the Northeast; this taiga contained southern Siberian species of conifers whose coexistence is observed today in a limited region including the settlements of Kirensk and Uchur, on the basis of which this flora has been called Uchur-Kirensk. It bears evidence of a quite severe continental climate similar to the present climate of southern Yakutia.

6. The numerous finds of mammoth remains and the remains of other animals comprising the "mammoth" or "Upper Paleolithic" faunistic complex in European U.S.S.R. are associated with finds of Uchur-Kirensk interglacial flora and with the deposits of the last glacial epoch. Further, a primitive form of mammoth, whose morphologic characteristics are similar to those of the *Elephas trogontherii*, is encountered in finds belonging to the initial stages of the interglacial.

These facts allow us to contend that the last interglacial epoch in the Northeast corresponds to the Riss-Würm interglacial epoch in the Alps or the Dnieper-Valday epoch of the Russian Plain, and the last major expansion of ice in the Northeast corresponds to the Würm (Valday) glacial epoch.

7. The remains of flora associated with deposits of the last glacial epoch indicate that the climate had become severe and continental by that time in the Northeast, resulting in the dominance of vegetation poor in species composition and differing only in details from the present vegetation of the Northeast, if the latter is examined in three dimensions.

*Or one of the first.
†With the exception of one pollen spectrum described above.

The vertical limit of the forest descended, on an average, 300–500 m, and the northern forest limit retreated southward, but the *Larix dahurica* reigned supreme among the forest vegetation groupings, as it does today. In the moraine belt of the last glaciation, found 700–800 m above sea level, apparently the mean July temperature was around +10° or +11° C.

Some characteristics of the species composition of the pollen spectra allow us to assume that a combination of larch forests, tundra, and xerophytic plant associations was characteristic of the vegetation of the last glacial epoch in the upper Kolyma, Yana, and Indigirka basins, which explains the existence of animals adapted to forest and open habitats among the fauna of that time. This combination of forest, tundra, and steppe elements in the flora and fauna is observed in the central parts of the Northeast even today, although the species composition of the fauna has undergone substantial changes since the end of the glacial epoch.

8. Observations by many researchers in the Northeast have established the northward extension of the tree limit during the postglacial thermal epoch. However, the forest expansion did not extend more than 100–200 km north of its present boundary and the southern Siberian conifers did not reach farther north than the Lena-Chaun mountain arc during the postglacial period.

9. From all that has been said, it follows that the Far Northeast of the U.S.S.R. experienced sharp climatic changes during the Quaternary, equivalent to and simultaneous with the climatic changes in Europe, and the climate of the Northeast, at least during the second half of the Quaternary period, was severe and continental. The greater continentality of the climate explains the relatively slight development of the ancient (and modern) glaciation in the Northeast. Therefore, we should reject once and for all the hypothesis that the Northeast had a mild and moist climate during the Pleistocene and that it gradually changed during that epoch.

Reproductions of the remains of coniferous and other plants mentioned in the text and which form the basis for many of the author's conclusions are shown in Plates I–V, appended to this article. Some of these are new species and are portrayed here for the first time.

---

### PLATE I

ITEM 1. Cone of *Picea obovata*. Terminal interglacial. The stream Levyy Promezhutochnyy, a tributary of the Elgi. Scale 1:1.

ITEM 2. (a) three fruit scales; (b) four seeds from this cone. Scale 1:1.

ITEM 3. Cones of *Larix sibirica* Ldb. From the same source as the cones of item 1. Scale 1:1.

ITEM 4. Fruit scale of this cone. Scale 1:1.

ITEM 5. Cone fragment of *Picea obovata*. From same source as item 1. Scale 1:1.

ITEM 6. Cone fragment of *Picea obovata*. From same source as item 1. Scale 1:1.

ITEM 7. (a) three fruit scales; (b) three fruit scales with alulae; (c) three seeds without alulae from the same cone. Scale 1:1.

ITEM 8. Fragment of *Larix Sukaczewii* Djil. cone. From same source as item 1. Scale 1:1.

ITEM 9. Cone of *Larix dahurica* Turcz. From same source as item 1. Scale 1:1.

ITEM 10. (a) fruit and bract scales; (b) three seeds with alulae; (c) one seed without alula from same cone. Scale 1:1.

ITEM 11. Cone of modern *Larix dahurica* Turcz. From the Trans-Baykal. Scale 1:1.

ITEM 12. Two seeds with alulae from this cone.

ITEM 13. *Larix* sp. ind. The stream Levyy Promezhutochnyy. Scale 1:1.

ITEM 14. Cone of modern *Picea obovata* Ldb. From the region of the mouth of the Kandyga river. Scale 1:1.

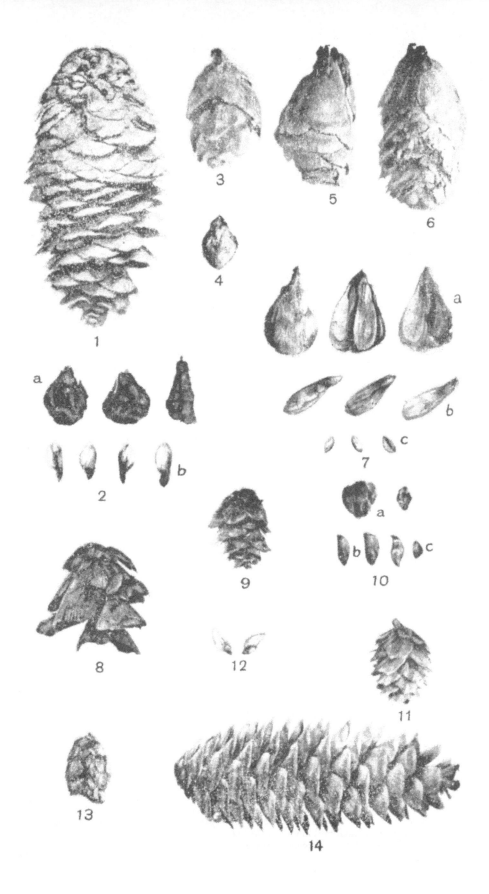

PLATE II

ITEM 15. Cone of modern *Picea obovata* Ldb. Yana river basin, valley of the stream Propashchey. Scale 1:1.

ITEM 16. Four fruit scales (upper right taken at base (cone)) and needle from the same tree as the cone. Scale 1:1.

ITEM 17. Fragment of *Pinus monticola* Dougl. cone. Lower Quaternary deposits, Khudzhakh river. Scale 1:1.

ITEM 18. Three fragments of fruit scales of the same cone. Scale 1:1.

ITEM 19. Cone of *Picea obovata* Ldb.

ITEM 20. Two fruit scales of the same cone. Scale 1:1.

ITEM 21. Cone of *Picea anadyrensis* Krysht. Scale 1:1.

ITEM 22. Fruit scale and seed with alula of the same cone. Scale 1:1.

ITEM 23. Cone of *Picea obovata* Ldb. First Interglacial, Berelyakh river, 110-m terrace, between the streams Kuranakh and Sosed. Scale 1:1.

ITEM 24. Two fruit scales of this cone. Scale 1:1.

15

17

16

18

19

20

21

22

23

24

PLATE III

ITEM 25. Cone of *Picea anadyrensis* Krysht. Second Interglacial. Berelyakh river between the streams Sennyy and Tengkelyakh, 60–80-m terrace. Side view. Scale 1:1.

ITEM 26. The same, view from base of cone. Scale 1:1.

ITEM 27. Cone of *Picea anadyrensis* Krysht. Same source as item 25. Scale 1:1.

ITEM 28. (*a*) three fruit scales; (*b*) two seeds with alulae; (*c*) three seeds without alulae from the same cone as above. Scale 1:1.

ITEM 29. Cone of *Picea obovata* Ldb. Same source as item 25. Scale 1:1.

ITEM 30. Four fruit scales from this cone.

ITEM 31. Two cones of *Larix Cajanderi* Mayr. Late Quaternary deposits of the Debin river, 30-m terrace at the village of Yagodnoye. Scale 1:1.

ITEM 32. Four seed casings ("nuts") of *Pinus pumila* Rgl. From same source. Scale 1:1.

ITEM 33. Five cones of *Picea hondoensis* Maur. var. *tripartita* Vassk. Lower Quaternary deposits, shore of Rekinniki Bay. Scale 1:1.

ITEM 34. Seed with casing, alula, and four bract scales of a *Picea hondoensis* cone; extreme left of item 33 on scale 5.5:1.

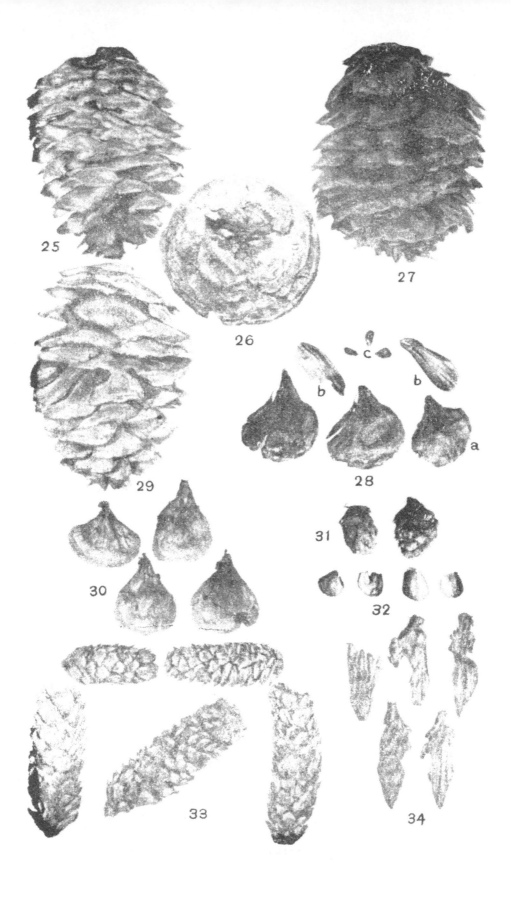

PLATE IV

ITEM 35. Three nuts of *Juglans cinerea* L. Upper Pliocene deposits of the northwest coast of Kamchatka. Scale 1:1.

ITEM 36. Cone of *Pinus itelmenorum* Vasskovsky. Same source. Scale 1:1.

ITEM 37. Three fruit scales of this same cone. Scale 1:1.

ITEM 38. Cone of *Pinus monticola* Dougl. Lower Quaternary deposits of the Upper Nera depression. Scale 1:1.

ITEM 39. Cone of *Larix occidentalis* Nutt. Lower Quaternary deposits near Magadan. Scale 1:1.

ITEM 40. Cone of *Pseudotsuga magadanica* Vasskovsky. Same source. Scale 1:1.

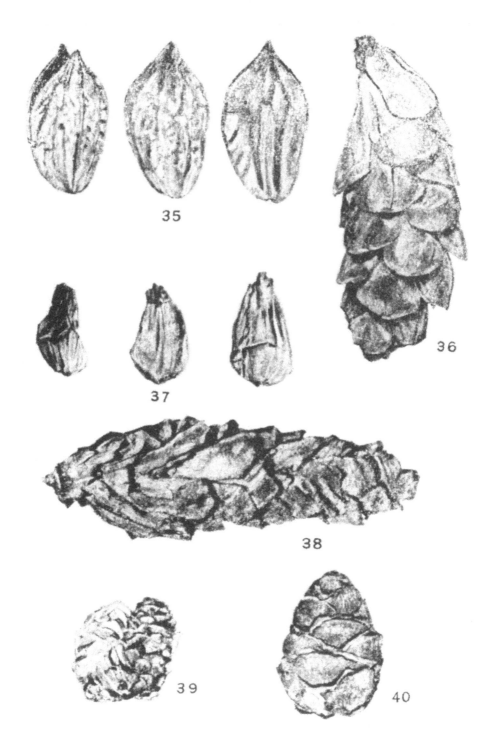

35

36

37

38

39          40

PLATE V

ITEM 41. Cone of *Picea Bilibinii* Vasskovsky. Lower Quaternary deposits near Magadan. Scale 1:1.

ITEM 42. Fragment of a *Picea Bilibinii* Vasskovsky cone. Lower Quaternary deposits of the western coast of Kamchatka. Scale 1:1.

ITEM 43. Fragment of a *Picea Bilibinii* Vasskovsky cone. Same source. Scale 1:1.

ITEM 44. Three fruit scales of *Picea Bilibinii* belonging to the cone depicted in item 42. Scale 1:1.

ITEM 45. Three fruit scales of the cone depicted in item 43. Scale 1:1.

ITEM 46. Three bract scales of *Picea Bilibinii* belonging to the cone depicted in item 42. Scale 2.5:1.

ITEM 47. Two cones of the *Tsuga minuta* Vasskovsky. Lower Quaternary deposits of the Dzelkan depression. Scale 1:1.

ITEM 48. Three cones of *Larix dahurica* Turcz. Late Quaternary deposits, 8-m aggradation terrace of the Bolshoy Taryn river. Slightly enlarged.

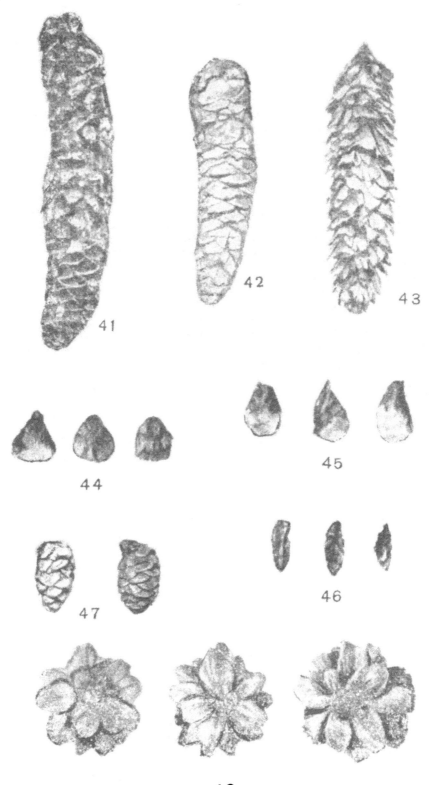

41

42

43

44

45

47

46

48

# BIBLIOGRAPHY

AREMBOVSKIY, I. V. 1951. K probleme stratigrafii antropogena Irkutskogo amfiteatra (Quaternary stratigraphy of the Irkutsk amphitheatre). *Trudy Irkutskogo gosudarstvennogo universiteta*, seriya geologicheskaya, vol. 5, no. 2.

BERG, L. S. 1913. K voprosu o peremeshchenii klimaticheskikh zon v poslelednikovoye vremya (The displacement of climatic zones in the postglacial period). *Pochvovedeniye*, no. 4.

————— 1938. *Osnovy klimatologii* (Principles of climatology). Moscow–Leningrad.

BOCH, S. G. 1937. Materialy k chetvertichnoy geologii Lyapinskogo, Nizhnesosvinskogo i Kondinskogo kraya Zapadno-Sibirskoy nizmennosti (Data on the Quaternary geology of the Lyapin, Lower Sosva, and Konda section of the West Siberian Lowland). *Trudy Komissii po izucheniyu chetvertichnogo perioda*, vol. 5.

BOGOLEPOV, K. V. 1955. K voprosu ob etapakh razvitiya tretichnoy rastitelnosti v Priangarskoy chasti Yeniseyskogo kryazha (The stages of development of Tertiary vegetation in the Angara portion of the Yenisey ridge). *Doklady Akademii nauk SSSR*, vol. 100, no. 5.

BOLI, A. 1948. *Severnaya Amerika* (North America). Moscow.

BOZHINSKIY, A. P. 1939. K istorii chetvertichnogo perioda doliny r. Angary (The Quaternary history of the Angara river valley). *Byulleten Moskovskogo obshchestva ispytately prirody*, otdel geologicheskiy, vol. 17, no. 6.

CHERSKIY, I. D. 1878–81. Predvaritelnyy otchet o geologicheskikh issledovaniyakh beregovoy polosy ozera Baykal (Preliminary survey of geologic investigations of the strandline of Lake Baykal). *Izvestiya Vostochno-Sibirskogo otdeleniya Russkogo geograficheskogo obshchestva*, vol. 9, no. 1–2. 1878; vol. 9, no. 5–6, 1878; vol. 11, no. 1–2, 1880; vol. 12, nos. 2 and 3, 1881.

————— 1891. Opisaniye kollektsii poslechetvertichnykh mlekopitayushchikh zhivotnykh, sobrannykh Novo-Sibirskoyu ekspeditsieyu 1885–1886 gg. (Description of a collection of post-Quaternary mammals collected by the New Siberian Expedition of 1885–1886). *Zapiski Akademii nauk*, vol. 65, Appendix 1.

DEREVYA I KUSTARNIKI SSSR (Trees and bushes of the U.S.S.R.), vol. 1. 1949. Moscow–Leningrad.

DICE, L. K. 1946. *The biotic provinces of North America*, Ann Arbor, Michigan.

DOROFEYEY, P. I. 1955. Neskolko zamechaniy k riss-vyurmskoy flore okrestnostey g. Galicha (Some notes on the Riss-Würm flora in the vicinity of the city of Galich). *Botanicheskiy zhurnal*, vol. 40, no. 3.

DUBROVO, I. A. 1954. O pervoy nakhodke primitivnogo slona *Elephas meridionalis* Nesti na severe Sibiri (The first find of the primitive elephant *Elephas meridionalis* Nesti in northern Siberia). *Byulleten Komissii po izucheniyu chetvertichnogo perioda*, no. 19.

DUMITRASHKO, N. V. 1948a. Osnovnyye voprosy geomorfologii i paleogeografii Baykalskoy gornoy oblasti (Fundamental problems of the geomorphology and paleogeography of the Baykal mountain region). *Trudy Instituta geografii Akademii nauk SSSR*, vol. 42.

————— 1948b. Molodost i drevnost relyefa yugo-vostochnoy Sibiri (Youth and antiquity of the relief of southeastern Siberia). *Trudy Instituta geografii Akademii nauk SSSR*, no. 39.

————— 1952. Geomorfologiya i paleogeografiya baykalskoy gornoy oblasti (Geomorphology and paleogeography of the Baykal mountain region). *Trudy Instituta geografii Akademii nauk SSSR*, vol. 55, no. 9.

DYAKOV, B. F. 1935. O melovykh otlozheniyakh poluostrova Kamchatki (The Cretaceous deposits of Kamchatka). *Problemy sovetskoy geologii*, no. 12.

—— 1936. Geologicheskiye issledovaniya na zapadnom beregu poluostrova Kamchatki (Geologic investigations on the western coast of Kamchatka). *Trudy Neftyanogo geologo-razvedochnogo instituta*, no. 83.

ELLIOT, N. A. 1938. *Forest trees of the Pacific coast.* New York.

FEDOROVA, R. V. 1952a. Rasprostraneniye pyltsy i spor tekuchimi vodami (Propagation of pollens and spores by flowing water). *Trudy Instituta geografii Akademii nauk SSSR*, vol. 52, no. 7.

—— 1952b. Kolichestvennyye zakonomernosti rasprostraneniya pyltsy drevesnykh porod vozdushnym putem (Quantitative laws of tree pollen propagation by air). *Trudy Instituta geografii Akademii nauk SSSR*, vol. 52, no. 7.

FLORA SSSR (Flora of the U.S.S.R.), vols. 1–22. 1934–55. Moscow–Leningrad.

GERASIMOV, I. P. 1952. Sovremennyye perezhitki pozdnelednikovykh yavleniy vblizi samoy kholodnoy oblasti mira (Modern relicts of late glacial phenomena near the coldest region of the earth). *Izvestiya Akademii nauk SSSR, seriya geograficheskaya*, no. 5.

GERASIMOV, I. P. and MARKOV, K. K. 1939. Lednikovyy period na territorii SSSR (The glacial epoch in the U.S.S.R.). *Trudy Instituta geografii Akademii nauk SSSR*, no. 33.

GRICHUK, M. P. 1955. K istorii rastitelnosti v basseyne Angary (On the history of the vegetation of the Angara basin). *Doklady Akademii nauk SSSR*, vol. 102, no. 2.

GRICHUK, V. P. 1950. Rastitelnost Russkoy ravniny v nizhne- i srednechetvertichnoye vremya (Vegetation of the Russian Plain in the Lower and Middle Quaternary periods). *Trudy Instituta geografii Akademii nauk SSSR*, vol. 46, no. 3.

—— 1952. Osnovnyye rezultaty mikropaleobotanicheskogo izucheniya chetvertichnykh otlozheniy Russkoy ravniny (Principal results of a micropaleobotanic study of the Quaternary deposits of the Russian Plain). In: *Materialy po chetvertichnomu periodu SSSR* (Data on the Quaternary period in the U.S.S.R.), no. 3. Moscow–Leningrad.

GROMOV, V. I. 1934. Materialy k izucheniyu chetvertichnykh otlozheniy v basseyne srednogo techeniya r. Obi (Materials for studying the Quaternary deposits of the middle Ob basin). *Trudy Komissii po izucheniyu chetvertichnogo perioda*, vol. 3, no. 2.

—— 1948. Paleontologicheskoye i arkheologicheskoye obosnovaniye stratigrafii kontinentalnykh otlozheniy chetvertichnogo perioda na territorii SSSR: mlekopitayushchiye, paleolit (The paleontologic and archaeological basis for the stratigraphy of the Quaternary continental deposits in the U.S.S.R.; Mammals, Paleolithic). *Trudy Instituta geologicheskikh nauk Akademii nauk SSSR*, no. 64, geologicheskaya seriya, no. 17.

IVANOV, G. A. 1928. Geologicheskiy ocherk iskopayemykh ugley srednego techeniya r. Leny; Kangalasskoye, Sangarskoye i Lunkhinskoye mestorozhdeniya YaASSR (Geologic sketch of the fossil coal of the middle course of the Lena; the Kangalassy, Sangar, and Lungkha deposits of the Yakut A.S.S.R.). *Materialy po obshchey i prikladnoy geologii*, no. 87.

KARAVAYEV, M. N. 1948. Rastitelnyy mir Yakutii dolednikovogo perioda (The plant world of Yakutia during the preglacial period). In: *Doklady na Pervoy nauchnoy sessii Yakutskoy bazy AN ASSR* (Reports of the First Scientific Session of the Yakut [Research] Base of the Academy of Sciences of the U.S.S.R.). Yakutsk.

KHAYKINA, S. L. 1948. Rezultaty sporovo-pyltsevogo analiza prob iz merzlykh sloyev chetvertichnykh otlozheniy na ruchye Struyka v basseyne Indigirki (Results of spore-pollen analysis of samples from the frozen layers of Quaternary deposits on the stream Struyka in the Indigirka basin). *Byulleten Komissii po izucheniyu chetvertichnogo perioda*, no. 13.

KOLOSOV, D. M. 1952. Razvitiye Tikhookeanskoy drevnelednikovoy provintsii SSSR (Development of the Pacific Ocean ancient glacial province of the U.S.S.R.). In:

*Materialy po chetvertichnomu periodu SSSR* (Data on the Quaternary period in the U.S.S.R.), no. 3. Moscow–Leningrad.

KOMAROV, V. L. 1926. Vvedeniye v izucheniye rastitelnosti Yakutii (Introduction to the study of the vegetation of Yakutia). *Trudy Komissii po izucheniyu Yakutskoy ASSR,* vol. 1.

KORENEVA, YE. V. 1955. Izucheniye sovremennykh morskikh otlozheniy metodom sporovo-pyltsevogo analiza (Study of contemporary marine deposits by the spore-pollen analysis method). *Trudy Instituta okeanologii Akademii nauk SSSR,* vol. 13.

KRASHENINNIKOV, I. M. 1946. Opyt filogeneticheskogo analiza nekotorykh yevraziatskikh grupp roda *Artemisia* L., v svyazi s osobennostyami paleogeografii Yevrazii (Attempt at a phylogenetic analysis of certain Eurasian groups of the genus *Artemisia* L., in connection with the paleogeographic characteristics of Eurasia). In *Materialy po istorii flory i rastitelnosti SSSR* (Data on the history of the flora and vegetation of the U.S.S.R.), no. 2. Moscow–Leningrad.

KRISHTOFOVICH, A. N. 1915. Amerikanskiy seryy orekh *Juglans cinerea* L. iz presnovod-nykh otlozheniy Yakutskoy oblasti (The American butternut *Juglans cinerea* L. from the freshwater deposits of Yakutia). *Trudy Geologicheskoy Komissii,* n.s., no. 124.

—— 1929. Vodyanoy orekh, *Trapa borealis* Heer, iz tretichnykh otlozheniy Tun-kinskoy doliny v Sayanakh (A water chestnut, *Trapa borealis* Heer, from the Tertiary deposits of the Tunka valley in the Sayan mountains). *Vestnik Geologiche-skoy komissii,* no. 9–10.

—— 1945. *Paleobotanika* (Paleobontany). Moscow.

KULIK, N. A. 1926. O severnom postpliotsene (The northern post-Pliocene). *Geologi-cheskiy vestnik,* vol. 5, no. 1–3.

LAMAKIN, V. V. 1935. Proshloye relyefoobrazovaniye v Tunkinskom Pribaykalye (Past formation of relief in the Tunka region of the Cis-Baykal). *Zemlevedeniye,* vol. 37, no. 1.

—— 1952. *Ushkani ostrova i problema proiskhozhdeniya Baykala* (The Ushkani Islands and the problem of the origin of Lake Baykal). Moscow.

LAVRENKO, YE. M. (Ed.). 1939. *Karta rastitelnosti SSSR; 1:4 000 000* (Vegetation map of the U.S.S.R.; scale 1:4,000,000). Moscow.

LAVRENKO, YE. M. and SOCHAVA, V. B. (Eds.). 1954. *Geobotanicheskaya karta SSSR* (Geobotanic map of the U.S.S.R.). Moscow.

LOPATIN, I. A. 1897. Dnevnik Turukhanskoy ekspeditsii 1866 g. (Log-book of the 1866 Turukhan expedition). *Zapiski Russkogo geograficheskogo obshchestva,* vol. 28, no. 2.

MALGINA, YE. A. 1950. Opyt sopostavleniya rasprostraneniya pyltsy nekotorykh vidov drevesnykh porod s ikh arealami v predelakh Yevropeyskoy chasti SSSR (Compari-son of the distribution of the pollen of certain tree species with their areal distribu-tion in the European part of the U.S.S.R.). *Trudy Instituta geografii Akademii nauk SSSR,* vol. 46, no. 3.

MARKOV, K. K. 1939. O soderzhanii ponyatiy "lednikovaya epokha" i "mezhlednikovaya epokha" (The substance of the concepts "glacial epoch" and "interglacial epoch"). *Izvestiya Vsesoyuznogo geograficheskogo obshchestva,* vol. 71, no. 7.

—— 1955. *Ocherki po geografii chetvertichnogo perioda* (Outlines of Quaternary geography). Moscow.

MASLOV, V. P. 1939. Sledy drevnego oledeneniya v severo-zapadnom Pribaykalye (Traces of ancient glaciation in the northwestern Baykal region). In: *Akademiku V. A. Obruchevu k 50-letiyu nauchnoy i pedagogicheskoy deyatelnosti* (On the fiftieth anniversary of the scientific and pedagogical acitvities of Academician V. A. Obruchev), vol. 2. Moscow.

MIKHAYLOV, N. I. 1947. Geomorfologicheskiye nablyudeniya v zapadnoy chasti gor Putorana (Geomorphologic observations in the western part of the Putorana moun-tains). *Voprosy geografii,* sbornik 3.

MIKI, SHIGERU. 1938. On the change of flora of Japan since the Upper Pliocene and the floral composition at the present. *Japanese Journal of Botany,* vol. 9, no. 2.

NAPALKOV, P. 1932. Ostrov Belyy (Belyy Island). *Zapiski po gidrografii*, no. 2.

NEYSHTADT, M. I. 1944. Yeshche k voprosu o pervichnosti listvennichno-sosnovoy taygi v Vostochnoy Sibiri (Additional information on the precedence of larch-pine taiga in eastern Siberia). *Sovetskaya botanika*, no. 4–5.

——— 1955. Stratigrafiya golotsenovykh otlozheniy na territorii SSSR (Stratigraphy of the Holocene deposits of the U.S.S.R.). *Trudy Instituta geografii Akademii nauk SSSR*, vol. 63.

NEYSHTADT, M. I. and TYULINA, L. N. 1936. K istorii chetvertichnoy i poslechetvertichnoy flory r. Mayn, pritoka Anadyrya (The history of the Quaternary and post-Quaternary flora of the Mayn river, a tributary of the Anadyr). *Trudy Arkticheskogo instituta*, vol. 40.

NIKITIN, P. A. 1938. Chetvertichnyye semennyye flory s nizovyev Irtysha (Quaternary seed-bearing flora from the lower Irtysh river). *Trudy Biologicheskogo nauchno-issledovatelskogo instituta Tomskogo gosudarstvennogo universiteta*, vol. 5 (Botany).

OBRUCHEV, S. V. 1932. Kolymsko-Indigirskiy rayon: Geomorfologicheskiy ocherk (The Kolyma-Indigirka region: A geomorphologic sketch). *Trudy Soveta po izucheniyu proizvoditelnykh sil Akademii nauk SSSR, seriya yakutskaya*, no. 1.

——— 1946. Orografiya i geomorfologiya vostochnoy poloviny Vostochnogo Sayana (Orography and geomorphology of the eastern half of the Eastern Sayan mountains). *Izvestiya Vsesoyuznogo geograficheskogo obshchestva*, vol. 78, no. 5–6.

OBRUCHEV, V. A. 1931. Priznaki lednikovogo perioda v severnoy i tsentralnoy Azii (Indications of the glacial period in northern and central Asia). *Byulleten Komissii po izucheniyu chetvertichnogo perioda*, no. 3.

——— 1938. *Geologiya Sibiri* (Geology of Siberia), vol. 3. Moscow–Leningrad.

——— 1951. *Izbrannyye raboty po geografii Azii* (Selected works on the geography of Asia), vol. 3. Moscow.

OKLADNIKOV, A. P. 1948. Paleontologicheskiye nakhodki na r. Lene (Paleontologic finds on the Lena river). *Byulleten Komissii po izucheniyu chetvertichnogo perioda*, no. 12.

——— 1950. Osvoyeniye paleontologicheskim chelovekom Sibiri (Conquest of Siberia by paleontologic man). In: *Materialy po chetvertichnomu periodu SSSR* (Data on the Quaternary period in the U.S.S.R.), no. 2. Moscow.

PALIBIN, I. V. 1936. Tretichnaya flora yugo-vostochnogo poberezhya Baykala i Tunkinskoy kotloviny (Tertiary flora of the southeastern shore of Lake Baykal and the Tunka depression). *Trudy Neftyanogo geologo-razvedochnogo instituta*, ser. A, no. 76.

PAVLOVSKIY, YE. V. 1948a. Geologicheskaya istoriya i geologicheskaya struktura Baykalskoy gornoy oblasti (Geologic history and geologic structure of the Baykal mountain region). *Trudy Instituta geologicheskikh nauk Akademii nauk SSSR*, no. 99, geologicheskaya seriya, no. 31.

——— 1948b. O chetvertichnom oledenenii Yuzhnogo Pribaykalya (The Quaternary glaciation of the southern Cis-Baykal region). *Izvestiya Akademii nauk SSSR, seriya geologicheskaya*, no. 5.

PERNOT, J. F. 1916. *Forests of Crater Lake National Park*. Washington, D.C.

PIDOPLICHKO, I. G. 1950. Po povodu raboty A. P. Vaskovskogo i A. P. Okladnikova o nakhodke dereva, obrabotannogo paleoliticheskim chelovekom (In reference to A. P. Vaskovskiy and A. P. Okladnikov's work on the discovery of wood worked by Paleolithic man). *Arkheologiya*, no. 3. Kiev.

PILIPENKO, P. P. 1930. Oledeneniye Yuzhnogo Pribaykalya (Glaciation of the southern Cis-Baykal region). *Mineralnoye syrye*, no. 7–8.

PIROZHNIKOV, P. L. 1931. K geograficheskomu poznaniyu oblasti, nakhodyashcheysya mezhdu Tazom i Yeniseyem (Contribution to the geographic knowledge of the region between the Taz and Yenisey rivers). *Zemlevedeniye*, vol. 33, no. 1–2.

POPOV, A. I. 1947. Vechnaya merzlota v Zapadnoy Sibiri i yeye izmeneniya v chetvertichnyy period (Permafrost in western Siberia and Quaternary changes in it). *Merzlotovedeniye*, vol. 2, no. 2.

——— 1949. Nekotoryye voprosy paleogeogrfii chetvertichnogo perioda v Zapadnoy

Sibiri (Some questions on the paleogeography of the Quaternary period in western Siberia). *Voprosy geografii*, sbornik 12.

———— 1952. Mŏrozoboynyye treshchiny i problema iskopayemykh ldov (Frost cracks and the problem of fossil ice). *Trudy Instituta merzlotovedeniya imeni V. A. Obrucheva Akademii nauk SSSR*, vol. 9.

———— 1953*a*. Osobennosti litogeneza allyuvialnykh ravnin v usloviyakh surovogo klimata (Lithogenic features of alluvial plains in a severe climate). *Izvestiya Akademii nauk SSSR*, seriya geograficheskaya, no. 2.

———— 1953*b*. *Vechnaya merzlota v Zapadnoy Sibiri* (Permafrost in western Siberia). Moscow.

———— 1955. Proiskhozhdeniye i razvitiye moshchnogo iskopayemogo lda (The origin and development of deep fossil ice). In: *Materialy k osnovam ucheniya o merzlykh zonakh zemnoy kory* (Contributions to the basic study of the frozen zones of the earth's crust), pt. 2. Moscow.

[Popov, M. G. 1953. O sisteme i filogeneticheskom razvitii roda *Mertensia* Roth (Boraginaceae) (On the system and phylogenetic development of the genus *Mertensia* Roth (Boraginaceae)). *Botanicheskiye materialy gerbariya*, vol. 25. Botanicheskiy institut Akademii nauk SSSR.]

Popov, Yu. N. 1947. O sovremernom oledenenii Severo-Vostoka Azii v svyazi s problemoy drevnego oledeneniya (The present glaciation of northeastern Asia in connection with the problem of ancient glaciation). *Izvestiya Vsesoyuznogo geograficheskogo obshchestva*, vol. 79, no. 3.

———— 1948. Nakhodki iskopayemykh trupov mlekopitayushchikh v merzlykh sloyakh pleystotsena severo-vostochnoy Sibiri (Finds of fossil carcasses of mammals in the frozen strata of the Pleistocene period in northeastern Siberia). *Byulleten Komissii po izucheniyu chetvertichnogo perioda*, no. 13.

Popova, A. I. 1955. Sporovo-pyltsevyye spektry chetvertichnykh otlozheniy tsentralnoy Yakutii v svyazi s istoriyey razvitiya rastitelnosti yeye v posletretichnoye vremya (Spore-pollen spectra of Quaternary deposits of central Yakutia in connection with the history of development of vegetation in post-Tertiary time). *Trudy Instituta biologii Yakutskogo filiala Akademii nauk SSSR*, no. 1.

Povarnitsyn, V. A. 1937. Pochvy i rastitelnost basseyna r. Verkhney Angary (Soils and vegetation of the Upper Angara basin). In: Buryat-Mongoliya, Trudy Buryat-Mongolskoy kompleksnoy ekspeditsii 1932 g. *Trudy Soveta po izucheniyu proizvoditelnykh sil, seriya Vostochno-sibirskaya*, no. 4.

Reverdatto, V. V. 1940. Osnovnyye momenty razvitiya posletretichnoy flory sredney Sibiri (Main factors in the development of the post-Tertiary flora of central Siberia). *Sovetskaya botanika*, no. 2.

Saks, V. N. 1936. O chetvertichnom oledenenii severa Sibiri (Quaternary glaciation of northern Siberia). *Arctica*, bk. 4.

———— 1947*a*. Klimaty proshlogo na severe SSSR (Climates of the past in the northern U.S.S.R.). *Priroda*, no. 12.

———— 1947*b*. Chetvertichnoye oledeneniye severa Sibiri (Quaternary glaciation of northern Siberia). *Priroda*, no. 4.

———— 1948. Chetvertichnyy period v Sovetskoy Arktike (The Quaternary period in the Soviet Arctic). *Trudy Arkticheskogo instituta*, no. 201.

———— 1955. Novyye dannyye po istorii geologicheskogo razvitiya Sibiri v chetvertichnyy period (New data on the history of the geologic development of Siberia during the Quaternary period). In: *Voprosy geologii Azii* (Problems of Asian geology), vol. 2. Moscow.

Schenk, C. A. 1939. *Fremdländische Wald- und Parkbäume*. vol. 1–3. Berlin.

Shanster, Ye. V. 1951. Allyuviy ravninnykh rek umerennogo poyasa i yego znacheniye dlya poznaniya zakonomernostey stroyeniya i formirovaniya allyuvialnykh svit (The alluvium of the lowland rivers of the temperate zone and its significance for determining the laws of the structure and formation of alluvial suites). *Trudy Instituta geologicheskikh nauk Akademii nauk SSSR*, no. 135, geologicheskaya seriya, no. 55.

SHCHERBAKOVA, YE. M. 1954a. O vozraste i razvitii Vostochnogo Sayana (The age and evolution of the Eastern Sayan mountains). In: *Materialy po paleogeogrfii*, no. 1. Moscow.

——— 1954b. Novaya nakhodka vintorogoy antilopy na territorii SSSR (New find of a screw-horned antelope, *Spiroceros kiaktensis*, in the U.S.S.R.). In: *Materialy po paleogeografii*, no. 1. Moscow.

SHELUDYAKOVA, V. A. 1938. Rastitelnost basseyna reki Indigirki (Vegetation of the Indigirka river basin). *Sovetskaya botanika*, no. 4–5.

SHEYNMAN, YU. M. 1948. O stepnykh landshaftakh na severnoy okraine Sibiriskogo ploskogorya (Steppe landscapes on the northern edge of the Siberian Plateau). *Izvestiya Vsesoyuznogo geograficheskogo obshchestva*, vol. 80, no. 5.

SHMIDT, F. B. 1872. O nanosakh v ustye r. Yeniseya (Alluvial deposits at the mouth of the Yenisey river). *Zapiski Akademii nauk.*

SOKOLOV, N. I. 1938. Terrasy r. Angary (The Angara river terraces). *Problemy fizicheskoy geografii*, vol. 4.

SOKOLOV, N. I. and TYUMENTSEV, N. V. 1949. K voprosu o nakhodke *Elephas trogontherii* Pall. v basseyne r. Angary (On the discovery of *Elephas trogontherii* Pall. in the Anagara river basin). *Doklady Akademii nauk SSSR*, vol. 69, no. 3.

SUKACHEV, V. N. 1910a. Nekotoryye dannyye o dolednikovoy flore severa Sibiri (Some data on the pre-glacial flora of northern Siberia). *Trudy Geologicheskogo muzeya Akademii nauk*, vol. 4, no. 4.

——— 1910b. O nakhodke iskopayemoy arkticheskoy flory na r. Irtyshe u s. Demyanskogo, Tobolskoy gubernii (Discovery of fossil arctic flora on the Irtysh river at the village of Demyanskoye, Province of Tobolsk). *Izvestiya Akademii nauk*, 6th ser. vol. 4, no. 6.

——— 1914. Issledovaniye rastitelnykh ostatkov iz zheludka mamonta, naydennogo na r. Berezovke Yakutskoy oblasti (Investigation of the plant remains from the stomach of a mammoth found on the Berezovka river, Yakutia). In: *Nauchnyye rezultaty ekspeditsii, snaryazhennoy Akademiey nauk dlya raskopki mamonta, naydennogo na r. Berezovke v 1901 g.* (Scientific results of the expedition equipped by the Academy of Sciences for excavation of a mommoth found on the Berezovka river in 1901), vol. 3. Petrograd.

——— 1932. Irtyshskaya fitopaleontologicheskaya ekspeditsiya (The Irtysh phytopaleontologic expedition). In: *Ekspeditsii Vsesoyuznoy Akademii nauk 1931 g.* (Expeditions of the All-Union Academy of Sciences, 1931). Leningrad.

——— 1936. Po Obi i Tymu (Along the Ob and the Tym rivers). In: *Ekspeditsii Akademii nauk SSSR 1934 g.* (Expeditions of the Academy of Sciences of the U.S.S.R., 1934). Moscow–Leningrad.

TIKHOMIROV, B.A. 1941. O lesnoy faze v poslelednikovoy istorii rastitelnosti severa Sibiri i yeye reliktakh v sovremennoy tundre (The forest phase in the postglacial history of the vegetation of northern Siberia and its relics in the tundra of today). In: *Materialy po istorii flory i rastitelnosti SSSR* (Data on the history of the flora and vegetation of the U.S.S.R.), vol. 1. Moscow–Leningrad.

TOLMACHEV, A. I. 1954. *K istorii vozniknoveniya i razvitiya temnokhvoynoy taygi* (History of the origin and development of the dark-coniferous taiga). Moscow–Leningrad.

TYUMENTSEV, N. V. 1941. K voprosu o geologicheskom vozraste stoyanki Buret (The geologic age of the site Buret). *Kratkiye soobshcheniya Instituta istorii materialnoy kultury*, no. 10.

VAKHRAMEYEV, V. A. and PUSHCHAROVSKIY, YU. M. 1954. O geologicheskoy istorii Vilyuyskoy vpadiny i prilegayushchey chasti Priverkhoyanskogo krayevogo progiba v mezozoyskoye vremya (The geologic history of the Vilyuy depression and the adjacent part of the Verkhoyansk marginal fault during the Mesozoic period). In: *Voprosy geologii Azii* (Prolems of Asian geology), vol. 1. Moscow.

VASILYEV, V. G. 1946. *Geologicheskoye stroyeniye severo-zapadnoy chasti Zapadno-Sibirskoy nizmennosti i yeye neftenosnost* (The geologic structure of the northwestern part of the West Siberian Lowland and its oil potential). Moscow.

VASKOVSKIY, A. P. 1950. Granitsa tundrovoy rastitelnoy zony na severnom poberezhye Okhotskogo morya (The boundary of the tundra vegetation zone on the northern coast of the Sea of Okhotsk). *Botanicheskiy zhurnal,* vol. 35, no. 3.

———— 1954. Ostatki serogo orekha i metasekvoyi v verkhnem pliotsene Kamchatki (Remains of the butternut and the metasequoia in the Upper Pliocene of Kamchatka). *Kolyma,* no. 8.

———— 1955. Razmery sovremennogo oledeneniya na Severo-Vostoke (The extent of the present glaciation of the Northeast). *Kolyma,* no. 10.

———— 1956. Novyye vidy iskopayemykh khvoynykh, naydennyye na Kraynem Severo-Vostoke Azii (New types of fossil conifers found in the far northeast of Asia). *Materialy po geologii i poleznym iskopayemym Severo-Vostoka SSSR,* no. 10.

VASKOVSKIY, A. P. and OKLADNIKOV, A. P. 1948. Nakhodka obrabotannogo chelovekom dereva na drevney terrase r. Susuman, basseyn Kolymy (The discovery of a wood worked by man on the ancient terrace of the Susuman river in the Kolyma basin). *Byulleten Komissii po izucheniyu chetvertichnogo perioda,* no. 13.

VASKOVSKIY, A. P. and TUCHKOV, I. I. 1953. Resheniye odnoy iz vazhnykh paleogeograficheskikh problem Mamontovoy gory na Aldane (Solution of one of the most important paleogeographic problems of Mamontova Gora on the Aldan river). *Kolyma,* [vol. 15] no. 9.

———— 1954. Resheniye odnoy iz vazhneyshikh paleogeograficheskikh problem Mamontova gory na Aldane (Solution of one of the most important paleogeographic problems of Mamontova Gora on the Aldan river). *Kolyma,* [vol. 16] no. 9.

VELIKOVSKAYA, YE. M. 1955. Krasnotsvetnyye otlozheniya pliotsena na territorii SSSR i zarubezhnoy Azii (Pliocene red beds in the U.S.S.R. and non-soviet Asia). *Doklady Akademii nauk SSSR,* vol. 100, no. 6.

VOLLOSOVICH, K. A. 1930. Geologicheskiye nablyudeniya v tundre mezhdu nizhnimi techeniyami rek Leny i Kolmy (Geologic observations in the tundra between the lower courses of the Lena and Kolyma rivers). *Trudy Komissii po izucheniyu Yakutskoy ASSR,* vol. 15.

VYSOTSKIY, N. K. 1896. Ocherk tretichnykh i posletretichnykh obrazovaniy Zapadnoy Sibiri (Outline of the Tertiary and post-Tertiary formations of western Siberia). In: *Geologicheskiye issledovaniya po linii Sibirskoy zheleznoy dorogi,* (Geologic investigations along the Siberian Railroad), no. 5. St. Petersburg.

YAROVOY, M. I. 1939. Rastitelnost basseyna r. Yany i Verkhoyanskogo khrebta (Vegetation of the Yana basin and the Verkhoyansk range). *Sovetskaya botanika,* no. 1.

YERMILOV, I. YA. 1935. Geologicheskiye issledovaniya na Gydanskom poluostrove v 1927 g. (Geologic investigations on the Gydan peninsula in 1927). *Trudy Polyarnoy komissii Akademii nauk SSSR,* no. 20.

YOKOYAMA, M. 1910. Climatic changes in Japan since the Pliocene epoch. *Journal of the Faculty of Science, Tokyo University,* vol. 32, article 5.

ZAKLINSKAYA, YE. D. 1950. Nekotoryye dannyye po pliotsenovoy flore Barguzinskoy doliny (Data on the Pliocene flora of the Barguzin valley). *Byulleten Komissii po izucheniyu chetvertichnogo perioda,* no. 15.

———— 1951. Materialy k izucheniyu sostava sovremennoy rastitelnosti i yeye sporovo-pyltsevykh spektrov dlya tseley biostratigrafii chetvertichnykh otlozheniy, shiroko-listvennyy i smeshannyy les (Materials for a study of the composition of the contemporary vegetation and its spore-pollen analysis for purposes of establishing the biostratigraphy of the Quaternary deposits, broadleaf and mixed forest). *Trudy Instituta geologicheskikh nauk Akademii nauk SSSR,* no. 127, geologicheskaya seriya, no. 48.